D0205624

Photonic and Optoelectronic Polymers

ACS SYMPOSIUM SERIES **672**

Photonic and Optoelectronic Polymers

Samson A. Jenekhe, EDITOR
University of Rochester

Kenneth J. Wynne, EDITOR
Office of Naval Research

Developed from a symposium sponsored by the
Pacific Polymer Federation at the Pacific Polymer Conference

American Chemical Society, Washington, DC

Library of Congress Cataloging-in-Publication Data

Symposium on Polymers for Advanced Optical Applications (1995: Kauai, Hawaii)

Photonic and optoelectronic polymers: developed from a symposium sponsored by the Pacific Polymer Federation at the Pacific Polymer Conference, Kauai, Hawaii, December 12–16, 1995/ Samson A. Jenekhe, editor, Kenneth J. Wynne, editor.

p. cm.—(ACS symposium series, ISSN 0097–6156; 672)

Includes bibliographical references and indexes.

ISBN 0–8412–3519–8

1. Photonics—Materials—Congresses. 2. Polymers—Optical properties—Congresses. 3. Optoelectronic devices—Materials—Congresses. 4. Polymers—Electric properties—Congresses.

I. Jenekhe, Samson A. II. Wynne, Kenneth J., 1940– . III. Pacific Polymer Federation. IV. Pacific Polymer Conference (4th: 1995: Kauai, Hawaii) V. Title. VI. Series.

TA1505.S986 1995
621.36—dc21

97–22245
CIP

This book is printed on acid-free, recycled paper.

PRINTED IN THE UNITED STATES OF AMERICA

Advisory Board

ACS Symposium Series

Foreword

THE ACS SYMPOSIUM SERIES was first published in 1974 to provide a mechanism for publishing symposia quickly in book form. The purpose of this series is to publish comprehensive books developed from symposia, which are usually "snapshots in time" of the current research being done on a topic, plus some review material on the topic. For this reason, it is necessary that the papers be published as quickly as possible.

Before a symposium-based book is put under contract, the proposed table of contents is reviewed for appropriateness to the topic and for comprehensiveness of the collection. Some papers are excluded at this point, and others are added to round out the scope of the volume. In addition, a draft of each paper is peer-reviewed prior to final acceptance or rejection. This anonymous review process is supervised by the organizer(s) of the symposium, who become the editor(s) of the book. The authors then revise their papers according to the recommendations of both the reviewers and the editors, prepare camera-ready copy, and submit the final papers to the editors, who check that all necessary revisions have been made.

As a rule, only original research papers and original review papers are included in the volumes. Verbatim reproductions of previously published papers are not accepted.

ACS BOOKS DEPARTMENT

Contents

Nonlinear Optical Properties and Applications

Optical Information Storage and Processing

Enhanced Properties and Device Performance Through Self-Assembly and Nanostructure Control

Indexes

Preface

EXPLOSIVE GROWTH OF INFORMATION in the world is fueling the need for efficient, reliable, and low-cost acquisition, storage, processing, transmission, and display technologies. In the last several years, research aimed at addressing this need has demonstrated the great potential of electroactive and photoactive polymers as photonic and optoelectronic materials. Recent discoveries in these areas include high-data-rate and low-loss polymer optical fibers, high-speed polymer-based electro-optic modulators, bright polymer light-emitting diodes, high-diffraction-efficiency photorefractive polymers, high-efficiency polymer photodetectors for visible–UV radiation, and organic thin-film transistors comparable to amorphous silicon devices. Although these recent achievements represent exciting opportunities for major innovations in information technologies, the commercial realization of many segments of polymer-based photonics and optoelectronics awaits further research advances in addressing many scientific and technical challenges.

This volume was developed from a symposium presented at the Pacific Polymer Conference (sponsored by the Pacific Polymer Federation), titled "Polymers for Advanced Optical Applications," in Kauai, Hawaii, December 12–16, 1995. We organized the symposium to provide an international forum to discuss recent advances and future prospects in the broad field of photonic and optoelectronic polymers. The symposium attracted more than 80 presentations and included 39 invited leading researchers from several countries. The chapters in this book were developed mostly from the invited presentations at the Kauai conference. In addition, five papers not presented at the symposium were invited for inclusion in the volume for the purpose of broadening the topical coverage of the book.

In the spirit of the symposium, this book provides a broad overview of research advances in several areas of photonic and optoelectronic polymers as well as their promising applications. Among the main topics covered are diverse polymers for digital and holographic information storage, including photorefractive polymers; electroluminescent polymers for light sources, high-speed transmission, high-power amplification, and high-speed modulation of optical signals in polymer waveguides and gradient-index fibers; thermally stable poled polymers for second-order nonlinear optics; self-assembly and nanostructure control as approaches to efficient photoelectronic and photonic properties; and thin-film transistors from organic semiconductors.

An important theme of this volume is the interrelationships among materials chemistry, photonic and optoelectronic properties, and device performance. The design and synthesis of novel polymer compositions and architectures aimed at enhanced properties are emphasized in some chapters. Other contributions feature the development of novel approaches to processing and fabrication of photonic and optoelectronic polymers into thin films, multilayers, fibers, waveguides, gratings, and device structures. These approaches, which emphasize polymer synthesis, processing, and device fabrication, are complementary and synergistic.

The interdisciplinary nature of many of the chapters suggests that chemists, chemical engineers, materials scientists, and others interested in the design, synthesis, and processing of diverse photonic and optoelectronic polymers will find this volume useful. This book will also be valuable to physicists, electrical engineers, optical engineers, and others concerned with the design, fabrication, and evaluation of polymer-based electronic, optoelectronic, and photonic devices and components.

Acknowledgments

The assembly of leading researchers for the symposium and the subsequent editing of this volume for publication would not be possible without the generous financial support of the Office of Naval Research. We also thank X. Linda Chen and Laura Devincentis at the University of Rochester, A. Ervin and M. Talukder at the Office of Naval Research, and Anne Wilson and Vanessa Evans-Johnson at ACS Books for their essential help with the book.

SAMSON A. JENEKHE
Departments of Chemical Engineering and Chemistry
University of Rochester
Rochester, NY 14627–0166

KENNETH J. WYNNE
Physical Sciences S&T Division 331
Office of Naval Research
U.S. Department of the Navy
800 North Quincy Street
Arlington, VA 22217–5660

May 2, 1997

WAVEGUIDING, LINEAR OPTICAL EFFECTS, AND APPLICATIONS

Chapter 1

Linear Optical Anisotropy in Aromatic Polyimide Films and Its Applications in Negative Birefringent Compensators of Liquid-Crystal Displays

Fuming Li, Edward P. Savitski, Jyh-Chien Chen, Yeocheol Yoon, Frank W. Harris, and Stephen Z. D. Cheng[1]

Department of Polymer Science and Maurice Morton Institute, University of Akron, Akron, OH 44325–3909

Samples of soluble aromatic polyimides of varying chemical structure and molecular weight were synthesized in refluxing *m*-cresol at elevated temperatures through a one-step polymerization route. Modifications of the dianhydride and diamine monomers were designed to prepare aromatic polyimides having new architectures based on the requirements of liquid crystal display (*LCD*) applications. The solution-cast films exhibit linear optical anisotropy (*LOA*), which is called *uniaxial negative birefringence* (*UNB*) and is characterized by the presence of a larger refractive index along the in-plane direction than in the out-of-plane direction. It is found that the *UNB* is critically associated with the backbone linearity and rigidity as well as the intrinsic polarizability of the polyimides. A specific polyimide synthesized from 2,2'-bis(3,4-dicarboxyphenyl)-hexafluoropropane and 2,2'-bis(trifluoro-methyl)-4,4'-diaminobiphenyl was used as an example to study the molecular weight effect on the *LOA*. For films having a fixed molecular weight, the refractive indices are constant for film thicknesses below 15 μm. They gradually change with further increase of the film thickness. On the other hand, the refractive index along the out-of-plane direction decreases while the in-plane refractive index increases when the polyimide molecular weight increases. The *LOA* is closely associated with the anisotropy of other second order parameters which are second derivatives with respect to the energy term. These parameters include the coefficient of thermal expansion, modulus, dielectric constant and refractive index. Films with this *UNB* can be used as negative birefringent compensators in twisted and super-twisted nematic *LCDs* to improve display viewing angles.

It has been recognized since the 1960s that aromatic polyimide films exhibit structural anisotropy in the directions parallel (in-plane) and perpendicular

[1]Corresponding author

(out-of-plane) to the film surface. This phenomenon has been defined as "in-plane orientation" *(1-5)*. Recently, it has been recognized that such anisotropic structure leads to anisotropic thermal, mechanical, dielectric and optical properties along the in-plane and out-of-plane directions *(5)*.

Our interests are particularly focused on the linear optical anisotropy (*LOA*) in solution-cast polyimide films and their applications. Since aromatic polyimide molecules commonly tend to align parallel to the film surface during the film forming process *(1-5)*, the in-plane refractive index is thus larger than the out-of-plane refractive index. The degree of in-plane orientation and the resultant extent of *LOA* in the films can be estimated readily using refractive index measurements. Therefore, the *LOA* can be expressed by the birefringence which is the difference in the refractive indices along the in-plane and out-of-plane directions. In the field of optics, this phenomenon is defined as *uniaxial negative birefringence* (*UNB*). One of the applications for the *LOA* in polyimide films is that they may be utilized to design negative birefringent compensators for twisted and super-twisted nematic liquid crystal displays (*TN*- and *STN-LCDs*). Such films can be used in both active and passive forms to improve *LCD* viewing angles *(6-8)*.

However, it is well known that aromatic polyimides are usually difficult to process since they do not melt flow before decomposition and are insoluble in conventional solvents. The traditional approach is to use a two-step polymerization route to make processing possible through the soluble poly(amic acid) precursors. For example, polyimide films are generally produced by solution-casting or spin-coating and then are either thermally or chemically imidized. The imidization history affects the ultimate structure, morphology and properties of the films *(9,10)*. Conventional aromatic polyimides synthesized *via* the two-step polymerization often possess strong optical absorption in the low *UV*-visible wavelength region (<450 nm). This is caused by the existence of electron conjugation states formed by the phenylene and imide groups and results in the dark yellow or brown color of the films. Also, crystalline and ordered structures in the films introduce density fluctuations which generate light scattering and, consequently, increase the optical loss. Different degrees of imidization in such films may also give rise to varying refractive indices along both the in-plane and out-of-plane directions. In the past, all these complications in aromatic polyimide films precluded them from applications in optical devices.

Over the past eight years a family of organo-soluble aromatic polyimides have been designed and synthesized in our laboratory *(4-8,11-13)*. These polyimides were synthesized in refluxing *m*-cresol at high temperatures through the one-step route in which the intermediate poly(amic acid)s were not isolated *(11-13)*. Most of the aromatic polyimides synthesized can be easily dissolved in common organic solvents such as acetone, chloroform, 2-pentanone, cyclopentanone, methyl ethyl ketone, tetrahydrofuran, dimethylformamide, N-methylpyrrolidinone and dimethyl sulfoxide. Solution of the imidized polymers can be directly processed to form films which are used as negative birefringent retardation compensators in *LCDs* *(6-8)*. In this publication we present our recent experimental results dealing with the effects of film thickness, molecular backbone structure and molecular weight on the structural and property anisotropy. Specifically, we examine changes in the *UNB* behavior of the aromatic polyimide films.

Principle of *UNB* Compensators in Liquid Crystal Displays

Thin-film-transistor (*TFT*) active matrix (*AM*) *TN* or *STN-LCDs* offer the advantages of high information content, color capability, broad gray scale and fast response. They have been widely used in televisions, lap-top computers and information-related displays. Most of the *TN*- or *STN-LCDs* are operated in the normally white (*NW*) mode. Namely, the *TN*- or *STN-LCD* is light-transmissive (white) under the undriven state (field-off state), and dark in the driven state (field-on state). One of the major drawbacks of the *NW-TN*- or *STN-LCDs* are the asymmetric viewing angle characteristics along the horizontal and vertical directions as well as the narrow range of viewing angles *(6-8,14)*. These are critical issues for applications such as in avionics information displays where cross-cockpit viewing is required. Loss of the contrast ratio at high viewing angles for *NW-TN*- or *STN-LCDs* is the result of light leakage in the dark state. From a molecular point of view, the liquid crystals are aligned in a *TN* orientation which is induced by surface alignment layers in the undriven state. In the driven state they are in a quasi-homeotropic state having a *uniaxial positive birefringence* (*UPB*). Without consideration of the pretilt angle in *AM-NW-TN*- or *STN-LCDs*, *LC* molecules are in a perfect homeotropic state under a large enough driving voltage. Therefore, under the driven state, a light beam traveling normal to the cell is parallel to the long optical axis of the *LC* molecules and does not see any birefringence. The cell shows in the black state. However, an off-axis light beam with an incident angle θ with respect to the normal direction of the *LCD* surface undergoes a retardation due to the contribution from both the ordinary (n_o) and extraordinary (n_e) refractive indices of the liquid crystal. Consequently, a light leakage occurs. The light leakage in the black state causes a low contrast ratio in the oblique direction. In order to eliminate the light transmission in the oblique viewing direction, an optical compensator (*OC*) can be inserted between the cross polarizers. The *OC* has a *UNB* value while the homeotropically aligned liquid crystal possesses a *UPB* value as shown in Figure 1. Therefore, the retardation of the *LC* (R_{LC}) is compensated by the opposite negative retardation of the *OC* (R_{OC}). Theoretically, without consideration of the pretilt angle and if the driving voltage is high enough to drive the *LC* into the perfect homeotropic state, the total retardation becomes $R = R_{LC} + R_{OC} = 0$. Therefore, the off-axis of the light transmission to the normal direction (leakage) under the driven state ideally approaches zero. Several techniques have been reported to generate such *UNB* compensators in the past ten years *(16-21)*. However, all these methods include either complicated processing procedures or lack uniformed *UNB* properties. Novel and simple methods are needed to develop compensating films with reproducible and controllable *UNB*.

Anisotropic Structures in Aromatic Polyimide Films

Several experiments have been customarily utilized to characterize the structural anisotropy in films. One of the most widely used methods is wide angle X-ray diffraction (*WAXD*) experiments *(1-4)*. Figure 2 shows a set of *WAXD* patterns under both the reflection and transmission modes on a crystalline aromatic polyimide films synthesized from 3,3',4,4'-biphenyltetracarboxylic dianhydride (BPDA) and 2,2'-

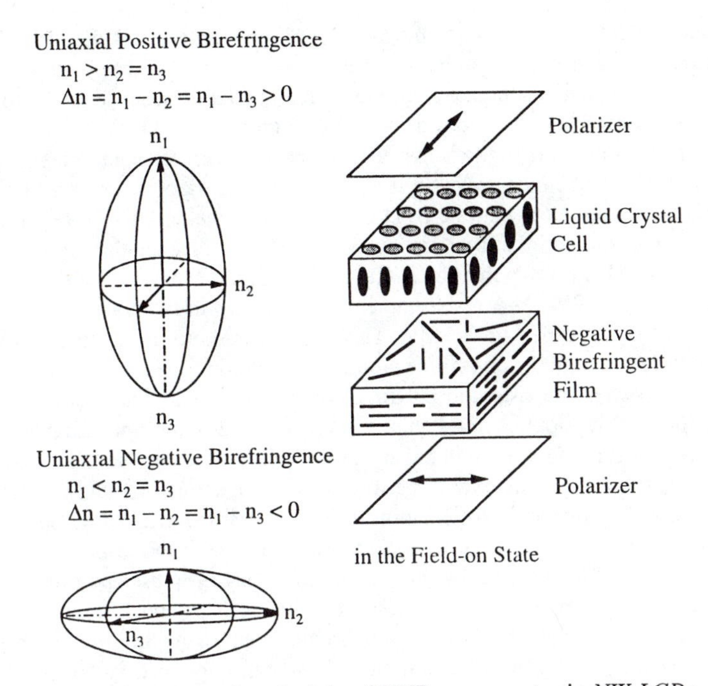

Figure 1. Schematic principle of *UNB* compensator in *NW-LCDs*

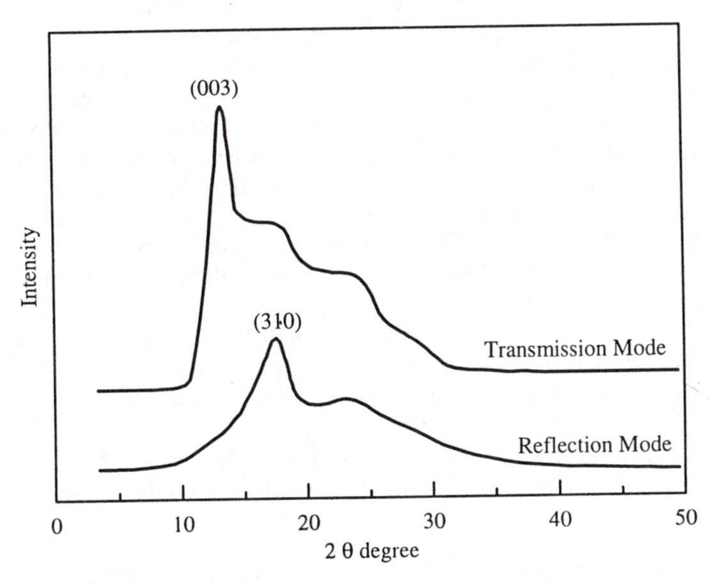

Figure 2. *WAXD* patterns for BPDA-PFMB films under both the reflection and transmission modes

bis(trifluoromethyl)-4,4'-diaminobiphenyl (PFMB) diamine (BPDA-PFMB) *(4)*. Detailed chemical structure can be found in Table I. It is clear that both patterns are different. This can be compared with the fiber patterns scanned along both the equatorial and meridian directions as shown in Figure 3. In fact, the pattern obtained from the reflection mode of the films correspond well with the equatorial scan of the fibers, while that obtained from the transmission mode are in agreement with the fiber meridian scan. Note that the BPDA-PFMB fiber equatorial scan represents *hk0* reflections and the meridian scan uncovers *00l* reflections since the crystal structure is monoclinic. This reveals that the *c*-axis of the BPDA-PFMB crystals is parallel to the film surface *(4,22)*. As a result, the chain molecules in the crystalline region are more or less laying down in the film. This method is quite useful to determine the anisotropic structure in the crystalline aromatic polyimide films.

However, when the aromatic polyimides are amorphous, which is an important feature for optical applications, this *WAXD* method may not provide definite conclusions. Furthermore, even in the crystalline polyimide, one may ask if the *c*-axis alignment is truly representative of the whole chain backbone orientation. One of the methods to detect molecular orientation is Fourier transform infrared spectroscopy (*FTIR*). It is known that several modes are associated with the in-plane and out-of-plane vibrations of the chemical groups such as carbonyl groups. The finger prints of the vibration modes are the carbonyl absorption bands at around 1783 cm^{-1} and 1727 cm^{-1}, which possess higher absorption frequencies than the imide absorption frequency range. The origin of this doublet has been explained in terms of the in-phase and out-of-phase coupled vibrations of the carbonyl groups *(23)*. The relative intensities of these two carbonyl absorptions can be used to characterize the anisotropy of polyimide films. The 1783 cm^{-1} absorbance, which possesses a transition moment vector parallel to the imide plane, is the symmetrical stretching band of the carbonyl groups (in-phase). In addition, it has a higher intensity compared with the intensity of the 1727 cm^{-1} absorbance which is the asymmetrical stretching band of the carbonyl groups (out-of-phase) *(24)*. Indeed, for many aromatic polyimides, we have found that the ratio of the intensities of these two bands is an accurate measure of the structural anisotropy in the thin films. Figure 4 shows an *IR* spectrum for 6FDA-PFMB (foe chemical structure, see Table I), which was synthesized from 2,2'-bis(3,4-dicarboxyphenyl)-hexafluoropropane (6FDA) and PFMB as well as BPDA-PFMB as two examples. Both absorbances are marked in the figure.

On the other hand, the anisotropy within the unoriented film can also be investigated *via* the ratio of two absorptions at 1783 cm^{-1} and 742 cm^{-1}, which represents the in and out-of-plane bending of the carbonyl group, respectively. Recently, it has been observed that the absorption of the *IR* beam yields different spectrum intensities at these two bands for BPDA-PFMB and their copolyimide films *(5)*, and the ratio between 1783 cm^{-1} and 742 cm^{-1} absorption bands is critically associated with the degree of in-plane orientation.

Chemical Backbone Structure Effect on Linear Optical Anisotropy

The molecular design of organo-soluble aromatic polyimides used in the optical applications has been focused on the issue to interrupt the electron conjugation along

Table I Chemical Structures of Dianhydrides, Diamines, and Polyimides

Dianhydride		Diamine	
BPDA		PFMB	
DBBPDA		DMB	
BTDA		DClB	
ODPA		DBrB	
6FDA		DIB	

Polyimide	
BPDA-PFMB	
6FDA-PFMB	

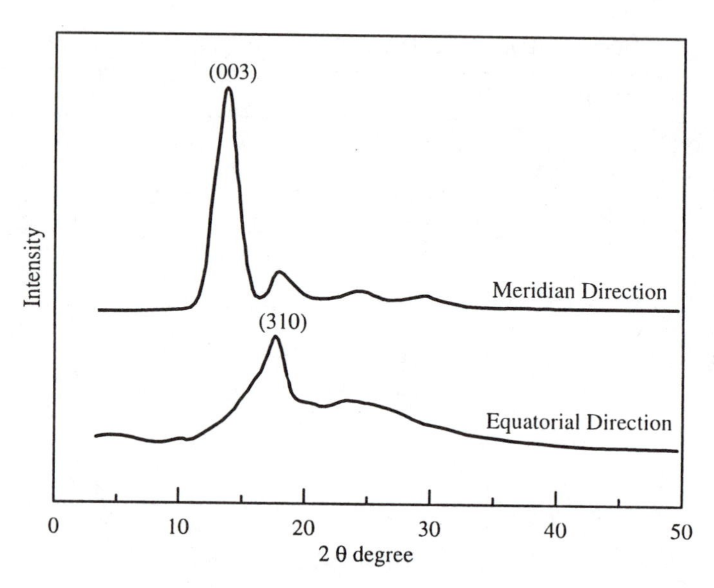

Figure 3. *WAXD* patterns of BPDA-PFMB fibers scans along both the equatorial and meridian directions

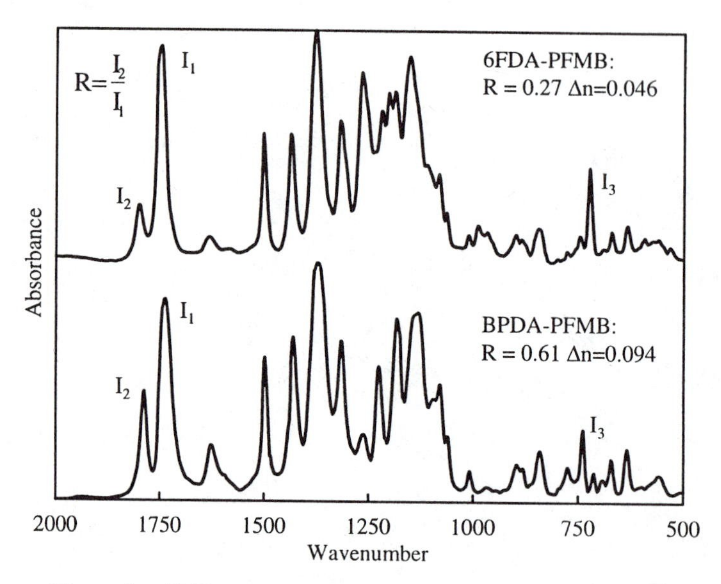

Figure 4. *FTIR* spectrum of 6FDA-PFMB and BPDA-PFMB

the chain backbones. This has been shown to increase solubility, light transmittance in the low *UV*-visible light wavelength region, decrease the amount of ordered structures and loosen the chain packing while the backbone chain linearity and rigidity are not scarified. One approach to disrupt the electron conjugation is introduce substituents at the 2- and 2'-positions of a 4,4'-diaminobiphenyl diamine. Typical substituents are methyl, trifluoromethyl, phenyl, cyano, halogen and other groups. Steric hindrances between these groups cause two phenylene groups in the diamine to be twisted at angles close to 90° from each other. This torsion between the phenylenes leads to nonplanar conformations. Using the PFMB diamine as an example, various aromatic polyimides involving PFMB have been prepared with different dianhydrides. They include BPDA, 6FDA, 3,3',4,4'-benzophenone-tetracarboxylic dianhydride (BTDA) and 4,4'-oxydiphthalic anhydride dianhydride (ODPA) having varying structure groups between the two phenylene-imide rings (for chemical structures, see Table I). Table II gives the in-plane (n_e) and out-of-plane (n_o) refractive indices and *UNB* values ($\Delta n = n_o - n_e$) determined at a wavelength of 632.8 nm. Their *UV*-visible transmittance (80%) is also included. It has been found that the refractive indices (1.516-1.639) as well as the *UNB* values (0.046-0.094) are critically affected by the structure of dianhydrides. Generally speaking, the introduction of groups containing N, Cl, Br and I increases the refractive indices while fluorine containing groups reduce the refractive indices. Furthermore, the chemical structures of linking groups between two phenylene-imide rings in the dianhydrides also critically affect the *UNB* values. For these organo-soluble aromatic polyimide films, their *UNB* values (Δn) are approximately one order of magnitude higher than those in the films solution-cast or spin-coated from coil macromolecules. This indicates that the molecular linearity, rigidity and intrinsic polarizability are important factors to determine structural and property anisotropy.

Table II Linear Optical Properties of Several Aromatic Polyimides Based on the PFMB Diamine

Polyimides	n_o	n_e	Δn	*80% trans. UV-Vis. (nm)*
BPDA-PFMB	1.540	1.634	0.094	445
BTDA-PFMB	1.576	1.639	0.063	426
ODPA-PFMB	1.587	1.639	0.052	408
6FDA-PFMB	1.516	1.558	0.046	388

Modifications can also be made to the dianhydride portions of these polyimides as well. For example, a similar approach can also be carried out for the BPDA dianhydride to twist the two phenylene-imide rings *via* synthesizing 2,2'-disubstituted-4,4',5,5'-biphenyltetra-carboxylic dianhydrides *(25)*. The substituent groups can also be methyl, trifluoromethyl, phenyl, cyano, halogen and other groups. For example, a new monomer, 2,2'-dibromo-4,4',5,5'-biphenyltetracarboxylic dianhydride (DBBPDA, see Table I), has been used to prepare a series of organo-soluble aromatic polyimides. It has been combined with different diamines which are all 2- and 2'- position substituted 4,4'-diamenobiphenyls with substituents such as methyl groups (DMB, see Table I) and halogen groups (Br, Cl and I, see Table I). Their n_e and n_o, *UNB* and 80% *UV*-visible transmittance values are listed in Table III. The same general conclusions can be achieved as in the case found in Table II.

Table III Linear Optical Properties of Several Aromatic Polyimides Based on DBBPDA Dianhydride

Polyimides	n_o	n_e	Δn	*80% trans. UV-Vis. (nm)*
DBBPDA-DMB	1. 606	1. 698	0.092	424
DBBPDA-DBrB	1. 635	1. 723	0.088	429
DBBTDA-DClB	1. 628	1. 713	0.085	462
DBBPDA-DIB	1. 656	1. 732	0.076	431
DBBPDA-PFMB	1. 564	1. 637	0.074	403

Differences in the linkages and conjugation of the dianhydrides and diamines also bring differences in crystallinity, organo-solubility, color and processability of the films. A complete relationship between the structures and properties can be established and utilized to design the polyimide architectures for various optical applications.

Optical birefringence dispersion is also an important factor in the design of compensation films. Figure 5 shows the optical birefringence dispersion of 6FDA-PFMB, which has been found to dissolve in a variety of common organic solvents and has been used to make *OC* in *NW-TN-* or *STN-LCDs*. The open circles in Figure 5 are experimental data and the solid line represents a Cauchy function which theoretically represents optical dispersion *(26)*. The dependence of the *UNB* values on wavelength for the 6FDA-PFMB film can also be compared with one of the liquid crystal mixtures developed for *LCD* application (Merck ZLI-1840, which is a mixture of several liquid crystal molecules) as shown in Figure 5. Although the magnitude of the Δn of 6FDA-PFMB films is about three times smaller, the Δn dispersion of this film is similar to that of *LC* mixtures. A good match of birefringence dispersion between the *OC* and the *LC* mixture over the entire visible range leads to a null phase retardation for all three primary colors (red, green and blue) employed in *LCDs*. As a result, a high contrast ratio can be achieved *(27,28)*.

Film Thickness Effect on Linear Optical Anisotropy

Aromatic polyimide films of varying thickness were prepared to study the effect of the thickness on the refractive indices. Figure 6 shows that the refractive indices are fairly independent of the film thickness in the range of 1 μm to 15 μm for 6FDA-PFMB. For thicknesses above 15 μm, the difference between the in-plane and out-of-plane refractive indices (n_e and n_o) gradually decreases with increasing the film thickness. This indicates that the thin film linear optical properties gradually approach the bulk properties which should be three-dimensionally isotropic.

From a microscopic standpoint, the molecules in the films are restricted in three-dimensional space between two walls having different characters: one is the silicon wafer surface (hard wall) and the other is the air-polymer surface (soft wall). The polyimide films form the *LOA* during the solvent evaporation process. Moreover, aromatic polyimides possess intrinsically larger differences in the refractive indices along and perpendicular to the molecular chain directions than flexible coil macromolecules. On the other hand, the film thickness in this study are at least one to several orders of magnitude greater than the molecular size. The *LOA* may thus not

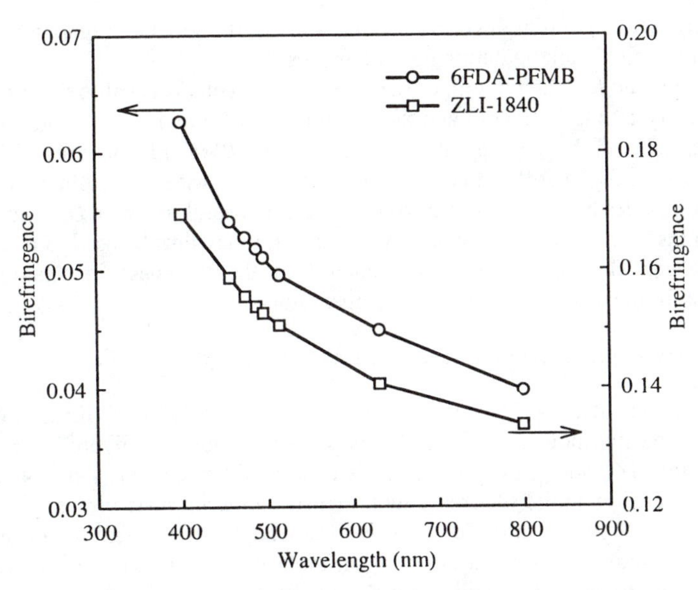

Figure 5. Comparison of birefringence dispersion of 6FDA-PFMB film with a liquid crystal mixture (Merck ZLI-1840) in the visible wavelength region

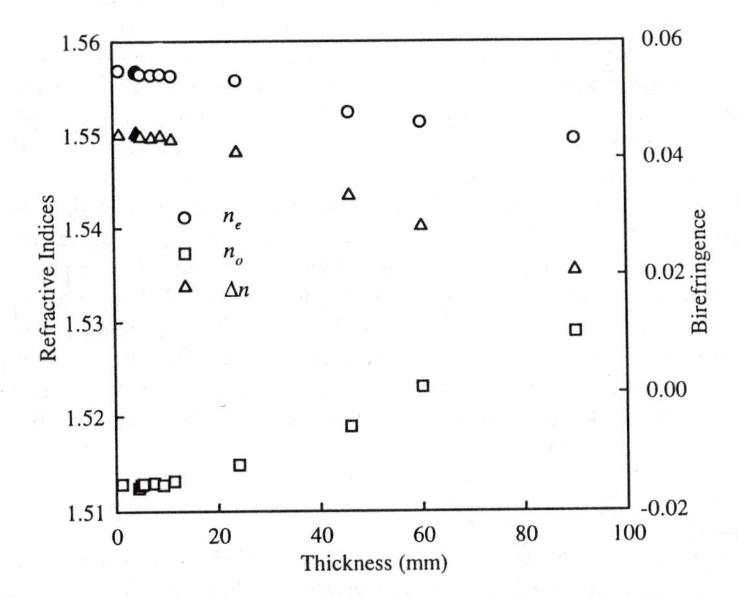

Figure 6. Film thickness effect on the refractive indices and *UNB* values for 6FDA-PFMB films

be directly caused *via* film thickness but is a result of the solvent evaporation and the internal force field built up during the film processing.

Although the retardation term of the liquid crystal cells is equal to the product of the liquid crystalline *UPB* (Δn) and the cell thickness (d) and it is around 1000 nm, this value should be compensated by another *UNB* retardation value. Overall, combined with the retardation contributed from two polarizers, the films with *UNB* retardation values between 50 nm to 200 nm are very useful in the *LCD*s. The 6FDA-PFMB films in the thickness range of 1 μm to 8 μm may thus be used to meet these requirements. The *UNB* value independence of the thickness in this range is important for the design of retardation compensators.

Molecular Weight Effect on Linear Optical Anisotropy

It is interesting that the *UNB* values of the aromatic polyimide films exhibit an intrinsic viscosity (molecular weights) dependence as shown in Figure 7 for 6FDA-PFMB films. The thickness of the films is controlled to be about 4 μm. The *UNB* values of the films initially increase with increasing the intrinsic viscosity and then level off when a critical intrinsic viscosity of around 1.1 is reached. As a result, the *LOA* of the films is dependent not only on molecular parameters such as rigidity, linearity and intrinsic polarizability, but also on the packing of the chain molecules in the film condensed state. Molecular weight differences in 6FDA-PFMB films may affect molecular packing due to changes in the chain end group density, which may possess a higher free volume fraction. When the molecular weight reaches a critical value, the effects of end groups on the packing becomes less important compared with other factors.

Molecular weights of the 6FDA-PFMB samples can be converted to different degrees of polymerization (*DP*) for the 6FDA-PFMB samples *(29)*. In Figure 8, the refractive indices along the in-plane and out-of-plane directions and the *UNB* values of the films are plotted as a function of $(DP)^{-1}$. Note that the parameter $(DP)^{-1}$ is roughly proportional to the end-group density, which approaches zero when the molecular weight increases to infinity. From this figure, it is evident that the refractive indices of the films along the in-plane direction (n_e) decrease, while the refractive indices along the out-of-plane direction (n_o) increase with increasing $(DP)^{-1}$ and both show the linear relationships. This enhances the *UNB* value of the films with increasing molecular weight (Figures 7 and 8). Since the refractive index is proportional to the polarizability per unit volume in an electromagnetic field, it seems that the different end group densities in chain molecules and, therefore, the uncoupled chain ends may significantly affect the polarizability along both the in-plane and the out-of-plane directions.

Correlation with Second Order Parameters

We have investigated many macroscopic properties of the organo-soluble aromatic polyimide films and all of them exhibit anisotropic behaviors which are closely associated with the anisotropic structure in the films. One question still remains is whether we can find a common physics background to describe these properties. If we consider the free energy (or internal energy or energy) term, the first derivative of

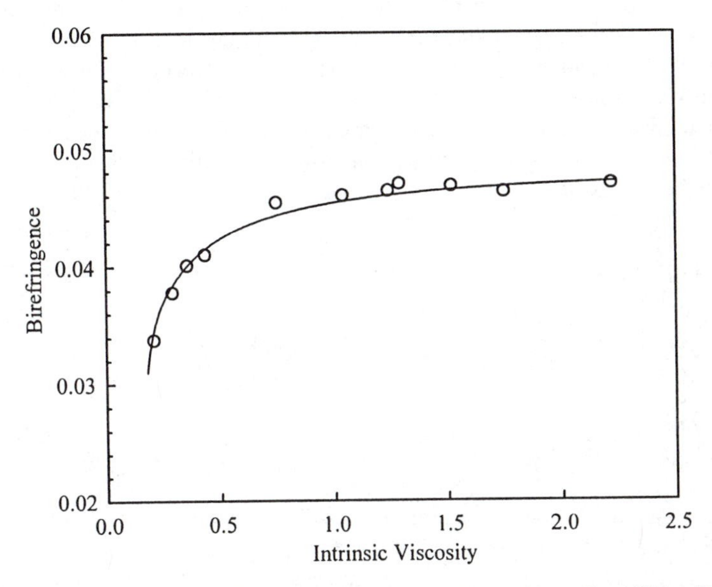

Figure 7. Intrinsic viscosity effect on the *UNB* values for 6FDA-PFMB polyimides

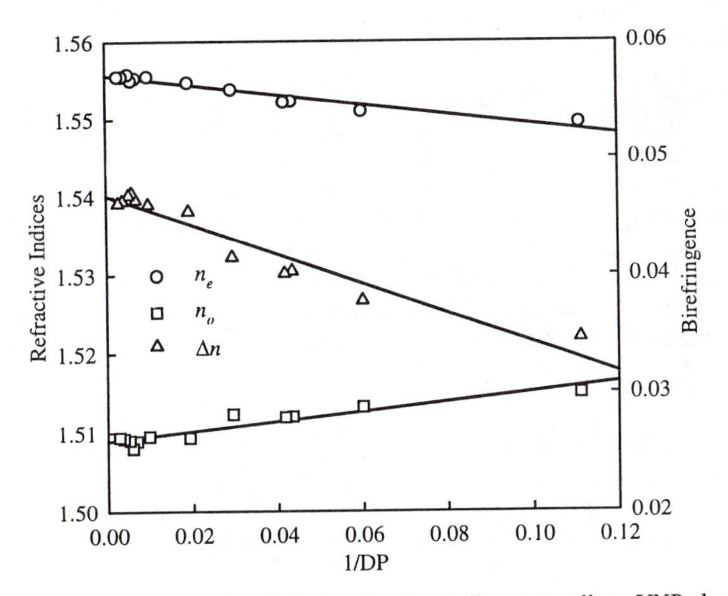

Figure 8. In-plane and out-of-plane refractive indices as well as *UNB* changes with reciprocal of degree of polymerization

the energy term with respect to pressure (p) is volume (V), that with respect to displacement (l) is stress (σ) and that with respect to the applied electric field (E) is the electric polarization (P) or dipole moment (μ). The second derivatives of these quantities are thus the coefficient of thermal expansion (CTE), modulus (E'), dielectric constant (ε) and refractive index (n). Note that these four parameters all belong to the second order type in physics. Mathematically and thermodynamically, they can be expressed as

$$\left(\frac{\partial G}{\partial p}\right)_T = V \qquad \frac{1}{V_0}\left(\frac{\partial V}{\partial T}\right)_p = CTE$$

$$\frac{1}{A_0}\left(\frac{\partial U}{\partial l}\right)_{T,p} = \sigma \qquad L_0\left(\frac{\partial \sigma}{\partial l}\right)_{T,p} = E'$$

$$\left(\frac{\partial U}{\partial E}\right)_T = \bar{\mu} = P/N \qquad \left(\frac{\partial P}{\partial E}\right)_T = \frac{\varepsilon}{(1+\chi)}\chi = \varepsilon_0(n^2-1).$$

Here, L_0, A_0, and V_0 are initial one, two and three dimensional size parameters, respectively. As a result, it is expected that they may share the common microscopic origin of the structural changes. Experimentally, it is also recognized that the refractive indices and *UNB* values of the films can be correlated with other parameters, such as *CTE*, E' and ε' behavior. Further research is necessary to obtain a quantitative relationships between these second order parameters.

Conclusion

In summary, we have briefly reviewed our understanding in the development of organo-soluble aromatic polyimides *via* molecular design and architecture control and the establishment of relationships between the anisotropic structure and properties in the films. These anisotropic structures and properties are also found to be film thickness and molecular weight dependent. In particular, the *LOA* behavior has led to a invention where these aromatic polyimide films can be used as *UNB* retardation compensators in *LCDs* to increase viewing angles. This technology initiated from our researches has been commercialized to produce compensators for wide applications in *LCDs*.

Acknowledgment

This work was supported by SZDC's Presidential Young Investigator Award from the National Science Foundation (NSF DMR-91-57738), and Advanced Liquid Crystal Optical Materials (ALCOM) NSF Science and Technology Center at Kent State University, Case Western Reserve University and The University of Akron.

Literature Cited

1. Ikeda, I. M. *J. Polym. Sci., Polym. Lett.* Ed. **1966,** *4,* 353.

2. Russell, T. P.; Gugger, H.; Swallen, J. D. *J. Polym. Sci., Polym. Phys. Ed.* **1983,** *21,* 1745.
3. Takahashi, N.; Yoon, D. Y.; Parrish, W. *Macromolecules* **1984,** *17,* 2583.
4. Cheng, S. Z. D.; Arnold, Jr., F. E.; Zhang, A.; Hsu, S. L.-C. Harris, F. W. *Macromolecules* **1991,** *24,* 5856.
5. Arnold, Jr, F. E.; D. Shen, D.; C. J. Lee, C. J.; Harris, F. W.; Cheng, S. Z. D.; Lau, S. F. *J. Mater. Chem.* **1993,** *3,* 353.
6. Harris, F. W.; Cheng, S. Z. D. *US patent* 5,344,916, 1995
7. Harris, F. W.; Cheng, S. Z. D. *US patent* 5,480,964, 1996.
8. Li, F.; Harris, F. W.; Cheng, S. Z. D. *Polymer* **1996,** *37,* xxxx.
9. Hoffman, D. A.; Ansari, H.; Frank, C. W. in *Material Science of High Temperature Polymers for Microelectronics*; Editor, Grubb, D. T.; Mita, I.; Yoon, D. Y. Eds.; MRS Symp. Ser.; Materials Research Society: Pittsburgh, Pennsylvania, 1991, Vol. 227, p. 125.
10. Hornak, L. A. in *Polymers for Lightwave and Integrated Optics: Technology and Applications*; Editor, Hornak, L. A. Ed.; Marcel Dekker: New York, New York, 1992, p. 207.
11. Harris, F. W.; Hsu, S. L.-C. *High Perform. Polym.* **1989,** *1,* 1.
12. Harris, F. W. *US patent* 5,071,997, 1991.
13. Harris, F. W.; Hsu, S. L.-C.; Tso, C. C. *ACS Polym. Preprints* **1990,** *31(1),* 342.
14. Lien, A.; Takano, H.; Suzuki, S.; Uchida, H. *Mol. Cryst. Liq. Cryst.* **1991,** *198,* 37.
15. Yamauchi, S.; Aizawa, M.; J.F.Clerc, J. F.; Uchida, T.; Duchene J. *SID'89 Digest* 1989, p. 378.
16. Hatoh, H.; Ishikawa, M.; Hisatake, Y.; Hirata, J.; Yamamoto, T. *SID'92 Digest* 1992, p. 401.
17. Ong, H. L. *Japan Display'92 Digest* 1992, p. 247.
18. Clerc, J. F. *SID'91 Digest* 1991, p. 758.
19. Kuo, C.-L.; Miyashita, T.; Suzuki, M.; Uchida, T. *SID'94 Digest* 1994, p. 927.
20. Yel, P. A.; Gunning, W. J.; Eblen, Jr. J. P.; Khoshnevisan, M. *US Patent* 5,196,953, 1993.
21. Eblen Jr., J. P.; Gunning, W. J.; Beedy, J.; Taber, D.; Hale, L.; P. A. Yel, Khoshnevisan. M. *SID'94 Digest* 1994, p. 245.
22. Cheng, S. Z. D.; Wu, Z.; Zhang, A.; Eashoo, M.; Hsu, S. L.-C.; Harris, F. W. *Polymer* **1991,** *32,* 1803.
23. Matsuo, T. *Bull. Chem. Soc. Japan* **1964,** *37,* 1844.
24. Dine-Hart R. A.; Wright, W. W. *Makromol. Chem.* **1971,** *143,* 189.
25. Harris, F. W.; Lin, S.-H.; Li, F.; Cheng, S. Z. D. *Polymer* **1966,** *37,* xxxx.
26. Pohl, L.; Finkenzeller, U. in *Liquid crystals. Applications and Uses*; Editor Bahadur, B. Ed; World Scientific: Singapore, 1990, Vol. 1, p. 139.
27. Wu, S.-T.; Lackner, A. M. *Appl. Phys.Lett.* **1994,** *64,* 2047.
28. Wu, S.-T. *Phys. Rev. A* **1986,** *33,* 1270; see also, *J. Appl. Phys.* **1991,** *69,* 2080.
29. Savitski, E. P., Ph.D. Dissertation, Department of Polymer Science, The University of Akron, Akron, Ohio 44325-3909, 1996.

Chapter 2

Transparent Zero-Birefringence Polymers

Shuichi Iwata[1,2], Hisashi Tsukahara[1,2], Eisuke Nihei[1], and Yasuhiro Koike[1,2]

[1]Department of Material Science, Faculty of Science and Technology, Keio University, 3–14–1 Hiyoshi, Kohoku-ku, Yokohama 223, Japan
[2]Kanagawa Academy of Science and Technology, 3–2–1 Sakado, Takatsu-ku, Kawasaki 213, Japan

Birefringence is caused by orientation of polymer chains during injection-molding and extrusion and it is a disadvantage in optical systems such as lenses, compact disks, waveguides. We propose new transparent zero-birefringence polymers for any degree of orientation of polymer chains, which are obtained by the following two methods. One is random copolymerization of negatively and positively birefringent monomers. The other is doping of a small amount of large anisotropic molecules into a polymer matrix to compensate the birefringence caused by the orientation of polymer chains. Zero-birefringence polymers synthesized by these two methods have low scattering loss, which is comparable with that of pure poly(methyl methacrylate) (PMMA), and had no observable microscopic heterogeneous structures.

Transparent amorphous polymers such as poly(methyl methacrylate) (PMMA) have been found to be useful materials for polymer optical fibers (POFs) (*1,2*), waveguides (*3*), lenses (*4*), optical disks (*5*), and other optical components because of their excellent mechanical properties and easy processing. Many recently developed optical applications utilizing polarization techniques need optical polymers for maintaining more accurate polarization. However, applications of optical polymers are limited by birefringence which occurs in the process of device fabrication.

Although a polymer may be composed of an anisotropic monomer, birefringence does not occur in it if the polymer structure forms random coils. Birefringence is, however, easily caused by both orientation of the polymer chains and photoelasticity which results from external mechanical stress. Orientational birefringence can be compensated by blending a negative and positive birefringent homopolymers and such pairs of polymer blends have been reported (*6,7*). However these polymer blends are inhomogeneous and not transparent, since they phase-separate, resulting in large-domain heterogeneous structures which dramatically increase light scattering. In addition, it is very difficult to blend polymers homogeneously in the extrusion process and injection-molding.

In order to circumvent this problem, we propose two methods for compensating the birefringence of polymers. One involves random copolymerization of positive and negative birefringent monomers and the other involves doping a small amount of large anisotropic molecules into a polymer matrix.

Compensation for Birefringence

Compensation by Random Copolymerization

Zero-birefringence copolymers synthesized from negative and positive birefringent monomers are more transparent than polymer blends. Perfect random copolymerization is achieved when the monomer reactivity ratios are equal to unity. As a result, the heterogeneous structure of the random copolymer is of the order of less than several monomer units which is much smaller than an optical wavelength, preventing light scattering. In the zero-birefringence copolymer, since the anisotropic polarization of the negative birefringent monomer units is compensated by the positive birefringent monomer units on the same polymer chain, the birefringence is zero for any degree of orientation.

In the copolymerization reaction between M_1 and M_2 monomers, the monomer reactivity ratios r_1 and r_2 are defined as (*8*):

$$r_1 = \frac{k_{11}}{k_{12}} \qquad r_2 = \frac{k_{22}}{k_{21}} \tag{1}$$

where k is the propagation rate constants in the following copolymerization reaction:

$$\cdots M_1^{\bullet} + M_1 \xrightarrow{k_{11}} \cdots M_1 M_1^{\bullet} \qquad \cdots M_2^{\bullet} + M_1 \xrightarrow{k_{21}} \cdots M_2 M_1^{\bullet}$$

$$\cdots M_1^{\bullet} + M_2 \xrightarrow{k_{12}} \cdots M_1 M_2^{\bullet} \qquad \cdots M_2^{\bullet} + M_2 \xrightarrow{k_{22}} \cdots M_2 M_2^{\bullet}$$

The monomer reactivity ratios r_1 and r_2 between monomers M_1 and M_2 are estimated by using Equation (2):

$$r_1 = \frac{Q_1}{Q_2}\exp\left[-e_1(e_1 - e_2)\right]$$
$$r_2 = \frac{Q_2}{Q_1}\exp\left[-e_2(e_2 - e_1)\right] \quad (2)$$

Here Q_1 (or Q_2) is the reactivity of the monomer M_1 (or M_2), and e_1 (or e_2) is the electrostatic interaction of the permanent charges on the substituents in polarizing the vinyl group of monomer M_1 (or M_2). When r_1 and r_2 of two monomers are unity, they can be randomly copolymerized perfectly.

Table I shows the Q and e values and the monomer reactivity ratios r_1 and r_2 for the radical copolymerization process. Since the monomer reactivity ratios between negatively birefringent methyl methacrylate (MMA) and positively birefringent 2,2,2-trifluoroethyl methacrylate (3FMA) and benzyl methacrylate (BzMA) are nearly equal to unity, these monomers can be randomly copolymerized, resulting in homogeneous and transparent copolymers.

Preparation of Copolymer Films. In order to eliminate inhibitors and impurities, MMA monomer, 3FMA monomer, and BzMA monomer were distilled at bp 46-47℃ / 100 mmHg, bp 56-57℃ / 110 mmHg, and bp 86℃ / 0.1 mmHg, respectively. A mixture of MMA and one of the positively birefringent monomers with a specified amount of initiator, benzoyl peroxide (BPO), chain transfer agent (*n*-butyl mercaptan (*n*BM)), and solvents (ethyl acetate) were placed in a glass tube and heated at 70℃. After polymerization, the polymer solution was filtered through a 0.2 μm membrane filter and precipitated into methanol. The polymer was dried under reduced pressure for about 48 hours. The polymer solution in ethyl acetate was cast onto a glass plate with a uniform film thickness (50-100 μm) by using a knife-coater. The polymer film was dried under vacuum.

Birefringence Measurement. The polymer film was uniaxially heat-drawn at 90℃ at a rate of *ca.* 6.6 mm/min. in hot silicone oil. Birefringence of the drawn film was determined by a measuring the optical path difference between the parallel and perpendicular directions to the draw, using a crossed sensitive color plate method (Toshiba Glass Co., SVP-30-II).

Zero-Birefringence Copolymers. Figure 1 shows birefringence Δn ($n_{//} - n_{\perp}$) of poly(MMA-co-3FMA) film sample as a function of the draw ratio. The subscripts "//" and "⊥" denote the parallel and perpendicular directions to the oriented polymer chains, respectively. The drawn PMMA has a negative birefringence ($\Delta n < 0$), while P3FMA has a positive birefringence ($\Delta n > 0$). The value and the sign of birefringence vary with the composition of copolymers. The copolymer of MMA/3FMA in the ratio of 45/55(wt./wt.) had a birefringence of zero at various draw ratios. The zero-birefringence of such a poly(MMA-co-3FMA) copolymer was independent of molecular weight from 1.0×10^5 to 3.2×10^5 and draw temperature (70-90℃).

The birefringence, Δn, of poly(MMA-co-BzMA) film sample as a function of the draw ratio is shown in Figure 2. Poly(benzyl methacrylate) (PBzMA) has a large positive birefringence ($\Delta n > 0$). However, since the PBzMA film was brittle, we could not show the data of PBzMA in Figure 2. The copolymer of MMA/BzMA in the ratio of 82/18 (wt./wt.) had the orientation birefringence of zero at various draw ratios. Zero-birefringent poly(MMA-co-BzMA) was independent of molecular weight from 1.0×10^5 to 5.5×10^5 and draw temperature from 70 to 90℃.

Compensation by Doping a Large Anisotropic Molecule into Polymer Matrix

In the random copolymerization method, the monomer reactivity ratios between negative and positive birefringent monomers must be nearly equal to unity in order to achieve random copolymerization and hence a homogeneous structure. Therefore, combinations of monomers to satisfy that condition are quite limited. Therefore, we propose another method involving doping of large anisotropic molecules into polymers to compensate the birefringence. The advantage of this method is that the birefringence can be compensated with many selections of doping molecules. If the doping molecules are miscible with the matrix of polymer, the molecules are randomly dispersed in the polymer without any aggregation and does not cause excess light scattering loss (*9*). The dopant molecules are oriented in proportion to the orientation degree of polymer chains. The long axis of the doping molecule tends to point to the stretched direction to minimize the enthalpy. In the case of the rod-like doping molecules such as *trans*-stilbene and diphenyl acetylene (tolan) used in this paper, they are easily oriented by the orientation of polymer chains when the polymer is heat-drawn or injection-molded. Polarization parallel to the long axis of the doping molecule is much larger than the polarization perpendicular to the axis and hence they have positive birefringence. Therefore, when these doping molecules in the PMMA matrix are oriented according to the orientation of MMA polymer chains, the negative birefringence caused by orientation of polymer chains is compensated by the orientation of these dopant molecules.

Birefringence of PMMA with Dopant Molecules. The film sample for investigating birefringence was prepared from a solution of PMMA and the dopant molecules. The methods of processing the films and measuring the birefringence were almost the same as in the random copolymerization method. Since the glass transition temperature (*Tg*) of the PMMA sample is decreased by the dopant molecules, the film was uniaxially heat-drawn at 70℃. Figures 3 and 4 show the birefringence of PMMA with dopant molecules, stilbene and tolan respectively, as

Table I. *Q* and *e* values of monomers

Monomer	*Q value*	*e value*	*Monomer reactivity ratio*	
MMA	0.74	0.40	$r_1 = 0.83$	$r_1 = 1.07$
3FMA	1.13	0.98	$r_2 = 0.86$	
BzMA	0.70	0.42		$r_2 = 0.94$

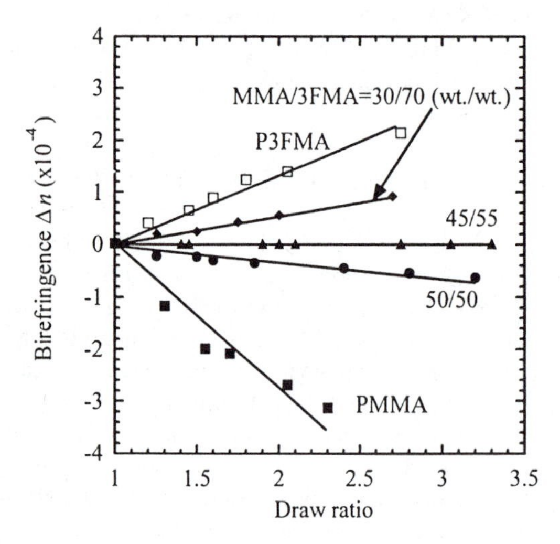

Figure 1. Birefringence of poly(MMA-co-3FMA) films, drawn at 90℃.

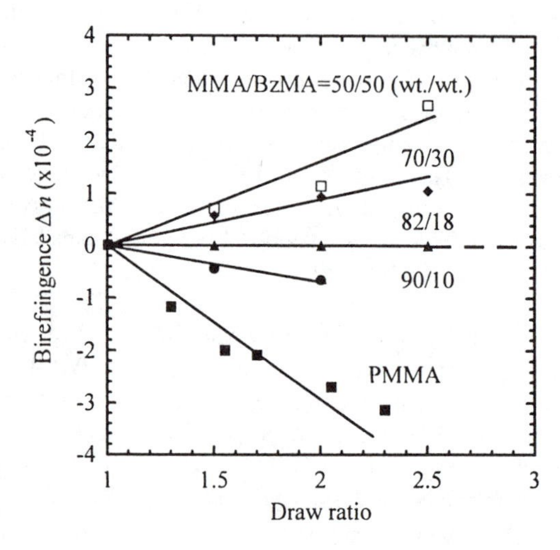

Figure 2. Birefringence of poly(MMA-co-BzMA) films, drawn at 90℃.

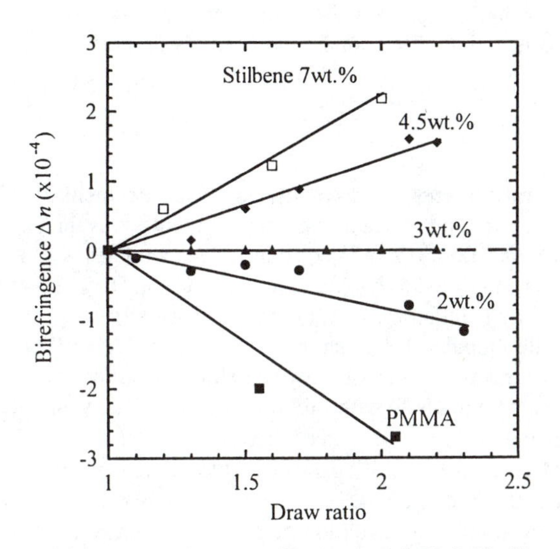

Figure 3. Birefringence of *trans*-stilbene doped PMMA films, drawn at 70℃.

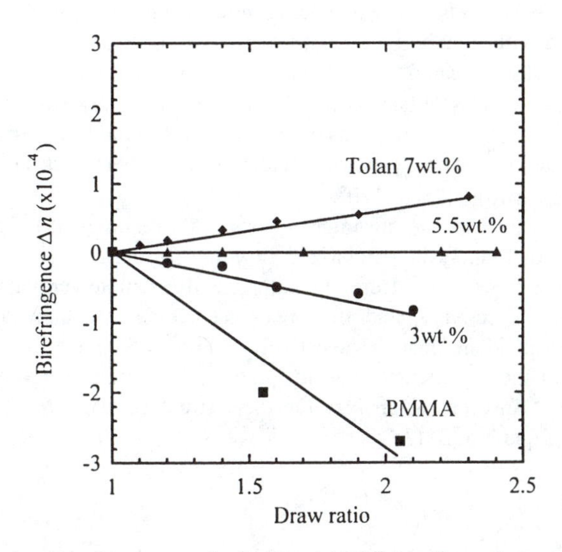

Figure 4. Birefringence of tolan doped PMMA films, drawn at 70℃.

a function of the draw ratio. Zero-birefringence in PMMA was achieved when it was doped with 3.0 wt.% stilbene (Figure 3) or 5.5 wt.% tolan (Figure 4). In both cases, zero-birefringence was always attained at any draw ratio, which proves that the degree of orientation of the dopant molecules is proportional to that of the host polymer chains.

Light Scattering Loss

Sample Preparation and Light Scattering Measurement (*9-12*). The bulk polymer used in the light scattering measurements was prepared as follows. Distilled MMA and BzMA were injected into clean ampules A and B, respectively, which were then connected to two clean ampules C and D. Di-*tert*-butyl peroxide (DBPO) used as an initiator and *n*BM as a chain-transfer agent were placed in ampule C, while ampule D was empty. MMA, BzMA, DBPO and *n*BM were degassed by several freeze-thaw cycles and slowly distilled into ampule D under vacuum by cooling ampule D with liquid nitrogen. Finally ampule D was sealed under vacuum and immersed in heated silicone oil at 130℃ for polymerization.

PMMA samples with dopant molecules were prepared by polymerizing MMA doped with either stilbene or tolan. The composition of the initiator and chain-transfer agent and the condition of polymerization were almost the same as the above copolymerization.

The scattered light intensity of the bulk polymer was measured by the apparatus described in detail elsewhere (*10*). The sample was placed in the center of a cylindrical glass cell. The gap between the sample and inner wall of the glass cell was filled with immersion oil with refractive index of 1.515. The glass cell was perpendicularly located at the center of the goniometer and a parallel beam of vertically polarized He-Ne laser (wavelength λ_0 = 632.8 nm) was injected from the side. Pure benzene was used as a calibration standard in order to estimate absolute intensity. The scattering intensity was measured at the range of scattering angle θ from 30° to 120°.

The average scattering intensity from 200 readings at a fixed angle was employed for data analysis. Polarized (V_V) and depolarized (H_V) scattered light intensities were measured. Here, the letter V and H denote vertical and horizontal polarizations, respectively, and the upper case letter and subscript denote the scattered and incident beam, respectively. Figure 5 shows a typical angular dependence in the V_V intensity due to large-scale heterogeneous structure inside bulk polymer. In order to analyze the local structure, we separated V_V into the two terms of Equation (3) (*13*):

$$V_V = V_{V1} + V_{V2} \qquad (3)$$

where V_{V1} denotes the background intensity independent of the scattering angle, and V_{V2} denotes excess scattering with angular dependence due to large-scale heterogeneities. The isotropic part (V_{V1}^{iso}) of V_{V1} is given by Equation (4).

$$V_{V1}^{iso} = V_{V1} - \frac{4}{3}H_V \quad (4)$$

The scattering loss was estimated by integrating the scattering intensities in all directions. The total scattering loss α_{total} (dB/km) contains three terms, i.e., α_1^{iso}, α_2^{iso} and α^{aniso}. Here α_1^{iso} is the loss due to the scattering V_{V1}^{iso} without angular dependence, α_2^{iso} is due to the isotropic V_{V2} scattering with angular dependence, and α^{aniso} is due to anisotropic scattering (H_V).

$$\alpha_1^{iso}(dB/km) = 1.16 \times 10^6 \pi V_{V1}^{iso} \quad (5)$$

$$\alpha_2^{iso}(dB/km) = \frac{1.35 \times 10^9 a^3 \langle \eta^2 \rangle}{n^4 \lambda^4} \times \left\{ \frac{(b+2)^2}{b^2(b+1)} - \frac{2(b+2)}{b^3} ln(b+1) \right\}$$

(6a)

$$b = \frac{16\pi a^2}{\lambda^2}$$

(6b)

$$\alpha^{aniso}(dB/km) = 3.86 \times 10^6 \pi H_V \quad (7)$$

Equation (6) is obtained by using Debye's theory (*14,15*). Here a (Å) is called the correlation length and is a measure of the size of the heterogeneous structure inside the polymer. The symbol $\langle\eta^2\rangle$ denotes the mean-square average of the fluctuation of all dielectric constants, n is the refractive index of medium, and λ is the wavelength of light in a medium.

Light Scattering Loss of Zero-Birefringence Copolymer. Table II shows the scattering parameters of zero-birefringence poly(MMA-co-BzMA) and those of the homopolymer glasses. The V_V and H_V intensities of all samples had no angular dependence. The intensity of the isotropic V_{V1}^{iso} scattering estimated from the thermal density fluctuation theory by Einstein (*16*) is written as,

$$V_{V1}^{iso} = \frac{\pi^2}{9\lambda_0^4}(n^2-1)^2(n^2+2)^2 kT\beta \quad (8)$$

where λ_0 is the wavelength of light in vacuum, k the Boltzman constant, T the absolute temperature, and β the isothermal compressibility. The experimental isotropic scattering loss α_1^{iso} of PMMA glass was 9.5 dB/km, which is close to the theoretical α_1^{iso} (*ca.* 9.7 dB/km) calculated using Equation (8) at 632.8 nm

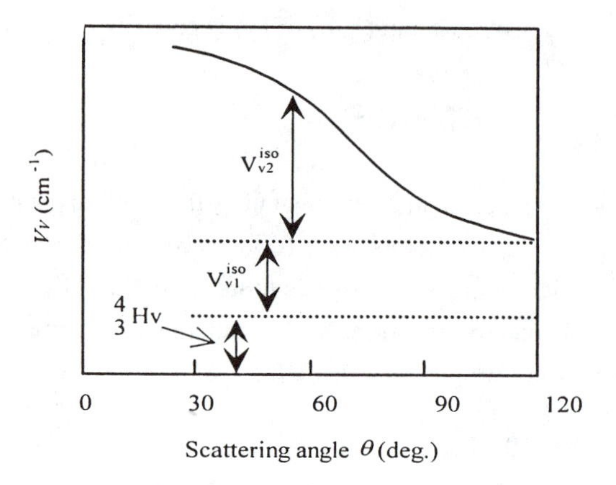

Figure 5. V_V scattering for polymer glass.

Table II. Scattering parameters of poly(MMA-co-BzMA) glasses

MMA/BzMA (*wt./wt.*)	*a* (*Å*)	$<\eta^2>$ ($\times 10^{-8}$)	α_1^{iso} (*dB/km*)	α_2^{iso} (*dB/km*)	α^{aniso} (*dB/km*)	α_{total} (*dB/km*)
100/0 (PMMA)	—	—	9.5	0	4.1	13.6
82/18	—	—	20.0	0	10.4	30.4
0/100 (PBzMA)	—	—	7.8	0	45.1	52.9

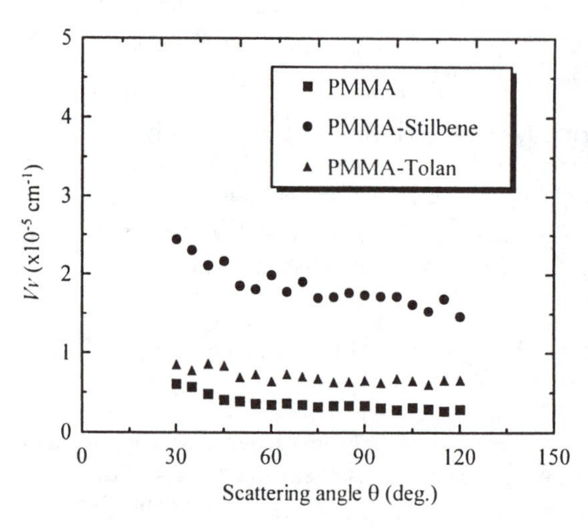

Figure 6. V_V scattering of PMMA with dopant molecules.

wavelength (*12*), whereas that of PBzMA glass was 52.9 dB/km and larger than that of PMMA. It could be mainly attributed to the large anisotropic scattering loss α^{aniso} due to the anisotropy of the phenyl groups of monomer units. On the other hand, the α_{total} of the zero-birefringence poly(MMA-co-BzMA) was 30.4 dB/km which is competitive with those of the homopolymers. Since the V_V intensities of zero-birefringence copolymers had no angular dependence, the α_2^{iso} became zero and heterogeneous structures were not observed. In the MMA-BzMA copolymerization process, monomer reactivity ratios, r_1 and r_2, are almost equal to unity, therefore, the maximum size of the heterogeneities would be the order of a few monomer units. Since the heterogeneities are much smaller than the wavelength of the beam, the excess scattering loss α_2^{iso} (caused by the V_{V2}) of Table II is almost zero even for poly(MMA-co-BzMA).

Light Scattering Loss of Zero-Birefringence PMMA with Dopant Molecules. V_V and H_V scattering intensities of PMMA and zero-birefringence PMMA with dopant molecules are shown in Figures 6 and 7, respectively. Scattering intensities increase with dopant molecules compared with pure PMMA. The increases in H_V intensities are due to large anisotropy of the dopant molecules (Figure 7) and angular dependencies in the V_V intensities indicate a heterogeneous structure (Figure 6). However, the value of α_{total} was less than 100 dB/km, which means that the materials are transparent enough for lenses, waveguides, compact disks, etc.

Injection-Molding

Preparation of Injection-Molded Plates. The injection-molded sample was made of polymer pellets which have a molecular weight of about 1×10^5 (Toshiba Co., IS-130-FII). The molding conditions of polymers having different compositions were almost the same and details are given in Table III. The dimensions of the samples used as heat distortion test pieces were 127 mm$\times$12.5 mm$\times$3.2 mm.

Birefringence of Injection-Molded Polymers. The birefringence of injection-molded poly(MMA-co-BzMA) plates were measured in monochromatic light at 589.3 nm with a Babinet compensation method and the birefringence distribution of samples are shown in Figure 8. The abscissa indicates the length from the injection gate. The conventional PMMA, MMA/BzMA copolymers with compositions 92/8 and 87/13 had birefringence of the order of 10^{-5}-10^{-6} and birefringence distribution. The birefringence of the samples increased on approaching the gate, which means that the orientation of polymer chains increases in the vicinity of the gate. It is noteworthy that the injection-molded zero-birefringence copolymers did not show any birefringence at all points.

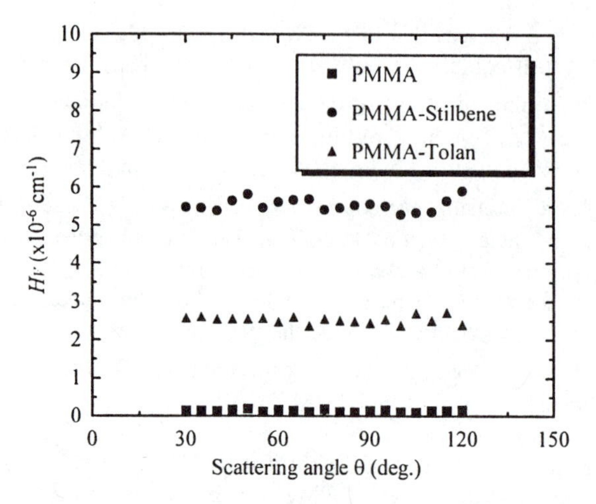

Figure 7. H_V scattering of PMMA with dopant molecules.

Table III. Conditions of injection-molding

Composition	*Temperature (℃)*		*Time (sec.)*	
MMA/BzMA (wt./wt.)	*Barrel*	*Mold*	*Injection*	*Cooling*
100/0 (PMMA)	240	60	15	30
92/8	230	60	15	30
87/13	230	60	15	30
82/18	230	60	15	30

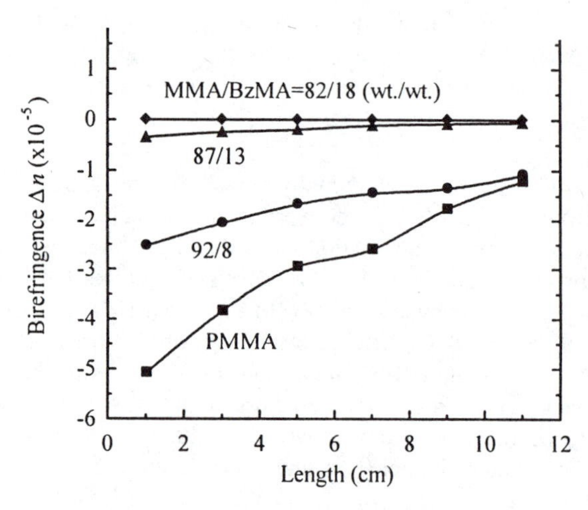

Figure 8. The birefringence distribution of the injection-molded polymers.

Figure 9 shows the photographs of the injection-molded polymers placed between crossed polarizers. Since linearly polarized incident light is converted to elliptically polarized light by the birefringence of a sample, bright areas of the samples in Figure 9 shows an evidence of the birefringence caused by orientation of polymer chains. The brightness gradually decreases from left side of PMMA sample (Figure 9a) because the birefringence decreases with increasing distance from the gate as shown in Figure 8. On the other hand, poly(MMA-co-BzMA) sample (Figure 9b) has no bright area and is quite dark, which means that the sample has no birefringence in the whole area of the sample plate.

Other Properties

The *Tg*, refractive indices (n_D) and Abbe numbers (ν_D) of zero-birefringence polymers are shown in Table Ⅳ. The zero-birefringence poly(MMA-co-3FMA) and poly(MMA-co-BzMA) have slightly lower Tg (the former 90.0℃, the latter 103.3℃) compared with PMMA, because the *Tg*s of P3FMA and PBzMA are 64 and 61℃, respectively. Furthermore, the *Tg* of zero-birefringence PMMA with dopant molecules was also decreased by plasticization.

The Abbe number (ν_D) measures the extent of wavelength dispersion of the refractive index. Polymers possessing a high Abbe number, such as PMMA, show low refractive index dispersion and are useful for lens and other optical elements. As shown in Table Ⅳ, the Abbe number of the zero-birefringence polymers (poly(MMA-co-BzMA), PMMA-Stilbene, and PMMA-Tolan) having phenyl groups were lower than that of PMMA. The Abbe number of a polymer is represented as,

$$\nu_D = \frac{6n_D}{\left(n_D^2+2\right)\left(n_D+1\right)} \cdot \frac{[R]}{[\Delta R]} \tag{9}$$

where n_D is the refractive index at the sodium *D* line (589 nm), $[R]$ the molecular refraction and $[\Delta R]$ the molecular dispersion. In Equation (9), since $[R]/[\Delta R]$ is almost a constant which is independent of the refractive index, polymers having high refractive indices usually have lower Abbe numbers. Therefore, the Abbe number of the zero-birefringence polymers decreases as the number of phenyl groups in the copolymer increases. On the other hand, the zero-birefringence poly(MMA-co-3FMA) (MMA/3FMA= 45/55) has a very high Abbe number ($\nu_D = 64.6$). The Abbe numbers of poly(MMA-co-3FMA) and P3FMA are higher than that of PMMA because of their lower polarizability (*11*) and lower refractive index.

CONCLUSIONS

Orientation zero-birefringence copolymers have been successfully synthesized by random copolymerization of negatively and positively birefringent monomers.

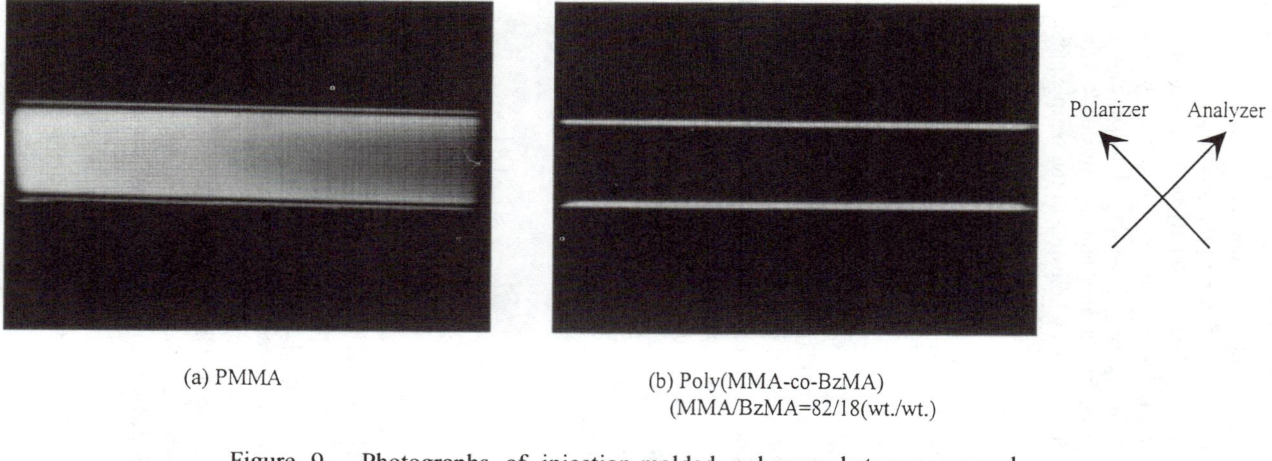

Figure 9. Photographs of injection-molded polymers between crossed polarizers.

The resulting zero-birefringence copolymers eliminate the birefringence caused by the process of the injection-molding while possessing enough transparency to be utilized as optical components. We have also proposed and demonstrated a new concept for compensating birefringence by doping large anisotropic molecules into polymers, resulting in zero-birefringence regardless of the orientation degree of polymer chains. The zero-birefringence polymers also had excellent transparency comparable to PMMA.

Table IV. Glass transition temperature (*Tg*), refractive indices and Abbe numbers of zero-birefringence polymer glasses

Composition (*wt./wt.*)	*Tg* (℃)	*Refractive-index* n_D	*Abbe number* ν_D
PMMA	109.1	1.492	58.9
MMA/3FMA=45/55	90.0	1.453	64.6
MMA/BzMA=82/18	103.3	1.506	47.6
PMMA/Stilbene=100/3	84.0	1.4930	50.7
PMMA/Tolan=100/5.5	82.0	1.5030	47.6

Literature Cited

(1) Ohtsuka, Y.; Koike, Y. *Appl. Opt.* **1985,** *24*, 4316.
(2) Kaino, T. *Appl. Phys. Lett.* **1986,** *48*, 757.
(3) Koike, Y.; Takezawa, Y.; Ohtsuka, Y. *Appl. Opt.* **1988,** *27*, 486.
(4) Koike, Y.; Tanio, N.; Nihei, E.; Ohtsuka, Y. *Polym. Eng. Sci.* **1989,** *29*, 1200.
(5) Bernacki, B. E.; Mansuripur, M. *Appl. Opt.* **1993,** *32*, 6547.
(6) Hahn, B. R.; Wendorff, J. H. *Polymer* **1985,** *26*, 1619.
(7) Saito, H.; Inoue, T. *J. Polym. Sci.*: *Part B* **1987,** *25*, 1629.
(8) Koike, Y. *Polymer* **1990,** *32*, 1737.
(9) Koike, Y.; Matsuoka, S.; Bair, H. E. *Macromolecules* **1992,** *25*, 4807.
(10) Koike, Y.; Tanio, N.; Ohtsuka, Y. *Macromolecules* **1989,** *22*, 1367.
(11) Tanio, N.; Koike, Y.; Ohtsuka, Y. *Polymer J.* **1989,** *21*, 259.
(12) Tanio, N.; Koike, Y.; Ohtsuka, Y. *Polymer J.* **1989,** *21*, 119.
(13) Meeten, G. H. Ed. *Optical Properties of Polymers*; Elsevier Applied Science Publishers: London, 1986.
(14) Debye, P.; Bueche, A. M. *J. App. Phys.* **1949,** *20*, 518.
(15) Debye, P.; Anderson, H. R.; Brumerger, H. *J. App. Phys.* **1957,** *28*, 679.
(16) Einstein, A. *Ann. Phys.* **1910,** *33*, 1275.

Chapter 3

Development of Polymer Optical Waveguides for Photonic Device Applications

Toshikuni Kaino[1], Itaru Yokohama[2], Satoru Tomaru[3], Michiyuki Amano[3], and Makoto Hikita[3]

[1]**Institute for Chemical Reaction Science, Tohoku University, 2–1–1 Katahira, Aoba-ku, Sendai-shi, Miyagi 980–77, Japan**
[2]**NTT Opto-electronics Laboratories, 3–1 Wakamiya, Morinosato, Atsugi-shi, Kanagawa 243–01, Japan**
[3]**NTT Opto-electronics Laboratories, 162 Shirakata, Tokai, Naka-gun, Ibaraki 319–11, Japan**

The development of nonlinear optical polymer waveguides directed toward practical photonic devices are discussed. Electro-optical polymer waveguides and third-order nonlinear optical polymer waveguides have been fabricated and characterized as optical modulators and optical switches. To make the most of polymer processability, hybrid structures of waveguides are proposed since such structures are anticipated to advance optical signal processing technologies. This hybrid fabrication technique has the potential to develop waveguides for optical sampling element and compact Kerr shutter switch applications.

There is a growing interest in the processing of information with optical devices for application in optical telecommunication systems. Optical signal transmission and processing require optical interconnection technology designed to overcome bottlenecks resulting from high circuit densities for high data transmission and processing rates from increased data quantities. These kinds of photonic devices have already attracted considerable attention as key elements in optical signal processing systems.

For device applications, waveguide structures can be effective for controlling optical signals with low power. Polymer waveguides offer the potential to create highly complex integrated optical devices and optical interconnects on a planar substrate because their excellent optical properties can be tailored by using different types of polymers(*1*). Optical processing requires highly nonlinear optical (NLO) materials to operate high data transmission rate signals(*2*). In this case, the trade-off between large optical nonlinearity, interaction length, and processability should be considered. Because high NLO chromophore (NLO-phore) concentrations are needed to achieve large optical nonlinearity, optical loss can be increased due to the $\pi-\pi^*$ transitional absorption tail of the polymer when the NLO-phore is added into the polymer. Thus, reduction of the attenuation loss of the polymer waveguide must be balanced against optical nonlinearity enhancement. Processability of the NLO polymer is also influenced by the concentration of NLO-phore; i.e., a highly functionalized polymer usually becomes difficult to process. The potential of combining an active processing function using NLO properties of materials with a passive transmission function is the most appealing prospect for waveguides made from NLO polymers.

In this paper, the development of nonlinear optical polymer waveguides directed toward practical photonic devices, i.e., electro-optical polymer waveguides and third-order nonlinear optical polymer waveguides, will be discussed. Hybrid polymer waveguides for future applications will also be presented.

Overview of Passive Polymer Waveguides

Recently, many kinds of polymer waveguides have been studied for constructing integrated optical devices and optical interconnections. They have attracted much attention because of their potential for applications. It should be emphasized that almost all the polymer optical waveguides developed so far have an optical loss of greater than 0.1 dB/cm except for poly (deuterated and fluorinated) methylmethacrylate (d-f- PMMA in short) core and fluorinated-polyimide core optical waveguides, detail of which will be presented later.

A silicon wafer area network using polymer integrated optics has been described(*3*). This device is based on a combination of an active reconfiguration function with a passive transmission function. Parallel optical link for Gbyte/s data communications using polymer waveguide was also reported(*4*). The applicability, advantages, and limitations for creating practical optical devices was related to polymer properties and polymer waveguide fabrication processes. Photolithographic techniques are commonly used to define the polymer waveguide patterns. However, optical waveguides are fabricated from polyimide or photo-crosslinking acrylate polymer by using direct writing with lasers or electron beams for fabricating optical waveguides. Using this method, simple straight multi-mode waveguides with propagation loss of 1 dB/cm at 670 nm have been fabricated(*4*). Polymer film layers are typically created by spinning or casting techniques on substrates such as silicon and SiO_2 glass. The process of creating a waveguide is essentially decided by the inherent properties of the selected polymer.

Among the various waveguide fabrication techniques, the method using reactive ion etching (RIE) to form waveguides is an excellent example of technology which can be compared to the solvent etching process. Waveguides created with etching techniques include two types of fabrication processes: (1) an etched groove backfilled with a high-index transparent polymer ; and (2)a ridge waveguide surround by a lower-refractive-index polymer to create a buried waveguide structure. A number of research groups have used these direct methods to make waveguides on silicon, fused silica glass, or organic glass substrates. For example, polyimide is spin-coated on substrates followed by RIE to make ridge waveguides(*5*). Film thickness and refractive index are usually measured by m-line spectroscopy. The refractive index at any wavelength was also determined by a single-term Sellmeier equation using the values at the wavelength of 1.3, 1.152, and 0.633 μm which were accurately determined by the prism coupling technique. Optical loss is measured by evaluating the scattered light from a streak pattern using a video scan technique (i.e., measurement of the light intensity from the waveguide along its length) and an optical multi-channel analyzer. This method can be applied for waveguide loss from 0.1 dB/cm to about 20 dB/cm. For lower losses the scattered intensity is too low, and for higher losses the streak becomes too short. Reliability, reproducibility, stability, acceptable cost performance, and compatibility with other optical systems are the important goal for polymer waveguide systems.

Deuterated and/or Fluorinated Polymer Waveguides

Passive waveguides need highly transparent polymers with excellent processabilities. PMMA and polystyrene (PS) are typical polymers that offer low loss and processability. When discussing the transparency of a polymer, one should consider the vibrational higher harmonics of the polymer(*6*). Figure 1 shows the higher harmonics of the C-H vibration of PMMA. In the visible wavelength region, PMMA-based polymer has the potential low loss of less than 1 dB/m. Although polymer optical waveguides have good processability and low manufacturing cost, they suffer from high optical loss in the near-IR region, 1.0-1.6 μm. For optical components, transparency in the near-IR region, rather than in the visible region, is needed because wavelengths of 1.3 and 1.55 μm are used in optical telecommunication. For the PMMA-based polymer, the loss in this near-IR region will be around 1 dB/cm. This loss can be reduced by fluorination and/or deuteration of hydrogen atoms in the polymer(*7*).

Few polymeric waveguides with a loss below 0.1 dB/cm have been reported for use at 1.3 and 1.55 μm. Among these are low loss deuterated and deuterated/fluorinated polymer core plastic optical fibers (POFs) which have been successfully fabricated(*8*). Figure 2 shows the loss spectra of a per-deuterated PMMA (P[MMA-d8]) core and a fluorinated and deuterated polystyrene (P-5F3DSt) core POFs. At around 1.3 μm wavelength, 0.015 dB/cm loss was obtained for the latter POF. Recently, poly-perfluorinated (butenyl ether) , (CYTOP as a brand name), was also used as a core of POF and its attenuation loss of 0.06 dB/m was obtained(*9*). Therefore, channel waveguides composed of polymers with deuterated methacrylate and deuterated fluoro-methacrylate monomers may be promising candidates for use in optical telecommunication systems. Imamura and co-workers fabricated a single-mode waveguide by using deuterated and fluorinated PMMA (d-f-PMMA for short) through a standard photo-process with RIE technique(*10*). The absorption loss limits of the polymers using these monomers are estimated to be 0.03 dB/cm at 1.3 μm and 0.40 dB/cm at 1.55 μm. A polymer was fabricated for use in waveguides by copolymerization of deuterated methacrylate and deuterated fluoromethacrylate monomers using the standard radical polymerization method. To fabricate single mode waveguides, it is also important to control the refractive index of the core and cladding polymers and to generate precise core patterns. The refractive index can be controlled by the composition of the copolymer.

An important factor in fabricating low-loss and stable waveguides is insuring no intermixing between core and cladding. Channel waveguides were fabricated as follows(*10*). First, a planar waveguide with core and buffer layers was fabricated on a substrate by spin coating. The thickness of the core layer was 8 μm and that of the buffer layer was 15 μm. RIE was then used to form the channel waveguide patterns by etching until the buffer layer surface was exposed. Finally, the core ridges were covered with a spin-coated cladding layer. The buffer and cladding layers had the same refractive index which was 0.40 % lower than that of the core. The loss value was measured by a cut-back method as a function of waveguide length, confirmed using a 1.3-μm-wavelength laser diode light source. Figure 3 shows the wavelength dependence of the loss of a straight waveguide(*10*). The spectrum has an absorption window at 1.3 μm. Attenuation loss of less than 0.1 dB/cm was obtained at 1.3 μm wavelength. Hida and co-workers had fabricated several optical circuits such as a 2x2 TO switch and a ring resonator using the polymer waveguide(*11,12*).

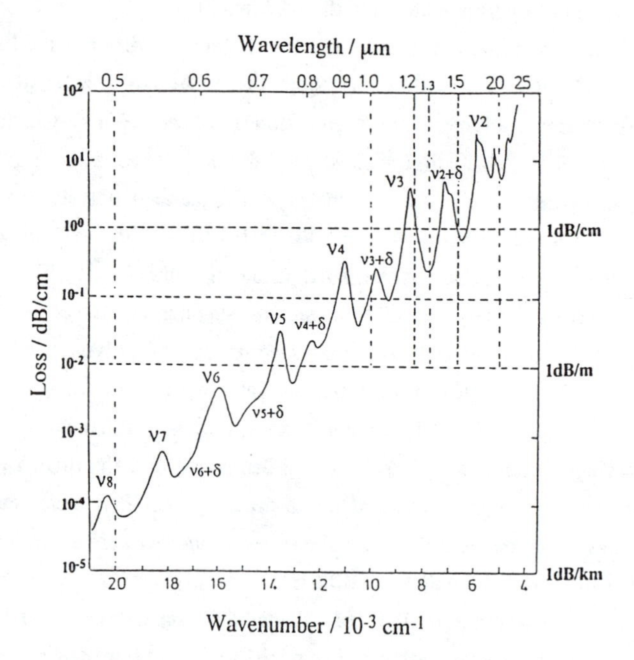

Figure 1. Absorption loss spectrum of PMMA.

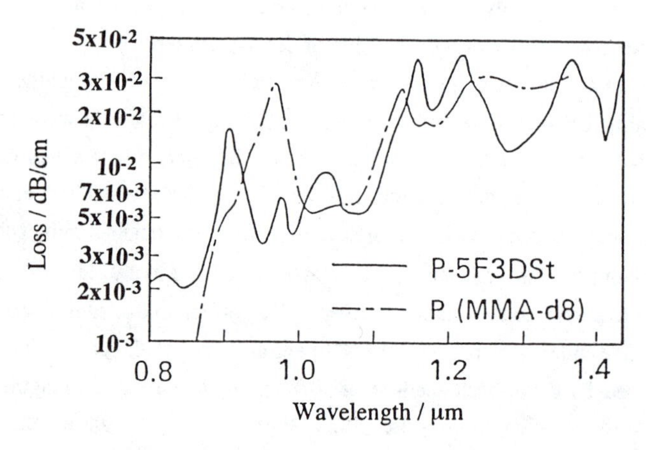

Figure 2. Loss spectrum of deuterated PMMA and fluorinated/deuterated PMMA core Plastic Optical Fibers (POFs).

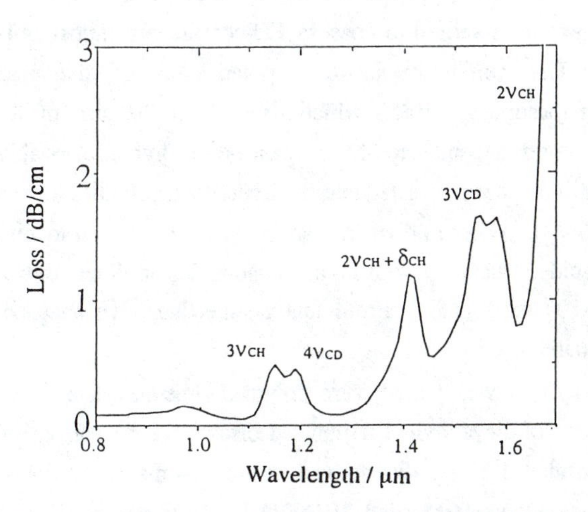

Figure 3. Loss spectrum of fluorinated/deuterated PMMA waveguide.

Sasaki and co-workers have also fabricated single-mode optical waveguides from fluorinated polyimides by using almost the same process as that for d-f-PMMA waveguides. A loss less than 0.5 dB/cm was obtained at 1.3 and 1.55 μm wavelengths(*13*). The structure has a large anisotropy in electronic polarizability, making polyimide a candidate for a birefringent material with large in-plane birefringence(*14*).

Second order Nonlinear Polymer Waveguides

Azo-dye attached poled polymers have been investigated as NLO materials, showing efficient electro-optical (EO) properties with low absorption losses(*15*). These poled azo-polymers have been shown to possess EO-coefficients almost as large as that for $LiNbO_3$(*16*). The nonlinear electronic polarization of dye-attached polymers originates from mesomeric effects which depend on the size of the p-conjugated systems. The length dependence of the second-order hyperpolarizabilities have been investigated and it has been revealed that the hyperpolarizabilities increase rapidly with the chain length(*17*). These polymers can be easily processed to make high-quality thin film waveguide structures, with high second-order nonlinear optical susceptibility ($\chi^{(2)}$) values (~10^{-7} esu) (*18*). From this perspective, dye attached polymers are promising $\chi^{(2)}$ materials.

The polymers we studied were polymethylmethacrylate copolymerized with methacrylate esters of dicyanovinyl-terminated bisazo dye derivative. Disperse Red 1 and the nitro-terminated bisazo dye derivatives are also discussed for comparison(*19*). These azo dyes are hereafter called 3RDCVXY, $2RNO_2$, and $3RNO_2$, respectively (Figure 4). Each dye molecule contains a dicyanovinyl group and a nitro group as an electron acceptor, and a diethylamino group as an electron donor. Donor-acceptor charge transfer will greatly contribute to second-order nonlinear optical molecular susceptibility, β, values. The 3RDCVXY and $3RNO_2$ dyes contain three benzene ring connected with azo groups, and their conjugated structure is longer than that of $2RNO_2$. The molecular structures of the copolymers and side chain azo dyes are shown in Figure 4 along with their $\chi^{(2)}$ wavelength dependence.

The Tgs of the synthesized polymers were 80, 100 and 135 ℃ for $2RNO_2$, $3RNO_2$ and 3RDCVXY, respectively. The maximum absorption wavelengths are 515 nm (2.41 eV) for 3RDCVXY, 500 nm (2.48 eV) for $3RNO_2$, and 470 nm (2.64 eV) for $2RNO_2$. The dimethyl-substitution of the 3RDCVXY effectively increases the dye concentration in the copolymer and as a result the content of the 3RDCVXY dye is nearly twice as large as that of the $3RNO_2$ dye.

$$-\!\left(CH_2 - \underset{\underset{\underset{O-R}{|}}{\underset{C=O}{|}}}{\overset{\overset{CH_3}{|}}{C}}\right)_m\!\!-\!\left(CH_2 - \underset{\underset{\underset{O-CH_3}{|}}{\underset{C=O}{|}}}{\overset{\overset{CH_3}{|}}{C}}\right)_n\!-$$

R= $-CH_2CH_2$ / CH_3CH_2 > N — C_6H_4 — N = N — C_6H_4 — NO_2

2RNO2 (DR-1)

$-CH_2CH_2$ / CH_3CH_2 > N — C_6H_4 — N = N — C_6H_4 — N = N — C_6H_4 — NO_2

3RNO2

$-CH_2CH_2$ / CH_3CH_2 > N — C_6H_4 — N = N — $C_6H_2(CH_3)_2$ — N = N — C_6H_4 — $CH=C(CN)_2$

3RDCVXY

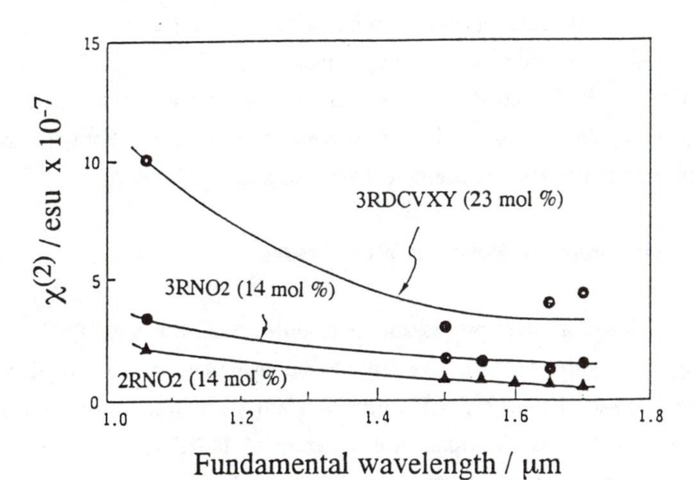

Figure 4. Chemical structure of azo-dye attached EO-polymers and their wavelength dependent $\chi^{(2)}$.

The polymer films on glass substrates were poled by parallel electrode poling. The poling direction was perpendicular to the film surface. The $\chi^{(2)}$ was determined using a standard procedure(*20*), assuming $\chi^{(2)}{}_{333}= \chi^{(2)}{}_{311}$ since the film was found to be isotropic. The $\chi^{(2)}$ increases as the fundamental wavelength decreases which corresponds to the absorption spectrum. This is caused by the $\chi^{(2)}$ enhancement effect of the second harmonic resonance near the absorption band. The $\chi^{(2)}$ measurement revealed that the maximum $\chi^{(2)}$ of the 3RDCVXY copolymer reaches 1.0×10^{-6} esu as shown in Figure 4, which is larger than those of $3RNO_2$ and $LiNbO^3$ by 3 and 7 times, respectively. A thermal aging test shows that the $\chi^{(2)}$ of 3RDCVXY is stable even at 80 °C for more than 6 months. This thermal and temporal stability makes 3RDCVXY a viable substitute for current inorganic EO materials.

As discussed in a previous section, deuteration or fluorination will be needed even for NLO waveguides. It is necessary to fabricate channel waveguides with low optical loss in order to make polymeric NLO switches that can be driven with low laser power(*21*). To this end, a deuterated 3RDCVXY polymer, as shown in Figure 5, was developed for EO applications(*22*). The polymer is composed of transparent copolymer units whose hydrogens are deuterated. Channel waveguides were fabricated using the deuterated 3RDCVXY polymer. Figure 6 shows a schematic of the channel waveguide fabrication process which is based on RIE technique. The selection of overcoat cladding layer is important to obtain flatness of the waveguide surface. A Mach-Zehnder interferometer using the 3RDCVXY EO-polymer was fabricated(*23*). The EO coefficient (r-coefficient) of the waveguide was 26 pm/V at about 70 MV/m poling voltage with a half-wave voltage of 12 V. For comparison, the $LiNbO_3$ interferometer has an r-coefficient value of around 32 pm/V.

Third Order Nonlinear Polymer Waveguides

We have developed a novel processable third-order NLO polymer, PSTF, in which tris-azo-dye NLO-phore was incorporated into the main chain of a polyurethane with a fluorinated alkyl backbone(*24*). Fluorinated alkyl units were used to reduce the polymer waveguide loss. The chemical structure of PSTF is shown in Figure 7. Since third-order NLO polymers generally have higher refractive indices than transparent polymers, the refractive index of the NLO polymers should be considered. We can apply the RIE process for PSTF polymer because the polymer has a moderate refractive index and the cladding material, UV cured fluorinated epoxy resin, has an

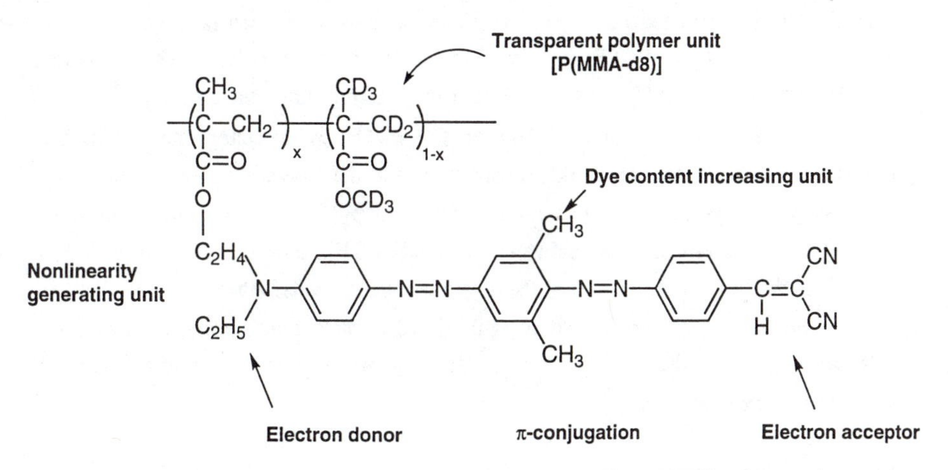

Figure 5. Chemical structure of deuterated 3RDCVXY polymer.

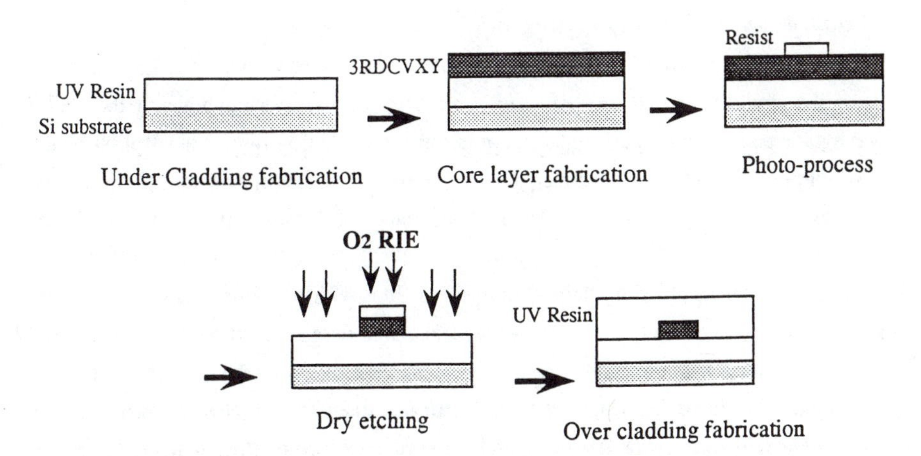

Figure 6. Fabrication method of 3RDCVXY polymer core channel waveguide.

appropriate refractive index for fabricating single-mode channel waveguides. The far-field pattern of the guided mode at a wavelength of 1.3 μm reveals that it is a quasi-single-mode waveguide. Because of the high NLO-phore content of the PSTF, the losses of the waveguide at 1.3 and 1.55 μm were 3.5 and 4.5 dB/cm, respectively, even though a fluorinated alkyl chain was used as part of the matrix polymer. The nonresonant $\chi^{(3)}$ of the PSTF polymer film measured by third harmonic generation (THG) was around 2×10^{-11} esu at 1.55 μm. Using this wavelength, we have confirmed the nonlinear optical effect of the PSTF by detecting the self phase modulation (SPM) of the waveguide. From the result, the nonlinear refractive index, n_2, of the polymer was calculated to be 2.8×10^{-14} cm^2/W(*25*). This value is about an order of magnitude smaller than the value obtained by THG measurement. It is due to the strong two-photon absorption of the polymer at that wavelength.

For an optical Kerr-switch using third-order optical nonlinearity, the switching gate power $P\pi$ for the π-phase shift of the signal beam in the nonlinear optical waveguide is expressed as:

$$P\pi = 3\lambda A / 4Ln_2 \qquad (1)$$

where λ is the signal beam wavelength, A is the core area, and L is the medium length. To reduce the gate power, the core area should be decreased or the medium length and n_2 of the medium should be increased. To date, optical switching using the PSTF waveguide could not be performed due to strong linear and nonlinear absorption in the waveguide. The absorption limits the wavelength at which optical switching can be operated. It is important to think about the trade-off between linear and nonlinear absorption and the $\chi^{(3)}$ value of polymer waveguides when the wavelength for switching is selected. For chalcogenide glass fiber with n_2 value of 2×10^{-14} cm^2/W, peak switching power of less than 3W was obtained using 1.2 meter fiber(*26*). For optical switching with the same gate power as the chalcogenide fiber switch, a polymer waveguide should be fabricated with two-order of magnitude higher n_2 or a polymer waveguide with one-order low loss and one-order higher n_2 than that of the PSTF waveguide.

Future Targets of Polymer Waveguides for Photonic Applications

Currently, several technical problems exist in 2nd and 3rd order NLO polymers. Questions such as when actual organic devices will be available or who will actually make these devices should be seriously considered. As mentioned above, the trade-off between $\chi^{(2)}$ and $\chi^{(3)}$ values of polymer waveguides, linear and nonlinear absorption, processability, and reliability should be carefully considered. To address these issues, novel hybrid waveguides are proposed, i.e., a waveguide with the combination of an active processing function with a passive transmission function. Hybrid waveguides with highly complex optical processing can be fabricated by constructing the waveguide with passive and active waveguides serially grafted or vertically integrated. Quasi-phase-matching (QPM) second-harmonic generation (SHG), or vertical coupling interferometric functions are expected using these waveguides.

The vertical integrated polymer device fabricated was a vertically stacked EO-polymer directional coupler (*27*). The nonlinear optical material used was a bisazo-dye functionalized EO-polymer, 3RDCVXY. A standard RIE process was used to fabricate the hybrid waveguides. Using the vertically stacked directional coupler, optical coupling similar to that of in-plane directional coupler was attained.

A QPM polymer waveguide was fabricated as the serially grafted polymer device (*28*). The QPM technique is very attractive for realizing frequency conversion devices, because it is easy to obtain a collimating beam and achieve frequency conversion with high efficiency. Many QPM studies have been undertaken but these mainly used inorganic waveguides(*29*). Poled polymers have also received particular attention because they are easy to process into optical waveguides(*30*). However, there have been few reports on QPM frequency conversion using the materials described here(*31*) because they are difficult to fabricate into waveguides with a precise periodic poling structure which is an important factor in realizing frequency conversion with high efficiency.

The serially grafting technique, where two different polymers, a UV cured resin and a poled polymer, are used, is an effective method for the fabrication of QPM waveguides with a precise periodic poled structure. Figure 8 shows the fabrication process of the polymer waveguide. The core materials in the waveguide were poled polymer and UV cured resin.

This poled polymer was 3RDCVXY which exhibits a d_{33} value of 80 pm/V and a optical loss of 0.65 dB/cm in the form of a single-mode waveguide operating at 1.3 μm. A UV cured epoxy resin is also used as a core. The refractive index of the UV cured resin can be easily controlled(*28*). This controllability makes it easy to form a periodic laminar structure composed of poled polymer and UV cured resin with almost the same refractive index. On top of the silicon substrate, a low refractive index UV cured epoxy resin is spin coated as an under-cladding layer of the waveguide. Core

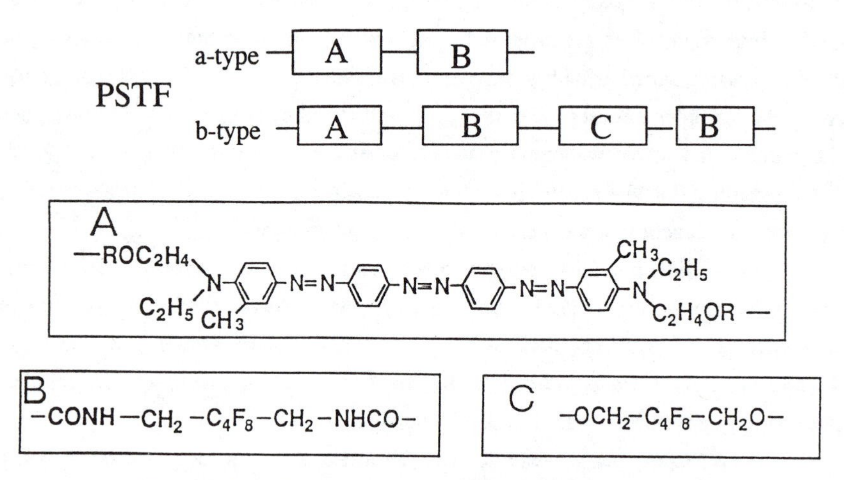

Figure 7. Chemical structure of PSTF $\chi^{(3)}$ -polymer.

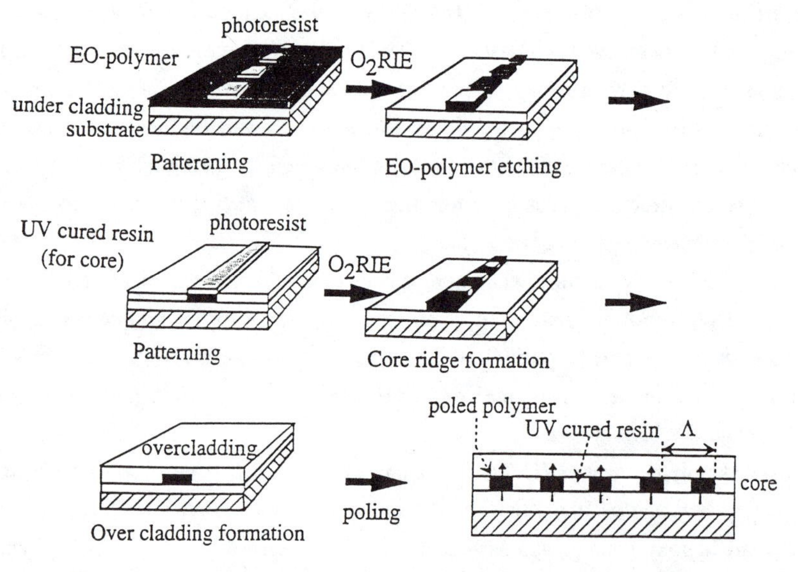

Figure 8. Schematic of periodic waveguide fabrication method.

ridge of EO-polymer and UV-cured epoxy resin whose refractive index difference is controlled to within 0.001 can be successfully fabricated as illustrated in the Figure 8. By covering the core ridge with a low refractive index UV cured resin, a channel waveguide with an excellent periodic structure was fabricated. Although the etching rates of these two polymers are different, overcladding layer can cover the groove formed due to etching speed difference. Finally, the NLO polymer in the waveguide is poled by supplying a high electric voltage through upper and lower electrodes. This poling process provides a strong second-order nonlinearity only to the periodical parts of poled polymer in the core layer. The periodic length (Λ) in the waveguide can be controlled easily by changing the mask pattern.

Using this technique, a QPM SHG polymer waveguide was successfully fabricated. The loss of the EO polymer and UV-cured resin were 0.67 dB/cm and 0.53 dB/cm, respectively, and that of the grafted interface was 0.005 dB/point. By changing the periodic length, in the 10 μm to 90 μm range at intervals of 1 μm, the phase matched wavelength for SHG can be controlled. SHG experiments were performed with 5 mm long QPM waveguides using an F center laser with a 1.48-1.65 μm fundamental wavelength. In this experiment, because there were many waveguides with different periodic lengths on the same substrate, a suitable QPM condition could be probed by choosing a waveguide with an appropriate periodic length. Figure 9 shows the relationship between the periodic length and the fundamental wavelength at the maximum SHG intensity. When the fundamental wavelength was 1.586 μm, the SHG intensity was strongest in the waveguide with a periodic length of 32 μm. The waveguide has a typical efficiency of about 4×10^{-1} %/W/cm^2 at a wavelength of approximately 1.55 μm(*32*). The phase matched SHG was controlled by changing the periodic length. This hybrid fabrication technique may be extended to the fabrication of frequency conversion devices. The QPM waveguide design can be applied not only for EO waveguides but also for Kerr shutter waveguides by considering the role of each waveguide.

Conclusion

We have fabricated and characterized NLO waveguides for optical modulators and optical switches. To make the most of polymer processability, we proposed hybrid waveguide structure. It is anticipated that this new concept of the application of hybrid waveguides will be a key for future advances in optical signal processing technologies. This hybrid fabrication technique can be applied for EO waveguides

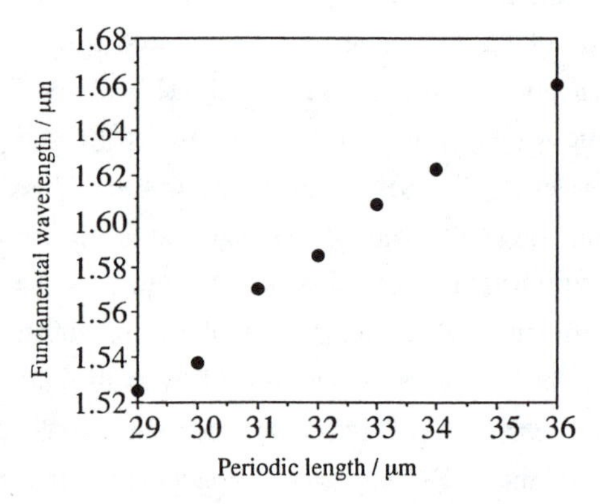

Figure 9. Relationship between periodic length and phase-matched wavelength.
(Reproduced with permission from ref. 32. Copyright 1996 American Institute of Physics)

and for compact Kerr shutter waveguides. The successful development of polymeric optical waveguiding devices will require much effort not only by NLO material and device researchers but also by optical system researchers.

Acknowledgment

We would like to thank M. Asobe, T. Watanabe, T. Kurihara, Y. Shuto, and S. Imamura at NTT Laboratories.

Literature Cited

(1) Beeson, K. W.; McFarland, M. J.; Pender, W. A.; Shan, J., Wu, C.; Yardley, J. T.*Proc. SPIE* **1992**, *1794*, 397.

(2)*Photonic Switching* 2, Toda, K.; Hinton, H. S., Eds.; Springer Series in Electronics and Photonics, Vol. 29; Springer-Verlag: Berlin, Heidelberg, 1990.

(3) Hornak, L. A.; Tewksbury, S. K.; Weidman, T. W.; Kwock, E. W.; Holland, W. R. ;Wolk, G. L., *Proc. SPIE* **1990**, *1337*, 12.

(4) Nutt, A. C. G. *Proc. SPIE* , **1992**, *1794*, 421.

(5) Matsuura, T.; Ando, S.; Matsui, S.; Sasaki, S.; Yamamoto, F. *Electron. Lett.*. **1993**, *29*, 2107.

(6) Kaino, T. In *Polymers for Lightwave and Integrated Optics*, Hornak, L. A., Ed.; Mercel Dekker: New York, 1992, Chap.1.

(7) Groh, W. *Makromol. Chem.* **1988**, *189*, 2861.

(8) Kaino, T. *Appl. Phys. Lett.* **1986**, *48*, 757.

(9) Oharu, K.; Sugiyama, N.; Nakamura, M.; Kaneko, I. *Reports Res. Lab. Asahi Glass Co. Ltd.* **1991**, *41*, 51.

(10) Imamura, S.; Yoshimura, R.; Izawa, T. *Electron. Lett.* **1991**, *27*, 1342.

(11) Hida, Y.; Onose, H.; Imamura, S. *IEEE Photonic Tech. Lett.* **1993**, *5*, 782.

(12) Hida, Y.; Imamura, S.; Izawa, T. *Electron. Lett.* **1992**,*28*, 1314.

(13) Sasaki, S., *Proceedings of Plastic Optical Fibers '94*, **1994** , p 157.

(14) Ando, S.; Sawada, T.; Inoue, Y. *Electron. Lett.* **1993**, *29*, 2143.

(15) Thackara, J.I.; Lipscomb, G.F.; Stiller, M.A.; Ticknor. A.J.; Lytel, R. *Appl. Phys. Lett.* **1988**, *52*, 1031.

(16) Singer, K.D.; Kuzyk, M.G.; Holland, W.R.; Sohn. J.E.; Lalama, S.J.; Comizzoli, R.B.; Katz. H.E.; Schilling, M.L. *Appl. Phys. Lett.* **1988**, *53*, 1800.

(17) Dalton, L.R.; Yu, L.P.U.; Chen, M.; Sapochak. L.S.; Xu, C *Synth. Met.*, **1993**, *54.*, 156.

(18) Shuto, Y.; Amano, M.; Kaino, T. *IEEE Photonic Technol. Lett.* **1991**, *3*, 1003.

(19) Amano, M.; Kaino, T., *Chem. Phys. Lett.* **1990**, *170*, 515.

(20) Man, H.-T.; Yoon, H.N.*Adv. Matter* **1992**, *4*, 159.

(21) Horsthuis,W.H.G.; Koerkamp, M.M.; Heideman, J.L.; Mertens. H.W.; Hams,B.H.*Proc. SPIE* **1993**, *2025*, 516.

(22) Amano, M.; Hikita, M. ; Shuto, Y.; Watanabe, T.; Tomaru, S.; Yaita, H.; Nagatsuma, T. *Proc. SPIE* **1994**, *2143*, 68.

(23) Shuto, Y.; Tomaru, S.; Amano, M.; Hikita, M. *IEEE QE* **1995**, *31*, 1451.

(24) Kurihara, T.; Tomaru, S.; Mori, Y.; Hikita, M.; Kaino, T. *Appl. Phys. Lett.* **1992**, *61*, 1901.

(25) Asobe, M.; Yokohama, I.; Kaino, T.; Tomaru, S.; Kurihara, T. *Proc. 42th Jpn. Applied Physics Society Related Annual Meeting* , **1995**, p 28p-ZQ-12.

(26) Asobe, M.; Kobayashi, H.; Itoh, H; Kanamori, T. *Opt. Lett.* **1993**, *18*, 1056.

(27) Hikita, M.; Shuto, Y.; Amano, M.; Yoshimura, R.; Tomaru, S.; Kozawaguchi, H. *Appl. Phys. Lett.* **1993**, *63*, 1161.

(28) Watanabe, T.; Amano, M.; Hikita, M.; Shuto, Y.; Tomaru, S. *Appl. Phys. Lett.* **1994**, *65*, 1205.

(29) Lim, E.; Fejer, M. M. ; Byer, R. L. *Electron. Lett.* **1989**, *25*, 174.

(30) Khanarian, G.; Norwood, R.A.; Haas, D.; Feuer. B.; Karim, D. *Appl. Phys. Lett.* **1990**, *57*, 977.

(31) Suhara, T.; Morimoto, T.; Nishihara, H. *Proceedings of Micro Optics Conference /GRaded Index '93*, **1993**, p 76.

(32) Tomaru, S.; Watanabe, Amano, M.; Hikita. M.; Yokohama, I.; Kaino, T. *Appl. Phys. Lett.* **1996**, *68*, 1760.

Chapter 4

High-Power Polymer Optical Fiber Amplifiers in the Visible Region

Takeyuki Kobayashi[1,2], Akihiro Tagaya[1,2], Shiro Nakatsuka[1,2], Shigehiro Teramoto[1], Eisuke Nihei[1], Keisuke Sasaki[1], and Yasuhiro Koike[1,2]

[1]Department of Material Science, Faculty of Science and Technology, Keio University, 3–14–1 Hiyoshi, Kohoku-ku, Yokohama 223, Japan
[2]Kanagawa Academy of Science and Technology, 3–2–1 Sakato, Takatsu-ku, Kawasaki 213, Japan

High-power (1200 W) and high-gain (36 dB, approximately 4000 times) amplification has been achieved in a graded-index (GI) polymer optical fiber amplifier (POFA) with 0.3 W input signal under optimized amplification conditions suggested from theoretical analysis. High gain of more than 25 dB (320 times) is expected through the wavelength region of 540~610 nm using Rhodamine 6G, Rhodamine B, and Rhodamine 101-doped GI POFA. The amplification at 649-nm wavelength where POF has a low loss window of transmission was achieved in an Oxazine 4 perchlorate-doped GI POFA.

The construction of optical fiber network systems is proceeding rapidly in recent years due to the emergence of an advanced information-oriented society. As long-haul communication media in such a multimedia society, conventional wire cables have been replaced by single-mode silica optical fibers because high-speed data transmission with a capacity of the order of giga bits/s (bps) is required. In short-distance communication, many junctions and connections of optical fibers would be necessary. The small core of the single-mode optical fiber requires high-accuracy alignment in the fiber connection and junction, which increases the cost of the whole system. Recently, a high bandwidth and low loss graded-index (GI) type polymer optical fiber (POF) having a large core diameter (such as 500 μm or more) was prepared at Keio University (*1-4*). The GI POF is one of the promising candidates to solve this problem of short-distance communication because of its easy processing and handling. Data communication in the GI POF network is expected to be carried out at a visible wavelength because the GI POF has a low loss region (around 650 nm wavelength). Therefore, optical amplifiers, couplers, and switches in such a wavelength region for the GI POF network are necessary.

Very recently, a GI type organic dye-doped polymer optical fiber amplifier (POFA) which has a gain in the visible region was proposed and demonstrated for the first time at Keio University (*5, 6*). High-power (more than 620 W) and high-gain (more than 620 times, 28 dB) nanosecond-pulse-light-amplification in a short length (0.5~2.0 m) of the GI POFA is possible due to the extremely large absorption and emission cross sections of the dye (approximately 10,000 times larger than those of rare-earth elements) and a large core diameter (more than 500 μm) (*7, 8*). As far as we know, such a high-power amplification with the signal and pump pulses of several nanosecond duration in the visible region was realized for the first time. Doping of organic dyes in the POF is possible because of its low heat-drawing temperature (180-220 ℃) compared with a silica optical fiber whose heat-drawing temperature is more than 1000 ℃. The availability of many organic dyes that can be used as dopants result in potential amplification over the whole visible region. These advantages offer attractive applications such as an amplifier in the GI POF network, a booster for a laser diode or other light sources, and so on.

In this paper, optimization of amplification conditions (signal wavelength, fiber length, and dye concentration) to achieve maximum gain in the GI POFA is described. In addition, amplification performance of GI POFA over a wide spectral range in the visible region is shown by theoretical analysis based on dye density distribution and cross section measurements. Actual amplification in the GI POFA at several wavelengths is demonstrated and discussed.

Preparation of POFA

Details of the preparation of the POFA have been described previously (*5-8*). In this section, the preparation of GI POFA is briefly described. A series of methyl methacrylate (MMA) solutions of organic dyes ranging in concentrations from 0.01 to 20 ppm were prepared, in which specified amounts of n-butyl mercaptan, benzyl n-butyl phthalate and dimethyl sulfoxide were dissolved. The n-butyl mercaptan is a chain transfer agent for controlling the molecular weight of polymer, the benzyl n-butyl phthalate is for obtaining the graded-index distribution (*9*), and the dimethyl sulfoxide is for enhancing the solubility of organic dyes in MMA. The organic dye/MMA solution, a t-butyl peroxy isopropylcarbonate initiator, was placed in a PMMA tube with an outer diameter of 10 mm and an inner diameter of 5 mm. The tube was placed in a furnace at 90-95 ℃. Polymerization was carried out for 24 hours followed by heat treatment at 110 ℃ for 60 hours.

A preform rod with a diameter of 10 mm prepared by this process was heat-drawn into a fiber at 190-250 ℃ by taking up reel. The preform rod moved down with a constant velocity V_1 (ca. 10 mm/min) and was heat-drawn by drive roll with a velocity V_2 (ca. 5 m/min). A GI POFA with a desired diameter can be obtained by controlling the ratio of V_2 to V_1.

An example of actual organic dye density distribution in a GI POFA preform rod is shown in Figure 1, where Rp is the radius of the preform rod and r is distance from the center axis. The dye density distribution was determined by two techniques: (1) fluorescence technique; (2) absorbance technique. In the first technique, the dye density was calculated from the fluorescence intensity measured with a fluorescence microscope. In the second technique, the dye density was obtained from the absorbance at 532 nm wavelength measured in the radial direction of disk-like sample (thickness: 0.1-1.0 mm) and the absorption cross section. From these two techniques, it was experimentally confirmed that the GI POFA had a quadratic dye density distribution which provides efficient amplification compared with homogeneous dye density distribution.

Figure 1 also shows a typical example of refractive index distribution in the GI POFA preform rod measured by an interferometric technique (*10, 11*), where n_0 and n mean the refractive indices at the center axis and at distance r, respectively. As shown in the curve of Figure 1, the preform rod has a cladding region coming from the PMMA tube and a quadratic index profile in the core region. The normalized refractive index distribution of the GI POFA was almost the same as that of this preform rod (*12*).

Theoretical Analysis

In order to analyze the amplification in a GI POFA, we assume the model shown in Figure 2 where the decay of the vibrational energy in the first excited singlet state S_1 is extremely fast. Thus the population densities of each level are described as follows:

$$N_3(t, z) \cong 0 \tag{1}$$

$$N_T = N_1(t, z) + N_2(t, z) \tag{2}$$

where N_i is the population density of level i at a position z along the fiber at a time t and N_T is the total density of dyes in the fiber. The rate equation for level 2 is written as

$$\frac{dN_2(t, z)}{dt} = \frac{\sigma_p^a N_1(t, z) P_p(t, z)}{h\nu_p} - \frac{N_2(t, z)}{\tau} - \frac{(\sigma_s^e N_2(t, z) - \sigma_s^a N_1(t, z)) P_s(t, z)}{h\nu_s} \tag{3}$$

Here, $P_p(t, z)$ and $P_s(t, z)$ are the pump and the signal power densities, $\sigma_p{}^a$ and $\sigma_s{}^a$ are absorption cross sections at the pump and the signal wavelengths, $\sigma_s{}^e$ is the emission cross section at the signal wavelength, ν_p and ν_s are frequencies for the pump and the signal, and h means Planck's constant. The signal and pump pulses

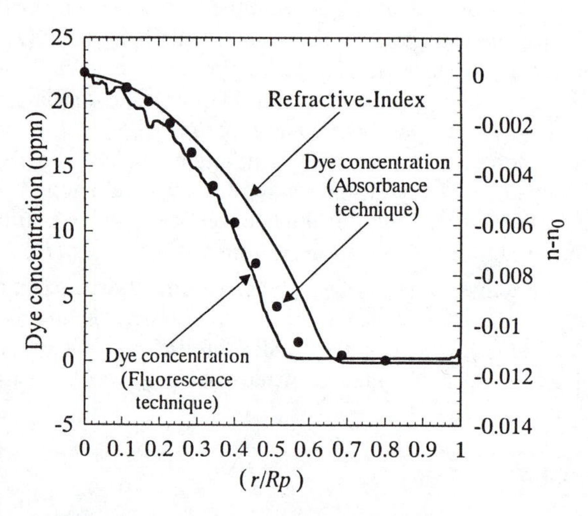

Figure 1. Dye density and refractive index distributions of Rhodamine B-doped GI POFA preform rod prepared from MMA solution of 20 ppm-Rhodamine B. Plots and solid line denote the dye density and refractive index values, respectively.

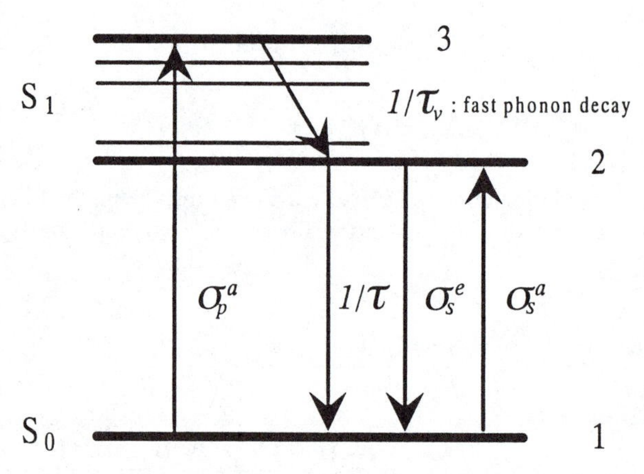

Figure 2. Schematic representation of amplifier model. $\sigma_p{}^a$, absorption cross section at pump wavelength; $\sigma_s{}^a$, absorption cross section at signal wavelength; $\sigma_s{}^e$, emission cross section at signal wavelength; τ, lifetime of level 2; τ_v, lifetime of level 3.

propagate in the positive direction of the fiber axis, according to:

$$\frac{dP_s(t,z)}{dz} = (\sigma_s^e N_2(t,z) - \sigma_s^a N_1(t,z))P_s(t,z) \tag{4}$$

$$\frac{dP_p(t,z)}{dz} = -\sigma_p^a N_1(t,z)P_p(t,z) \tag{5}$$

Taking into account the radial distributions of the optical power density and the dye density, the functions of power density distribution $\Psi(r)$ and the dye density distribution $\Theta(r)$ are defined as

$$I_{s,p}(t,z,r) = P_{s,p}(t,z)\Psi(r) \tag{6}$$

$$n_{1,2}(t,z,r) = N_{1,2}(t,z)\Theta(r) \tag{7}$$

By integrating the overlap between the optical power density and the dye density distributions, equations (3)~(5) become

$$\begin{aligned}\frac{dN_2(t,z)}{dt} = {} & \frac{2\pi\sigma_p^a N_1(t,z)P_p(t,z)}{h\nu_p}\int_0^{a_0}\Theta(r)\Psi(r)rdr - \frac{N_2(t,z)}{\tau} \\ & - \frac{2\pi(\sigma_s^e N_2(t,z) - \sigma_s^a N_1(t,z))P_s(t,z)}{h\nu_s}\int_0^{a_0}\Theta(r)\Psi(r)rdr\end{aligned} \tag{8}$$

$$\frac{dP_s(t,z)}{dz} = 2\pi(\sigma_s^e N_2(t,z) - \sigma_s^a N_1(t,z))P_s(t,z)\int_0^{a_0}\Theta(r)\Psi(r)rdr \tag{9}$$

$$\frac{dP_p(t,z)}{dz} = -2\pi\sigma_p^a N_1(t,z)P_p(t,z)\int_0^{a_0}\Theta(r)\Psi(r)rdr \tag{10}$$

where a_0 is the core radius. Here, $\Psi(r)$ is defined as the Gaussian function, in which 95 % of the total optical energy is in the core region. $\Theta(r)$ is defined as the quadratic function that is similar to the refractive index profile. Equations (8)~(10) are solved by numerical integration, where Δt = 20 ps and Δz = 0.01 mm.

Cross Section Measurements

Absorption Cross Section. Absorption cross section $\sigma^a(\lambda)$ is related to absorbance ABS(λ) by equation (11).

$$\sigma^a(\lambda) = \frac{ABS(\lambda)}{0.4343 N_1 z} \tag{11}$$

Here, N_1 means the dye density and z means the optical path. The absorption cross section of a variety of dyes in bulk PMMA was obtained by measuring ABS(λ), dye density, and optical path.

Emission Cross Section. In general, quantum yield of fluorescence is defined as

$$\Phi = \frac{\int \frac{K(\lambda)\lambda}{hc} d\lambda}{\int \frac{A(\lambda)\lambda}{hc} d\lambda} = \int E(\lambda) d\lambda \leq 1 \tag{12}$$

where $K(\lambda)$ and $A(\lambda)$ are fluorescence and absorption intensities at wavelength λ, and $E(\lambda)$ is the line shape function. The emission cross section $\sigma^e(\lambda)$ is related to $E(\lambda)$ by

$$\sigma^e(\lambda) = \frac{\lambda^4 E(\lambda)}{8\pi c n^2 \tau} \tag{13}$$

Therefore, the emission cross section for a variety of organic dyes can be obtained from the measured quantum yield, fluorescence spectrum, and lifetime data.

Figure 3 shows the absorption and emission cross section spectra obtained for Rhodamine 101 perchlorate (R101) in bulk PMMA. These cross sections ($\sigma_{a\ max} = 3.2 \times 10^{-20}$ m^2, $\sigma_{e\ max} = 2.3 \times 10^{-20}$ m^2) are approximately 10,000 times lager than those of Er^{3+} in GeO_2-SiO_2 glass ($\sigma_{a\ max} = 8.0 \times 10^{-25}$ m^2, $\sigma_{e\ max} = 7.0 \times 10^{-25}$ m^2) (*13*). The optical amplification in the GI POFA was simulated using these cross section data. Further details of the cross section measurements were described elsewhere (*7, 8*).

Amplification Performance of POFA

Optimization for maximum signal gain. Calculated signal gain against fiber length in a R101-doped GI POFA is shown in Figure 4 for a series of launched pump powers. Such a calculation enables us to optimize fiber length and dye concentration for each launched pump power to achieve maximum signal gain.

Signal gain against signal wavelength. Calculated signal gain versus signal wavelength for a variety of organic dye-doped GI POFA are shown in Figure 5. These signal gains were calculated under the optimized conditions (fiber length and dye concentration) mentioned in the previous section. High gain of more than 25

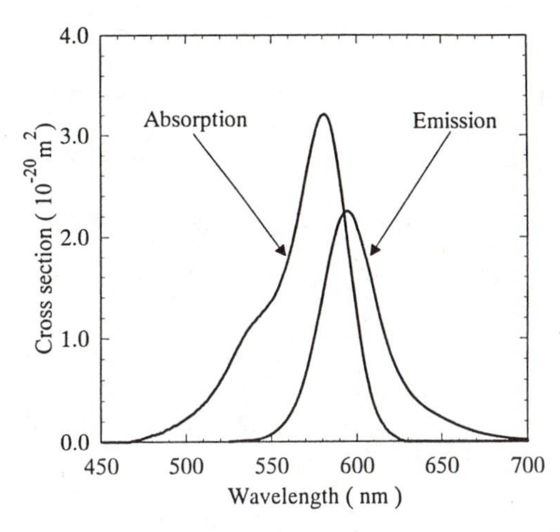

Figure 3. Absorption and emission cross section spectra of Rhodamine 101 perchlorate in bulk PMMA.

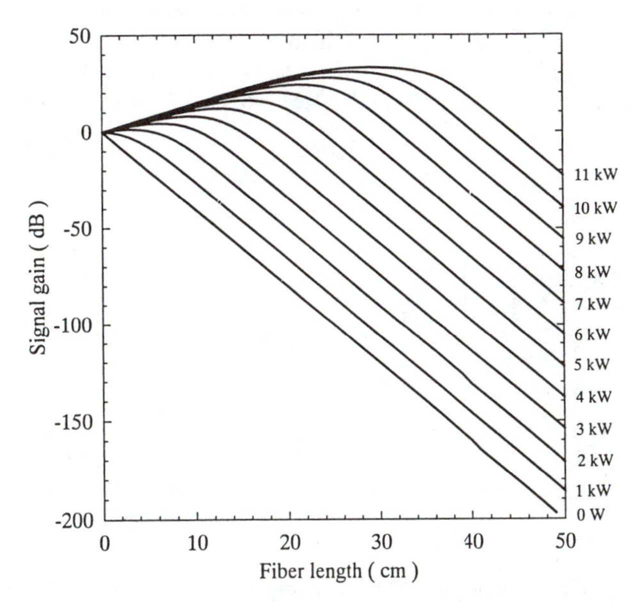

Figure 4. Signal gain versus fiber length for Rhodamine 101 perchlorate-doped GI POFA at a series of launched pump powers. Rhodamine 101 perchlorate concentration = 1.0 ppm. Core diameter = 500 μm. Launched signal power (at 598 nm, FWHM = 3.5 ns) = 1.0 W. Pump wavelength (FWHM = 6.0 ns) = 532 nm.

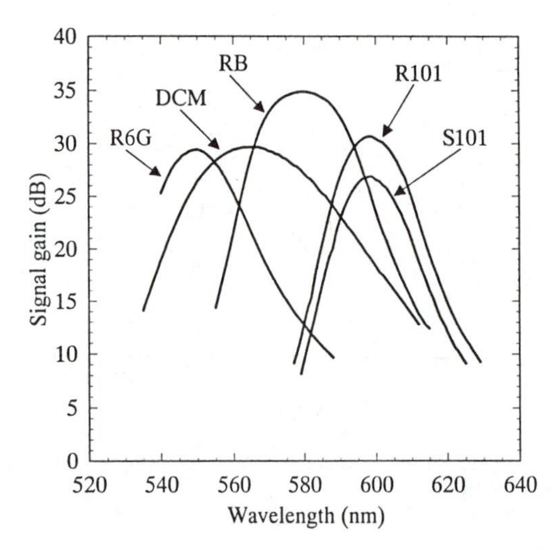

Figure 5. Calculated signal gain versus signal wavelength for Rhodamine B (RB), Rhodamine 6G (R6G), DCM, Rhodamine 101 (R101), and Sulforhodamine 101 (S101)-doped GI POFA. Launched signal power (FWHM = 3.5 ns) = 1.0 W. Launched pump power (at 532 nm, FWHM = 6.0 ns) = 10 kW. Core diameter = 500 μm.

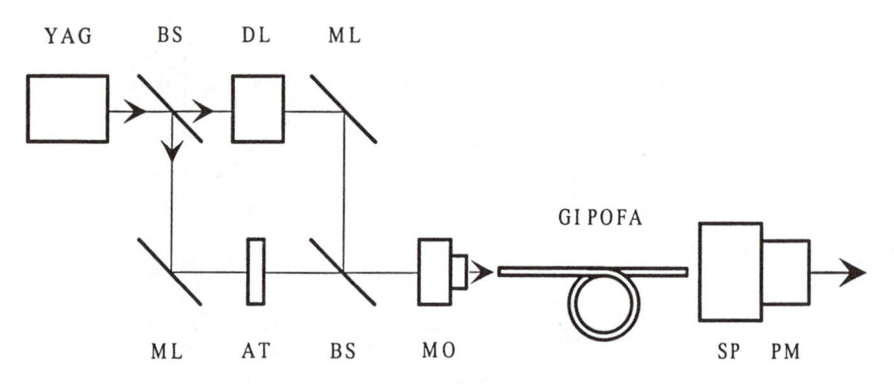

Figure 6. Experimental setup for investigating output spectra of the GI POFA. YAG, frequency-doubled Q-switched Nd:YAG laser; DL, dye laser; BS, beamsplitter; ML, mirror; AT, attenuator; MO, microscope objective; SP, spectroscope; PM, photomultiplier tube.

dB (320 times) can be obtained throughout the spectral region of approximately 70-nm wavelength (540~610 nm) by using Rhodamine 6G, Rhodamine B (RB), and R101-doped GI POFA. In particular, a high gain of 35 dB (3200 times) can be achieved around 580 nm. These results show the possibility of amplification covering the whole visible region in the GI POFA by doping with a variety of organic dyes.

Amplification Experiments

Amplification experiment was carried out under the optimum conditions (signal wavelength, fiber length, and dye concentration) in order to achieve the maximum signal gain of 35 dB by using RB-doped GI POFA. The setup for the amplification experiment is shown in Figure 6. The pump source was a frequency-doubled Q-switched Nd:YAG laser at 532-nm wavelength and the signal source was a dye laser at 580-nm wavelength. The pump and the signal lights were combined by means of a beamsplitter and were coaxially launched into the RB-doped GI POFA. After passing through a spectroscope, the output from the RB-doped GI POFA with a launched signal power of 0.3 W was detected by a photomultiplier tube connected to an oscilloscope. Full width at half maximum for the pump and signal lights were approximately 6.0 ns and 3.5 ns, respectively. The launched pump power and signal power were estimated from the pump and signal coupling efficiencies of the RB-doped GI POFA.

The signal gain of the RB-doped GI POFA at 580-nm wavelength plotted against the launched pump power at 532 nm wavelength is shown in Figure 7. The maximum signal gain of 36 dB (4000 times) was obtained with the launched pump power of 3.5 kW, in which 0.3 W of input signal was amplified up to 1200 W. As far as we know, such a high-power amplification of the signal and the pump pulses of several nanosecond duration in the visible region is realized for the first time. Agreement between the experimental data and the calculated values shows the validity of the optimization based on the theoretical analysis. Other results of the amplification experiments are summarized in Table I. It should be pointed out that the amplification at 649 nm wavelength where GI POF has a low loss region was achieved in an Oxazine 4 perchlorate-doped POFA. Except for the case of RB-doped GI POFA, higher gains than the results shown in Table I are expected because the amplification conditions have not been optimized (signal wavelength, fiber length, and dye concentration) yet.

Conclusions

Polymer optical fiber amplifiers which have a gain in the visible region have been demonstrated for the first time at Keio University. The high-gain (36 dB, approximately 4000 times) and high-power (1200 W) amplification was achieved experimentally under the optimized amplification conditions based on a theoretical analysis using the absorption and emission cross section data. A high gain of more than 25 dB (320 times) is expected throughout the spectral region of

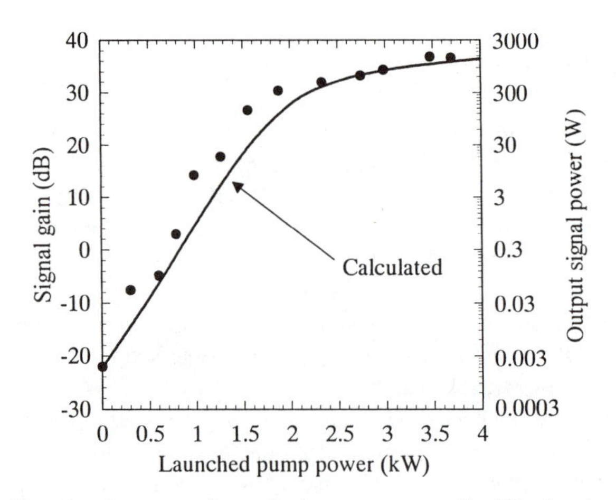

Figure 7. Signal gain versus launched pump power for Rhodamine B-doped GI POFA. Launched signal power (at 580 nm, FWHM = 3.5 ns) = 0.3 W. Pump wavelength = 532 nm. Fiber length = 90 cm. Core diameter = 300 μm. Assumed dye concentration = 0.17 ppm.

Table I. The experimental results of gain measurements

Dye	Gain (dB/times)	Signal Wavelength (nm)	Fiber Length (m)
Rhodamine B	36/4000	580	0.9
	28/620	591	1.0
Rhodamine 6G	26/400	572	1.2
Rhodamine 101 P. C.	13/20	598	2.2
Pyrromethene 567	14/25	567	1.5
Perylene red	20/100	597	1.6
Oxazine 4 P.C.	18/63	649	1.0

approximately 70-nm wavelength (540~610 nm) by using Rhodamine 6G, RB, and R101-doped GI POFA. Amplification at 649-nm wavelength where GI POF has a low loss region, on the other hand, was achieved in the Oxazine 4 perchlorate-doped GI POFA. These results demonstrate the potential of GI POFA to cover the whole visible spectral region.

Literature Cited

(1) Koike, Y.; Ishigure, T.; Nihei, E. *J. Lightwave Technol.* **1995**, *13*, 1475.
(2) Ishigure, T.; Nihei, E.; Koike, Y. *Appl. Opt.* **1994**, *33*, 4261.
(3) Ishigure, T.; Nihei, E.; Koike, Y. In *Conference on Lasers and Electro-Optics in Europe (CLEO EUROPE 94), Technical Digest*, **1994**, *CThD5*.
(4) Koike, Y. In *Third Int. Conference on Plastic Optical Fibers and Applications (POF 94)*, Conference Proceedings, **1994**.
(5) Tagaya, A.; Koike, Y.; Kinoshita, T.; Nihei, E.; Yamamoto, T.; Sasaki, K. *Appl. Phys. Lett.* **1993**, *63*, 883.
(6) Tagaya, A.; Koike, Y.; Nihei, E.; Teramoto, S.; Fujii, K.; Yamamoto, T.; Sasaki, K. *Appl. Opt.* **1995,** *34*, 988.
(7) Tagaya, A.; Teramoto, S.; Yamamoto, T.; Fujii, K.; Nihei, E.; Koike, Y.; Sasaki, K. *IEEE J. Quantum Electron.* **1995,** *31*, 2215.
(8) Tagaya, A.; Kobayashi, T.; Nakatsuka, S.; Nihei, E.; Sasaki, K.; Koike, Y. *Jpn. J. Appl. Phys.*, submitted.
(9) Koike, Y. In *Proc. ECOC'92*, **1992**.
(10) Koike, Y.; Sumi, Y.; Ohtsuka, Y. *Appl. Opt.* **1986**, *25*, 3356.
(11) Ohtsuka, Y.; Koike, Y. *Appl. Opt.* **1980**, *19*, 2866.
(12) Koike, Y. *Polymer*, **1991**, *32*, 1737.
(13) Barnes, W. L.; Laming, R. I.; Tarbox, E. J.; Morkel, P. R. *IEEE J. Quantum Electron.*, **1991**, *27*, 1004.

Chapter 5

Optimization of Modal and Material Dispersions in High-Bandwidth Graded-Index Polymer Optical Fibers

Eisuke Nihei[1], T. Ishigure[2], and Yasuhiro Koike[1,2]

[1]Department of Material Science, Faculty of Science and Technology, Keio University, 3–14–1 Hiyoshi, Kohoku-ku, Yokohama 223, Japan
[2]Kanagawa Academy of Science and Technology, Yokohama, Kanagawa 236, Japan

The optimum refractive-index distribution of the high bandwidth graded-index polymer optical fiber (GI POF) was clarified by consideration of both modal and material dispersions. The ultimate bandwidth achieved by the POF is investigated by a quantitative estimation of the material dispersion as well as the modal dispersion. The results indicated that even if the refractive-index distribution is tightly controlled, the bandwidth of the poly (methyl methacrylate) (PMMA)-base GI POF was dominated by the material dispersion when the required bit rate becomes larger than 4 Gb/s for 100-m transmission. It was also confirmed that the material dispersion strongly depends on the matrix polymer and that the fluorinated polymer whose material dispersion (-0.078 ns/nm·km) is lower than that of poly methyl methacrylate (-0.305 ns/nm·km) allows for a 10 Gb/s transmission rate in 100 m link.

Recent developments in optical fiber technology brings new life to the prospect of optical telecommunications and has triggered broad efforts to develop the science and technology essential to realize new communication concepts such as information super highway.

Single mode glass fiber is widely used in the trunk area of the telecommunication system and the optical fiber system is expected to extend to the subscriber area and the network in building, offices, and homes. However, use of the single mode glass fiber system in short distance network is not necessarily suitable because it requires the precise handling and connection due to the small core size (5-10 μm) and thus it would be expensive in broadband residential area network, ATM-LAN, interconnection, etc. In contrast, polymer optical fiber (POF) which has more than 500 μm of core is promising candidate to solve such problems.

Since all commercially available POF have been of the step-index (SI) type, the modal dispersion limits the possible bit rate of POF links to less than 100 megabit per second (Mb/s). Because of this, it has been thought that POF cannot be utilized for high speed transmission medium. Recently, however, we proposed a large core, low loss, and high bandwidth graded-index polymer optical fiber (GI POF) (*1,2*) for the first time and we confirmed that 2.5 Gb/s signal transmission for 100 m distance was possible in the GI POF (*2,3*).

It is well known that the dispersion in the optical fibers is divided into three parts, modal dispersion, material dispersion, and waveguide dispersion. In the case of the SI POF, the modal dispersion is so large that the other two dispersions can be approximated to be almost zero. However, the quadratic refractive-index distribution in the GI POF can dramatically decrease the modal dispersion. We have succeeded in controlling the refractive-index profile of the GI POF to be almost a quadratic distribution by the interfacial-gel polymerization technique (*2*). Therefore, in order to analyze the ultimate bandwidth characteristics of the GI POF in this paper the optimum refractive index profile is investigated by taking into account not only the modal dispersion but also the material dispersion.

Experimental Section

In order to investigate the material dispersion of the optical fiber, we use several measurements methods that have already been reported (*4,5*). The material dispersion of glass fibers can usually be measured directly by measurement of the delay time of light pulse of several wavelengths through the fiber. However, this method cannot be applied to POF because the higher attenuation of POF than glass optical fiber limits the transmission distance of light pulses and much higher accuracy is required to detect the pulse delay time. Therefore, dispersion measurement in the bulk samples of the polymer used to prepare the GI POF was adopted.

Sample Preparation. The material dispersion which is caused by the wavelength dependence of the refractive index was approximated by the direct measurement of the refractive indices of the polymer at several wavelengths. The PMMA samples were prepared by bulk polymerization of their monomers. We also succeeded in preparing low loss fluorinated polymer-base GI POF (*6*). For a comparison of the material dispersion between PMMA and the fluorinated polymer, poly (hexafluoroisopropyl 2-fluoroacrylate) (PHFIP 2-FA) sample was prepared. The PHFIP 2-FA is the partially fluorinated polymers whose monomer unit contains only three carbon-hydrogen bonds. As dopants to form the graded-index profile in the GI POF (*2*), benzyl benzoate (BEN) for PMMA and dibutyl phthalate (DBP) for PHFIP-2-FA were selected. The dopant feed ratio to the monomer (monomer / dopant) was 5/1 (wt./wt.). The monomer and dopant mixture with initiator and chain transfer agent were placed in a glass ampule with an inner diameter of 18 mm and was polymerized at 90 °C for 24 hours. The method of polymerization of the polymer rod has been described in detail elsewhere (7,8).

Refractive Index. The refractive indices of the bulk polymers were measured by using an Abbe's refractometer at room temperature. In order to measure the refractive indices at several wavelengths, an interference filter was inserted in the refractometer. Two or three measurements were made at each wavelength and averaged. Usually, the refractive-index data were fit to a three-term Sellmeier dispersion equation of the form (*9*):

$$n^2 - 1 = \sum_{i=1}^{3} \frac{A_i \lambda^2}{\lambda^2 - \lambda_i^2} \tag{1}$$

where n is the refractive index of polymer sample, A_i is the oscillator strength, λ_i is the oscillator wavelength, and λ is the wavelength of light.

Results and Discussion

Refractive Index. Average refractive indices of the PMMA, BEN doped PMMA, PHFIP-2-FA, and DBP doped PHFIP 2-FA are summarized in Figure 1 for different wavelengths. The solid lines are the best fitted curves computed by a least-square method for Equation (1). The coefficients resulting from the Sellmeier fitting of the data are given in Table I. It should be noted that the slopes of the refractive index versus wavelength curves indicate that the PHFIP 2-FA has the smaller dispersion of the refractive-index with respect to wavelength than PMMA.

Table I. Coefficients of the Sellmeier Equation Fit of Data

	A_1	λ_1	A_2	λ_2	A_3	λ_3
PMMA	0.4963	71.80	0.6965	117.4	0.3223	9237
PMMA-BEN	0.4855	104.3	0.7555	114.7	0.4252	49340
PHFIP 2-FA	0.4200	58.74	0.0461	87.85	0.3484	92.71
PHFIP 2-FA-DBP	0.2680	79.13	0.3513	83.81	0.2498	106.2

Material Dispersion. The material delay distortion D_{mat} for propagation of a pulse in the step-index (SI) type and single-mode type optical fiber can be obtained from the following equation(*9*):

$$D_{mat} = -\left(\frac{\lambda \delta\lambda}{c}\right)\left(\frac{d^2 n}{d\lambda^2}\right) L \tag{2}$$

where $\delta\lambda$ is the root mean square spectral width of the light source, λ is the wavelength of transmitted light, c is the velocity of light, $d^2n/d\lambda^2$ is the second-order dispersion, and L is length of the fiber.

The material dispersions of the PMMA, BEN doped PMMA, PHFIP 2-FA, and DBP doped PHFIP 2-FA are shown in Figure 2. The material dispersion in Figure 2 were calculated from the results of Sellmeier fitting of Equations (1) and (2). The material dispersion of silica taken from Ref. *9* is also shown in Figure 2. The material dispersion of PMMA is larger than that of the silica from the visible to the near infra-red region. On the other hand, it was confirmed that the substitution of the hydrogen atoms in the polymer with fluorine decreases the material dispersion.

In the case of the fluorinated polymer matrix, a little amount of additives such as DBP or BEN which has a large material dispersion increases the material dispersion of the doped polymer. Therefore, a fluorinated dopant should be used for fluorinated polymer in order to keep the material dispersion low.

Refractive-Index Profile. The refractive-index profile was approximated by the conventional power law. The output pulse width from the GI POF was calculated by the Wentzel-Kramers-Brillouin (WKB) method (*10*) in which both modal and material dispersions were taken into account as shown in Equations (3), (4), and (5). Here, $\sigma_{\text{Intermodal}}$, $\sigma_{\text{Intramodal}}$, and σ_{total} signify the root mean square pulse width due to the modal dispersion, intramodal (material) dispersion, and both dispersions, respectively.

$$\sigma_{\text{Intermodal}} = \frac{LN_1\Delta}{2c}\cdot\frac{g}{g+1}\cdot\left(\frac{g+2}{3g+2}\right)^{\frac{1}{2}} \cdot\left[C_1^2+\frac{4C_1C_2\Delta(g+1)}{2g+1}+\frac{4\Delta^2C_2^2(2g+2)^2}{(5g+2)(3g+2)}\right]^{\frac{1}{2}} \quad (3)$$

$$\sigma_{\text{Intramodal}} = \frac{\sigma_s L}{\lambda}\left[\left(-\lambda^2\frac{d^2n_1}{d\lambda^2}\right)^2 - 2\lambda^2\frac{d^2n_1}{d\lambda^2}(N_1\Delta) \cdot C_1\cdot\left(\frac{2g}{2g+2}\right)+(N_1\Delta)^2\left(\frac{g-2-\varepsilon}{g+2}\right)^2\cdot\frac{2g}{3g+2}\right]^{\frac{1}{2}} \quad (4)$$

where,

$$C_1 = \frac{g-2-\varepsilon}{g+2}$$

$$C_2 = \frac{3g-2-2\varepsilon}{2(g+2)}$$

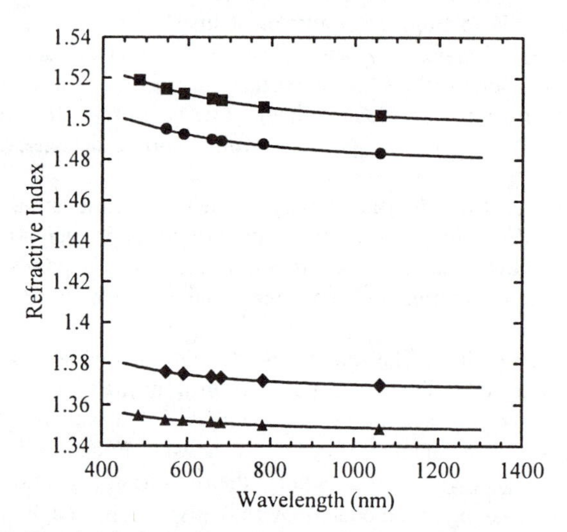

Figure 1. Refractive-index dependence of the polymers on wavelength. ■: benzyl benzoate doped PMMA; ●: PMMA; ◆: dibutyl phthalate doped partially fluorinated polymer; ▲: partially fluorinated polymer.
Reproduced with permission from reference 13. Copyright 1996 Optical Society of America.

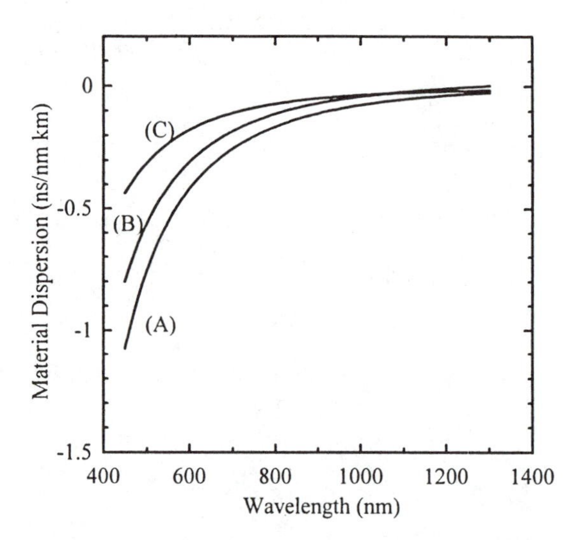

Figure 2. Comparison of the material dispersion among PMMA, PHFIP 2-FA, and silica. (A): PMMA; (B): Silica; (C):PHFIP 2-FA.

$$\varepsilon = \frac{-2n_1}{N_1} \cdot \frac{\lambda}{\Delta} \cdot \frac{d\Delta}{d\lambda}$$

$$N_1 = n_1 - \lambda \cdot \frac{dn_1}{d\lambda}$$

Here, σ_s is root mean square spectral width of the light source (in nanometer) and L is the fiber length.

$$\sigma_{\text{total}} = \left[\left(\sigma_{\text{Intermodal}}\right)^2 + \left(\sigma_{\text{Intramodal}}\right)^2 \right]^{\frac{1}{2}} \qquad (5)$$

Calculated results in PMMA-base and PHFIP 2-FA-base GI POFs are shown in Figures 3 and 4, respectively. Here, it was assumed that the fiber length and the spectral width of the light source (laser diode) were 100 m and 2 nm, respectively. When only modal dispersion was taken into account, the pulse width was minimized when g equaled 1.97 in both PMMA-base and PHFIP 2-FA-base GI POFs. On the other hand, when the material dispersion at 650-nm wavelength was taken into account in Figure 3, the total pulse width (σ_{total}) becomes 45 times broader, and the index exponent giving the smallest pulse width (σ_{total}) was shifted to 2.33. It is quite noteworthy that the effect of the material dispersion on the pulse width in PHFIP 2-FA-base GI POF in Figure 4 is much smaller that that in PMMA-base GI POF in Figure 3. Based on the results shown in Figures 3 and 4, we estimated the possible bit rate in the GI POF link when the rms spectral width of the light source was assumed to be 2 nm. It has been reported by Personick (*11*) that the power penalty of the receiver reaches 1 dB when the pulse width exceeds one fourth of the bit period (1/B [in second] where B is the bit rate [in bit per second]). Therefore, the possible bit rate $B_p$was defined by Equation (6)

$$B_p = \frac{1}{4\sigma_{\text{total}}} \qquad (6)$$

Calculated possible bit rate in PMMA-base and PHFIP 2-FA-base GI POF links is shown in Figures 5 and 6, respectively. The results indicate that maximum bit rates of 25.7 Gb/s and 5.48 Gb/s can be achieved at 780-nm wavelength in PHFIP 2-FA-base GI POF and PMMA-base GI POF links, respectively. Also, the maximum bit rates at 780-nm wavelength are higher than those at 650-nm wavelength in both PHFIP 2-FA-base and PMMA-base GI POF links, because the material dispersion becomes small with increase in wavelength of light as shown in Figure 2. The attenuation spectra of light transmission for both PMMA-base and

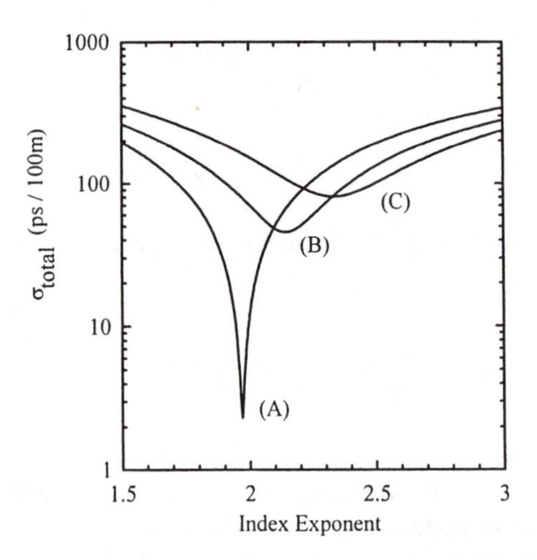

Figure 3. Pulse width (σ_{total}) versus index exponent of PMMA-base GI POF assuming equal power in all modes and light source having rms spectral width of 2 nm.
(A): Only modal dispersion is considered at 650-nm wavelength.
(B): Both modal and material dispersions at 780-nm wavelength are considered.
(C): Both modal and material dispersions at 650-nm wavelength are considered.
Reproduced with permission from reference 14. Copyright 1996 The Society of Polymer Science, Japan.

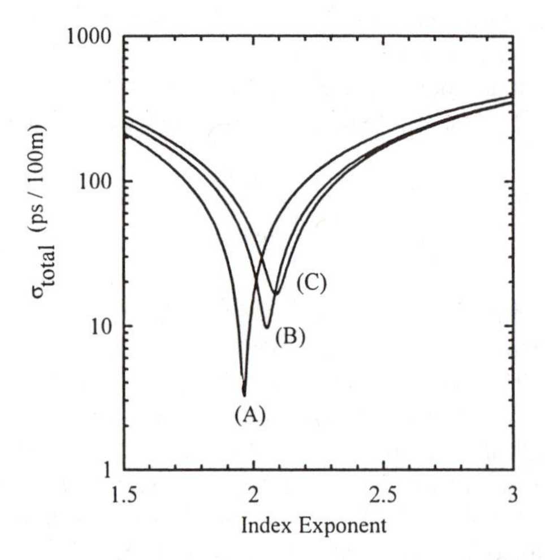

Figure 4. Pulse width (σ_{total}) versus index exponent of PHFIP 2-FA-base GI POF assuming equal power in all modes and light source having rms spectral width of 2 nm. (A): Only modal dispersion is considered at 780-nm wavelength.
(B): Both modal and material dispersions at 780-nm wavelength are considered.
(C): Both modal and material dispersions at 650-nm wavelength are considered.
Reproduced with permission from reference 14. Copyright 1996 The Society of Polymer Science, Japan.

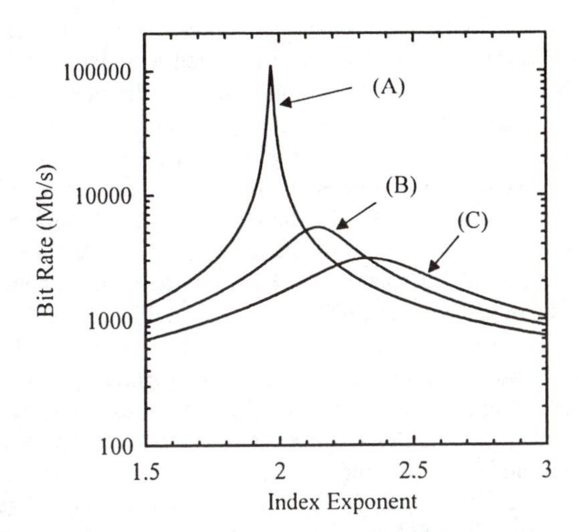

Figure 5. Relationship between index exponent g and possible bit rate in 100-m length PMMA-base GI POF link. Source width was assumed to be 2 nm.
(A): Only modal dispersion is considered at 650-nm wavelength.
(B): Both modal and material dispersions at 780-nm wavelength are considered.
(C): Both modal and material dispersions at 650-nm wavelength are considered.

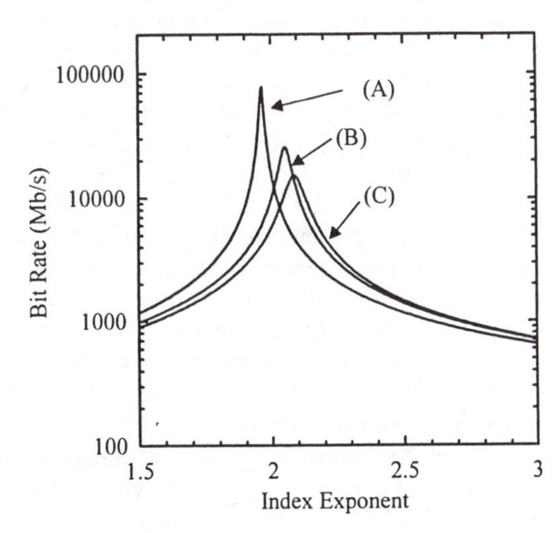

Figure 6. Relationship between index exponent g and possible bit rate in 100-m length PHFIP 2-FA-base GI POF link. Source width was assumed to be 2 nm.
(A): Only modal dispersion is considered at 780-nm wavelength.
(B): Both modal and material dispersions at 780-nm wavelength are considered.
(C): Both modal and material dispersions at 650-nm wavelength are considered.

PHFIP 2-FA-base GI POFs are shown in Figure 7. As most carbon-hydrogen bondings in the PMMA-base GI POF are substituted with carbon-fluorine bondings in the PHFIP 2-FA-base POF, the absorption loss in the near infrared to infrared region due to carbon-hydrogen stretching vibration can be eliminated. Consequently, the total attenuation at the near infrared region of the PHFIP 2-FA-base GI POF is lower than that of PMMA-base GI POF as shown in Figure 7. For instance, the attenuation of PHFIP 2-FA-base GI POF at 780-nm wavelength is 135 dB/km, while 840.5 dB/km in PMMA-base GI POF. This low attenuation at near infrared region of PHFIP 2-FA-base GI POF enables the use of inexpensive laser diode (LD) for compact disc as the light source in PHFIP 2-FA-base POF link. Therefore, the fluorinated polymer base GI POF is advantageous because of the low attenuation at near infra-red region and of the low material dispersion (*6*).

In the POF link, it has been proposed that inexpensive LED can be utilized as the light source (*12*). However, since the LED has large spectral width (15-20 nm) than the laser diode (1-3 nm) the effect of the spectral width on the GI POF link by the LED should be investigated. The possible bit rate in 100 m PMMA-base and PHFIP 2-FA-base GI POF links is shown in Figures 8 and 9, respectively when the spectral widths of the light source are assumed to be 0, 1, 2, 5, and 20 nm. The wavelengths of light sources in Figures 8 and 9 are 650 nm and 780 nm, respectively. In the case of the PMMA-base GI POF, the maximum bit rate dramatically decreases with an increase in the source spectral width increases as shown in Figure 8. When the source width is 20 nm, the bit rate becomes almost independent of the index exponent as shown in Figure 8E. Therefore, the GI POF with almost the quadratic refractive index distribution is more effective for high speed transmission when a light source with narrow spectral width such as laser diode (LD). It was confirmed that it is impossible to cover more than 1 Gbit/s of bit rate unless a light source with less than 5 nm spectral width can be used and the refractive-index profile can be controlled in the order of 2 to 3 of the index exponent g. On the other hand, in the case of the PHFIP 2-FA-base GI POF, the low material dispersion enables to transmit more the transmission of more than 1 Gb/s even if the spectral width of the light source is 10 to 20 nm.

The relationship between the possible bit rate versus wavelength when the index exponent g is fixed in both PMMA and PHFIP 2-FA-base GI POF links is shown in Figures 10 and 11, respectively. In the case of PMMA-base GI POF, when the index exponent g is 2.33, the possible bit rate in 100 m link increases with increasing wavelength. When the index exponent is larger than 3, there is no wavelength dependence of the possible bit rate. In this case, as the index profile becomes close to step-index, the modal dispersion is dominant and the effect of the material dispersion is negligible even if the wavelength changes. On the other hand, in the case of PHFIP 2-FA-base GI POF, it is noteworthy that by changing the index exponent, the possible bit rate can be maximized at desired wavelength in the range of 500 nm to 1000 nm.

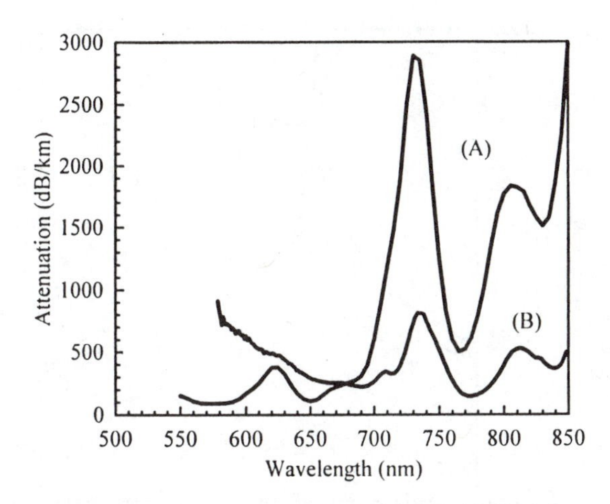

Figure 7. Total attenuation spectra of the GI POF. (A): PMMA-base; (B): PHFIP 2 FA-base.

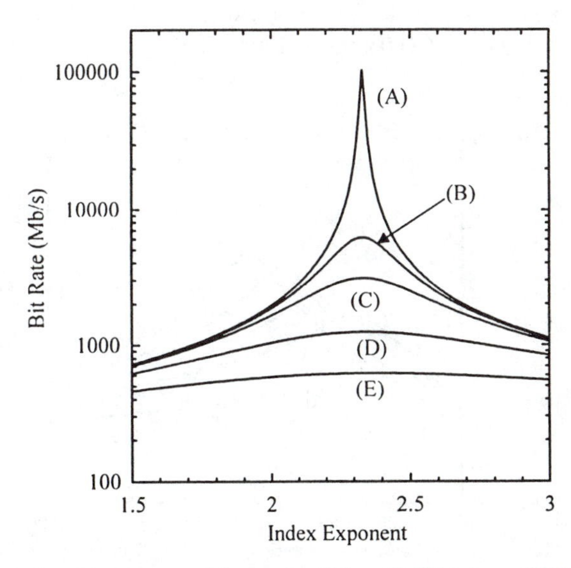

Figure 8. Relationship between bit rate in 100-m PMMA-base GI POF link and index exponent at 650-nm wavelength when the light source has a finite spectral width as follows: (A): 0 nm; (B):1 nm; (C): 2 nm; (D): 5 nm; (E): 20 nm.

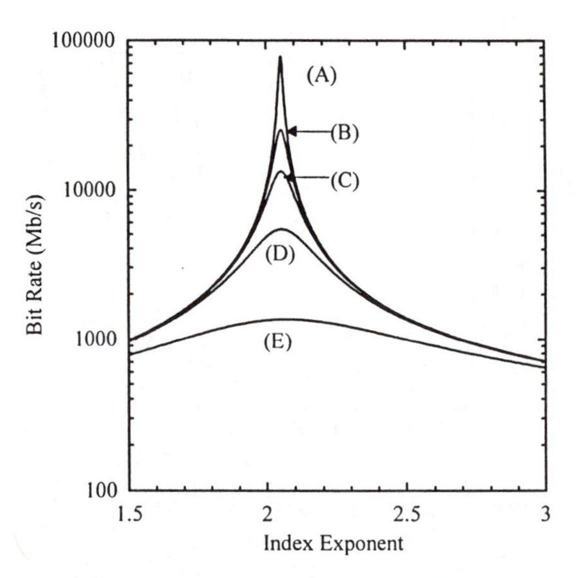

Figure 9. Relationship between bit rate in 100-m PHFIP 2-FA-base GI POF link and index exponent at 780-nm wavelength when the light source has a finite spectral width as follows: (A):0 nm; (B):1 nm; (C): 2 nm; (D): 5 nm; (E): 20 nm.

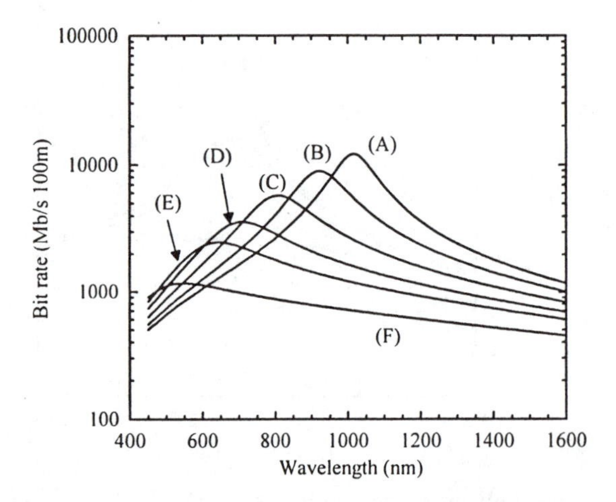

Figure 10. Wavelength and index exponent g dependencies of possible bit rate in 100-m PMMA-base GI POF link at 650-nm wavelength. (A): g= 1.9; (B): g=2.0; (C): g=2.15; (D): g=2.33; (E): g=2.5; (F): g=3.0.

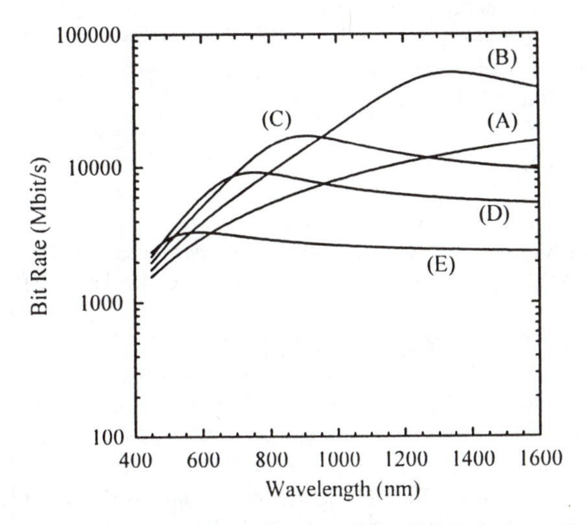

Figure 11 Wavelength and index exponent g dependencies of possible bit rate in 100-m PHFIP 2-FA-base GI POF link at 780-nm wavelength. (A): g= 1.95; (B): g=2.0; (C): g=2.05; (D): g=2.1; (E): g=2.25.

Conclusion

The optimum refractive-index profile of GI POF was investigated by taking both modal and material dispersions into account. The material dispersion seriously affects the bit rate performance of GI POF link when the signal spectral width is larger than the order of 1 nm. In order to achieve bit rates of more than 1 Gb/s in 100-m POF link, the source width should be less than 5 nm and the index profile of the PMMA-base GI POF should be controlled in the range from 2 to 3 of the index exponent g. The PHFIP 2-FA-base GI POF can transmit higher bit rates than PMMA-base GI POF even if the source width is as broad as 20 nm because the fluorinated polymer has a low material dispersion. Furthermore, the low absorption loss of PHFIP 2-FA-base GI POF at the near infra-red region enables to use the longer wavelength in the POF link. It is concluded that fluorinated polymer-base GI POF is superior to the PMMA-base GI POF in both attenuation of light transmission and total bit rate at near infrared region where several inexpensive LEDs and LDs exist.

Literature Cited

(1) Ishigure, T.; Nihei, E.; Koike, Y. *Appl. Opt.* **1994**, *33*, 4261.
(2) Koike, Y.; Ishigure, T.; Nihei, E. *IEEE J. Lightwave Technol.* **1995**, *13*, 1475.
(3) Ishigure, T.; Nihei, E.; Yamazaki, S.; Kobayashi, K.; Koike, Y. *Electron. Lett.* **1995**, *31*, 467.
(4) Horiguchi, M.; Ohmori, Y.; Miya, T. *Appl. Opt.* **1979**, *18*, 2223.
(5) Heckert, M. J. *IEEE Photon. Technol Lett*. **1992**, *4*, 198.
(6) Ishigure, T.; Nihei, E.; Koike, Y.; Forbes, C. E.; LaNieve, L.; Straff, R.; Deckers, H. A. *IEEE Photon. Technol. Lett.* **1995**, *7*, 403.
(7) Koike, Y.; Tanio, N.; Ohtsuka, Y. *Macromolecules.* **1989**, *22*, 1367.
(8) Koike, Y.; Matsuoka, S.; Bair, H. E. *Macromolecules.* **1992**, *25*, 4809.
(9) Fleming, J. W. *J. Am. Ceram. Soc*. **1976**, *59*, 503.
(10) Olshansky, R.; Keck, D. B. *Appl. Opt.* **1976**, *15*, 483.
(11) Personick, S. D. *Bell Syst. Technol. J.* **1973**, *52*, 875.
(12) Yoshimura, T.; Nakamura, K.; Okita, A.; Nyu, T.; Yamazaki, S.; Dutta A. K. *Conference Proceeding of Plastic Optical Fibers & Applications, Boston*, October **1995**, 119
(13) Ishigure, T.; Nihei, E.; Koike, Y. *Polymer J.* **1996**, *28*, 272.
(14) Ishigure, T.; Nihei, E.; Koike, Y. *Appl. Opt.* **1996**, *35*, 2048.

Chapter 6

Preparation and Characterization of Gradient-Index Polymer Fibers

W. C. Chen[1], J. H. Chen[2], S. Y. Yang[2], J. J. Chen[2], Y. H. Chang[2], B. C. Ho[2], and T. W. Tseng[2]

[1]**Department of Chemical Engineering, National Taiwan University, Taipei, Taiwan, Republic of China**
[2]**Union Chemical Laboratories, Industrial Technology Research Institute, Hsinchu, Taiwan, Republic of China**

Gradient-index (GRIN) polymer fibers with a quadratic distribution of the refractive index are prepared by a novel closed extrusion process. The effects of the die geometry and the length of the diffusion zone on the refractive index distribution of the polymer fibers are studied. The results show that the non-quadratic refractive index portion of the GRIN polymer fiber can be removed by a core-shell separation die design. Also, the spatial refractive index distribution (Δn) values can be increased by adding a low refractive index monomer in the shell reactant mixture. The GRIN polymer fibers have potential applications in imaging and optical fiber communication.

Gradient-index (GRIN) polymer fibers with a parabolic refractive index distribution have attracted extensive interest in light of their promising potential in optical fiber communication and imaging (*1-9*). High bandwidth GRIN polymer optical fiber has been successfully obtained, which can exhibit 2.5 Gbit/s signal transmission over 100 m using a 650-nm high speed laser diode (*7*). An input signal of 0.85 W at 591 nm was amplified to 420 W (27-dB gain) by injection of 690 W of pump power at 532 nm into GRIN polymer optical fiber amplifier (POFA) with a 0.5 m length (*9*). Microlenses made from GRIN polymer optical fibers have been used as the essential components of the Selfoc lens array (SLA), which is currently used in the imaging parts of commercially available fax machines (*4*).

Several approaches have been developed to prepare GRIN polymer fibers in our laboratories, e.g., initiator diffusion technique (*10*), interfacial-gel copolymerization (*11*), and extrusion technique (*12,13*). Our previous study successfully used a closed extrusion process to prepare GRIN polymer fibers (*12,13*). This new process has also overcomed several drawbacks in conventional extrusion processes, e.g., poor reproducibility and low production rate. The refractive index profiles of the GRIN polymer fibers prepared by the new process are affected by the following two factors: (1) the formulation of the reactant mixtures, e.g., different types of monomer combination and variation of host polymers; and (2) the parameters of the extrusion process, e.g., the die geometry , the temperature and the length of the diffusion zone. In this study, the effects of the die geometry and the length of the diffusion zone on the refractive index profiles of the prepared GRIN polymer fibers are reported. Also, imaging applications of GRIN polymer fibers are discussed.

Experimental Section

Materials. Methyl methacrylate (MMA, 99%, Janssen Chimica), benzyl methacrylate (BzMA, 98%, TCI), and 2,2,3,3-tetrafluoropropyl methacrylate (4FMA, Echo Chemical Co.) were purified by vacuum distillation. PMMA1 (Chi-Mei, Co., $M_w = 8.3x10^4$, $M_n = 4.0x10^4$), PMMA2 (Asahi, DELPET 70H, $M_w = 13x10^4$, $M_n = 8x10^4$), 1-hydroxycyclohexyl phenyl ketone (HCPK, TCI) and hydroquinone (HQ, 99%, Janssen Chimica) were used without further purification. The refractive indices of the homopolymers PMMA, PBzMA, and P4FMA measured at 589 nm are 1.49, 1.57, and 1.42, respectively.

Preparation of GRIN Polymer Fibers. The closed extrusion process for preparing GRIN polymer fibers was described in our previous reports(*12, 13*). Figure 1 shows a schematic diagram of the apparatus of the extrusion process used for preparing GRIN polymer fibers. A material supply tank(A) contained the solution of a polymer (1a) and at least one monomer (2a) (may also contain another monomer (2b); meanwhile, a material supply tank (B) contained the solution of a polymer (1b) and at least one other monomer (2c). These two solutions were heated at 60^0C. Gear pumps (C) and (D) with a speed of 0.5 and 2 cc/min were used to feed the two reactant mixtures A (**core**) and B (**shell**) with a volume ratio of 1:3 into a concentric die (E). Next, a dual-layer composite fiber was extruded out of the orifice of the die and fed into an enclosed zone (F), which was maintained at a constant temperature of 90 ^0C. While the fiber went through the diffusion zone, the monomer (2a) and/or (2b) in the inner layer and the monomer (2c) in the outer layer diffused into each other, subsequently producing the effect of a continuous distribution of refractive index in the fiber. At the end of the diffusion zone, the fiber was extruded through a smaller orifice. For imaging application, 40% of the outermost portion of the fiber was removed through a core-shell separation die design (G). The fiber was then fed through a hardening zone (H) where it was hardened by four UV lamps of 60 W/cm each. Thus, a polymer fiber with a parabolic distribution of refractive index was taken up through rolls by a take-up roll (I). The formulations of the reactant mixtures **I** - **III** are listed in Table I. Two different kinds of PMMA polymers, i.e., PMMA1 and PMMA2, were used for adjusting the viscosity of the reactant mixtures suitable for gear pumps in the extrusion process. The polymers fibers **I-III** were prepared from the reactant mixtures **I-III**, respectively.

Characterization. The refractive index profile of the GRIN polymer fibers was measured using an Interphako interference microscopy (Carl Zeiss Jena), in which a matching oil with a refractive index of 1.492 was used as the reference. The instrument accuracy in measuring the refractive index was in the range of $\pm$ 0.0002. The polymer thin films for the measurement of refractive index distribution were prepared by cutting the polymer fiber using a Microtome HM 350 (Microm). The film thickness was 30 μm. A polymer film was placed on a slide and covered by a small piece of microcover glass. A drop of embedding liquid was placed on the slide and the oil was diffused all over the surface of the microcover glass. The refractive index varies within the GRIN cylindrical material according to in the equation:

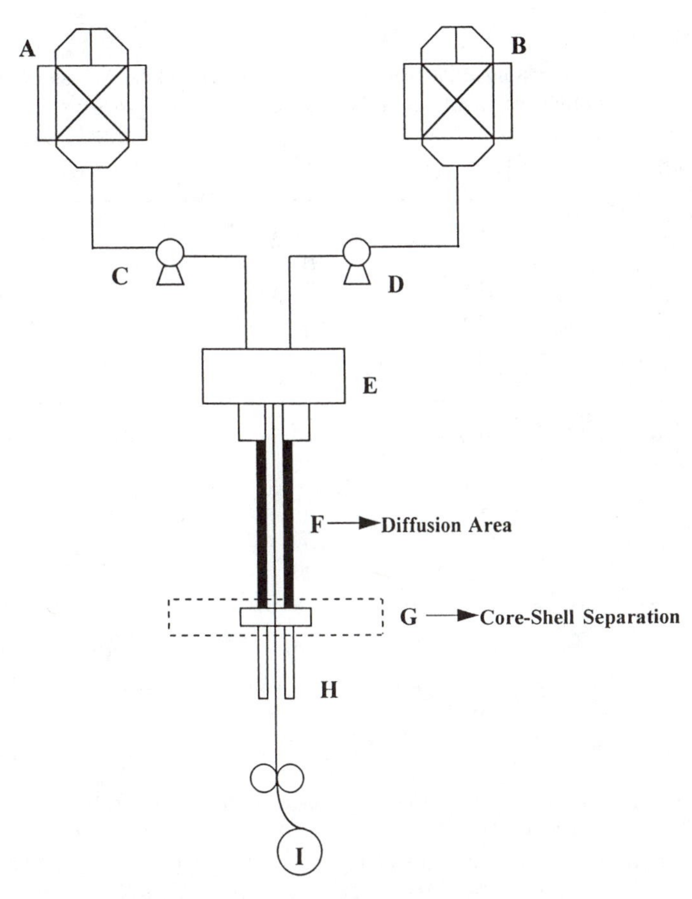

Figure 1. Experimental setup of the closed extrusion process for preparing GRIN polymer fibers. A and B, material supply tanks; C and D, gear pumps; E, a concentric die; F, an enclosed zone; G, (a) without core-shell separation die, (b) with core-shell separation die; H, a hardening zone; I, rolls.

Table I. The formulations of the reactant mixtures **I-III** and the Δn values of the resulting polymer fibers **I-III** from the extrusion process

	I[a,b]	**II**[a]			**III**[b]
Mixture 1 (Core)					
PMMA1(%)	57.7	56.2			32.8
PMMA2 (%)	0	0			20.9
BzMA (%)	27.9	36.8			36.8
MMA (%)	13.9	6.5			9.0
HCPK (%)	0.5	0.5			0.5
HQ (ppm)	50	50			50
Mixture 2 (Shell)					
PMMA1 (%)	59.7	59.7			33.9
PMMA2 (%)	0	0			16.9
MMA (%)	39.8	39.8			37.8
4FMA (%)	0	0			10.9
HCPK (%)	0.5	0.5			0.5
HQ (ppm)	50	50			50
Diffusion Zone Length (cm)	45	30[c]	55[c]	80[c]	45
Δn	0.018	0.030	0.026	0.017	0.055

[a]Polymer fibers prepared by the extrusion process without a core-shell separation die design.

[b]Polymer fibers prepared by the extrusion process with a core-shell separation die design.

[c]Polymer fiber **II** prepared by the extrusion process with a different diffusion zone length.

$$n(r) = n_c\,[1-(1/2)(A)(r/R_p)^2] \quad (1)$$

where n_c and n(r) are the refractive indices at the center axis and at the distance r from the center, respectively, R_p is the radius of the fiber, and A is the distribution constant. The relationships between Δn, (r/R_p), and $(r/R_p)^2$ were examined by quadratic curve-fitting and line-fitting methods, respectively. Here, Δn is the difference between the refractive index of the center (n_c) and periphery (n_p) of the GRIN polymer fiber. The distribution constant A was determined from the line-fitting of Δn with $(r/R_p)^2$.

Results and Discussion

Control of the Refractive Index Distribution of GRIN Polymer Fibers. There are two main processes for preparing GRIN polymer fibers. The first process is an open system process, which is based on internal diffusion and surface evaporation of monomers in a liquid jet (*4*). This process can produce polymer fibers with a quadratic distribution of the refractive index. However, the difficulty of controlling gas blowing rate resulted in poor reproducibility. Also, the low gas blowing rate and the long length of the diffusion zone led to low production rate. We have successfully developed a closed extrusion process which overcomes these drawbacks in the open system process. However, the polymer fibers prepared by the closed extrusion process reported previously had only 62% portion of the quadratic refractive index distribution no matter how we adjusted the monomer mixture composition and process parameters(*13*). Recently, a GRIN polymer fiber within a 100% quadratic distribution of the refractive index has been successfully prepared by modifying the die geometry with the core-shell separation die design in our closed extrusion process. Figure 2 shows the relationships between Δn and r/R_p for the polymer fibers **Ia** and **Ib** from the closed extrusion process. The non-quadratic index portion of the polymer fiber **Ia** was successfully removed by the core-shell separation die design. Hence, this kind of GRIN polymer fibers can be used for imaging applications which require a fiber with a completely quadratic distribution of the refractive index.

The closed extrusion process is a liquid (core mixture)-liquid (shell mixture) diffusion control process. The key parameters for controlling the refractive index distribution in this process are the compositions of the reactant mixtures, pump speed, the temperature and the length of the diffusion zone. In our previous study, we have shown that the refractive index distribution of the GRIN polymer fibers was significantly affected by the monomer composition of the reactant mixtures and the temperature of the diffusion zone (*13*). The refractive index difference between the center and the periphery (Δn) increased with increasing composition of the high refractive index monomers in the reactant mixtures and decreased with increasing temperature of the diffusion zone (*13*). In the present study, the effect of the length of the diffusion zone on the refractive index distribution was investigated. From Table 1, the Δn values of the GRIN polymer fibers **II** were 0.03, 0.026, and 0.017 for the diffusion zone lengths of 30 cm, 55 cm, and 80 cm, respectively. The mutual diffusion of BzMA and MMA monomers increased with increasing length of the diffusion zone and resulted in enhancing the distribution uniformity of the monomers in the polymer fiber. Thus, increasing length of the diffusion zone resulted in a decreasing Δn value. This result indicates that the selection of the diffusion length is very critical to the design of the refractive index distribution of GRIN polymer fibers.

Preparation of GRIN Polymer Fibers for Imaging Applications. GRIN polymer fibers for imaging applications require not only a completely quadratic index distribution but the Δn values must be adjusted to a wide range for various applications. For example, the Δn values for the imaging lens in a home-style fax machine could be as large as 0.052 (*14*). Hence, the adjustment of the composition of the reactant mixture from that of the GRIN polymer fibers **I** and **II** would be necessary to achieve such a large Δn value. One way to enlarge the Δn value could be by reducing the refractive index of the fiber periphery by adding the low refractive index monomer 2,2,3,3-tetrafluoropropyl methacrylate (4FMA), which has the refractive index of 1.42, in the shell reactant mixture. Figure 3 shows the relationships between Δn with (r/R_p) and $(r/R_p)^2$ for the polymer fiber **III**. The addition of the low refractive index monomer 4FMA largely increased the Δn value of the polymer fiber **III** to 0.055. Thus, it can be used as an imaging lens for fax machines. The linear portion of the relationship between Δn and $(r/R_p)^2$ and the distribution constant A estimated from Figure 3 are 70% and 0.11, respectively. This means that the quadratic refractive index portion of the GRIN polymer fiber **III** is only 70% even with the core-shell separation die design. This result suggests that the diffusion behavior of the monomer in polymer has been changed by adding the monomer 4FMA. Thus, the original core-shell separation die design cannot be used to fully remove the non-quadratic refractive index portion of the polymer fiber. A preliminary study of the fluid mechanics of the reactant mixture including the 4FMA monomer showed that both the viscosity and the elasticity were quite different from the rheological behavior of the reactant mixture without the 4FMA monomer. This change in rheological behavior may change the diffusion characteristics of monomers in the reactant mixture. A further study of the diffusion behavior of the 4FMA monomer in polymer combined with the effects of the die geometry, pump speed, the temperature and length of the diffusion zone will be very critical for preparing GRIN polymer fibers with a large Δn value.

Conclusions

Gradient-index polymer fibers with a quadratic distribution of the refractive index were successfully prepared by a closed extrusion process. The die geometry and the length of the diffusion zone showed significant effects on the refractive index distributions of the polymer fibers. The non-quadratic refractive index portion of polymer fiber **I** was removed by a core-shell separation die design. The refractive index difference between the center and the periphery (Δn) of polymer fiber **II** decreased from 0.03 to 0.017 when the length of the diffusion zone was increased from 30 cm to 80 cm. The Δn value of polymer fiber **III** was as large as 0.055, achieved by adding the low refractive index monomer 4FMA into the shell reactant mixture. This kind of polymer fiber with a large Δn value has a potential application as an imaging lens for fax machine. However, addition of the 4FMA monomer changed the diffusion behavior of monomers in the reactant mixture. Further study of how to achieve GRIN polymer fiber with a large Δn value is warranted.

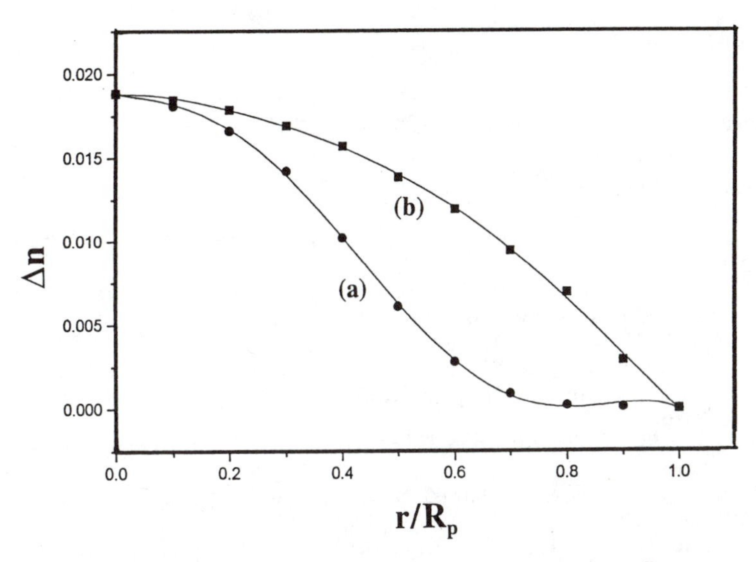

Figure 2. Relationships between Δn with (r/R_p) of polymer fibers **I** (a) prepared from the extrusion process without a core-shell separation die design (b) with a core-shell separation die design. The length and temperature of the diffusion zone were 45 cm and 90^0C, respectively.

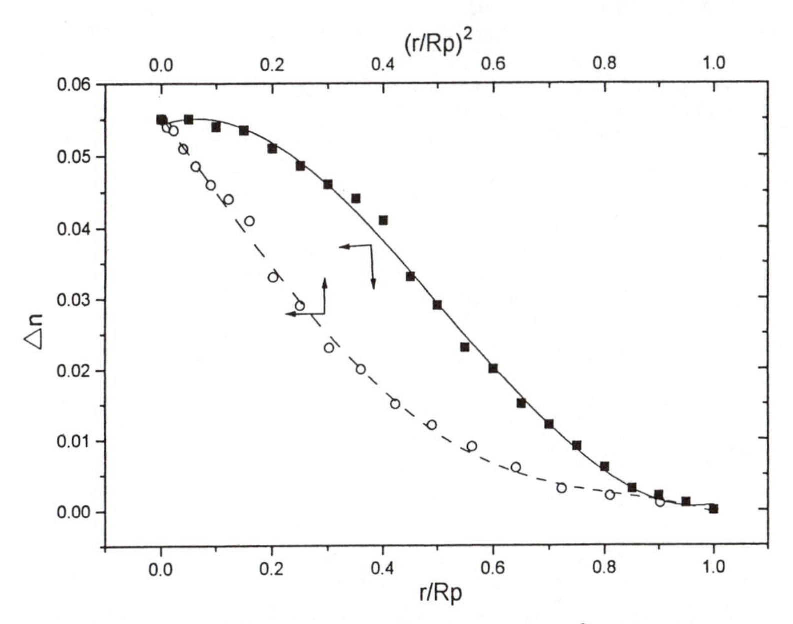

Figure 3. Relationships between Δn with (r/R_p) and $(r/R_p)^2$ for the polymer fiber **III** prepared from the extrusion process with a core-shell separation die design. The temperature of the diffusion zone was 90^0C.

Acknowledgment

The authors would like to thank the Ministry of Economic Affairs of Taiwan, R.O.C. for financial support of this work under Contract No.84-EC-2-A-15-0085.

Literature Cited

(1) Atkinson, L. G.; Kindred, D. S.; Zinter, J. B. *Opt. Photonics News* **1994**, *5*, 28.
(2) Koike, Y. In *Polymers for Lightwave and Integrated Optics* ; Hornak, L. A., Ed.; Marcel Dekker, Inc., New York, 1992, 71.
(3) Moore, D. T. *Appl. Opt.* **1980**, *19*, 1035.
(4) Yamamoto, T.; Mishina, Y.; Oda, M. *US Patent* **1989**, 4,852,982.
(5) Ishigure, T.; Horibe, A.; Nihei, E.; Koike, Y. *J. Lightwave Technol.* **1995**, *13*, 1686.
(6) Ishigure, T.; Horibe, A.; Nihei, E.; Koike, Y. *J. Lightwave Technol.* **1995**, *13*, 1475.
(7) Ishigure, T.; Nihei, E.; Yamazaki, S.; Kobayashi, K.; Koike, Y. *Electro. Lett.* **1995**, *31*, 467.
(8) Ishigure, T.; Nihei, E.; Koike, Y.; Forbes, C.; Lanieve, L.; Straff, R.; Deckers, H. *IEEE Photonics Technol. Letters* **1995**, *7*, 403.
(9) Tagaya, A.; Koike, Y.; Nihei, E.; Teramoto, S.; Fujii, K.; Yamamoto, T.; Sasaki, K. *Appl. Opt.* **1995**, *34*, 988.
(10) Yang, S. Y.; Chang, Y. H.; Ho, B. C.; Chen, W. C.; Tseng, T. W. *Polym. Bull.* **1995**, *34*, 87.
(11) Yang, S. Y.; Chang, Y. H.; Ho, B. C. Chen, W. C.; Tseng, T. W. *J. Appl. Polym. Sci.* **1995**, *56*, 1179.
(12) Ho, B. C.; Chen, J. H.; Chen, W. C.; Chang, Y. H.; Yang, S. Y.; Chen, J. J.; Tseng, T. W. *Polym. J.* **1995**, *27*, 310.
(13) Chen, W. C.; Chen, J. H.; Yang, S. Y.; Cherng, J. Y.; Chang, Y. H.; Ho, B. C. *J. Appl. Polym. Sci.* **1996**, *60*, 1379.
(14) *Technical Information on Selfoc Lens Array*, NSG American, Inc., 1991.

Chapter 7

Dipole–Dipole Interactions in Controlled-Refractive-Index Polymers

Patricia R. H. Bertolucci and Julie P. Harmon[1]

Department of Chemistry, University of South Florida, 4202 East Fowler Avenue, Tampa, FL 33620–5250

This study compared methacrylate and acrylate polymers to structural analogs with fluorinated ester groups. Two types of relaxations were characterized, the primary relaxation associated with the glass transition and secondary relaxations associated with side group motion and localized segmental motion. Dielectric analysis was used to characterize the response of dipoles to an electric field as a function of temperature. Mechanical properties were analyzed via dynamic mechanical analysis and stress relaxation measurements. Relaxation behavior was interpreted in terms of intermolecular and intramolecular mechanisms.

In this study, the specific polymers of interest are poly(2,2,2-trifluoroethyl methacrylate), PTFEMA, poly(2,2,2-trifluoroethyl acrylate), PTFEA, and poly(hexafluoroisopropyl acrylate), PHFiPA, with the respective non-fluorinated counterparts, poly(ethyl methacrylate), PEMA, poly(ethyl acrylate), PEA, and poly(isopropyl acrylate), PiPA.

Optical properties of fluoroalkyl acrylate and methacrylate polymers are attracting attention due to the expanding interest in polymer optical fiber applications. These polymers are transparent, and their refractive indices can be lowered in a predictable way by increasing the amount of fluorine substituents (*1*). Replacing hydrogen substituents with fluorine also increases near-IR transparency, since C-H stretch overtones are the primary near-IR absorptions in these polymers (*2*). Transparent, low refractive index polymers are used as fiber cladding materials (*3*). These polymers are also useful as moisture barriers, as low-surface free energy materials, and in contact lens formulations (*4,5*). The latter is due to the fact that fluorination imparts the polymer with enhanced oxygen permeability (*4,6*). Although these polymers are of interest in current applications, physical and chemical analysis of these materials is not widely discussed in the literature.

[1]Corresponding author

Numerous synthetic techniques for poly(alkyl methacrylates) and poly(alkyl acrylates) have been developed. Among these are anionic (*7*), ultraviolet (*8*) and bulk, free-radical (*9,10*) polymerization techniques. Many dielectric spectroscopy and dynamic mechanical spectroscopy studies have been conducted on polymethacrylates and polyacrylates and are summarized in reference 11 (*11*). Smith and Boyd have developed a molecular model in methyl acrylate polymers for the determination of the strength of the dielectric beta relaxation (*12*). The dynamic mechanical behavior of poly(methyl methacrylate), PMMA, is well known and several review papers have been written describing the relaxations in this polymer (*13-16*). Maps of the modulus mechanism have also been constructed for PMMA (*17*). Jones and Johnson have studied the mechanical properties of PMMA and copolymers formed with ethyl methacrylate and butyl methacrylate for use as bone cements (*18*). The dynamic mechanical and electrical properties of several alpha-substituted methacrylate polymers have been reported by Hoff and co-workers (*19*). Investigations of polyalkyl methacrylates and acrylates were conducted via physical aging studies. Beiner et al. have considered the influence of aging on the dynamic shear moduli of two poly(alkyl methacrylate)s at the onset of their glass transitions (*20*). Reaction-cast methacrylates and acrylates were characterized by thermo-mechanical analysis after aging. Allen et al. noted that all samples exhibited physical aging effects at their glass transition except those that were heavily cross-linked (*21*).

Studies characterizing the physical properties of the fluorinated counterparts of the polymethacrylates and polyacrylates are sparse. Ishiwari et al. characterized the dynamic mechanical Young's modulus of a series of fluorinated methacrylate and acrylate polymers (*22*). The data for poly(trifluoroethyl methacrylate), PTFEMA, of interest here, however, was limited to temperature-dependent storage modulus at one frequency, and no secondary transitions were observed. Koizumi et al. characterized poly(fluoroalkyl methacrylate)s and poly(fluoroalkyl α-fluoroacrylate)s by dielectric analysis (*3,22*). PTFEMA was characterized, but the α and β relaxations were not separated and the dielectric strength analysis was limited to the γ transition region. Shimizu characterized ordered structures in poly(fluoroalkyl α-fluoroacrylate)s with ester groups of six or more carbon atoms in length (*23*). X-ray analysis and calorimetric methods have been employed in the determination of the mesomorphic character of fluoroalkyl methacrylate and acrylate polymers (*24*). Mathais et al. have characterized semifluorinated polymethacrylates via NMR, FTIR, GPC, DSC, intrinsic viscosity and optical microscopy methods (*25*). Poly(2,2,2-trifluoroethyl methacrylate) was characterized by ^{13}C NMR (*26*). Poly (ethyl α- fluoroacrylate) has been characterized by ^{13}C NMR, ^{19}F NMR, TGA, GPC and the dielectric permittivity has also been determined (*27*).

In this paper, we analyze the effect of fluorine substitution in the polymers listed above by dielectric analysis (DEA), dynamic mechanical analysis (DMA) and stress relaxation measurements. The effect of fluorination on the α relaxation was characterized by fitting dielectric data and stress data to the Williams, Landel and Ferry (WLF) equation. Secondary relaxations were characterized by Arrhenius analysis of DEA and DMA data. The "quasi-equilibrium" approach to dielectric strength analysis was used to interpret the effect of fluorination on "complete" dipole relaxation for a given process via Cole-Cole plots. Differential scanning calorimetry (DSC) and FTIR data were also briefly discussed.

Experimental Section

2,2,2-Trifluoroethyl methacrylate, 2,2,2-trifluoroethyl acrylate and hexafluoro-isopropyl acrylate were obtained from PCR, Inc. (Gainesville, Florida). Ethyl methacrylate, ethyl acrylate and isopropyl acrylate were obtained from Scientific Polymer Products (Ontario, New York). These liquid monomers were first washed with 1% activated carbon, purchased from Calgon, to remove the inhibitor, methoxyhydroquinone. The monomer was isolated from the carbon and inhibitor via vacuum filtration. The monomer was warmed to minimize absorption of moisture from the atmosphere and placed in a polyethylene bag with 1% by weight UV initiator of isobutylbenzoinether. The solution was degassed with nitrogen, and an impulse sealer was used to seal the bag. After shaking lightly to dissolve the initiator, the assembly was placed in a UV reactor of 3500 μwatts/in^2 for approximately one hour.

Dynamic Mechanical Analysis and Stress Relaxation Behavior. Samples were compression molded into bars of the dimensions 38.x12.5x0.78 ± 0.007 mm and 65.x9.7x1.7 ± 0.007 mm in a Carver laboratory hot press model C. A TA Instruments 983 DMA, which was operated in the fixed frequency mode, was used to characterize the storage and loss moduli as a function of temperature. Samples were scanned at frequencies from 0.05 to 10.0 Hz over a temperature range from -150°C to above the glass transition temperature. The displacement was 0.4 - 0.6 mm. Stress relaxation curves were determined for the same size samples at a constant strain. The sample was displaced for 10.0 minutes and then allowed to recover for 10.0 minutes. The stress data were taken in five degree increments. A microprocessor controlled Liquid Nitrogen Cooling Accessory (LNCA) was used for sub-ambient operations.

Dielectric Analysis. Samples of 0.80 ± 0.01 mm thickness were compression molded into disks one inch in diameter. Permittivity, e', and the loss factor, e'', were determined at temperatures from -150°C to 30°C above the glass transition temperature of the samples using a TA Instruments 2970 DEA. The frequency range scanned was from 10^{-2} to 10^5 Hz.

Fourier Transform Infrared Spectrophotometry. Thin films of the polymers were prepared from 5% by weight solutions of the polymers in methylene chloride via a doctor blade. A Nicolet Magna-FTIR Spectrometer 550 was used to record the transmission spectra of thin films.

Refractive Index. A Bausch and Lomb ABBE-3L refractometer was used to determine the refractive indices of thin films of the polymers. A refractive index standard of 1.440 ± 0.0005 was used for calibration. A smooth surface polymer sample was placed in the refractometer and three drops of a 1.700 ± 0.0002 refractive index standard was used to form a layer of liquid between the film and the prisms.

Differential Scanning Calorimetry. A TA Instruments 2920 MDSC was used to determine the glass transition temperatures. Samples of 9-12 mg were scanned under a 25 ml/min compressed helium purge at a scanning rate of 3°C/min.

Data Analysis. PeakFit version 2.0 from AISN software, Jandel Scientific (Corte Madera, California) was used to separate overlapping transitions in tan(δ) and *e″* versus temperature plots. TA Instruments model 2000 thermal analyzer was used with version 4.1 time-temperature superposition software to analyze the stress data.

Gel Permeation Chromatography. Poly(methyl methacrylate) equivalent molecular weights were determined for dilute solutions of the polymers dissolved in tetrahydrofuran. The solvent was degassed by sonication. A Waters model 590 pump system was used with a Perkin-Elmer LC-25 RI detector and a Spectroflow 757 Absorbance detector. Phenomenex Phenogel columns were used ranging in molecular weight detection between 5K-300K. The pore size ranged from 50Å-10^6Å and the spherical particles were 5μ in diameter.

Results and Discussion

The structures of the polymers examined in this set of experiments are depicted in Figure 1 along with the refractive indices, η_D. A detailed discussion of the effect of fluorination on the refractive index of methacrylate polymers has been presented elsewhere (*1*). All of the polymers examined had poly(methyl methacrylate) equivalent molecular weights greater than 300,000 g/mol.

Table I. Properties of Fluorinated Polymers and Their Non-Fluorinated Counterparts

Polymer	T_g(°C)	λ^{-1}(cm^{-1})
PEMA	52.44	1728
PTFEMA	58.06	1750
PEA	-40.05	1735
PTFEA	-6.75	1760
PiPA	-27.01	1735
PHFiPA	-8.35	1783

It was evident that fluorination increased the T_g of all polymers studied as shown in Table I. The effect was more pronounced in the acrylate polymers than in the methacrylate polymers. Fluorination also affected the position of the carbonyl stretch in FTIR. Table I shows that in the fluorinated polymers the carbonyl stretch was

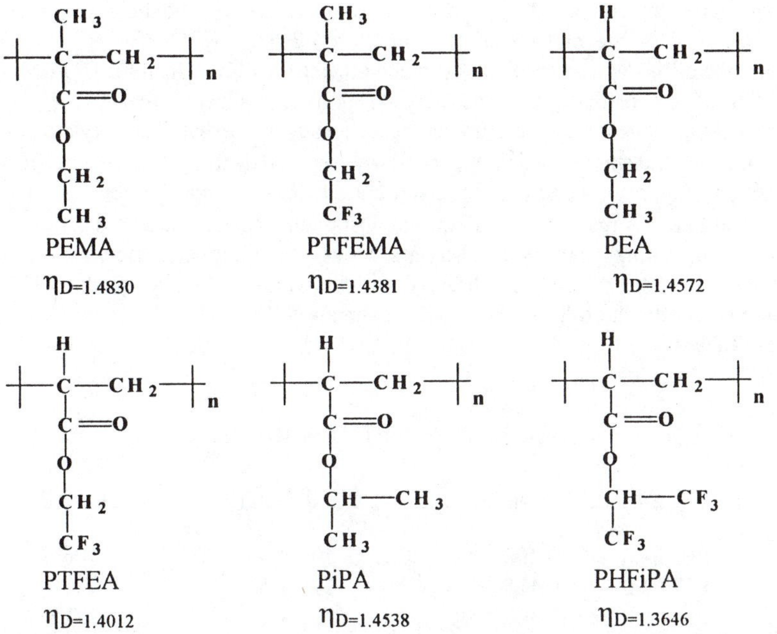

PEMA η_D=1.4830

PTFEMA η_D=1.4381

PEA η_D=1.4572

PTFEA η_D=1.4012

PiPA η_D=1.4538

PHFiPA η_D=1.3646

Figure 1. Methacrylate and Acrylate Polymers With Their Corresponding Refractive Indices.

shifted to higher energies. The effect was most pronounced in PHFiPA with six fluorine substituents. The effect of fluorination on the viscoelastic response of these polymers was probed by DEA, DMA, and stress relaxation in these studies.

Temperature and Frequency Dependence of Dielectric Response. Dielectric data were used to characterize the alpha transition in all of the polymers investigated. Methacrylate polymers with alkyl groups longer than one carbon atom exhibited strong beta transitions which overlapped the alpha transition region (*3,11*). In Figures 2 a and b for PEMA and 3 a and b for PTFEMA, the dielectric permittivy, e', and the dielectric loss, e'', are plotted versus temperature. Three observations were made concerning this data: 1) Fluorine substituents shifted the glass transition to higher temperatures as compared to hydrogen substituents, 2) The beta transition was overlapped with the alpha transition in both PEMA and PTFEMA, and 3) The magnitude of the alpha transition was much greater for PEMA than for PTFEMA.

All of the methacrylate and acrylate polymers showed that the e'' maxima increased in height and temperature as the frequency increased. The acrylate polymers exhibited α transitions which were clearly separated from the region of the β transition. Figures 4 a and b and 5 a and b are e' and e'' versus temperature plots for PiPA and PHFiPA respectively. Fluorine substituents, again, shifted the α transition to higher temperatures, but in the acrylate series, the fluorinated polymers exhibited narrow transition regions as observed in the width of the e'' peak. This was characterized by the full width at half maximum (FWHM) in e'' versus temperature plots (Table II).

Table II. Dielectric Data at 10,000 Hz

Polymer	e'' max	FWHM(°C)	T_g (°C)
PEMA	0.400	30	106
PTFEMA	0.125	27	121
PEA	0.400	20	-14.8
PTFEA	0.431	17	28.5
PiPA	0.629	21	-4.17
PHFiPA	0.316	15	56.6

FWHM was narrowest in the polymer with six fluorine substituents, PHFiPA. PTFEMA and PTFEA, which each contain three fluorine substituents, exhibited e'' narrowing of 3°C as compared to their non-fluorinated counterparts. This narrowing was due to a more uniform motion involved in dipole alignment in the fluorinated polymers–the spectrum of relaxation times was narrower in the fluorinated polymers. The non-fluorinated polymers appeared to be more highly polarizable, hence, more energy was dissipated in dipole alignment. The intramolecular interactions which were disturbed during dipole alignment must have been more extensive and diverse in nature in the non-fluorinated polymers. The amount of dipole alignment in the electric field is

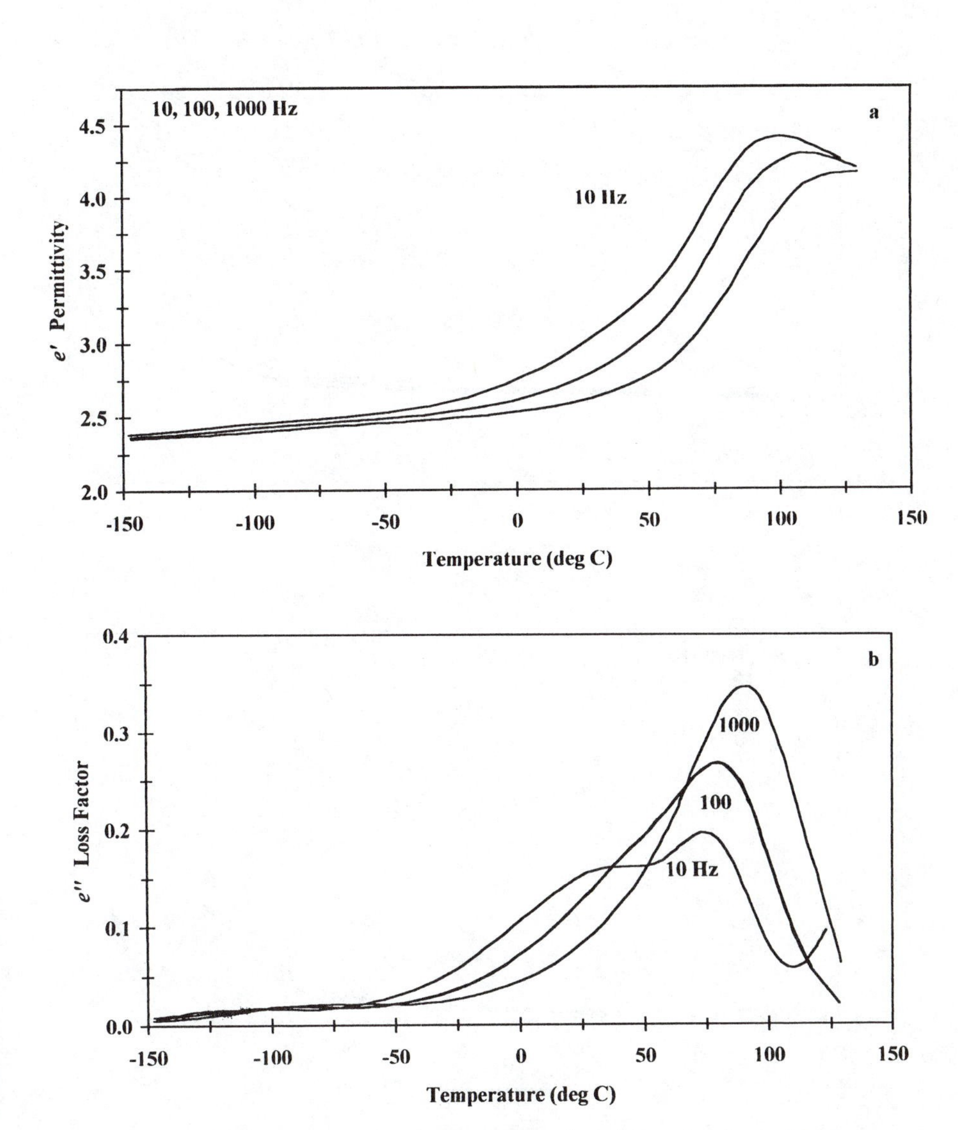

Figure 2. Transition Behavior in PEMA Showing a) Permittivity and b) Loss Factor Versus Temperature.

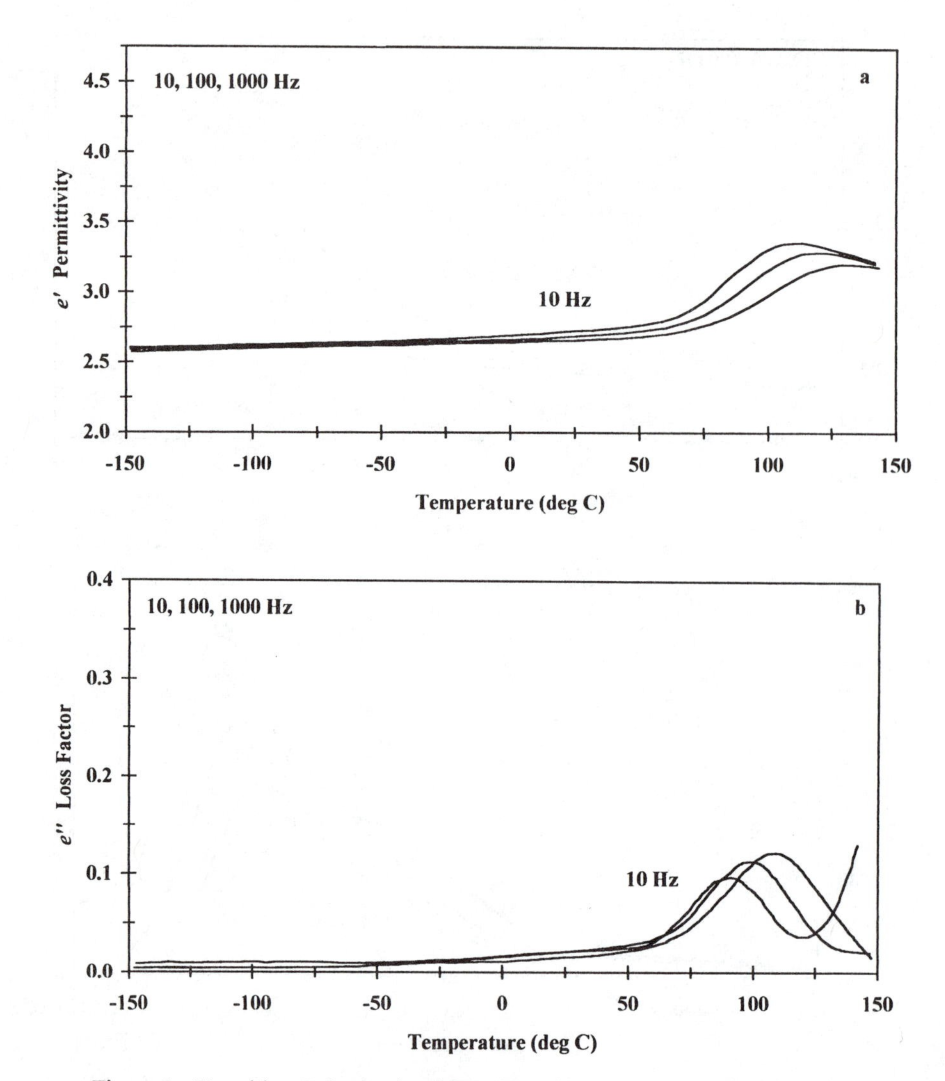

Figure 3. Transition Behavior in of PTFEMA Showing a) Permittivity and b) Loss Factor Versus Temperature.

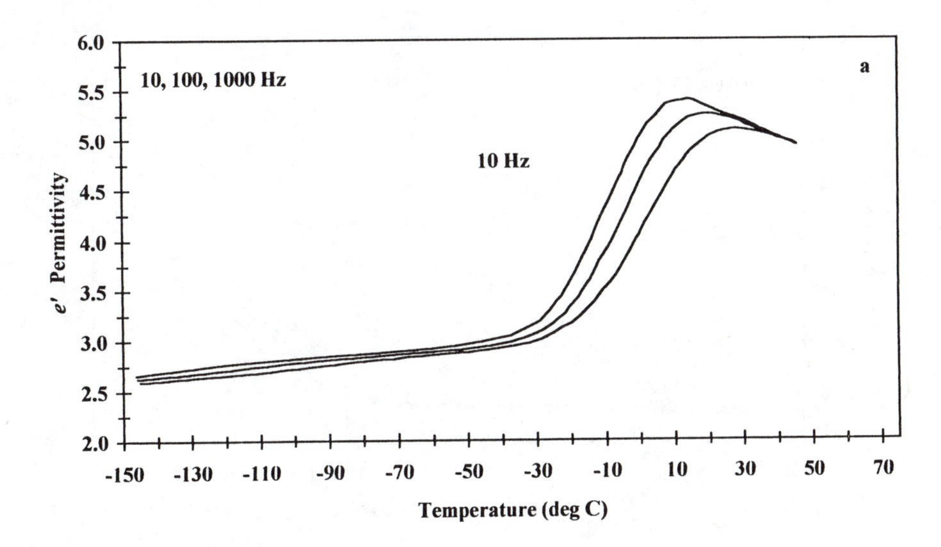

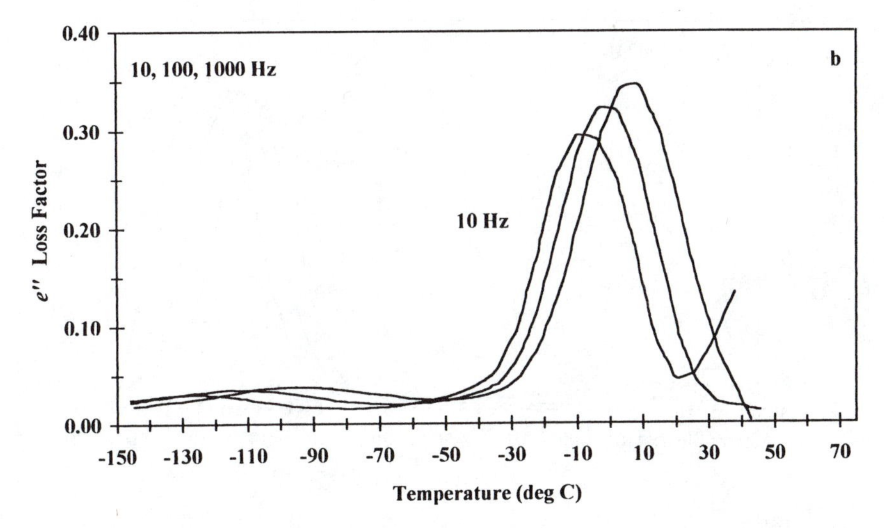

Figure 4. Transition Behavior in PiPA Showing a) Permittivity and b) Loss Factor Versus Temperature.

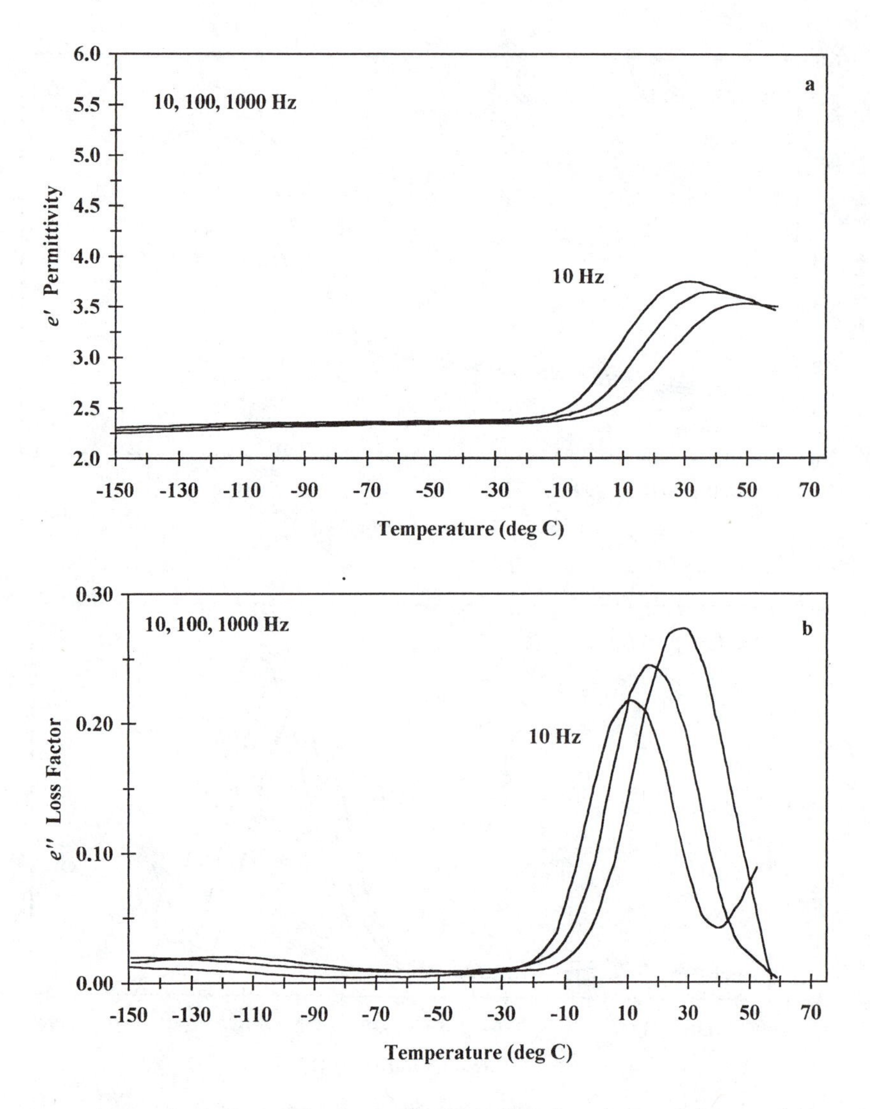

Figure 5. Transition Behavior in PHFiPA Showing a) Permittivity and b) Loss Factor Versus Temperature.

represented as e'. Characteristically, e' is smaller at low temperatures where hindered mobility relates to dipole alignment. At the glass transition, e' increased dramatically (*28*). Figure 6 shows plots of e' versus temperature at 10 Hz for PiPA and PHFiPA. The e' maximum above T_g was 1.7 units higher for PiPA than for PHFiPA. Fluorination showed similar effects for the other polymers studied herein. This demonstrated that the decrease in e' step height in fluorinated samples may have been due to a steric effect caused by the large fluorine substituents restricting dipole alignment.

An analysis of the temperature dependence of the α transition revealed, as expected, adherence to the WLF theory (*22,29,30*). That is, Arrhenius plots of the glass transition data were not linear, and the data fit the WLF equation :

$$\ln a_T = \frac{-C_1(T - T_g)}{C_2 + (T - T_g)} \tag{1}$$

where a_T is the shift factor, C_1 and C_2 are the WLF constants, and T_g represents the reference temperature. The data for the methacrylate polymers were analyzed in two ways. First, a peakfit program was used to separate the α and β transitions. The appropriate C_1 and C_2 constants were determined. In the second analysis, the data with overlapping transitions were fit to equation 1. The parameters in equation 1 were similarly determined for acrylate polymers. A representative fit of PTFEMA is shown in Figure 7.

Table III. WLF Constants, Reference Temperature and Activation Energy

Polymer	C_1	C_2	T_g (°C)	E_{app}(kcal/mole)	f_g
PEMA	10.2	64.7	60	80	0.043
PEMA (PeakFit)	15.2	45.5	75	93	0.029
PTFEMA	10.4	59.2	83	103	0.042
PTFEMA (PeakFit)	12.0	54.4	87	131	0.036
PEA	13.1	63.2	-24.5	59	0.033
PTFEA	9.8	41.4	3.1	83	0.044
PiPA	11.3	51.0	-31.0	59	0.038
PHFiPA	11.0	51.2	28.0	89	0.039

The fractional free volume at the glass transition, f_g, was calculated from C_1 (*30*). The apparent activation energy was determined from the following equations (*27,31,32*):

$$\Delta E_a = 2.303R\left[\frac{d\log A_T}{d(1/T)}\right] = -RT^2\left[\frac{d\log A_T}{dT}\right] \tag{2}$$

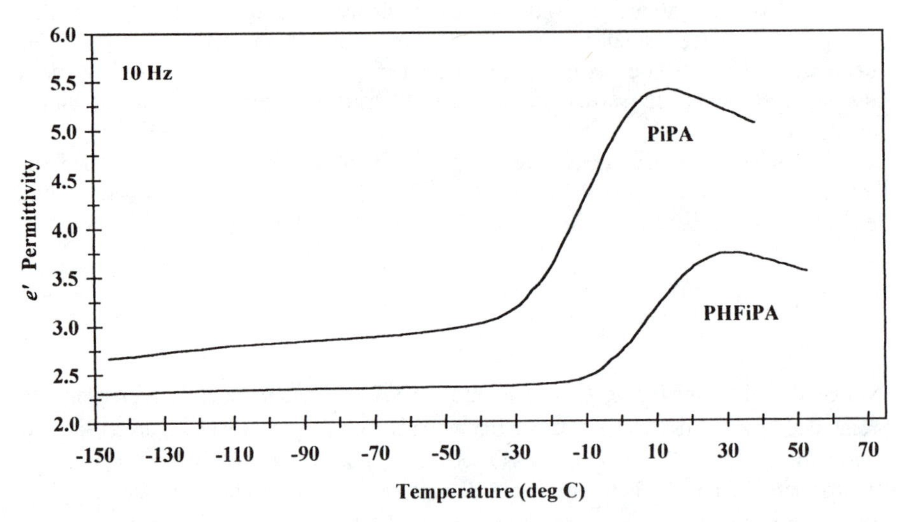

Figure 6. PiPA and PHFiPA Given for Frequency of 10.0 Hz.

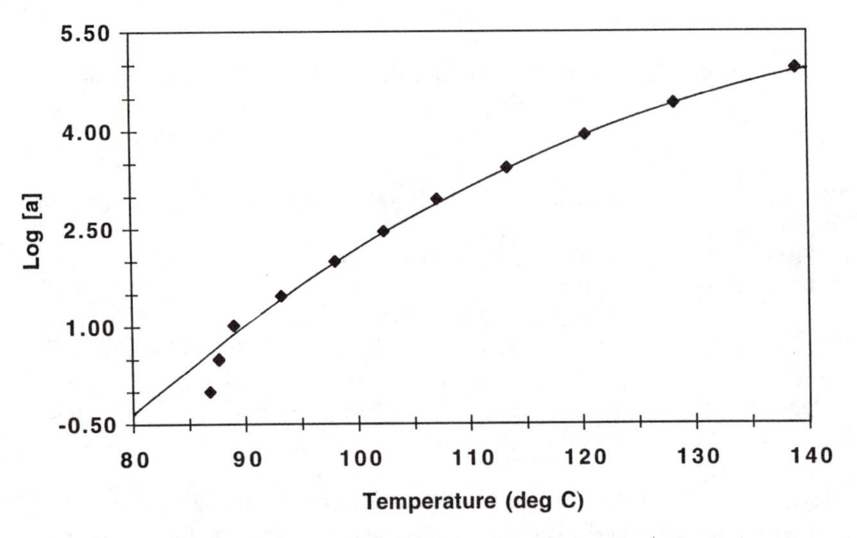

Figure 7. Representative Plot of WLF Constants.

at T_g equation 2 becomes:

$$\Delta E_a = 2.303\left(\frac{C_1}{C_2}\right)RT_g^{\,2} \tag{3}$$

The results are shown in Table III which lists the constants, the reference T_g and the apparent activation energy. The universal WLF constants, as listed by Aklonis (*29*), were $C_1 = 17.44$ and $C_2 = 51.6$. The data for the methacrylate polymers displayed constants near the WLF universal constants when the "as is" overlapping transition data were used, and the results, $C_1 = 10.4$ and $C_2 = 59.2$ for PTFEA, were close to the literature values, 11.6 and 66 (*3*). This indicated that the molecular motion associated with the glass transition was a cooperative type motion involving the β transition. An attempt to fit data obtained from the PeakFit program, which separated the α transition from the β transition for methacrylate polymers, resulted in a poor fit to the WLF equation. The overall effect of fluorine substituents on the glass transition temperature was evident. PHFiPA exhibited a T_g that was 59°C higher than PiPA. Similarly, fluorination increased the apparent activation energy for the α transition. The orientation polarization behavior for samples in the glass transition region, as analyzed by the WLF equation, pointed to the fact that the fluorine substituents diminished dipole alignment. The higher apparent activation energies for the glass transition in fluorinated polymers were reasonable when one considers (Figure 6) the difference in the extent of dipole alignment between the fluorinated and non-fluorinated polymers.

Dielectric data revealed secondary transitions in all of the polymers studied. Secondary relaxations were all much broader than those of the α transitions. A representative plot of e" versus temperature for the β relaxation in PHFiPA is shown in Figure 8. It is well known that poly(alkyl methacrylate)s exhibit both β and γ transitions. The β transition has been interpreted as due to the hindered motion of the -COOR group about the carbon-carbon bond which links the side group to the main chain. The γ transition is thought to be due to local molecular motion of alkyl groups in the side chain (*3*). The β transitions in PEMA and PTFEMA overlapped the α, T_g, transitions. An attempt to resolve the β transition with the PeakFit program resulted in tan(δ) and *e"* versus temperature curves which exhibited anomalous frequency effects.

Two facts emerged concerning γ transition behavior in the methacrylate polymers: first, the activation energy for the γ process in PTFEMA, 6.6 kcal/mol, was lower than that in PEMA, 8.4 kcal/mol; and secondly, the transition temperature was lower in PTFEMA than in PEMA, for example, at 10 Hz the values were -132°C and -107°C respectively. This was an indication that these groups reorient independently of the main chain (*12*). A pattern emerged which was also apparent in the acrylate polymers as discussed below. The activation energy for the γ tansition in secondary relaxations of PEMA, 8.4 kcal/mole, agreed with that reported in the literature, 9.0 kcal/mole (*11*).

In all acrylate polymers, only one sub-T_g transition, the β transition, was observed at temperatures down to -150°C. This transition has been attributed to ester group motion as in methacrylate polymers (*12*). The activation energies for this process, along with the transition temperature at 100 Hz, were: PEA -97°C and 11.5 kcal/mol, PTFEA -122°C and 7.1 kcal/mol, PiPA -82°C and 8.4 kcal/mol, and PHFiPA -144°C and 5.4 kcal/mol. Transition temperatures and activation energies were lower in the fluorinated than those in non-fluorinated acrylates. This was the same trend noted for the secondary γ transitions in PTFEMA and PEMA. This contrast to the behavior of the α transitions indicated that reorientation was likely intramolecular in secondary transitions and independent of the main chain.

Another method of analyzing dielectric data was via Cole-Cole plots (*31,33*). Here, e'' is plotted versus e' for a series of frequencies at a constant temperature. The plots form a semicircle which intersects the x-axis at two points. The first intersection point, ε_∞, represents the high frequency dielectric constant which characterizes the situation where frequencies are so high that dipoles do not relax within the time frame of the electrical oscillation. The second intersection point, ε_0, is the low frequency, static or relaxed dielectric constant. The ε_0 represents the dielectric constant of the oriented system. The difference between the relaxed and unrelaxed dielectric constant, $\Delta\varepsilon$, determined graphically, characterizes the strength of the relaxation. By changing the temperature, it was possible to the obtain relaxation strengths for all α and most secondary transitions (Figures 9 and 10). A representative Cole-Cole plot for the α transition of PHFiPA is depicted in Figure 10. In reviewing the data for the glass transition (Table IV), it was noted in every case that the fluorinated polymers exhibited lower relaxation strengths. The $\Delta\varepsilon$ was proportional to the amount and strength of dipoles per unit volume which absorb energy and reorient in the electric field (*32*). The fluorine substituents appeared to hinder dipole alignment associated with the glass transition temperature. This effect was reasonable if one considers the large scale segmental motion associated with the glass transition; the bulky fluorine substituents hinder this motion. This resistance to motion was also observed in the higher values for ΔE_{app} obtained from the WLF fit at T_g for fluorinated polymers. The increase in T_g with fluorine content indicated that more thermal energy was needed to induce segmental slippage.

Table IV. Relaxation Strength for the Glass and Beta Transition

	T_g			β			
Polymer	ε_0	ε_∞	$\Delta\varepsilon$	ε_0	ε_∞	$\Delta\varepsilon$	μ(D)
PEMA	2.32	4.02	1.70	2.27	2.59	0.32	0.647
PTFEMA	2.52	3.51	0.99	2.52	2.62	0.10	0.738
PEA	3.60	7.20	3.60	2.62	2.74	0.12	0.643
PTFEA	3.60	6.30	2.70	3.10	3.47	0.37	0.741
PiPA	2.80	6.10	3.30	2.56	2.68	0.12	0.647
PHFiPA	2.45	3.75	1.20	2.09	2.35	0.26	0.826

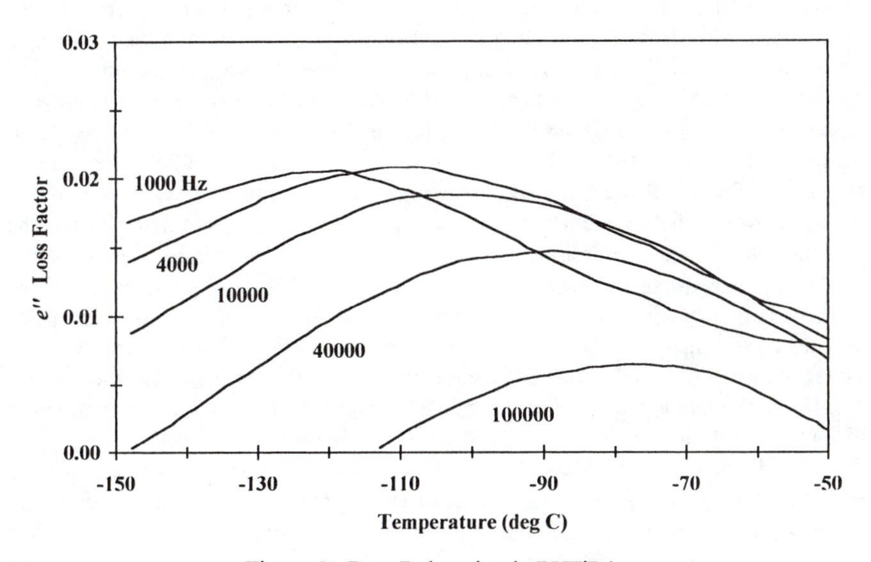

Figure 8. Beta Relaxation in PHFiPA .

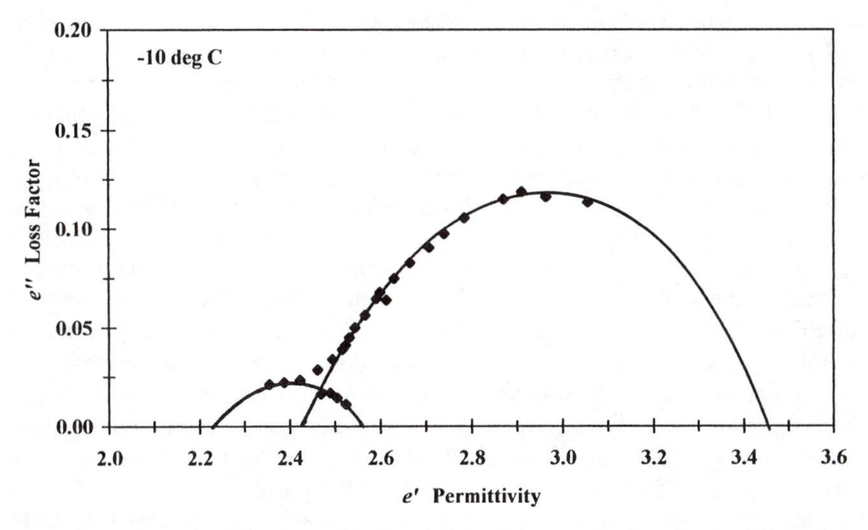

Figure 9. Cole-Cole Plot With Resolution of the Beta Relaxation in PEMA .

In contrast, Table IV shows that for acrylate polymers, the fluorine substituents increase the relaxation strength for the beta relaxation of the polymers. It can be envisioned that small scale motion (rotation) of groups more effectively changed the dipole moment of the fluorinated groups than that of the non-fluorinated groups. Thus, the relaxation strength was higher in the fluorinated acrylate polymers. The relaxation strengths were as much as ten times higher for the α process. The acrylate data were also consistent with the fact that for the β relaxation, fluorination decreased the transition temperature and the activation energy. The intramolecular motion responsible for this β relaxation was in contrast to intermolecular motion at the α transition where steric hindrance was important. The methacrylate data showed the reverse trend for the β transition. However, the cooperative nature of α-β motion rendered it impossible to separate α-β effects. The $\Delta\varepsilon$ for the γ transitions in the methacrylate polymers were unresolvable. Dipole moment calculations for polar polymers are complex. Systems consisting of polymer molecules do not comply to the Onsager theory (*34*), however, the Debye equation is the most reliable method for calculating the dipole moments of the individual repeat units as noted in the literature (*35-37*). The Debye equation was used in the dipole calculations that we made here for the individual monomer units. These calculations were useful for interpreting β transitions. In Table IV, μ, given in units of Debye, represents the dipole moment of the individual monomer repeat units. In all cases, fluorination increased μ. This increase in μ correlated with the increase in the dielectric strength of the β transition noted in the fluorinated polymer.

Dynamic Mechanical Spectroscopy. PEMA and PTFEMA were characterized by DMA in the fixed frequency mode. Figure 11 shows a plot of E' and E" versus T for the samples. Although the instrument recorded data through the glass transition, there was irreversible deformation in the sample due to flow, and the α transition could not be analyzed via DMA. It was noted, however, that as in dielectric analysis, the β transition was overlapped with the α transition; this overlapping was more pronounced in PEMA than in PTFEMA. The modulus of PEMA in the glassy state was somewhat higher than that of PTFEMA, while both polymers exhibited a continued decrease in modulus with temperature. The transition temperatures at 5 Hz and activation energies for the β relaxation determined by peak fitting were 24.0°C and 18.3 kcal/mol for PEMA and 24.5°C and 21.1 kcal/mol for PTFEMA. McCrum reported an activation energy of 31 kcal/mole for the β transition in PEMA, but noted that there were inconsistencies in values reported by various authors because of the merging α and β transitions (*11*). We believe that any rigorous analysis of β transition behavior in these polymers was limited by the overlapping nature of the α and β processes. At 5 Hz, the γ relaxation occurred at -120.4°C for PEMA and -124.6°C for PTFEMA with activation energies of 11.1 and 6.6 kcal/mol respectively. The data correlated well with the dielectric data we obtained for the γ relaxation behavior indicating that the events involved in dipole alignment was similar to the motion encountered in mechanical relaxation.

Stress Relaxation Studies. Time-temperature superposition was used to generate stress relaxation profiles through the α transition for PEMA and PTFEMA. The WLF

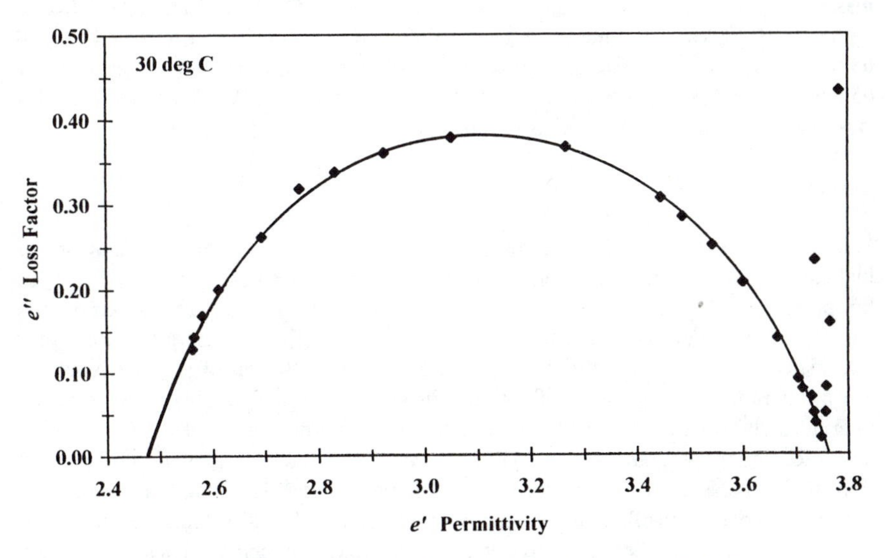

Figure 10. Cole-Cole Plot of the Alpha Relaxation in PHFiPA .

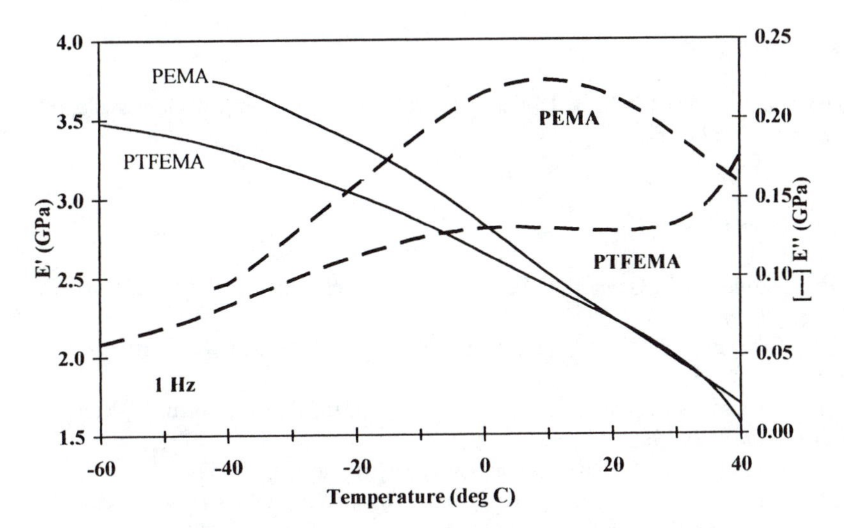

Figure 11. E' and E" Versus Temperature in PEMA and PTFEMA.

fitting routine determined glass transition temperatures of 60°C for PEMA and 70°C for PTFEMA. The T_g for PEMA reported in the literature for a similar experiment was 62°C (*38*). Values for C_1 and C_2 were 12.67 and 68.33 for PEMA and 12.78 and 68.03 for PTFEMA. Aklonis et al. reported C_1 and C_2 values for PEMA of 17.6 and 65.5 respectively (*38*). The activation energies for the glass transition determined by equation 3 were 94.18 and 101.2 kcal/mol in PEMA and PTFEMA respectively. This was in agreement with DEA and DSC studies.

Summary

Fluorine substituents have interesting effects on the relaxation behavior of methacrylate and acrylate polymers. In general, fluorine substituents increase the T_g of the polymers. This was evidenced from DEA, DSC and DMA. The dielectric and stress relaxation data also indicated an increase in the activation energy at the T_g for fluorinated polymers. We attributed this effect to the stiffening of the matrix and to steric hindrance induced by the fluorine substituents. All of this results in complex intermolecular interactions encountered in molecular motion at T_g. Discounting anomalous effects encountered in PTFEMA and PEMA due to the overlap of α and β transitions, the secondary transition behavior indicated that fluorinated polymers have lower secondary transition temperatures and lowered activation energies for these transitions than for the respective non-fluorinated polymers. These secondary relaxations are intramolecular in nature and steric effects, due to fluorination, are not as pronounced as those noted in the glass transition region.

Acknowledgment

The authors wish to thank Optical Polymer Research, Inc., in Gainesville, FL for support of this work.

Literature Cited

(1) Harmon, J.P.; Gaynor, J.; Schueneman, G.; Schuman, P. *J. Appl. Polym. Sci.* **1993**, *50*, 1645.
(2) Takezawa, Y.; Tanno, S.; Taketani, N.; Ohara, S.; Asano, H. *J. Appl. Polym. Sci.* **1991**, *42*, 3145.
(3) Koizumi, S.; Tadano, K.; Tanaka, Y.; Shimidzu, T.; Kutsumizu, S.; Yano, S. *Macromolecules* **1992**, *25*, 6563.
(4) Biagtan, E. Ph.D. Dissertation, University of Florida, 1995.
(5) Park, I.J.; Lee, S.; Choi, C.K. *J. Appl. Polym. Sci.* **1994**, *54*, 1449.
(6) Koβmehl, G.; Volkheimer, J.; Schafer, H. *Acta Polymer* **1992**, *43*, 335.
(7) Narita, T.; Hagiwara, T.; Hamana, H. *Makromol. Chem. Rapid Commun.* **1985**, *6*, 175.
(8) Allen, N.S. *Photopolymerization and Photoimaging Science and Technology*; Elsevier Science: New York, NY, 1989; pp 93-95.
(9) Gupta, D.S.; Reddy, V.S.; Cassidy, P.E.; Fitch, J.W. *Abstracts of Papers of the American Chemical Society* **1993**, *205*, POLY 342.

(10) Antonucci, J.M.; Stransbury, J.W.; Cheng, G.W. *Abstracts of Papers of the American Chemical Society* **1990**, *199*, POLY 111.
(11) McCrum, N.G.; Read, B.E.; Williams, G. *Anelastic and Dielectric Effects in Polymeric Solids*, Dover Publications, Inc.: Mineola, NY, 1967; pp 251-273.
(12) Smith, G.D.; Boyd, R.H. *Macromolecules* **1991**, *24*, 2731.
(13) Ferry, J.D. *Viscoelastic Properties of Polymers*; Wiley, New York, NY, 1980; pp 445.
(14) Meares, P. *Polymers: Structures and Bulk Properties*; Applied Science Publishers, London, England, 1965; pp 251-260.
(15) Ward, I.M.; Hadley, D.W. *Mechanical Properties of Solid Polymers*, Wiley, New York, NY, 1990; pp 173.
(16) Povolo, F.; Goyanes, S.N. *Polymer J.* **1994**, *26*, 1054.
(17) Gilbert, D.G.; Ahshby, M.F.; Beaumont, P.W.R. *J. Mater. Sci.* **1986**, *21*, 3194.
(18) Johnson, J.A.; Jones, D.W. *J. Mater. Sci.* **1994**, *29*, 870.
(19) Hoff, E.A.W.; Robinson, D.W.; Willbourn, A.H. *J. Polym. Sci.* **1955**, *18*, 161.
(20) Beiner, M.; Garwe, K.; Schroter, K.; Donth, E. *Polymer* **1994**, *35*, 4127.
(21) Allen, P.E.M.; Lai, C.H.; Williams, D.R.G. *Eur. Polym. J.* **1993**, *29*, 1293.
(22) Ishwari, K.; Ohmori, A.; Koizumi, S. *Nippon Kagaku Zasshi* **1985**, *10*, 1924.
(23) Shimizu, T.; Tanaka, Y.; Kutsumizu, S.; Yano, S. *Macromolecules* **1993**, *26*, 6694.
(24) Budovskaya, L.D.; Ivanova, V.N.; Oskar, L.N.; Lukasov, S.V.; Baklagina, Y.G.; Sidorovich, A.V.; Nasledov, D.M. *Vysokomol. Soedin., Ser. A.* **1990**, *32*, 561.
(25) Mathias, L.J.; Jariwala, C.P. *Macromolecules* **1993**, *26*, 5129.
(26) Passaglia, E.; Aglietto, M.; Ciardelli, F.; Mendez, B. *Polymer J.* **1994**, *26*, 1118.
(27) Victor, M.W.; Saffariannour, M.; Reynolds, J.R. *J.M.S.- Pure Appl. Chem.* **1994**, *A31*(6), 721.
(28) Hartwig, G. *Polymer Properties at Room and Cryogenic Temperatures*; Plenum Press: New York, NY, 1994; pp 176.
(29) Aklonis, J.J.; MacKnight, W.J.; Shen, M. *Introduction to Polymer Viscoelasticity*; John Wiley & Sons: New York, NY, 1972; pp 60-85.
(30) Sperling, L.H. *Introduction to Physical Polymer Science*; John Wiley & Sons: New York, NY, 1986; pp 261-266.
(31) Havriliak, S., Jr.; Havriliak, S. *J. Polym. Sci. Part B: Polym. Phys.* **1995**, *33*, 2245.
(32) Huo, P.; Cebe, P. *Macromolecules* **1992**, *25*, 902.
(33) McCrum, N.G.; Read, B.E.; Williams, G. *Anelastic and Dielectric Effects in Polymeric Solids*, Dover Publications, Inc.: Mineola, NY, 1967; pp 116-125.
(34) McCrum, N.G.; Read, B.E.; Williams, G. *Anelastic and Dielectric Effects in Polymeric Solids*, Dover Publications, Inc.: Mineola, NY, 1967; pp 83.
(35) VanKrevelen, D.W.; Hoftyzer, P.J. *Properties of Polymers*, Elsevier Scientific: New York, NY, 1976; pp 233.
(36) Riande, E.; Saiz, E. *Dipole Moments and Birefringence of Polymers*, Prentice Hall: Englewood Cliffs, NJ, 1992; pp 7.
(37) Salom, C.; Riande, E.; Hernandez-Fuentes, I.; Diaz-Calleja, R. *J. Polym. Sci. Part B: Polym. Phys.* **1993**, 31, 1591.
(38) Aklonis, J.J.; MacKnight, W.J.; Shen, M. *Introduction to Polymer Viscoelasticity*; John Wiley & Sons: New York, NY, 1972; pp 51.

Nonlinear Optical Properties and Applications

Chapter 8

High-Temperature Nonlinear Optical Chromophores and Polymers

R. D. Miller[1], D. M. Burland[1], M. Jurich[1], V. Y. Lee[1], P. M. Lundquist[1], C. R. Moylan[1], R. J. Twieg[1], J. I. Thackara[1], T. Verbiest[1], Z. Sekkat[2], J. Wood[2], E. F. Aust[2], and W. Knoll[2]

[1]IBM Almaden Research Center, 650 Harry Road, San Jose, CA 95120–6099
[2]Max Planck Institute for Polymer Research, D–55128 Mainz, Germany

Poled polymers for fully integrated modulator and switching applications will require orientational stability under device operating conditions and thermal stability of the chromophores and polymers during poling and integration. The latter may require heating to 250°C or more for brief periods. We describe here the preparation of a number of NLO polyimide derivatives which meet these thermal requirements. Derivatives containing the NLO chromophore donor substituents embedded in the polymer main chain show exceptional thermal and orientational stability. We also describe preliminary results on the efficient photoassisted electric field poling of high Tg azo-containing polymers at room temperature. This process may alleviate some of the difficulties inherent to poling at very high temperatures.

Organic nonlinear optical materials (NLO) have attracted attention recently as potentially fast and efficient components of optical communication and computing systems. Although they have been studied in many forms including crystals, clathrates, organic glasses, sol-gel composites, vapor deposited films, Langmuir-Blodgett structures, etc., poled polymer films provide the best opportunity for near-term applications, particularly in the area of electro-optic modulators and switches (*1*). In this regard, there are two possible scenarios: (i) external applications separate from the silicon device circuitry and (ii) fully integrated electro-optic modulators and switches. Each of these configurations, while sharing certain common characteristics, differ significantly in the thermal stability requirements. For purposes of discussion, we have designated the applications as type I and type II respectively. For each application, long term stability of the polymers (chemical, oxidative and orientational) will be required at operating temperatures which may vary between 80–125°C depending on the configuration and application. Orientational stability in poled samples will require either extensive crosslinking and/or polymers with high glass transition temperatures (*1,2*). Regarding the latter, Prêtre et al. (*3*) have suggested that polymer glass temperatures (Tg) in excess of 170°C will be required to maintain acceptable orientational stability for operation at 100°C over a period of several years. While in situ polymer and/or chromophore crosslinking has been demonstrated to greatly enhance the orientational stability in poled polymer systems (*1,2*), there is a price to be paid in terms of reproducibility and manufacturability.

In type II integrated electro-optic applications, the thermal stability requirements are more demanding. The polymer and chromophore must not only survive the poling and elevated operating temperature conditions, but must also be compatible with current semiconductor processing procedures. This may involve brief temperature excursions in the 250–350°C range associated with integration, hermetic sealing, chip connection, etc. (*4*). Since applications of this type will certainly require waveguide configurations, the optical losses of the NLO polymer should also be as low as possible and not increase substantially upon poling or processing. A real question for fully integrated electro-optical devices is whether integration and poling of the active polymeric devices can be accomplished without adversely effecting the electrical properties of the operating semiconductor devices. Some progress in this important area has recently been reported by Steier et al. (*5*).

While much effort has been devoted to the stabilization of polar order in polymer systems, the search for thermally stable NLO chromophores is a much more recent phenomenon (*1,2*). Indeed, even the definition of thermal stability for chromophore candidates and comparative assessment is nontrivial. Accordingly, we have developed a protocol useful for screening prospective candidates for high temperature NLO applications and have demonstrated its utility for a number of examples (*6*). The procedure involves (i) preliminary screening of candidates by thermal gravimetric analysis (TGA) and differential scanning calorimetry (DSC), (ii) variable temperature UV-visible studies of the chromophores contained as a guest in a soluble polyimide host, (iii) the study by electric field induced second harmonic generation (EFISH) of the effects of heating on chromophore nonlinearity, (iv) solution UV-visible studies on chromophores which have been heated to high temperatures, and finally (v) incorporation of promising candidates into polyimide polymers and study of the thermal and oxidative stability of the bonded chromophore by variable temperature UV-visible spectroscopy. Recently, Goldfarb et al. have further refined the DSC technique for studying the thermal stability of NLO chromophores using sealed microcapillary techniques (*7*). These authors have calculated the temperatures where 10% of the initial melting point endotherm or the decomposition exotherm of the NLO chromophore is lost for over 30 samples, including some of ours. Our results agree reasonably well with theirs obtained from the decomposition exotherms (vida infra), supporting the validity of our thermal protocol.

Thermal Stability of NLO Chromophores

The simplest technique for screening the thermal stability of large numbers of potential NLO candidate is thermal analysis. We routinely use thermal gravimetric analysis

(TGA) and differential scanning calorimetry for this purpose (*6*). In the initial screening of classes of compounds, a heating rate of 20°/minute is employed. From the TGA studies, we record the temperature at which 5% (T_5) of the original sample weight is lost. Since this technique measures only the loss of volatile fragments, it is difficult to distinguish between sample volatility and decomposition of the chromophore producing volatile fragments. For the DSC technique, we report an onset decomposition temperature defined as the intersection between the tangent of the exothermic or endothermic decomposition and the adjusted baseline (T_d). This temperature is strongly dependent on the heating rate. This dependency is shown for the common chromophore DR1(**1**) in Figure 1. For most chromophores, the curves, as shown in Figure 1, may be extrapolated to zero heating rate to produce a quantity which we define as the equilibrium onset decomposition temperature (T_d°). T_d° is *always* lower than T_d values obtained at higher heating rates, sometimes by as much as 50–80°C (*8*). While T_d values obtained at heating rates of 20°/minute are applicable for determining the relative stability of large numbers of materials, the equilibrium onset decomposition temperatures T_d° values are more useful for predicting the use temperature of NLO chromophores for high temperature poled polymer applications! Application of the complete thermal protocol, while time consuming, provides useful data for prediction of the utility of NLO chromophores in high temperature poled polymer applications. While it is clear from our studies that there is an inverse relationship between chromophore thermal stability and molecular nonlinearity as determined by electric field induced second harmonic generation studies, the relationship varies with structure, and reasonable compromises of the thermal stability-nonlinearity tradeoff may be achieved (*9–17*). In this regard, we have observed that for amino donor substituents, higher thermal stabilities are always observed for N-aryl substituents than for hydrogen or alkyl substitution. This appears to be true regardless of the nature of the acceptor groups or the bridging substituents (*10*).

The data in Table I were measured for a number of representative NLO azo containing chromophores by the application of our thermal protocol (*6*). Also included in the table (column 6) are the temperatures for which 10% of the chromophore decomposition exotherm is lost after 30 minutes of heating in a sealed capillary. These data were determined by Goldfarb et al. (*7*), and we thank these authors for the use of these data prior to publication. The limited data provided in the last column were determined by variable temperature UV-visible studies of the decrease in absorbance at the long wavelength λ_{max} of the chromophore when it was incorporated into a polyimide via a thermally stable flexible tether (vide infra) (*6b*). While there is some scatter in the data, it is obvious that the thermal protocol has predictive utility in assessing the utility of NLO chromophores for high temperature poled polymer applications.

Tethered NLO Polyimides

Given the improved thermal stability of diarylamine substituted chromophores, we have developed a number of routes to materials of this type. For tethered substituents, the chromophores may be attached to the oxygen of 3,5-diaminophenol by either SN-2

Table I. Thermal stability comparisons of a variety of NLO chromophores using thermal protocol techniques. (a) Temperature for which 10% of the original chromophore nonlinearity is lost after 0.5h heating. (b) Temperature where 10% of the original molar extinction coefficient at the long wavelength λ_{max} is lost after 0.5h. (c) Temperature where 10% of the measured heat of the exothermic DSC decomposition is lost after 0.5h. (d) Temperature where 10% of the original absorbance at the λ_{max} is lost upon heating for 0.5h for chromophores tethered to a polyimide backbone.

Chromophore	T_d (°C)	T_d^0 (°C)	$T_{1/2}$ (10%)[a] EFISH (°C)	$T_{1/2}$ (10%)[b] UV-VIS (°C)	$T_{1/2}$ (10%)[c] DSC (°C)	$T_{1/2}$ (10%)[d] UV-VIS (Polyimide)
N, N=N, NO_2 **2**	322	270	242	254	251	263
Ph_2N, N=N, NO_2 **3**	393	352	342	328	312	325
Ph_2N, N=N, NC, CN, CN **4**	364	249	257	260	227	--
Ph_2N, N=N, CN, CN, H **5**	383	310	273	275	261	--
Ph_2N, N=N, N, S, NO_2 **6**	356	291	276	276	254	282

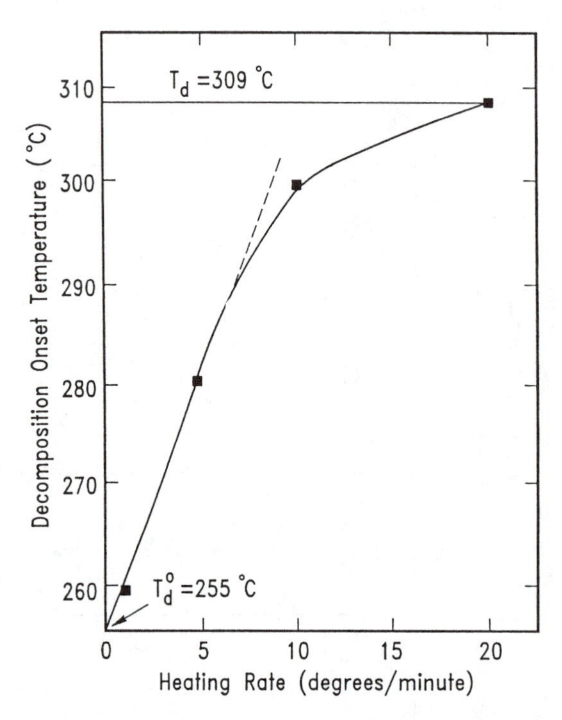

Figure 1. Variation of the onset decomposition temperature of DR1(**1**) with heating rate.

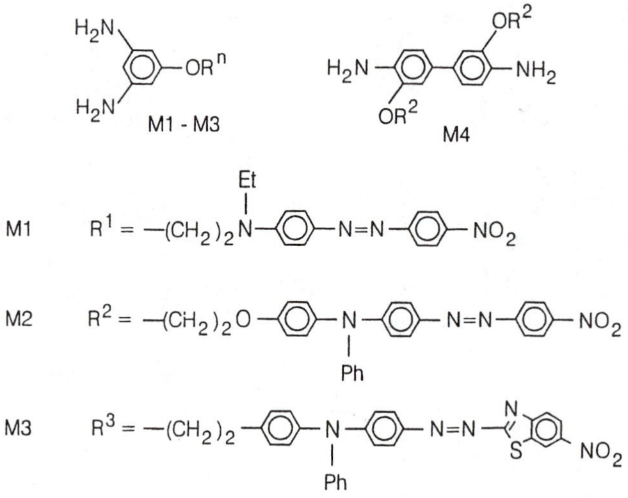

H_2N, H_2N, OR^n — M1 - M3

H_2N, NH_2, OR^2, OR^2 — M4

M1 $R^1 = -(CH_2)_2N(Et)-C_6H_4-N{=}N-C_6H_4-NO_2$

M2 $R^2 = -(CH_2)_2O-C_6H_4-N(Ph)-C_6H_4-N{=}N-C_6H_4-NO_2$

M3 $R^3 = -(CH_2)_2-C_6H_4-N(Ph)-C_6H_4-N{=}N-$(benzothiazolyl, N, S)$-NO_2$

Figure 2. Diamine monomers for the generation of NLO tethered polyimides.

displacement from N-tosyloxyalkyl substituted chromophores (*18*) or by Mitsunobu coupling (*18,19*) of the hydroxyalkyl or hydroxyalkoxy substituted structures (M1-M4, Figure 2). In the latter case, the amino groups of the phenol were first protected as the bis-trifluoroacetamide, and the amino substituents liberated by treatment with K_2CO_3-methanol-water after coupling. The monomer M1 is included as an example containing an alkyl amino donor substituent, while the others contain triarylamino substituted donors. M4 contains two NLO groups/diamine in an effort to increase the chromophore density in the polymer. Alkyl tethers seem not to significantly effect the stability of the chromophore as long as they are not directly attached to the donor nitrogen.

The desired polyimide derivatives were prepared by condensation polymerization as shown in Scheme 1. The initial polyamic acids (**PAA 1-4**) could be spun onto a substrate and cured by ramping the temperature to 260°C. For those examples where X=$(CF_3)_2C$, the polyimide derivatives were soluble in organic solvents. These materials were usually spun in the preimidized form. The improved thermal stability of the arylamino substituted chromophores tethered to the polyimide backbone is obvious from the DSC analysis. The onset decomposition temperature measured by DSC of the aryl substituted polymer **PI-2b** is almost 40°C higher than that of **PI-1b** (T_d = 376 vs 338°C). Likewise, this stabilizing effect is apparent by variable temperature, UV-visible spectroscopic studies of the polymer films. In this case, the absorbance at the λ_{max} of the chromophore was monitored versus time at a variety of temperatures (*18*). While the absorbance of PI-2b (λ_{max} 498 nm) was little changed after one hour at 275°C, that of **PI-1b** (λ_{max} 474 nm) decreased by almost 30%. In each case, the heating was conducted under an inert atmosphere. Studies such as these are the origin of the data shown in the last column of Table I.

Nonlinearity Measurements. The electro-optic coefficients (r_{33}) for the polymers **PI-1** through **PI-4** were measured by one of three procedures depending on the poling technique, and the results are shown in Table II. For samples that were poled by corona discharge, an attenuated total reflection (ATR) technique was employed (*20*). An ellipsometry technique was utilized when single films were electrode poled (*21,22*). A heterodyne technique (*23,24*) was used to determine r_{33} of polymers poled in modulator geometries. Using the ellipsometric technique, the films were poled with a rather low field of 75 V/μm. The values of the electro-optic coefficient (r_{33}) obtained in these electrode poling experiments would be expected to increase linearly with field (*1*). In the case of **PI-2b** which was studied extensively, a triple-stack phase modulator composed of two crosslinked acrylate buffer layers sandwiching the NLO polymer layer was constructed (E ~ 250 V/μm). The r_{33} value measured in this modulator was similar to those obtained both by corona poling and by single polymer layer ellipsometry, after extrapolating the latter to the modulator poling voltage. In the case of two related NLO polyimides, the r_{33} values measured either at 250 V/μm and/or extrapolated to this poling field compared quite favorably with those predicted by calculation (*17,25*) (i.e., 8.8 pm/V for **PI-2a** and 8.3 pm/V for **PI-2b**, respectively). However, the measured r_{33} value for the 6-nitrobenzothiazole polymer **PI-3** seems anomalously low based on the substantially larger hyperpolarizability of

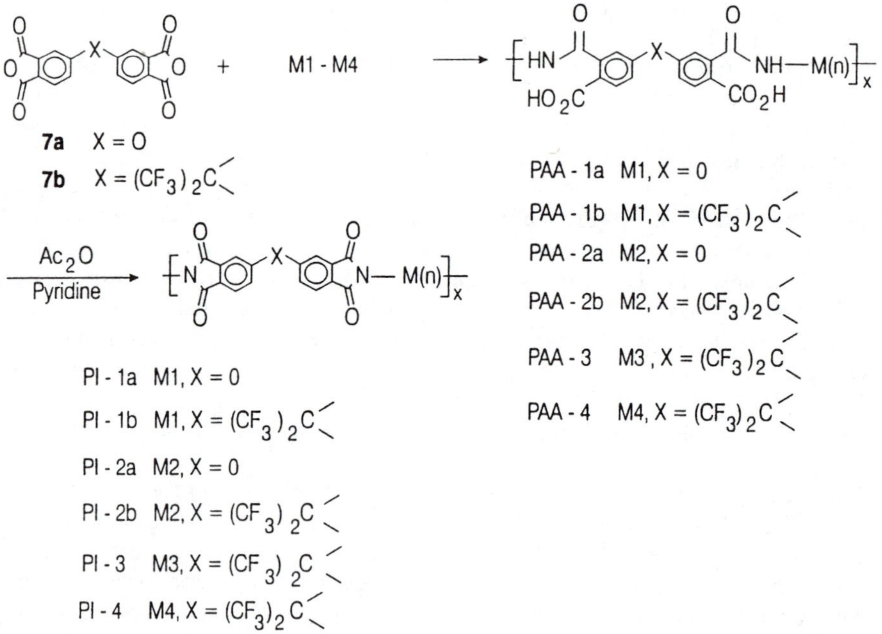

7a X = O
7b X = $(CF_3)_2C<$

+ M1 - M4 →

PAA - 1a M1, X = O
PAA - 1b M1, X = $(CF_3)_2C<$
PAA - 2a M2, X = O
PAA - 2b M2, X = $(CF_3)_2C<$
PAA - 3 M3, X = $(CF_3)_2C<$
PAA - 4 M4, X = $(CF_3)_2C<$

Ac_2O / Pyridine →

PI - 1a M1, X = O
PI - 1b M1, X = $(CF_3)_2C<$
PI - 2a M2, X = O
PI - 2b M2, X = $(CF_3)_2C<$
PI - 3 M3, X = $(CF_3)_2C<$
PI - 4 M4, X = $(CF_3)_2C<$

Scheme 1. Preparation of tethered NLO polyimides by condensation polymerization. In the polymer structures, –Mn is derived from the diamines shown in Figure 2.

Table II. The thermal, linear and nonlinear optical properties of a variety of NLO functionalized side chain polyimides: (a) poling temperature; (b) polymer glass transition temperature as measured by DSC analysis at a heating rate of 20°/minute; (c) electrode poling (75 V/μm), r_{33} value measured by ellipsometry; (d) corona poling (~ 230 V/μm), r_{33} measured at 1.3 μm by attenuated total reflection (ATR); (e) corona poling (~ 200 V/μm); (f) measured in a stacked three-level waveguide phase modulator; (g) corona poling (~ 230 V/μm), r_{33} measured by ATR at 633 nm and extrapolated to (1.3 μm) assuming the validity of the two-state model.

Polyimide	T_{pol} [a] (°C)	T_g [b] (°C)	λ_{max} (nm)	Chromophore Wt. %	r_{33} (1.3 µm) (pm/V)
PI-1b	205	213	474	36	3.75 [c]
PI-2a	210	223	500	47	8.2-10.0 [d]
PI-2b	220	228	498	45	7.0 [e] 8.1 [f] 3.1 [c]
PI-3	220	225	548	53	2.45 [c]
PI-4	210	210	496	58	13.0 [g]

the chromophore (*10*). It is possible that the poling is less efficient for this more extended chromophore or that **PI-3** films are more conducting.

Greatly improved orientational stability for the polyimides **PI-1–PI-4** would be anticipated by virtue of their relatively high glass transition temperatures (see Table II). Relaxation data were obtained by monitoring the intensity of the second harmonic signal of films of **PI-2a** and **PI-2b** poled in a corona field (~ 250 V/μm, poling temperatures 210–220°C) (*18*). After a small initial decay in the signal (< 10%), each sample was stable for over 1000h at 100°C. In addition, preliminary results suggest that thermal annealing of **PI-2b** at 175°C for ~ 3 hrs with the poling field on increases the characteristic relaxation time τ (derived from a stretched exponential fit (*1*) of the decay data) by ~ 50%. The characteristic relaxation time is defined as that where the second harmonic signal has decayed to 1/e of its initial value.

Embedded Donor Polyimides

Thus far we have been discussing only NLO polyimide derivatives which contained the active functionality tethered to the polymer main chain as a pendant substituent. Although the thermal and oxidative stability of these materials are relatively good and the orientational stability is adequate at 100°C , the Tg's of these materials are all below 230°C. Since the processing temperatures for complete on-chip integration will often exceed 250°C, NLO polymers with Tg values in excess of this temperature will be required. It seems unlikely that this goal will be achieved with any processable NLO polyimides containing the NLO chromophore attached via a flexible tether. To date, no polyimide derivatives of this type with a Tg value exceeding 250°C has been reported and most NLO polyimide derivatives cluster in the 200–240°C range (*18,26*). In an effort to raise the glass transition temperature of NLO polyimides, we considered derivatives where the chromophore was incorporated into the main chain via either the donor or acceptor functionality. This type of bonding should increase the polymer rigidity but still maintain adequate polymer flexibility to allow poling. It was anticipated that polyimides containing the chromophore incorporated into the main chain through both donor *and* acceptor bonding would result in main chain polymers which would be difficult to pole (*1,2*).

Initially, it was decided that the donor substituent should contain the aromatic diamine functionality necessary for polymerization (*27*). However, the target monomer **M-5** proved impossible to prepare by classical electrophilic substitution reactions. All attempts to couple triarylamines substituted with peripheral electron donor groups with diazonium salts failed and resulted instead in dearylation (*18*). The desired monomer **M-5** was finally prepared by the Mills reaction of tris-p-aminophenylamine with p-nitronitrosobenzene **12** as shown in Scheme 2. This is an exceptionally versatile synthetic procedure limited only by the ability to prepare substituted nitrosobenzene partners.

The embedded donor polyimides were prepared by the classical condensation route shown in Scheme 3 (*28*). The flexibility of the polymer could be tuned somewhat by the selection of the dianhydride. Although a wide variety of soluble polyamic acids could be prepared by this method, in practice most of the polyimide derivatives were insoluble in common organic solvents. The two derivatives (**PI-5** and

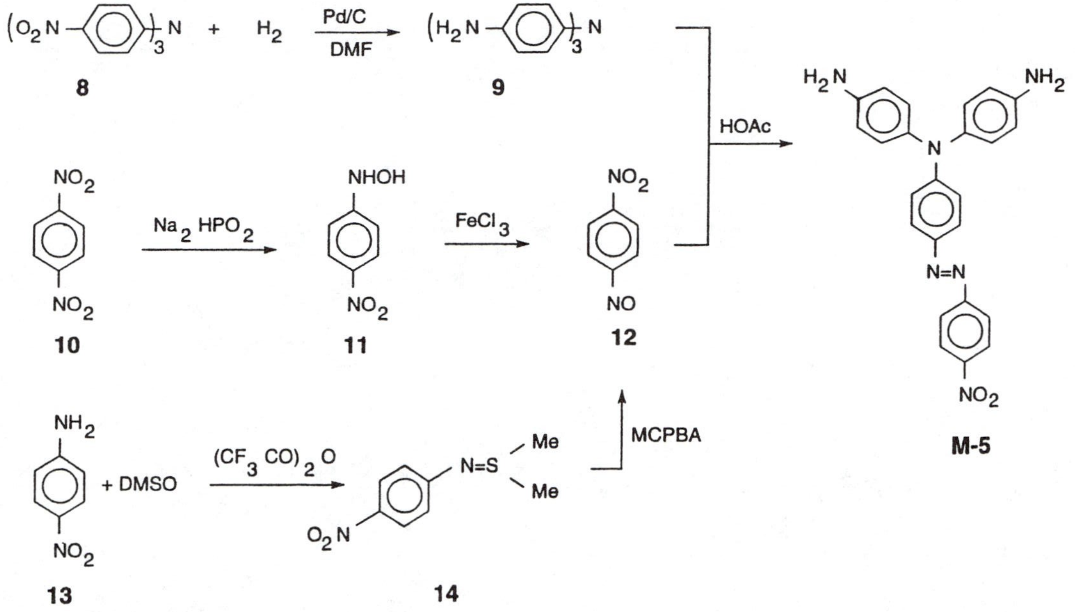

Scheme 2. The synthesis of the monomer **M-5** used in the preparation of embedded donor type NLO polyimides.

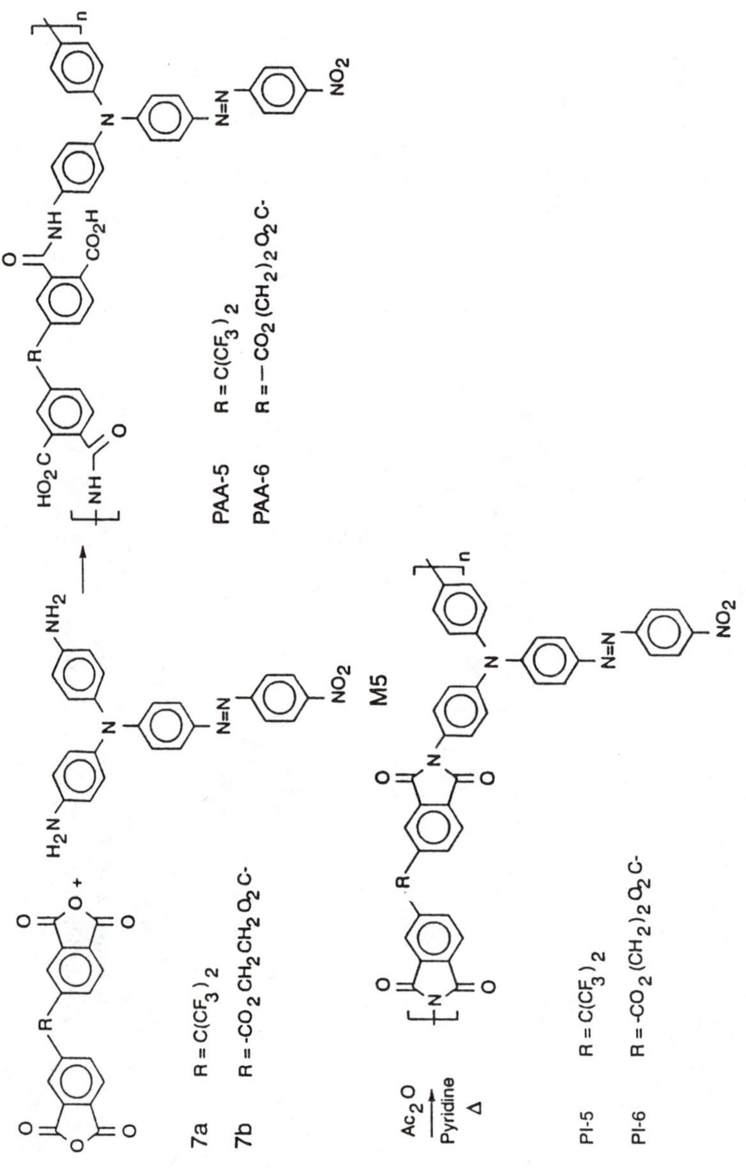

7a $R = C(CF_3)_2$

7b $R = -CO_2CH_2CH_2O_2C-$

M5

PAA-5 $R = C(CF_3)_2$

PAA-6 $R = -CO_2(CH_2)_2O_2C-$

$\xrightarrow[\text{Pyridine, } \Delta]{Ac_2O}$

PI-5 $R = C(CF_3)_2$

PI-6 $R = -CO_2(CH_2)_2O_2C-$

Scheme 3. Preparation of embedded donor type NLO polyimides by condensation polymerization.

PI-6) were both processable in the imidized form. The increased flexibility provided by the use of the anhydride **7b** was evidenced by the lower Tg of **PI-6** (255°C) compared with the more rigid derivative **PI-5** (350°C).

The NLO chromophore in **PI-5** was exceptionally thermally stable as shown by the DSC data presented in Figure 3. For comparison, related data for two tethered examples (**PI-1b** and **PI-2b**) are included in the figure. The films of **PI-5** were also exceptionally stable and no change in the absorbance at the λ_{max} (474 nm) of the chromophore was evident even after several hours at 350°C. Figure 4 shows the stability of **PI-5** monitored by UV-visible spectroscopy in comparison with other tethered NLO polyimides. The more flexible derivative **PI-6** was also very stable, albeit somewhat less so than **PI-5**. Variable temperature UV-visible spectroscopy of **PI-6** showed less than 10% decrease at the λ_{max} after 1h at 300°C. Significant loss in absorbance (22%) was observed after 0.5h at 330°C. The results from the DSC studies corroborated the UV-visible findings. It is believed that the alkyl spacer group and the ester functionality in **PI-6** are the ultimate source of the instability.

Both the **PAA-5** and the imidized form could be oriented by electrode poling. The former was poled by ramping the temperature slowly from 150–280°C (*28*). The decay of the orientated samples while heating at 3°/minute is shown in Figure 5. The lower orientational stability of the sample derived from the polyamic acid relative to the polyimide itself is believed to result from the lower cure temperature used for the former. Extrapolation of the decay curve from **PI-5** poled at 300°C intersects the temperature axis around 360°C which is close to the measured polymer Tg (i.e., 350°C).

The thermal stability of the polar order of **PI-5** is unprecedented for a thermoplastic material. The decay at a variety of temperatures is shown in Figure 6a. The material is stable indefinitely at 250°C and shows only about 12% loss after 0.5h at 300°C. Above 300°C the loss of polar order occurs rapidly as the Tg of the polymer is approached. **PI-6** is less orientationally stable (see Figure 6b), as expected from its lower Tg value (255°C). This material is stable indefinitely at 200°C and shows only 12% loss after 0.5h at 220°C. Approximately 50% of the original nonlinearity remains after 0.5h at 240°, a temperature only 15°C below the Tg. The measured r_{33} values (electrode poling) for **PI-5** (poled at 300°C) and **PI-6** (poled at 255°C) were 5 and 6 pm/V respectively (1.3 μm) (*28*). The poling electric field strengths varied from 160–180 V/μm. Relaxation of the poled NLO chromophores was followed by monitoring the decrease as a result of the decay in the poled order in the second harmonic signal emanating from the samples. The observed decay for both embedded donor polyimides can be fit quite well by a stretched exponential function (*1*). The variation of the characteristic decay time τ (point at which the second harmonic coefficient (d) has decayed to 1/e of its initial value) with temperature for **PI-5**, **PI-6** and the tethered derivatives **PI-2a** and **PI-2b** is shown in Figure 7. The improved stabilities of the embedded derivatives are evident from the much larger values of τ relative to **PI-2a** and **PI-2b** at elevated temperatures.

Electric field poling at very high temperatures is exceedingly demanding and requires highly purified polymer samples. In the case of **PI-5**, this required multiple solvent reprecipitations, isolation and drying. The samples, after spinning from cyclopentanone, were dried in a vacuum oven at 200°C for several hours. Even after

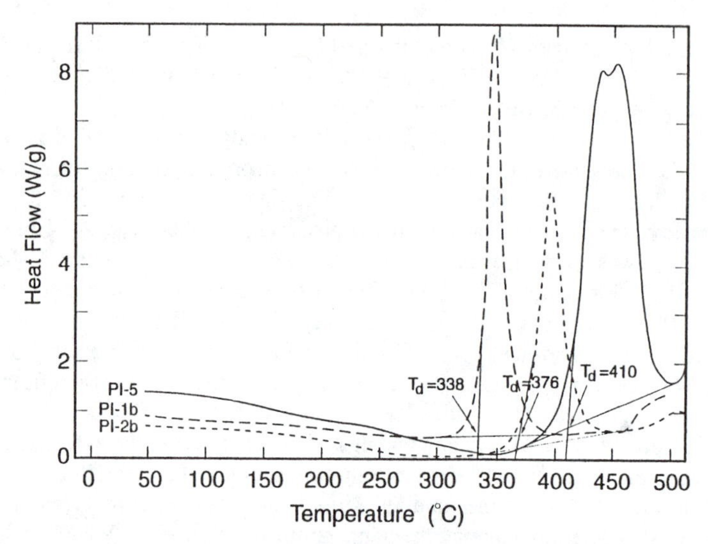

Figure 3. Thermal stability of NLO polyimides by DSC analysis (scanning rate 20/minute): (—) **PI-5**, (---) **PI-1b**, (···) **PI-2b**.

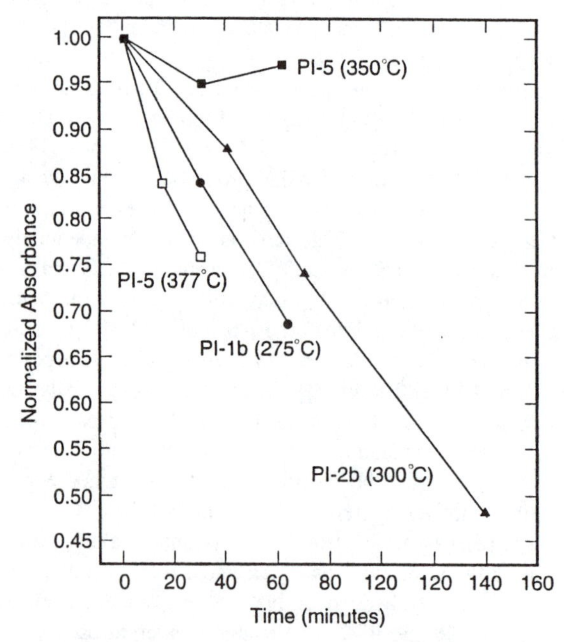

Figure 4. Variable temperature UV-visible studies of a variety of NLO polyimides: (-■-) **PI-5** (350°C), (-▲-) **PI-2b** (300°C), (-●-) **PI-1b** (275°C), (-□-) **PI-5** (377°C). Temperatures were measured on the wafer and the absorbance was recorded at the λ_{max}.

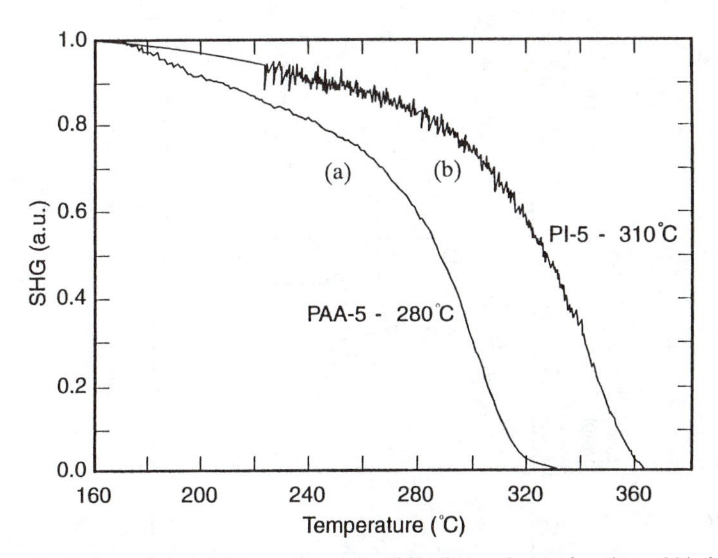

Figure 5. Polar order stability as determined by thermal ramping (rate 3°/minute); laser frequency fundamental 1.047 μm. (a) film prepared from the poly(amic acid) **PAA-5** heated to a maximum temperature of 280°C in the corona field; (b) chemically imidized film of **PI-5** poled at 310°C (corona field).

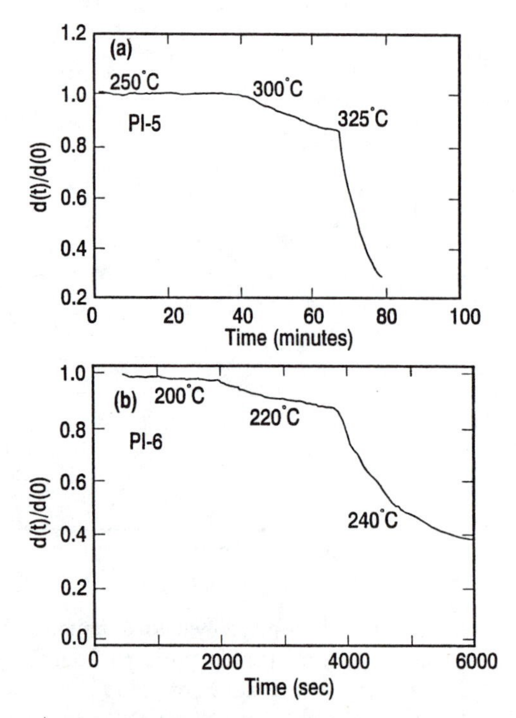

Figure 6. Polar order stability of **PI-5(a)** and **PI-6(b)** at various operating temperatures. The time interval at each temperature was 30 minutes.

rigorous purification, the conductivity of the samples increased according to the Arrenhius relation as the poling temperature exceeded 300°C. These data are shown in Figure 8. Prolonged heating above 300°C led to substantial Joule heating due to the increasing electrical current, leading ultimately to a thermal run away and sample breakdown. The intensity of the second harmonic signal for a sample of **PI-5** heated to the glass transition temperature varied quadratically with the applied field as expected. These data are shown in Figure 9. Also shown in the same figure is the power dissipated (mW) with increasing electric field. In this particular experimental run, breakdown and shorting occurred at a field of 220 V/μm, resulting in a catastrophic drop in the second harmonic signal due to rapid orientational relaxation, since the polymer is held near the Tg in the absence of a significant poling field.

We have studied the high temperature electrode poling behavior of **PI-5** in some detail. Figure 10 shows the generated second harmonic signal as a function of temperature (heating rate ~ 8°/min.). At the onset of the experiment, the sample temperature was 250°C and the second harmonic signal increases with temperature until the sample reached 350°C. At this point, where the current increases significantly, the heater was turned off. Over a space of 2–3 minutes, the second harmonic signal increased by a factor of two and the sample shorted. This increase in signal is apparently due to the increased applied electric field which occurs when the sample conductivity decreases as the sample begins to cool. Some of the increased signal upon cooling is undoubtedly due to the increased contribution from third-order processes as the field increases. After the sample is shorted, it is not possible to measure the electro-optic coefficient, but the second harmonic coefficient can still be estimated, since it does not require a field across the sample. The r_{33} value just prior to shorting was ~ 4 pm/V (1.3 μm). If we assume a normalized second harmonic coefficient of 1.0 at this point, we estimate that this value increases by a factor of 3–4 immediately after shorting. This would suggest that the r_{33} value at this point, if it could be measured, could be as high as 12 pm/V (1.3 μm) This is not an unreasonable value since, based on the measured nonlinearity of the chromophore, a weight % of 47% for the NLO chromophore in the polymer, and using standard values for the refractive index and the dielectric constant based on those of an isotropic model polyimide such as ULTEM (*29*), a calculated r_{33} value (17,25) at 250°C (250 V/μm) was estimated to be around 11 pm/V.

We have demonstrated that poling nonlinear polymers at very high temperatures near the glass transition temperature of the polymer can be very difficult due to conductivity effects which result in a decreased field across the NLO layer and cause significant resistive heating. For the construction of waveguide devices from high Tg polymers, the situation is also exacerbated by the need to find buffer layers with thermal and mechanical stability at the poling temperature which also have the proper electrical characteristics. In this regard, the buffer layers must be somewhat conducting, but the resistance must be such at the polymer poling temperature that most of the applied voltage is dropped across the NLO layer. This is a serious materials challenge for the construction of operating devices from high Tg thermoplastic NLO polymers.

Another difficulty which will be encountered in trying to make waveguiding modulators and switches from high Tg NLO polymers is that the optical losses

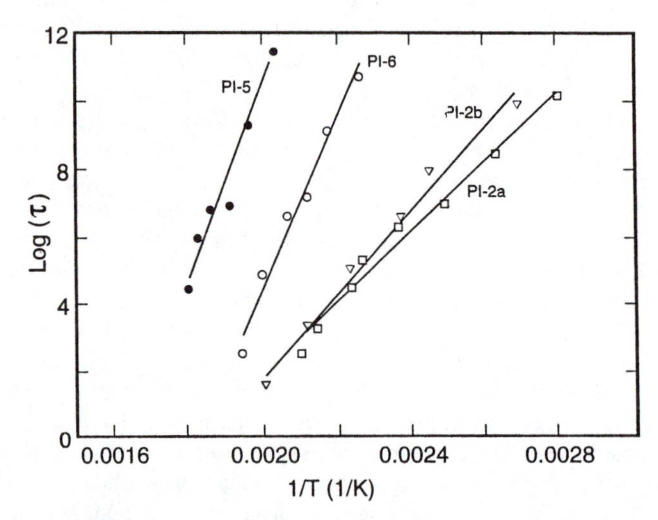

Figure 7. Arrhenius plots for some tethered and embedded donor NLO polyimides: (-●-) **PI-5,** (-○-) **PI-6,** (-▽-) **PI-2b,** (-□-) **PI-2a.**

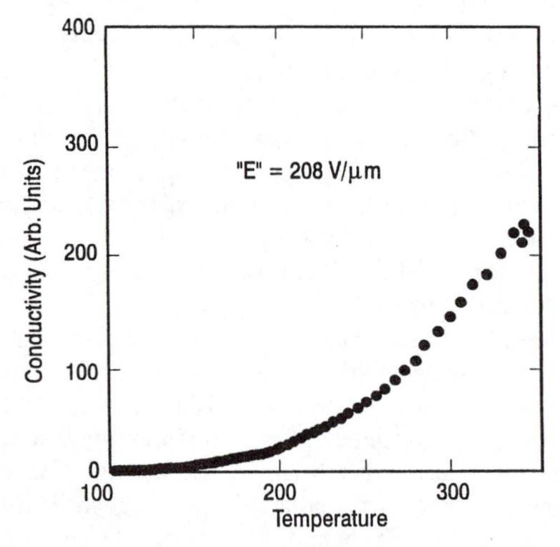

Figure 8. Conductivity of a **PI-5** film as a function of temperature (electric field = 208 V/μm).

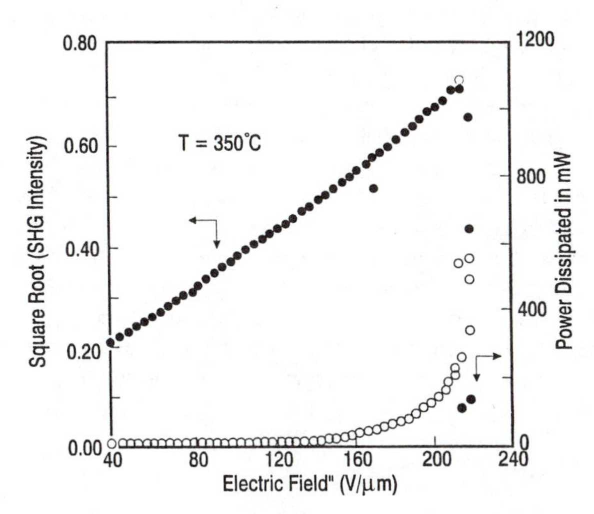

Figure 9. Second harmony intensity and power dissipated (**PI-5**) as a function of applied field at 350°C.

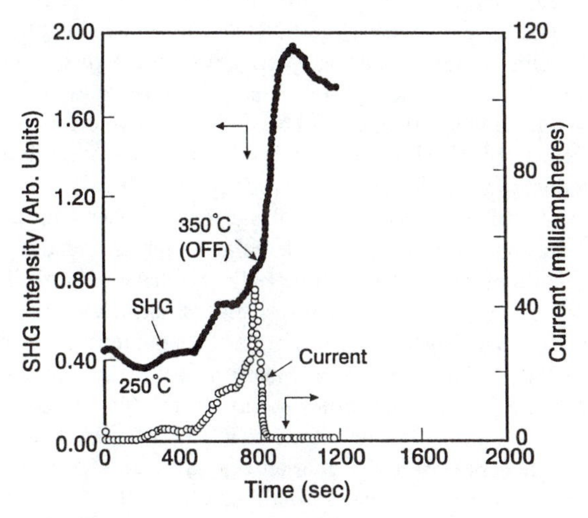

Figure 10. Plot of SHG and current for **PI-5** as a function of temperature ramping (8°C/minute). The heater was turned off at 350°C when the current began to increase rapidly.

(scattering and absorption) must remain low after poling. This is certainly a nontrivial problem, and we find that the losses always increase, sometimes significantly so, upon heating to high temperatures even without poling. These effects are not always detectable by monitoring the polymer UV-visible absorption spectrum before and after heating. This effect is shown in Figure 11 for **PI-3**. The top portion shows the UV-visible spectra of the polymer before and after heating to 250°C. This result is consistent with the earlier observation that the polymer film shows only a 10% decrease in the absorbance at the λ_{max} even after 0.5h at 282°C. On the other hand, part b of Figure 11 shows the photothermal deflection spectra (PDS) of the same polymer after heating at various temperatures for 0.5h. At 1300 nm (see vertical line), the absorbance has already changed by almost two orders of magnitude after heating to 250°C and significant changes are observed even at lower temperatures. Although **PI-3** may be somewhat more sensitive than many high temperature NLO polymers, we believe that the results presented are not atypical and one can expect increased losses upon heating to the elevated temperatures necessary for electric field poling in high Tg NLO polymers.

Photoassisted Electric Field Poling

The difficulties associated with polymer poling at high temperatures (i.e., increased conductivities, lack of suitable buffer layers and increased optical losses) has led us to explore other techniques that might be useful for poling high temperature polymers. We now report (30) very preliminary results on the photoassisted electropoling (PEP) of a high Tg NLO polyimide which suggest that the technique might have some utility.

The photoisomerization of trans azo derivatives to the cis isomer is facile and is both photochemically and thermally reversible. The latter is particularly fast for donor-acceptor substituted derivatives. The rapid, reversible isomerization leads to substantial movement of the chromophore from its initial position in the film prior to irradiation. Irradiation of a film with polarized light (linear or circularly) leads to the rapid development of considerable birefringence. When an electric field is simultaneously applied across the film, a polar order is induced parallel to the propagating direction of the irradiating light (*31*). We have performed one study with a high Tg NLO polymer to look for photo-orientation far below the polymer Tg (*30*). For this we have chosen **PI-1b** which has a glass temperature of 210°C. Irradiation perpendicular to the film surface with circularly polarized light at room temperature results in the development of polar order normal to the film surface. The results of this very preliminary experiment are shown in Figure 12. After irradiation at room temperature at applied field of 135 V/μm, an r_{33} value of 31.5 pm/V was measured (633 nm). Extrapolation of this electro-optic coefficient to 1.3 μm would predict a value of ~ 9 pm/V. This value is actually slightly larger than that obtained by electrode poling when the actual measured value (75 V/μm, 1.3 μm) is linearly extrapolated to a poling field of 135 V/μm. In the case of **PI-1b**, efficient photoassisted poling was observed at a temperature almost 190°C below the polymer Tg! This exciting, albeit very preliminary, result suggests that PEP techniques might be useful for poling very high Tg azo-containing NLO polymers at modest temperatures far below the polymer

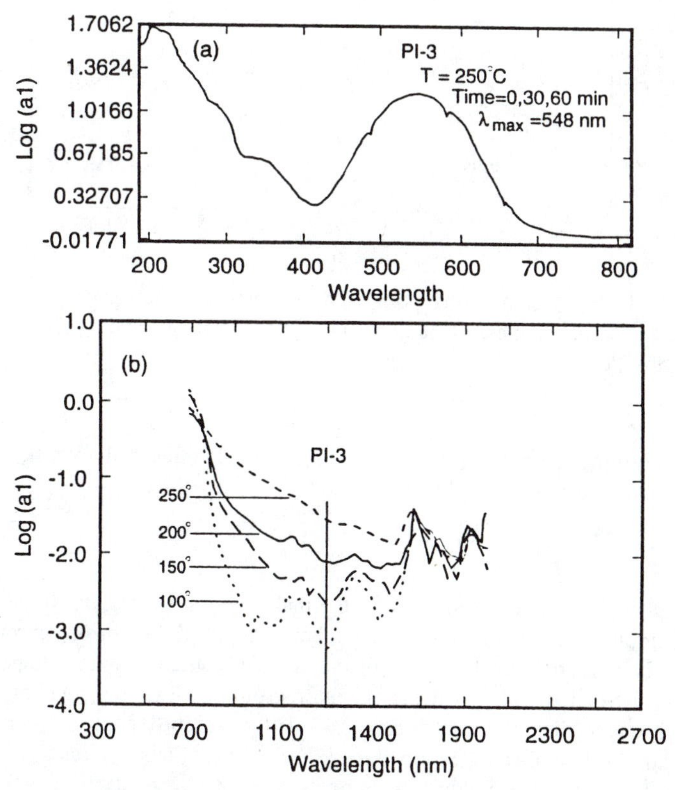

Figure 11. Optical losses upon heating **PI-3**: (a) Absorption spectra of **PI-3** after heating to 250°C; (b) PTD spectra of **PI-3** upon heating.

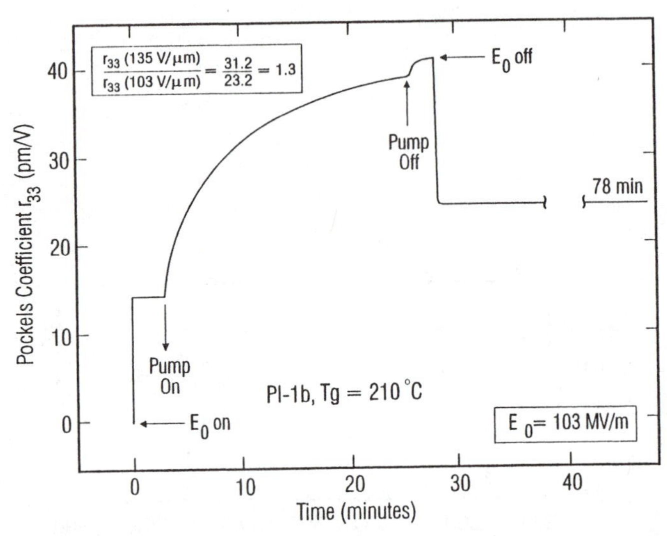

Figure 12. PEP profile of **PI-1b** pumped at 633 nm with circularly polarized light propagating perpendicular to the film surface. Poling field = 103 MV/m, temperature 25°C.

glass temperature, thereby avoiding some of the significant drawbacks of high temperature thermal electric field poling.

Summary

In summary, we have developed a useful thermal protocol for testing the viability of NLO chromophores for high temperature polymer applications. In our chromophore studies, we have found that the nature of the donor substituent is more important for determining thermal stability than the acceptor group, at least for standard electron transmitting connecting bridges. In this regard, for amino donors, aryl substitution *always* leads to significant improvements in thermal stability relative to alkyl substituted derivatives. Using this screening protocol, we have identified a number of highly nonlinear chromophores with acceptable thermal stabilities. A number of these were incorporated into polyimide derivatives as tethered substituents to yield high Tg NLO polymers. The orientational stabilities of these tethered materials at 100°C were adequate for device applications. NLO polyimides with much higher Tg's ranging from 250–350°C were prepared by embedding the donor substituent into the polymer main chain. The thermal and oxidative stability of these materials were extraordinary (300–350°C) and orientational stability up to 300°C was observed for one of these materials. This greatly exceeds that of any NLO poled polymer reported, including examples which have been heavily crosslinked. Finally, preliminary results suggest that photoassisted electric field poling (PEP) of polymers containing azo chromophores at temperatures far below the polymer glass temperature may provide a solution to the difficulties inherent to thermal poling at very high temperatures.

Acknowledgments

We gratefully acknowledge partial suppport of this work by the Air Force Office of Scientific Research Contract F49620-92-C-0025 and by the NSF Center for Polymeric Interfaces and Macromolecular Assemblies (CPIMA) under cooperative agreement NSF-DMR 9400354.

Literature Cited

(1) Burland, D. M.; Miller, R. D.; Walsh, C. A. *Chem. Rev.* 1994, *94,* p. 31.
(2) Dalton, L. R.; Harper, A. W.; Ghosn, R.; Steier, W. H.; Ziari, M.; Letterman, H.; Shi, X.; Mustacich, R. V. Jen, A. K.-Y.; Shea, K. J. *Chem. Mater.* 1995, *7,* p. 1060.
(3) Prétre, P.; Kaatz, P.; Bohren, A.; Günter, P.; Zysset, B.; Olalheim, M.; Stähelin, M.; Lehr, F. *Macromolecules* 1994, *27,* p. 5476.
(4) Lipscomb, G. F.; Lytel, R. *Mol. Cryst. Liq. Cryst. Sci. Technol. B., Nonlinear Optics,* 1992, *3,* p. 41.
(5) Kalluri, S.; Chen, A.; Chuyanov, V.; Biari, M.; Steier, W. H.; Dalton, L. R. *Proc. SPIE,* 1995, *2527,* p. 375.
(6) (a) Miller, R. D.; Betterton, K. M.; Burland, D. M.; Lee, V. Y.; Moylan, C. R.; Twieg, R. J.; Walsh, C. A.; Volksen, W. *Proc. SPIE* 1994, *2042,* p. 354. (b) Miller, R. D.; Lee, V. Y.; Moylan, C. R.; Siemans, R.; Twieg, R. J.; Volksen, W.; Walsh, C. A. *Mol. Cryst. Liq. Cryst. Sci., Technol. B, Nonlinear-Optics* 1996 (submitted).
(7) Goldfarb, I. SBIR-Phase II Research Prime Contract F49620-93-C-0051, Subcontract 771-602, 1996.
(8) Moylan, C. R.; Twieg, R. J.; Lee, V. Y.; Miller, R. D.; Volksen, W.; Thackara, J. I.; Walsh, C. A. *Proc. SPIE* 1994, *2285,* p. 17.
(9) Moylan, C. R.; Miller, R. D.; Twieg, R. J.; Betterton, K. M.; Lee, V. Y.; Matray, T. J.; Nguyen, C. *Chem. Mater.* 1993, *5,* p. 1499.
(10) Moylan, C. R.; Twieg, R. J.; Lee, V. Y.; Swanson, S. A.; Betterton, K. M.; Miller, R. D. *J. Am. Chem. Soc.* 1993, *115,* p. 12599.
(11) Twieg, R. J.; Betterton, K. M.; Burland, D. M.; Lee, V. Y.; Miller, R. D.; Moylan, C. R.; Volksen, W.; Walsh, C. A. *Proc. SPIE* 1993, *2025,* p. 94.
(12) Twieg, R. J.; Burland, D. M.; Hedrick, J.; Lee, V. Y.; Miller, R. D.; Moylan, C. R.; Seymour, C. M.; Volksen, W.; Walsh, C. A. *Proc. SPIE* 1994, *2143,* p. 2.
(13) Moylan, C. R.; Miller, R. D.; Twieg, R. J.; Lee, V. Y. *Second-Order Non-Linear Optics;* Lindsey, G. A., Singer, K. D., Eds.; ACS Symposium Series 601; American Chemical Society: Washington, D.C., 1995.
(14) Shi, R. E.; Wu, M. H.; Yamada, S.; Cai, Y. M.; Garito, A. F. *Appl. Phys. Lett.* 1993, *63,* p. 1173.
(15) Ermer, S.; Leung, D. S.; Lovejoy, S. M.; Valley, J. F.; Stiller, M. *Proc. Organic Thin Films for Photonic Applications 1993, Tech. Digest*; American and Optical Society of America; Vol. 17, p. 50.
(16) Rao, V. P.; Wong, K. Y.; Jen, A. K.-Y.; Drost, K. J. *Chem. Mater.* 1994, *6,* p. 2210.
(17) Moylan, C. R.; Swanson, S. A.; Walsh, C. A.; Thackara, J. I.; Twieg, R. J.; Miller, R. D.; Lee, V. Y. *Proc. SPIE* 1993, *2025,* p. 192.
(18) Miller, R. D.; Burland, D. M.; Jurich, M.; Lee, V. Y.; Moylan, C. R.; Twieg, R. J.; Thackara, J.; Verbiest, T.; Volksen, W.; Walsh, C. A. In *ACS Symposium Series 601*; Lindsey, G. A., Singer, K. D., Eds.; American Chemical Society: Washington, D.C., 1995; Chap. 10.

(19) Miller, R. D.; Hawker, C.; Lee, V. Y. *Tet. Lett.* 1995, *36*, p. 4393.
(20) Moricheré, D.; Dentan, V.; Kazjar, F.; Robin, P.; Levy, Y.; Dumont, M. *Optics. Commun.* 1989, *74*, p. 69.
(21) Teng, C.; Man, H. *Appl. Phys. Lett.* 1991, *58*, p. 435.
(22) Chollet, P.-A.; Gadret, G.; Kajzar, F.; Raimond, P. *Thin Solid Films* 1994, *242*, p. 132.
(23) Valley, J. F.; Wu, J. W.; Valencia, C. L. *Appl. Phys. Lett.* 1990, *57*, p. 1084.
(24) Thackara, J. I.; Jurich, M.; Swalen, J. D. *J. Opt. Soc. Am. B* 1994, *11*, p. 835.
(25) Moylan, C. R.; Swanson, S. A.; Walsh, C. A.; Thackara, J. I.; Twieg, R. J.; Miller, R. D.; Lee, V. Y. *Proc. SPIE* 1993, *2025*, p. 192.
(26) (a) Moylan, C. R.; Twieg, R. J.; Lee, V. Y.; Miller, R. D.; Volksen, W.; Thackara, J. I.; Walsh, C. A. *Proc. SPIE* 1994, *2285*, p. 17. (b) Verbiest, T.; Burland, D. M.; Jurich, M. C.; Lee, V. Y.; Miller, R. D.; Volksen, W. *Macromolecules* 1995, *28*, p. 3005. (c) Yu, D.; Gharavi, A.; Yu, L. *Appl. Phys. Lett.* 1995, *66(9)*, p. 1050. (d) Yang, S.; Peng, Z.; Yu, L. *Macromolecules* 1994, *27*, p. 5858. (e) Sotoyama, W.; Tatusuura, S.; Yoshimura, T. *Appl. Phys. Lett.* 1994, *64(17)*, p. 2197. (f) Becker, M. W.; Sopochak, L. S.; Ghosen, R.; Xu, C.; Dalton, L. R.; Steier, W. H. Shi, Y.; Jen, A. K.-Y. *Chem. Mater.* 1994, *6*, p. 104. (g) Chen, T.-A.; Jen, A. K.-Y.; Cai, Y. *J. Am. Chem. Soc.* 1995, *117*, p. 7295. (h) Prétre, P.; Kaatz, P.; Bohren, A.; Gunter, P.; Zysset, B.; Ahlheim, M.; Staehlin, M.; Lehr, F. *Macromolecules* 1994, *27*, p. 5476.
(27) For an example of a donor-embedded addition polyimide polymer, see: Lon, J. T.; Hubbard, M. A.; Marks, T. J.; Lin, W.; Wong, G. K. *Chem. Mater.* 1992, *4*, p. 1148.
(28) (a) Verbiest, T.; Burland, D. M.; Jurich, M. C.; Lee, V. Y.; Miller, R. D.; Volksen, W. *Science* 1995, *268*, p. 1604. (b) Miller, R. D.; Burland, D. M.; Jurich, M.; Lee, V. Y.; Moylan, C. R.; Thackara, J. I.; Twieg, R. J.; Verbiest, T.; Volksen, W. *Macromolecules* 1995, *28*, p. 4970. (c) Miller, R. D.; Burland, D. M.; Jurich, M.; Lee, V. Y.; Moylan, C. R.; Thackara, J. I.; Twieg, R. J.; Verbiest, T.; Volksen, W. *Mol. Cryst. Liq. Cryst., Technol. B, Nonlinear Optics* 1996, *15*, p. 343.
(29) Philipp, H. R.; Le Grand, D. G.; Cole, H. S.; Liu, Y. S. *Polym. Eng. Sci.* 1989, *29*, p. 1574.
(30) Sekkat, Z.; Wood, J.; Aust, E. F.; Knoll, W.; Volksen, W.; Miller, R. D. *J. Opt. Soc. B* 1996, *13*, p. 1713.
(31) (a) Sekkat, Z.; Dumont, M. *Appl. Phys.* 1992, *B45*, p. 486. (b) Sekkat, Z.; Dumont, H. *Mol. Cryst. Liq. Cryst. Sci. Technol. B: Nonlinear Optics* 1992, *2*, p. 359. (c) Sekkat, Z.; Kang, C.-S.; Aust, C. F.; Wegner, G.; Knoll, W. *Chem. Mater.* 1995, *7*, p. 142. (d) Kang, C.-S.; Winkelhahn, H.-J.; Schultze, M.; Neher, D.; Wegner, G. *Chem. Mater.* 1994, *6*, p. 2159.

Chapter 9

Highly Stable Copolyimides for Second-Order Nonlinear Optics

Dong Yu, Wenjie Li, Ali Gharavi, and Luping Yu[1]

Department of Chemistry and James Frank Institute, University of Chicago, 5735 South Ellis Avenue, Chicago, IL 60637

Nonlinear optical (NLO) copolyimides exhibiting high glass transition temperatures (T_g) and high thermal stabilities were synthesized. The relationship between the chromophore loading level and the physical properties of the copolymers was studied. It was found that decreasing the chromophore loading level increased the glass transition temperature more than 40 °C. As a result, the thermal stability of the electric field induced dipole orientation of the nonlinear optical chromophores tethered in the polyimides was enhanced. At 180 °C, the second harmonic generation (SHG) signal maintained 85 % of the initial value after 700 h. Sizable second harmonic coefficients (d_{33}: 146~32 pm/V) and electro-optic coefficients (r_{33}: 3~25 pm/V at 1300 nm) were obtained.

Polyimides, especially aromatic polyimides, are of growing interest in the investigation of second-order nonlinear optical (NLO) materials because of their high glass transition temperatures which can be utilized to stabilize the dipole orientation of the NLO chromophore at high temperatures (*1-5*). Another attractive feature is that the high thermal stability of the polyimides enables them to survive at the elevated temperatures during the fabrication and operation in the integrated optoelectric devices. In previous work, we described a general approach to synthesize a series of aromatic polyimides functionalized with pendant NLO chromophores (*2-3*). The structure-property relationship of the polyimides has been systematically studied. We observed that the polyimides with the same NLO chromophore but different dianhydrides have distinct physical properties, such as thermal properties and optical nonlinearities. It was also found that these functionalized polyimides exhibit glass transition temperatures which are consistently lower than the corresponding non-functionalized polyimides. It can be expected that if the common diamino monomers, for example, 1, 4-phenylenediamine, are used together with diamino monomers bearing NLO chromophore and corresponding dianhydrides, copolyimides with NLO properties can be synthesized. This copolymerization approach will offer the opportunity to study the effects of the chromophore loading level on the glass transition temperature (T_g), the thermal stability and the optical nonlinearity (the chromophore loading level is defined as the weight percentage of the

[1]Corresponding author

chromophore unit in the polymer). Thorough understanding of these relationships is important to further enhance the performance of NLO materials. The copolyimides exhibited higher glass transition temperatures and better thermal performances of the NLO properties than our previous polyimides. However, the NLO properties diminished due to the reduced NLO chromophore loading level. The results indicate that the copolyimides are the materials of choice for high thermal stability within a tolerable trade-off range in optical nonlinearity. In this paper, we discuss the detailed experimental results.

Experimental Section

The diamino monomer with the chromophore of disperse red 1 was synthesized by an approach described previously (*2,3,6*). The 1, 4-phenylenediamine was purified by recrystallization from methanol. All other chemicals were purchased from Aldrich Chemical Co. and used as received.

Polymerization. The synthesis of polyimide PI 1 was described in the previous paper (*2-3*). The following polymerization procedure exemplifies the synthesis of copolyimides PI 2-6. To a solution of diamino monomer 1 (0.1825 g, 0.434 mmol) and 1, 4-phenylenediamine (0.0156 g, 0.145 mmol) in 5 mL NMP at 0 °C was added 4, 4'-(hexafluoroisopropylidene)diphthalic anhydride (6FDA) (0.2571 g, 0.579 mmol) under nitrogen. After stirring for 1 h and then at room temperature for another 1 h, the resulting poly(amic acid) was imidized using a mixture of acetic anhydride and pyridine (10 mL/ 5 mL, 2 : 1). At room temperature the mixture was stirred for 2 h and then heated to 90 °C for another 3 h. The polymer was precipitated into methanol (100 mL) and collected by suction filtration. The solid was further purified by repeated dissolution in NMP and precipitation into methanol. The collected polymer was washed with methanol in a Soxhlet extractor for 2 days and then dried under a vacuum at 60 °C for 2 days. The ^{1}H NMR of PI 2-6 are similar except for the relative magnitudes of the peaks.

PI 1. ^{1}H NMR ($CDCl_3$, ppm): δ 1.06 (s, -CH_2**CH_3**, 3 H), 3.34 (s, -**CH_2**CH_3, 2 H), 3.71 (s, -**CH_2**CH_2O-, 2 H), 4.23 (s, -CH_2**CH_2**O-, 2 H), 6.62 (s, ArH, 2 H), 7.16 (s, ArH, 1 H), 7.19 (s, ArH, 1 H), 7.34 (s, ArH, 1 H), 7.67 (s, ArH, 2 H), 7.90 (m, ArH), 8.19 (s, ArH, 2 H). Anal. Calcd for $C_{41}H_{26}F_6N_6O_7$: C, 59.43; H, 3.16; N, 10.14. Found: C, 58.54; H, 3.28; N, 10.03.

PI 2. ^{1}H NMR ($CDCl_3$, ppm): δ 1.06 (s, -CH_2**CH_3**), 3.34 (s, -**CH_2**CH_3), 3.72 (s, -**CH_2**CH_2O-), 4.23 (s, -CH_2**CH_2**O-), 6.62 (s, ArH), 7.16 (s, ArH), 7.19 (s, ArH), 7.34 (s, ArH), 7.57 (s, ArH), 7.67 (s, ArH), 7.90 (m, ArH), 7.98 (m, ArH), 8.19 (s, ArH). Anal. Calcd for $C_{37}H_{22}F_6N_5O_{6.25}$: C, 59.20; H, 2.93; N, 9.33. Found: C, 58.98; H, 3.02; N, 9.40.

PI 3. Anal. Calcd for $C_{33}H_{18}F_6N_4O_{5.5}$: C, 58.93; H, 2.68; N, 8.33. Found: C, 58.92; H, 2.73; N, 8.30.

PI 4. Anal. Calcd for $C_{91}H_{46}F_{18}N_{10}O_{15}$: C, 58.70; H, 2.47; N, 7.53. Found: C, 58.53; H, 2.70; N, 7.48.

PI 5. Anal. Calcd for $C_{29}H_{14}F_6N_3O_{4.75}$: C, 58.59; H, 2.36; N, 7.07. Found: C, 58.36; H, 2.53; N, 7.00.

PI 6. Anal. Calcd for $C_{28.2}H_{13.2}F_6N_{2.8}O_{4.6}$: C, 58.51; H, 2.28; N, 6.78. Found: C, 57.94; H, 2.54; N, 6.70.

The ^{1}H NMR spectra were collected on a Varian 500-MHz NMR spectrometer. The GPC measurements were performed on a Waters RI system equipped with an UV detector, a differential refractometer detector, and an Ultrastyragel linear column at 35 °C using THF (HPLC grade; Aldrich) as an eluant.

The molecular weight and the molecular weight distribution were calculated based on monodispersed polystyrene standards. The FTIR spectra were recorded on a Nicolet 20 SXB FTIR spectrometer. A Perkin-Elmer Lambda 6 UV/Vis spectrophotometer was used to record the UV/VIS spectra. Thermal analyses were performed using the DSC-10 and TGA-50 systems from TA Instruments. Elemental analyses were performed by Atlantic Microlab, Inc. A Metricon Model 2010 prism coupler was used to measure the refractive indices of the polyimides. A He-Ne laser (632.8 nm) and two diode lasers (780 nm and 1300 nm) were used as the light sources.

The polyimide films were prepared by casting or spin-coating the polymer solution in tetrachloroethane (TCE) onto the glass slide or Indium-Tin-Oxide (ITO) glass slide. The thickness of the spin-coated film was around 0.5 μm - 2 μm depending upon the concentration and viscosity of the polymer solution. And then the films were coated with a thin layer of polystyrene. The copolyimides were poled using the corona discharge method. The distance from the needle electrode to the surface of the polymer film was fixed at 1 cm and the applied electric field was applied as high as 6000 V. After poling, the polystyrene buffer layer was removed by using benzene.

The polyimide films were characterized by a second harmonic generation experiment using a mode-locked Nd:YAG laser (Continuum PY61C-10, 10-Hz repetition rate) as the light source. The second harmonic signal generated by the fundamental wave (1064 nm) was detected by a photomultiplier tube (PMT). After amplification it was averaged in a boxcar integrator. A quartz crystal was used as the reference sample. The temperature dependent second harmonic generation was conducted by the same apparatus.

Films for the electro-optic coefficient measurements were prepared by corona poling, on which a silver electrode of 1000 Å was then deposited by vacuum evaporation. The electro-optic coefficients (r_{33}) were determined by the simple reflection technique at 1300 nm (*7*).

Results and Discussion

The copolyimides were synthesized via the poly(amic acid)s followed by chemical imidization (Scheme 1). They are soluble in solvents such as THF, chloroform, NMP, tetrachloroethane, DMF and DMSO. The polymer with a relatively low chromophore loading level possesses poorer solubility in THF and chloroform.

These polyimides exhibit the number-averaged molecular weights and the weight-averaged molecular weights in the range of M_n = 9,000 to 16,000 daltons and M_w = 28,000 to 49,000 daltons (Table I). The elemental analyses gave an excellent match with the theoretical values.

The ^1H NMR spectra of copolyimides PI 2-6 all show one peak at 7.6 ppm due to the four aromatic protons in the phenyl group without NLO chromophores. When the ratio of the phenylenediamine over the NLO diamine increased, the integration of the peak at 7.6 ppm gradually increased and was correlated well with the compositions of the copolyimides by comparing with the integration of the peaks at the aliphatic region which were solely due to the NLO chromophore.

The FTIR spectra of the polymers exhibited two characteristic absorption bands at 1785 and 1730 cm^{-1} due to the stretching modes of the imide rings. The absorption peaks of the nitro group occurred at 1518 and 1340 cm^{-1}. These results support the imide structure and indicate the existence of the NLO chromophore.

Thermal Properties. DSC studies indicated that as the NLO chromophore loading level decreased, the glass transition temperatures (T_g) of the copolyimides gradually increased from 235 °C to 278 °C (Table II). The high glass transition

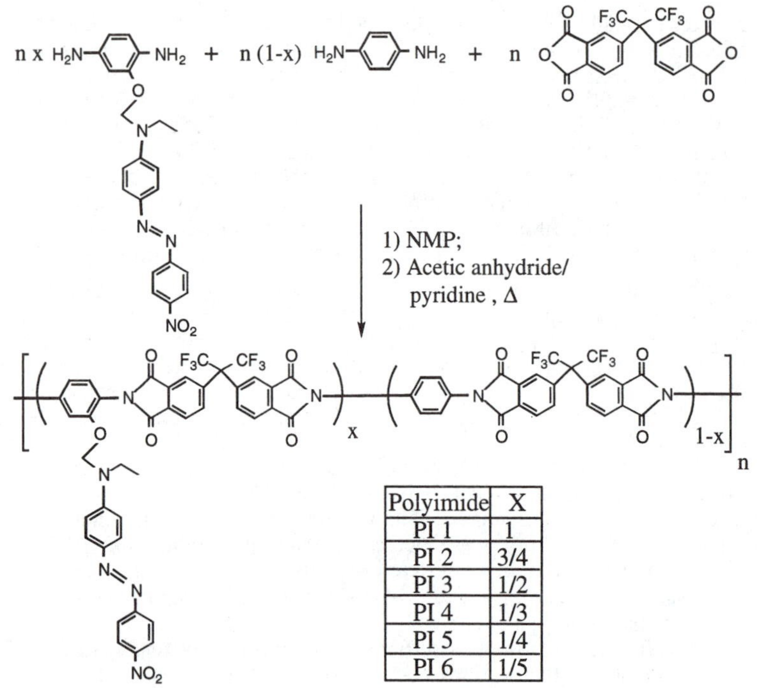

Polyimide	X
PI 1	1
PI 2	3/4
PI 3	1/2
PI 4	1/3
PI 5	1/4
PI 6	1/5

Scheme 1. Synthesis of aromatic copolyimides

Table I. Molecular weights of the copolyimides

Polyimide	PI1	PI2	PI3	PI4	PI5	PI6
x	1	3/4	1/2	1/3	1/4	1/5
M_n	14,000	16,000	15,000	12,000	10,000	9,000
M_w	35,000	38,000	43,000	49,000	33,000	28,000
Polydipersity	2.5	2.4	2.9	4.2	3.3	3.3

temperatures of these polymers require a high thermal stability of an NLO chromophore because the polymer film has to be poled around its glass transition temperature to induce optical nonlinearity. The TGA studies (Figure 1) showed that the decomposition temperatures (the onset point at which the sample starts the weight loss) of these copolyimides were higher than the corresponding glass transition temperatures (320 °C vs. 235 °C for PI 1 and 345 °C vs. 278 °C for PI 6). It indicated that high temperature poling for a short term is feasible without damaging the NLO chromophores. Isothermal experiments for PI 6, by ramping the temperature to 300 °C and maintaining the temperature at 300 °C for 2 h either in the nitrogen or in the air flow, showed only a slight decrease in the TGA trace (Figure 2). It is also interesting to note that the temperature corresponding to the first weight-loss, which is related to the decomposition of the NLO chromophore, gradually increased as the chromophore loading level decreased while all of the copolyimides have almost the same second weight loss at 520 °C due to the decomposition of the backbone of the copolyimides. This phenomena might be explained by a T_g effect. The chromophore may not decompose until the local viscosity is sufficiently low to permit molecular motion.

Table II. Physical properties of PI1-PI6

Polyimide	PI1	PI2	PI3	PI4	PI5	PI6
x	1	3/4	1/2	1/3	1/4	1/5
Chromophore loadng level	32.3%	26.8%	19.9%	14.4%	11.3%	9.3%
T_g (°C)	235	236	257	267	277	278
T_d (°C)	319	327	342	344	348	354
λ_{max} (nm)	477	477	477	478	477	476
Φ	0.18	0.19	0.18	0.17	0.21	0.19
d_{33} (532nm) (pm/V)	146	115	70	51	38	32
d_{33} (∞)	16	14	9	7	5	4
r_{33} (1300nm) (pm/V)	25	19	11	10	4	3

Linear optical properties. The refractive indices of both TE and TM modes were measured at three different wavelengths (632.8 nm, 780 nm and 1300 nm) (Table III). The Sellmeier equation ($n^2 - 1 = \frac{A}{\lambda_0^{-2} - \lambda^{-2}} + B$) was employed to deduce the refractive indices at the laser wavelengths of interest such as 1,064 nm and 532 nm, where λ_0 is the absorption maximum wavelength, λ is the wavelength at which the refractive index is to be determined, A is a constant proportional to the chromophore oscillator strength, and B accounts for all other absorption contributions (*8*). When the chromophore loading level decreased, the refractive indices decreased linearly. These polymers are birefringent in their pristine films as evidenced by the difference between the refractive indices of the TE and TM modes. The refractive indices of the TE mode (in-plane) are larger than those of the TM mode (out-of-plane) which implies that the polyimide backbone might lie in the plane of the film. The refractive indices of the TE mode did not change after corona electrical poling. The refractive indices of the TM mode became larger after poling and the birefringence became smaller simultaneously. It is caused by the alignment of the molecules of the chromophore along the direction of the applied poling field which is perpendicular to the polymer film plane (out-of-plane).

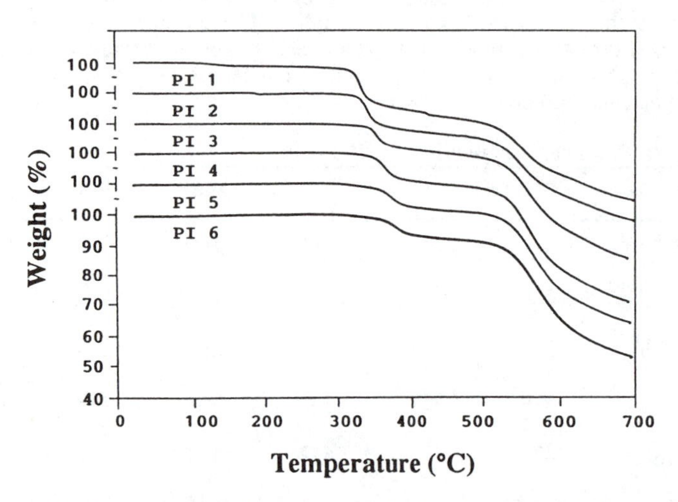

Figure 1. TGA traces of copolyimides with the heating rate 10 °C / min under a nitrogen atmosphere.

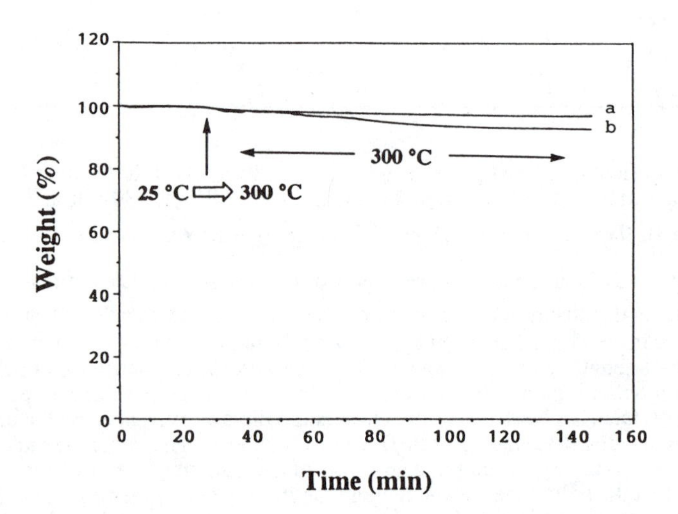

Figure 2. TGA traces of PI 6 isothermal at 300 °C for 2 h. a) in the N_2 flow; b) in the air flow.

The absorption peak in the UV/Vis spectra after poling decreased and the position of the peak slightly blue-shifted, which are typical phenomena for the poled polyimide sample. An important parameter which is used to characterize the efficiency of the poling is the order parameter, defined as $\Phi = 1 - \frac{A_1}{A_0}$, where A_0 and A_1 are the absorbancies of the polymer film before and after corona poling, respectively. To make the physical data comparable among different samples, the order parameters of the copolyimides are maintained around 0.19.

Table III. Refractive indices of TE and TM modes at different wavelengths

Polymer	Wavelength (nm)	unpoled sample n(TE)	unpoled sample n(TM)	poled sample n(TE)	poled sample n(TM)
PI 1	632.8	1.7399	1.7172	1.7421	1.7410
	780	1.6741	1.6489	1.6729	1.6699
	1300	1.6390	1.6159	1.6403	1.6173
PI 2	632.8	1.7297	1.6934	1.7170	1.7113
	780	1.6662	1.6358	1.6586	1.6527
	1300	1.6238	1.6113	1.6226	1.6165
PI 3	632.8	1.6856	1.6555	1.6839	1.6801
	780	1.6494	1.6180	1.6415	1.6374
	1300	1.6178	1.5874	1.6114	1.5910
PI 4	632.8	1.6639	1.6398	1.6669	1.6421
	780	1.6252	1.6010	1.6317	1.6040
	1300	1.5963	1.5730	1.6020	1.5799
PI 5	632.8	1.6428	1.6111	1.6427	1.6224
	780	1.6140	1.5858	1.6154	1.5939
	1300	1.5918	1.5663	1.5911	1.5702
PI 6	632.8	1.6376	1.6196	1.6407	1.6144
	780	1.6112	1.5856	1.6117	1.5890
	1300	1.5899	1.5660	1.5901	1.5661

UV/Vis measurement is also one of the techniques we used to evaluate the thermal stability and the stability of the optical nonlinearity. Control experiments have been done for two polyimide films of PI 6 (Figure 3). These two films were spin-coated from the same batch of the polymer solution so that they had similar thickness. Sample A was poled and sample B was unpoled. The absorption peak of sample A decreased after poling. A minor rise later occurred after one hour and a minor fall after one week when the sample was aged at 180 °C in the air. The minor rise for the poled sample, which corresponds to the relaxation process of the dipole alignment, accounts for the removal of the surface charge after poling. The decrease in the absorption peak indicates that the instability of the nonlinearity at high temperatures primarily comes from the thermal decomposition of the chromophore as opposed to the thermodynamic decay of the polar order. Otherwise, if only the relaxation process of the dipole alignment happens, the absorption peak in the UV/Vis spectra should increase instead of decrease. For the unpoled sample, only a successive minor reduction in absorbance was observed after one week which was probably due to the minor thermal damage of the chromophore at such a high temperature.

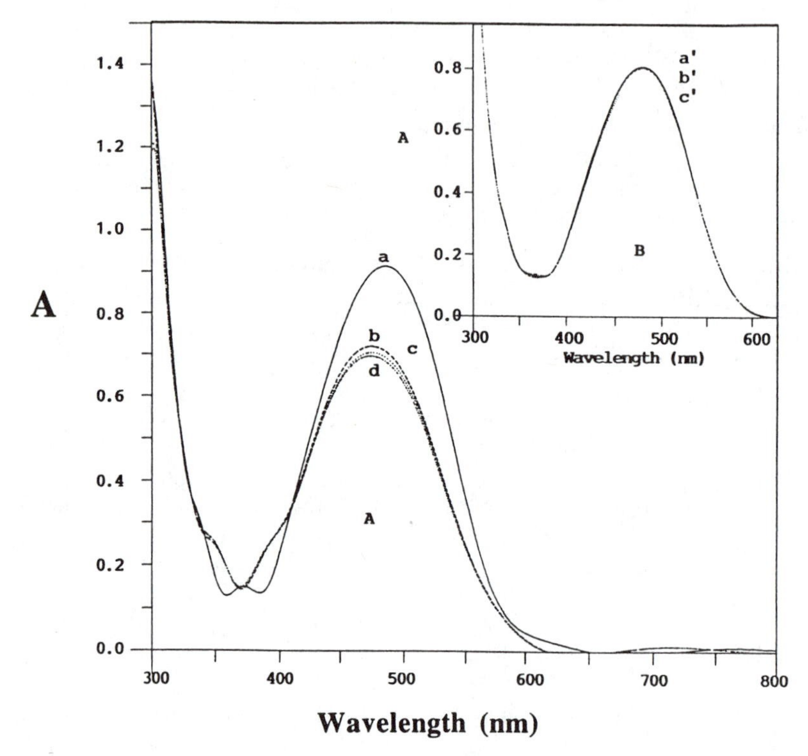

Figure 3. UV/Vis spectra of PI 6 for both poled (A) and unpoled (B) samples. a: before poling; b: right after poling; c: after 1 h curing at 180 °C; d: after 1 week curing at 180 °C; a': before curing; b': after 1 h curing at 180 °C; c': after 1 week curing at 180 °C.

Nonlinear optical properties. The optical nonlinearity was characterized by two experiments: second harmonic generation (SHG) and electro-optic measurements. All of the results are listed in Table II. The second harmonic generation (SHG) for the poled polyimides was conducted to determine the second harmonic generation coefficients (d_{33}). PI 1 possessed the highest value of SHG coefficient at 532 nm (d_{33} = 146 pm/V) and the nonresonant value at ∞ nm (determined based upon the approximate two-level model) is 16 pm/V (*9*). When the chromophore loading level was reduced, both the d_{33} and d_{33} (∞) values decreased (see Table II). This relationship is obvious because the d_{33} values are linearly proportional to the concentration of the NLO chromophore.

The electro-optic coefficients (r_{33}) were measured using the simple reflection technique at 1300 nm which is beyond the absorption band of the chromophore. A large value of r_{33} was obtained for PI 1 (r_{33} = 25 pm/V at 1300 nm). A similar trend as for d_{33} exists for r_{33} values in which it became smaller as the chromophore loading level became smaller.

The temporal stability of SHG was measured at 180 °C. The copolyimides exhibit a much improved stability, compared with the original PI 1 and other polyimides we synthesized (*2*). Few polymers can survive at such high temperatures for a long time. The polymer with higher T_g possessed a higher stability. Figure 4

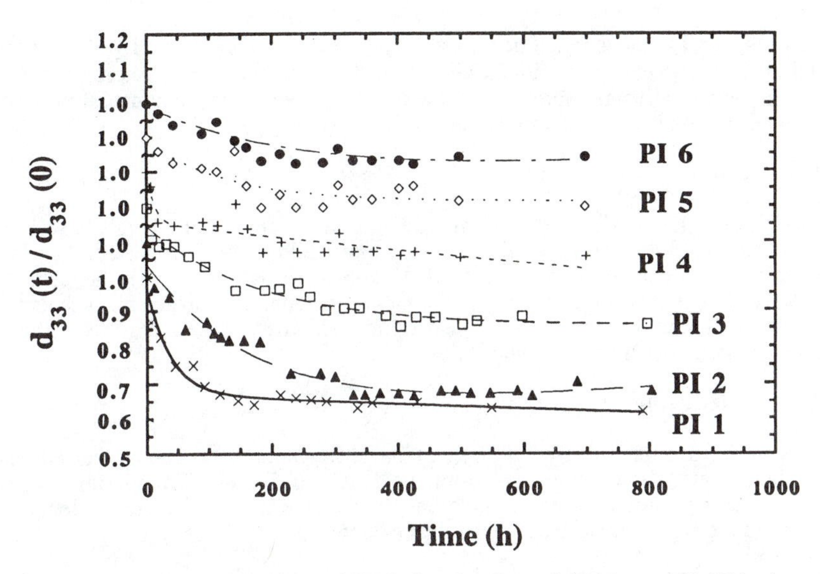

Figure 4. Temporal stability of SHG signals of copolyimides at 180 °C in the air.

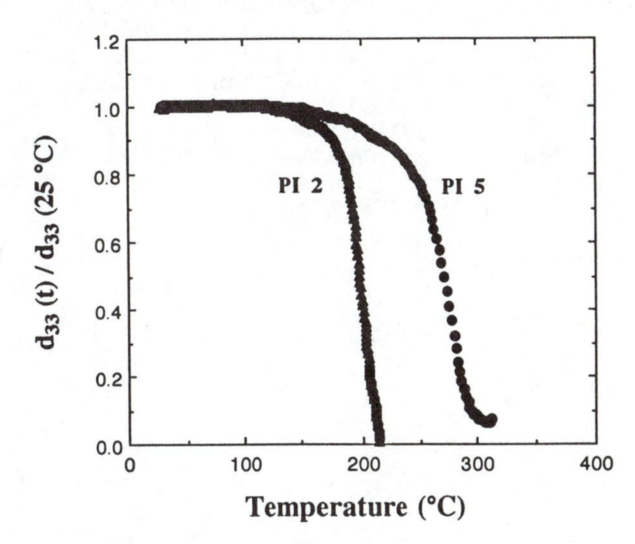

Figure 5. SHG signals of PI 2 and PI 5 as a function of temperature.

shows that PI 6 with the highest T_g stabilized around 85 % of its initial value after 700 h at 180 °C, while PI 1 with the lowest T_g only exhibited 60 % of the original SHG signal after 800 h at the same temperature.

The temperature-dependent SHG was carried out in situ by monitoring the SHG signal while the temperature was ramped in 4 °C / min (Figure 5). PI 5 which has a higher Tg shows much better stability than PI 2 with a lower T_g. Compared with the onset decay temperature (200 °C) for PI 2, the onset temperature of 250 °C for PI 5 is a significant improvement (the onset temperature is the temperature at the intersection of the first and second tangents).

Conclusions

The NLO copolyimides with varying NLO chromophore loading levels have been synthesized. As expected, as the chromophore loading level decreased, the glass transition temperature increased as well as the thermal performance of the resulting polymers. The copolymerization provides a general approach to the enhancement of the thermal stability and the nonlinear optical stability of the materials within a tolerable trade-off in optical nonlinearity.

Acknowledgements

This work was supported by the Office of Naval Research grants and by the National Science Foundation. Supports from the National Science Foundation Young Investigator program and the Arnold and Mabel Beckman Foundation (Beckman Young Investigator Award) are gratefully acknowledged.

Literature Cited

1. Yu, D.; Yu, L. P. *Macromolecules* **1994**, *27,* 6718.
2. Yu, D.; Gharavi, A.; Yu, L. P. *Macromolecules* **1995**, *28*, 784.
3. Yu, D.; Gharavi, A.; Yu, L. P. *J. Am. Chem. Soc.* **1995**, *117*, 11680.
4. Chen, T.; Jen. A. K.-Y.; Cai, Y. *J. Am. Chem. Soc.* **1995**, *117*, 7295.
5. Verbiest, T.; Burland, D. M.; Jurich, M. C.; Lee, V. Y.; Miller, R. D.; Volksen, W. *Science* **1995**, *268*, 1604.
6. Mitsunobu, O. *Synthesis*, **1981**, 1.
7. Teng, C. C.; Man, H. T. *Appl. Phys. Lett* **1990**, *56*, 1754.
8. Born, M.; Wolf, E. *Principles of Optics* ; Pergamon: Oxford, U.K., 1980.
9. Oudar, J. L. *J. Chem. Phys.* **1977**, *67*, 446.

Chapter 10

Accordion Polymers for Nonlinear Optical Applications

Geoffrey A. Lindsay[1], Kenneth J. Wynne[2], John D. Stenger-Smith[1], Andrew P. Chafin[1], Richard A. Hollins[1], M. Joseph Roberts[1], Larry H. Merwin[1], and Warren N. Herman[3]

[1]Chemistry and Materials Branch, Naval Air Warfare Center Weapons Division, 4B2200D, China Lake, CA 93555
[2]Physical Sciences S&T Division 331, Office of Naval Research, U.S. Department of the Navy, 800 North Quincy Street, Arlington, VA 22217–5660
[3]Avionics Branch, Naval Air Warfare Center Aircraft Division, 455150R, Patuxent River, MD 20670

The design and synthesis of mainchain polymers with chromophores in the syndioregic (head-to-head) configuration are reviewed, and recent unpublished results are included. Polymers were designed for both electric-field poling and Langmuir-Blodgett-Kuhn (LBK) deposition. Corona poling in the dark was shown to give excellent results for polycinnamamides, whereas poling in room light at elevated temperatures resulted in photo-oxidative degradation. A Mach-Zehnder intensity modulator was fabricated and the measured electro-optic coefficient (r_{33}) was 9 pm/V at a wavelength of 1.3 μm. New mainchain syndioregic polymers containing isomers of phenylene diacetonitrile as bridging groups were synthesized. Preliminary d_{33} and r_{33} measurements on one promising polymer were close to the expected values. In an alternative approach which avoids electric-field poling polar multilayer -$(AB)_n$- films were fabricated by the LBK technique. Two complementary amphiphilic polymers were alternatively deposited by Y-type deposition. In Polymer A, the chromophore's electron accepting end is connected to a hydrophobic bridging unit, and its electron donating end is connected to a hydrophilic bridging unit. The converse is true for Polymer B. These multilayer thin films have second order nonlinear optical properties which are stable at ambient temperature in the absence of oxygen. Microstructural information on a 92-bilayer polymer film was obtained from polarized optical measurements.

Polar Order in The Organic Solid State

Electrically polarizable polymer films which are macroscopically noncentrosymmetric (i.e., those containing bulk polar order) are of interest for electro-optic switches and signal modulators for high speed data processing (*1*). Incorporation of molecules with high nonlinear optical (NLO) activity into polymers (NLOPs) has attracted considerable interest for photonics applications due to their fast NLO response time, low cost, ease of fabrication, and large nonlinear second-order optical susceptibility, $\chi^{(2)}$.

Experience has shown that these films must have a high glass transition temperature, typically > 180°C, to insure thermal stability for many years. An important focus of current research is therefore directed to the enhancement of the thermal and temporal stability of $\chi^{(2)}$. The stability of polar order in the thermodynamically non-equilibrium glassy state varies from polymer to polymer. Chemical attachment of the chromophore to the polymer as a sidechain (*2*), as part of the mainchain (*3*), or in a crosslinked structure (*4*), has been found to retard the relaxation of alignment.

Polar order is rare in organic solids, as dipoles tend to align anti-parallel. However, polar order is readily achieved by applying an electric field across the film at or above the glass transition temperature. The field aligns the chromophore ground state dipole moments and then the film is cooled with the field on freezing in the order. It is important to achieve a high degree of polar alignment of the chromophores in order to maximize the second-order NLO properties. Typically, poling is carried out with a corona discharge above the film and a ground plane beneath the film, or with contacting electrodes on both sides of the film.

The poled polymer approach is state-of-the-art in terms of the application of polymers to electro-optic devices such as modulators and waveguides. Chromophore-bound polymers for second-order nonlinear optical applications with glass transition temperatures greater than 320°C have been reported (*5*). However, the electro-optic coefficients are modest, typically less than 10 pm/V at 1.3 μm. Part of the poling problem is caused by electrical conductivity at high poling temperatures, which reduces the poling field, or worse, sometimes causes dielectric breakdown.

In electric-field poling, the degree of polar alignment is related to $\mu E/kT$, where μ is the ground state dipole moment of the chromophore, E is the applied electric field, k is Boltzmann's constant, and T is the poling temperature. Because the polymer must be heated to the glass transition temperature for the chromophores to have adequate mobility for alignment, the high thermal vibrational energy of the chromophores works against the torque of the electric field. The poling process has drawbacks. There is a possibility of damaging the surface of the film by charge-injection (fields up to 250 V/μm are used for poling). It may be difficult to achieve a high degree of uniformity over large areas by corona poling (e.g., poling a polymer film on a 6" silicon wafer). For contacting poling, in order to place a large number of devices on a 6" integrated polymer-on-silicon wafer, it may require depositing and removing special electrodes.

An alternative to the poled polymer approach is the use of an intrinsically polar organic solid. As noted above, such structures are uncommon. A chemical approach to the synthesis of thin hybrid organic/inorganic polar films by alternate deposition of chromophore and siliceous phase in a geometrically controlled fashion has produced stable intrinsically polar materials with attractive nonlinear optical properties (*6*). In addition, it is known that polar polymer films may be prepared by the use of the Langmuir- Blodgett-Kuhn (LBK) process (*7*). The fabrication of micron-thick films needed for guiding optical signals takes days by the LBK process, while the process for achieving stable polar order by electric field poling takes only a few hours. However, the LBK method offers low temperature deposition, and may yield more ordered films than electric-field poling, due to less thermal disordering during processing and better control of thickness.

Ashwell, et al., have shown that small molecules with large molecular hyperpolarizabilities can be used to build up relatively thick NLO films which have large second harmonic coefficients (*8*). Penner, et al, demonstrated that 0.5-micron thick NLO films of high optical quality could be fabricated by LBK heterolayer deposition of side chain polymers (*9*). Recently, Lindsay, et al., reported the use of two mainchain chromophoric polymers to fabricate LBK heterolayer NLO films (*10*).

This paper explores both electric field poling of polymers and LBK deposition of intrinsically polar polymer films containing the chromophore in the main-chain, or "accordion" architecture (*3a*). Mainchain syndioregic polymers and poled films constitute the first topic, followed by a second section that describes a bilayer approach to intrinsically polar films.

Mainchain Syndioregic Polymers and Poled Films

With reference to Figure 1, mainchain NLOPs are categorized relative to the orientation of the chromophore dipole along the main chain as isoregic (head-to-tail), syndioregic (head-to-head and tail-to-tail) and aregic (random). Provided all chains were oriented in the extended chain conformation, isoregic polymers would yield a very polar solid state structure. However, this degree of alignment is difficult to achieve (*11*). In the folded chain conformation shown in Figure 1, only syndioregic order has the optimum topology to generate a polar solid.

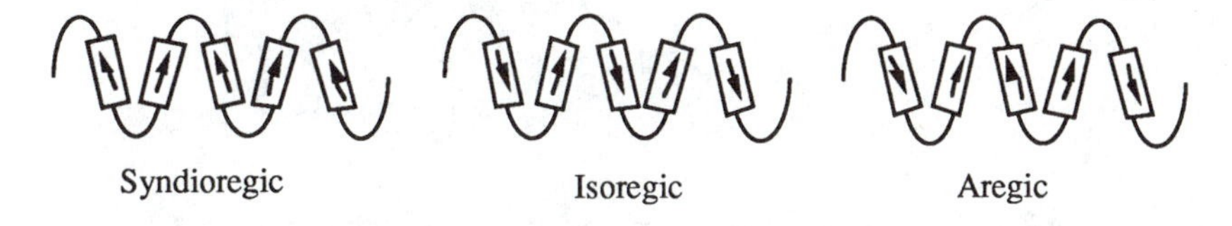

Figure 1. Representation of three chromophore orientations in polymers containing mainchain topology.

Synthesis. The most facile synthetic scheme found for making the accordion polymers involves a Knoevenagel condensation polymerization of two precursors. One is a bis-aldehyde precursor which will form the electron-donating ends of a pair of linked chromophores, e.g.:

OCH — C6H4 — N(C_2H_5) — R — N(C_2H_5) — C6H4 — CHO

which is prepared, for example, by condensing ethyl aniline with α,α'-dibromo xylene then adding the aldehydes via the Vilsmeier reaction. For improved thermal-oxidative stability, R would preferably be entirely aromatic and this is a goal of work in progress. The ethyl group is preferable to the methyl group because it increases solubility and is marginally more stable to oxidation.

The other precursor is a compound which will form the electron-accepting ends of the pair of chromophores, e.g.:

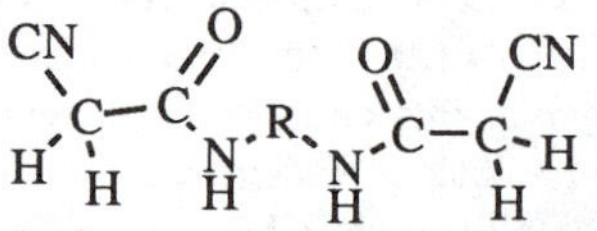

which is prepared, for example, by reacting ethyl α-cyanoacetate with *trans*-1,2-diaminocyclohexane (R). Once again an entirely aromatic R may be preferred and is the subject of current work.

The polymerization conditions typically employ refluxing the precursors in pyridine with catalytic amounts of piperidine for several days (*12a*). The structure of the *trans*-cyclohexyl ethyl amino syndioregic main chain cinnamamide polymer, **1**, (Tg = 208°C) is given in Figure 2.

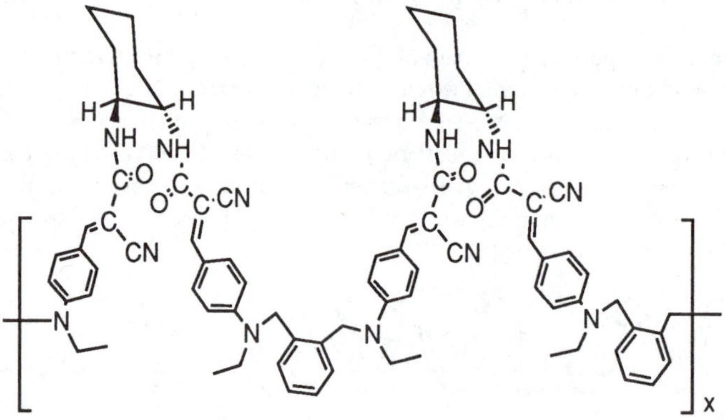

Figure 2. Two repeat units of the original *trans*-accordion polymer, **1**

Molecular weights typically range from 10,000 to 50,000 Daltons. Depending on the molecular weight, the configuration of the cyclohexane ring, and the substituents on the amine donor, the glass transition temperature of the accordion polymers ranges from 173° to 212°C (*12*). Comparing polymers at the same molecular weight (DP = 60), the *cis*-cyclohexyl ethyl amino syndioregic main chain cinnamamide polymer, **2**, had a lower glass transition temperature (185°) and poorer solubility than **1**. The *trans*--cyclohexyl methyl amino accordion polymer, **3**, has the highest glass transition temperature (212°), however it was difficult to obtain molecular weights of **3** higher than about 12,000 g/mol (degree of polymerization, DP = 20).

NMR Characterization. To identify the *cis* and *trans* isomers, solution NMR spectroscopic techniques were used (chemical shifts, the magnitude of the C-H coupling constants, and the dynamic exchange of the cyclohexane ring which exists for the *cis* isomer but not for the *trans* isomer). The *trans* materials have two axial protons in the 1,2 position and the *cis* materials have one axial and one equatorial proton. The *cis* materials thus undergo chair-chair interconversions, while the *trans* materials are relatively rigid since a chair-chair interconversion would lead to two sterically disfavored axial groups. Heteronuclear Multiple Quantum Coherence (HMQC) and Heteronuclear Multiple Bond Coherence (HMBC) measurements were also made on *cis* and *trans* diamide monomers as well as on model compounds and on polymers (*13*).

Using solid state ^{13}C NMR, a comparison of room temperature T_1 relaxation data for **1** and **2**, provides molecular level evidence for the higher T_g and stability of the *trans* isomer. Thus, certain ^{13}C chromophore resonances in **1** show larger T_1 values than the corresponding T_1 values for **2**, demonstrating lower molecular motion and increased rigidity for **1**.

An attempt was made to correlate the temperature dependence of molecular level motions of the polymers with the thermal stability of NLO activity in poled polymer films. To observe changes in local motion with temperature, CP/MAS spectra of the *trans* (**1**) and *cis* (**2**) accordion polymers were run at temperatures up to 110°C (*14*), the limit of our instrument. Over that temperature range, there was no significant change in T_1, T_{1p} or line width for **1** and **2**. Thus, although **1** and **2** have differing degrees of rigidity, there respective degrees of rigidity are constant over the temperature range investigated, indicating that physical alignment in these polymers should be stable up to at least 110°C.

Film Casting, Corona Poling, Optical Measurements and Stability. To evaluate optical behavior, thin films of **1** were bladed onto microscope slides from 8-10% solutions of polymer in solvent, air dried under a Petri dish, then heated in vacuum above the glass transition temperature for 18 hours to remove residual solvent. Both pyridine and a mixture of 5% to 10% chlorobenzene in pyridine were used as solvents. The addition of chlorobenzene tended to inhibit cracking. These films were corona poled above T_g for 10-30 minutes under nitrogen, then slowly cooled to room temperature with the poling field on. Preliminary measurements on **1** indicated unexpectedly low nonlinear optical properties. This problem was subsequently attributed to exposure to UV and visible light during poling at 200°C (*15*), and has been rectified as described below.

Dispersion in the refractive indices of the NLO polymer films was determined by the prism coupled waveguiding technique (*16*). The experimental setup was identical to that reported previously (*12b*). A gadolinium-gallium-garnet prism was used in conjunction with polarized light a from He-Ne laser (543 nm and 633 nm), a solid state laser (836 nm), and a Nd:YAG laser (1064 nm). The refractive indices at these four wavelengths were fitted to a single oscillator Sellmeier equation of the form shown in Figure 3 where λ_{max} = 400 nm was obtained from the UV-VIS spectra.

Figure 3a shows the change in indexes before and after poling in the dark in N_2. The index of refraction of the mode having its electric field is in the plane of the film (transverse electric) is n_{TE}. The index of refraction of the mode having its electric field is nearly perpendicular to the plane of the film (transverse magnetic) is n_{TM}. Unpoled films that were bladed from the pyridine/chlorobenzene solutions showed a small negative birefringence (n_{TM} - n_{TE} = -0.002). This is consistent with a majority of chromophores lying in the plane of the as cast film. After poling, the polymer showed a large positive birefringence (n_{TM} - n_{TE} = +0.022). In contrast to the as cast film, these data indicate that a majority of the chromophores in the poled film are oriented perpendicular to the plane of the film. Protected from air and light, the birefringence is thermally stable.

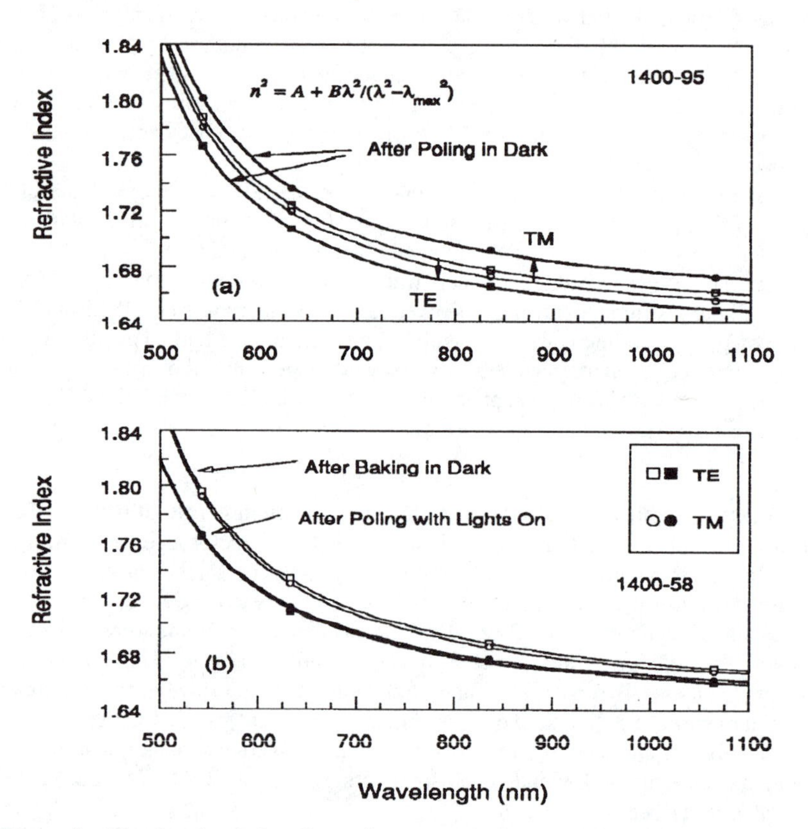

Figure 3: (a) refractive index dispersion curves before and after poling **1** in the dark under nitrogen purge: open squares for the TE mode unpoled film, open circles for the TM mode unpoled film, filled squares for the TE mode poled film, filled circles for the TM mode unpoled film; (b) refractive index dispersion curves before and after poling **1** in air with exposure to a fluorescent desk lamp at 125°C.

Figure 3b shows the effect on the refractive index of the combination of visible-UV light and air on films of **1** at high temperature. The presence of air and light at high temperature bleaches the chromophore and the polymer loses what little birefringence it had.

The nonlinear optical d-coefficients were measured using the experimental setup described previously (12b, 17). The poling field was removed when the film cooled to ~45 °C, and the second harmonic data were obtained 17 hours after poling a film in the dark in nitrogen. To aid in eliminating charge effects due to poling, an aluminum plate, shorted to the ground plane, was placed on the film surface. As a check against charge injection effects, second harmonic data taken four days after poling showed no measurable difference from that taken 17 hours after poling. A 0.4 OD neutral density filter was used in the collection of the p-p data. The d-coefficients were then obtained from a computer fit to an expression (18) for the second harmonic power, using the measured refractive index and thickness data. Assuming Kleinman symmetry, since the UV-VIS spectrum of **1** indicated negligible absorption at 532 nm, we obtained $d_{33} = 30 \pm 5$ pm/V with $d_{31}/d_{33} = 0.35$. No correction was made for absorption.

Molecular orbital calculations using MOPAC 6.0 with the PM3 Hamiltonian on model compounds for **1** containing the *trans*-cyclohexyl bridge, indicate that the di-equatorial conformation is approximately 7 kcal/mole lower in energy than the di-axial conformation (*19*). The zero frequency hyperpolarizability, β_0, calculated for the di-equatorial and di-axial conformations was 24×10^{-30} cm^5/esu and 12×10^{-30} cm^5/esu, respectively (*19*). To obtain a rough estimate of the agreement of our SHG measurements with the molecular hyperpolarizabilities, we used a two-level model (*20*) (with $\lambda_{max} = 400$ nm), to estimate a dispersion-corrected β at 1064 nm. Next, we estimated the order parameter $<\cos^3\theta>$ using the method of Page et al. (*21*), which gave a value of 0.35 ($\Phi = 0.027$). Finally the rigid oriented gas model (*22*) was used to calculate d_{33} at 1064 nm. The calculated d_{33} values of 32 pm/V, using the calculated β value and density for a single chromophore, and 26 pm/V, using the calculated β value and density for a pair of chromophores, give good estimates of the measured d_{33} of a poled film of **1**.

To evaluate orientational and chemical stability, poled films of **1** were placed on a hot stage in a setup designed to measure the second harmonic decay during isothermal aging (*12a*). With the temperature maintained at 125 °C, p-polarized light at 1047 nm from a diode-pumped Nd:YLF laser was incident on the sample at an angle of incidence of 52 degrees. The pulse width was 12 ns and repetition rate was 1 kHz. The fundamental beam in this case was not focused. The results of this experiment for a time period of 120 hours at 125°C in air showed fairly good stability (85% retention of original SHG). When a UV-VIS light was turned on in air at 125°C, an additional rapid decay process was introduced.

Waveguide construction and r_{33} measurement. Refractive index and propagation loss measurements were made by preparing a 1.4 μm film of **1** on a glass substrate. Prism coupling was used to excite the guided modes at 0.859 and 1.3 μm wavelengths. Measurements were made for an unbleached film and a film bleached at room temperature for 4 hrs with a UV light source (17 mw/cm @360 nm in air). The difference in index of refraction before and after bleaching was 0.052 at 0.859 μm and 0.003 at 1.3 μm, and optical loss was < 1 dB/cm (*12a*).

A Mach-Zehnder intensity modulator was constructed with **1** as the core layer (2.3 μm) of a three layer stack (*12a*). The cladding material (3.5 μm) was a UV curable epoxy (Norland NOA81). A gold ground plane formed the lower electrode along with a patterned gold layer for the upper electrode. Channel waveguides were formed by photobleaching with a 5 μm wide mask. Poling was accomplished using the device electrodes with an average field of 107 V/μm at 200°C. The device was overcoated with a polymer layer and cut/polished for end fire coupling. The measured electro-optic (EO) coefficient (r_{33}) was found to be 8.5 pm/V and is stable at room temperature for at least 129 days. Using equation 38 in reference (*23*), d_{33} = 30 pm/V from SHG measurements, the index values from Figure 3 (extrapolating to 1.32 μm), and correcting for dispersion and local field effects, one obtains a calculated r_{33} value of 6.5 pm/V.

Cyclization vs. Polymerization. One of the simple variations on **1** is the substitution of the *ortho*-xylylene bridging group with the *meta*-xylylene group. Upon doing this, it was surprising to find that no polymer formed -- only the cyclomer **4** shown below in Figure 4. This work will be reported elsewhere (*24*) and is summarized here.

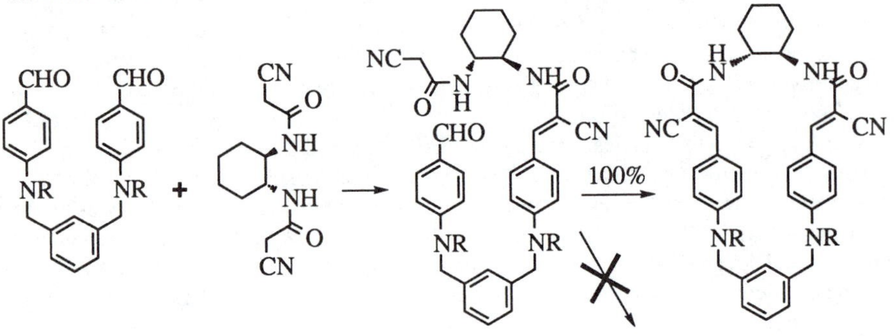

Figure 4. Formation of cyclomer **4** (vs. polymer) in the Knoevenagel condensation.

The rationale for cyclization instead of polymerization of *meta* vs. *ortho* isomers may be explained by the difference in strain energies for the cyclic compounds. The strain energy of a cyclic molecule can be estimated by means of an isodesmic reaction scheme in which the kinds and number of bonds are kept the same for both sides of the reactants-to-products equation (i.e., the number of C-H, N-H, C=O, etc. bonds). The difference in energy between the two sides (reactants - products) is the deviation from simple bond additivity, i.e., the strain energy. The lowest energy conformation for the open chain *meta* compound (the hypothetical polymer) was nearly the same in energy as for the cyclic **4** (less than 1 kcal/mole difference), therefore, essentially strainless. For comparison, the cyclic *ortho* compound was calculated to have a strain energy of 10 kcal/mole greater than the open chain. Indeed, during the preparation of **1**, no evidence of cyclic product was found by GPC. For the *meta* isomer, the entropy difference between cyclomer and polymer may overcome the slight energy difference favoring the linear polymer.

Design and Synthesis of New Syndioregic Polymers for Electric Field Poling.

Promising results for **1** encouraged us to try new chromophores with higher glass transition temperatures, hyperpolarizabilities, and improved photothermal stability. The commercially available electron-accepting groups, *ortho-*, *meta-*, and *para*-phenylene diacetonitrile which contain activated methylenes, lend themselves nicely for use in the Knoevenagel condensation polymerizations with our bis-aldehyde electron donating precursors. We have investigated the *ortho* and *meta* isomers, but not yet the *para*. The structures of these polymers are shown in Figure 5.

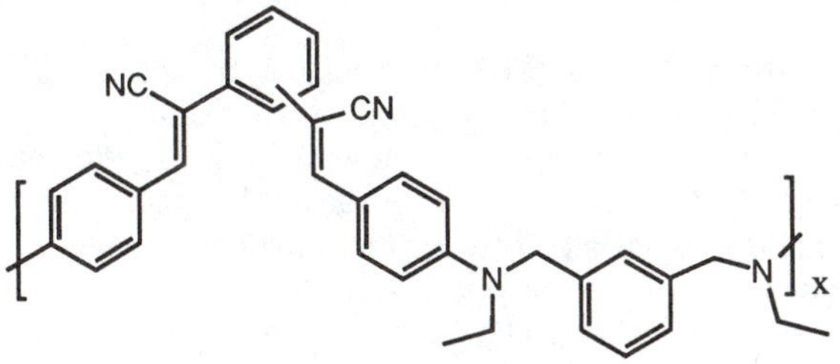

Figure 5. Polymers containing the phenylene diacetonitrile acceptor bridge: *ortho*, **5**, and *meta*, **6**.

The characterization data for **1, 5** and **6** are given below:

Polymer	Yield (%)	T_g(°C)	T_d(°C)(a)	M_n(b)
1	80	205	330	~25,000
5	75	175	283	~4,000
6	74	165	369	~90,000

(a) 2% weight loss in N_2 at 10°C/min heating rate by TGA. (b) by ^{1}H NMR end group analysis.

The polymerization of the *meta*-diacetonitrile isomer yields polymer, **6**, with a Mn of 90,000 g/mol. In our initial attempt, the *ortho*-isomer yielded only low molecular weight polymer **5**. It is interesting that the Tg of **5** is 10° higher than for **6**, even though it has much lower molecular weight. The thermal stability of **5** is over 80° lower than that for **6** (T_d = 283° vs. 369°C). We do not know the reason for the lower stability, but the proximity of the vinylenes in **5** may make them susceptible to an intramolecular chemical reaction.

With presently accessable Mn, the the glass transition temperatures of polymers **5** and **6** are lower than that for **1**. The contributing factors are: (1) lack of hydrogen bonding, (2) the energy to rotate about the single bond connecting the phenylene to the cyanovinylene group is only about 3kcal/mol (from MOPAC® calculations), and (3) the meta-substituted bridging groups seem to lead to a lower lower glass transition temperature than do the ortho-substituted bridging groups.

The d_{33} coefficient of a lower molecular weight (13,000 g/mol) version of **6** was measured using the methods described in the previous section (*25*). After poling at 170°C with a similar time-temperature protocol, the d_{33} was found to be 27 pm/V -- close to the value found for **1.** The electro-optic coefficient of **6,** as measured in a three layer Mach-Zehnder interferometer, is essentially the same as that found for **1** (8.2 pm/V) (*26*).

We are now in the process of extending the length of the chromophore in **1** and **6** with phenylene vinylene, thienylene vinylene, and vinylene groups.

Heterolayer $\chi^{(2)}$ LBK Films from Accordion Polymer Bilayers

The Langmuir-Blodgett-Kuhn (LBK) molecular layer deposition process has been utilized with both small molecules and polymers to give multilayer dipole-aligned films (*7, 8*). The polymers designed for the present work have a syndioregic mainchain NLO polymer topology (*3a*) in which hydrophilic and lipophilic bridging groups are located on alternating bridging sites between the chromophores in order to bring about polar self-ordering at an air-water interface (*27*). As shown in Figure 6, the electron accepting end of the chromophore in Polymer A is connected to a relatively hydrophilic bridging unit, and its electron donating end is connected to a relatively hydrophobic bridging unit. The converse is true for Polymer B.

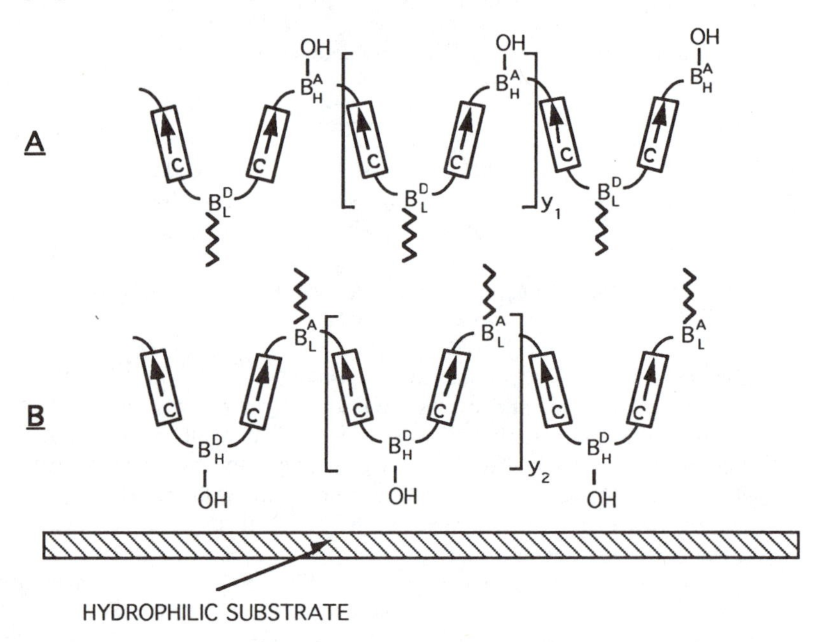

Figure 6. Idealized structure of a complementary pair of accordion polymers in a polar bilayer.

The polymers are designed such that the backbones could fold and position the relatively hydrophilic bridging groups near the water interface, and the relatively hydrophobic bridging groups near the air interface. As multiple layers are deposited by the Y-type method, in alternating $(AB)_n$ fashion, the hydrophilic surfaces of adjacent layers are in contact, as are the hydrophobic surfaces of adjacent layers. This would likely be the most thermodynamically stable conformation.

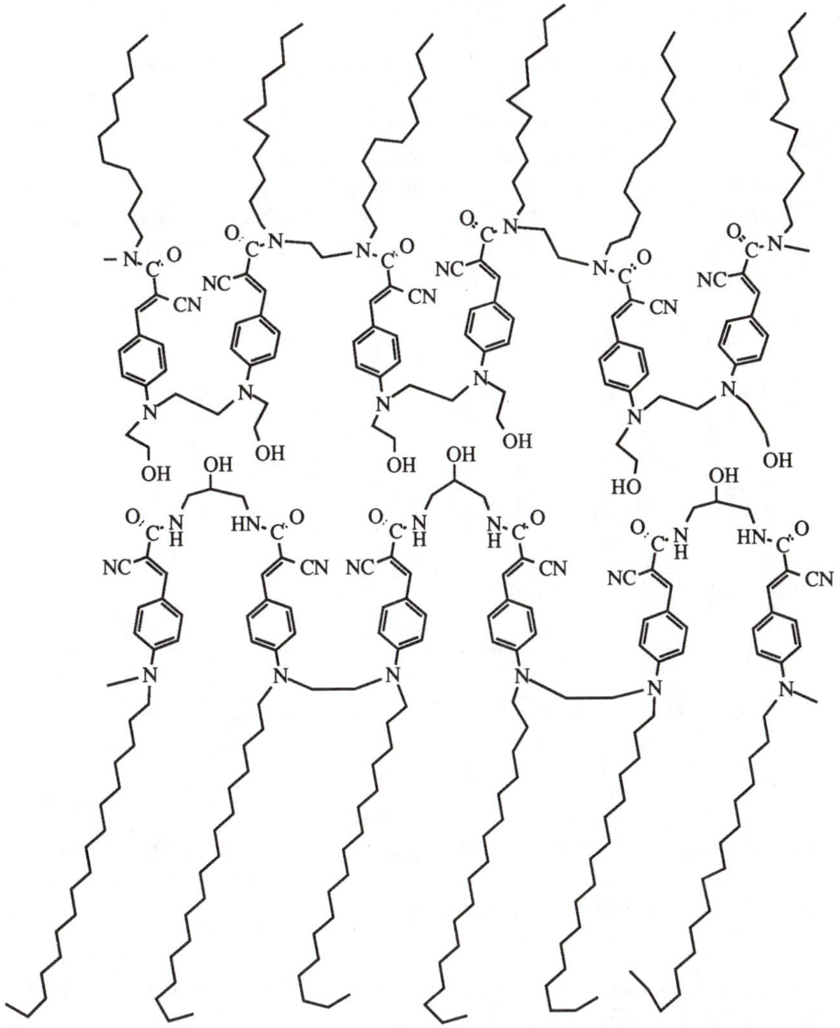

Figure 7. A schematic bilayer of **7**(A) on the bottom, and **8**(B) on the top.

The synthesis of polymer **7**(A) (*3a*) and polymer **8**(B) (*10*), shown in Figure 7 (*28*), was similar to the procedure used for **1**. An estimate of the number average molecular weight (Mn) of these polymers was made from the proton NMR spectra. A narrow Mn fraction (55,000 g/mol) of **7**(A) was obtained. The Mn of **8**(B) was about 8,000 g/mol by NMR end-group analysis. These results are in agreement with measurements by gel permeation chromatography on similar polymers, and no indication of cyclic oligomers were observed.

Differential scanning calorimetry showed that both **7**(A) and **8**(B) undergo two glass transitions, a low one at about 18 to 20°C (perhaps due to the fatty chains being phase separated), and a higher one at about 92° for **7**(A) and about 66° for **8**(B).

Film Deposition. Chloroform solutions of **7**(A) (typically 0.5 mg/ml) and chloroform/pyridine (95/5 v/v) solutions of **8**(B) (typically 0.6 mg/mL) were spread on pure water (18 MegOhm, Barnstead Nanopure system) in the two compartment circular NIMA film balance. The solvents were allowed to dry from the films in each compartment for several minutes, then the films were compressed at a rate of about 10 cm^2/min. until a surface pressure of about 20 mN/m was obtained. The compressed films were aged at 20 mN/m for about 20 min. to allow densification of the monolayers and to assure that there was no leakage of polymer from the compression area.

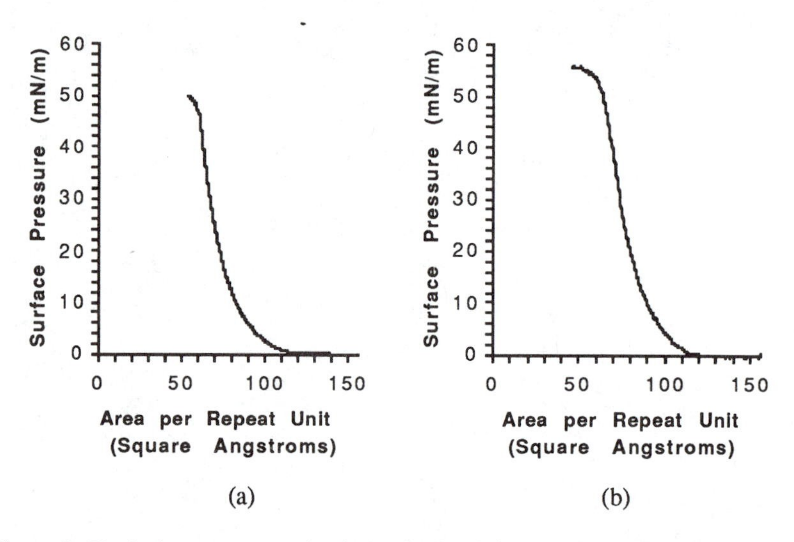

Figure 8. Typical pressure-area isotherms at the air/water interface (one repeat unit contains two chromophores): (a) **7**(A) was compressed at 9 cm^2/min at 23°C; and (b) **8**(B) was compressed at 10 cm^2/min. at 24°C.

Multilayer films were deposited onto hydrophobic microscope slides at a rate of about 2 to 3 mm/min (**7**(A) on the down stroke). The glass slides were made hydrophobic by exposure to refluxing hexamethyldisilizane (HMDS) for about 30 minutes. During the deposition process, the change in surface area of each compartment was monitored as the transfer of film was made at constant surface pressure.

An interesting potential advantage to this approach is that the dipole direction at some point within the multilayer film can be inverted, for example, at the point that maximizes the overlap integrals of fundamental and second harmonic for phase matching in frequency doubling applications (9). This may be effected by depositing one of the polymers twice in succession, then continuing with the alternating pattern.

Optical Measurements. Linear absorption and second harmonic generation (SHG) intensity measurements were made in transmission on freshly prepared films of various thicknesses. The linear absorption measurements were made normal to the films with a Cary 5E UV-VIS spectrometer. The SHG signal was generated by transmission of a fundamental beam from a Q-switched Nd:YAG laser (pulse width of 10 ns and repetition rate of 10 Hz) at an angle of about 55° from normal incidence to the film. The SHG intensity was measured with an array of intensified Si-photodiodes. The quadratic increase in SHG intensity as a function of thickness and the linear increase of absorption with thickness indicate the films have a high degree of uniformity of thickness and polar alignment. These data for 19-bilayer and 92-bilayer films are plotted in Figure 9 (each point on Figure 9 is an average of 1500 laser pulses).

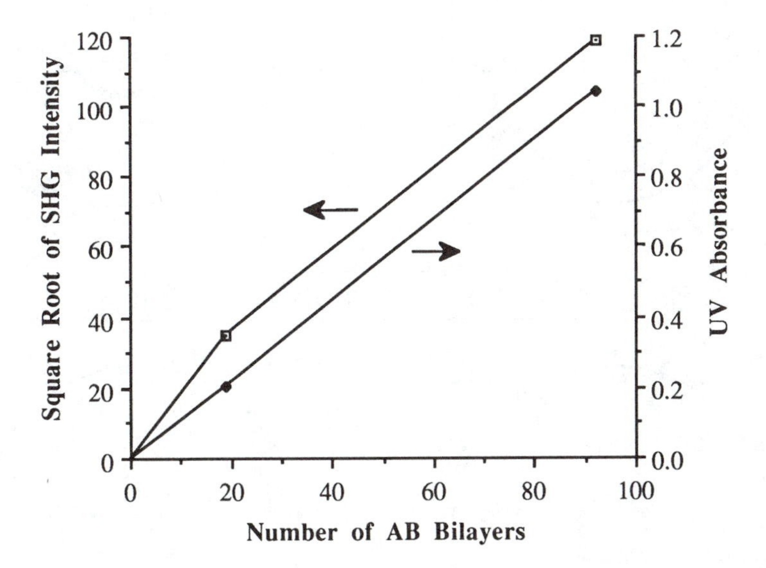

Figure 9. LBK bilayers of **7**(A) deposited on the downstrokes and **8**(B) deposited on the upstrokes (arbitrary units for square root of SHG intensity and UV absorbance).

The experimental setup for observing polarized SHG about the azimuthal angle for these films is as follows: The sample was at the waist of a focused (100 mm lens) beam (1064 nm) from a Q-switched Nd:YAG laser. The pulse width was 150 ns and the repetition rate was 1 KHz. The fundamental beam was polarized inside the laser cavity and a halfwave plate was used to control the polarization of the beam incident on the sample. The second harmonic signal was detected by a Hamatsu R928 photomultiplier tube in conjunction with a Stanford Research SR250 boxcar averager using active baseline subtraction. To account for laser power fluctuations, the second harmonic was normalized by monitoring the fundamental signal. The sample was mounted on a Oriel rotation stage and data was collected at 5° increments of the azimuthal angle (ϕ). The angle of incidence of the laser beam was fixed at 56°.

Complete 360° SHG scans were recorded for 6 different polarization combinations of the fundamental and second harmonic. The s-polarized second harmonic from an s-polarized fundamental (s→s) was zero, the s-polarized second harmonic from an p-polarized fundamental (p→s) was nearly in the noise, the s→p and 45°→s were weak as expected, and the other 2 combinations (p→p and 45°→p) gave significant second harmonic. The polar plots of these data are shown in Figure 10 for the 92 bilayer film. Theoretical fits to these data included reflections of the second harmonic (*29*).

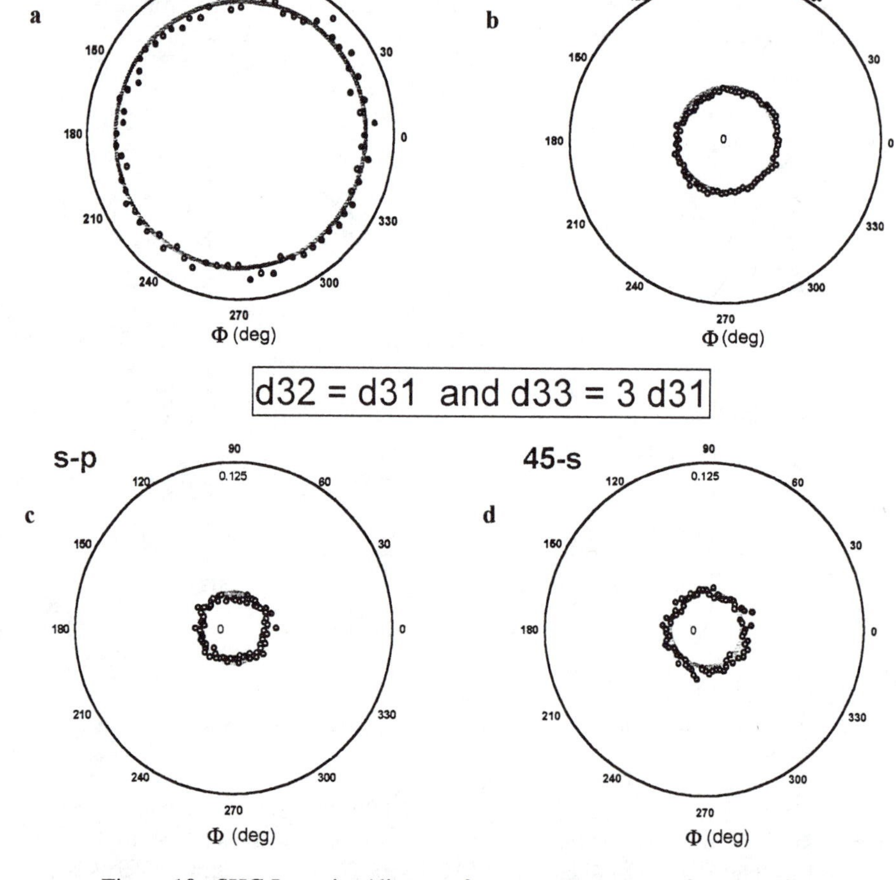

Figure 10. SHG Intensity (distance from center, a.u.) as a function of azimuthal angle for the 92 bilayer film: (a) p→p; (b) 45°→p; (c) s→p; and (d) 45°→s (the fundamental laser beam at an angle of 56° from incidence).

Using our first 24-bilayer film made from **7**(A) and **8**(B), an asymmetry in the SHG intensity about the azimuthal angle was noted (*10*). The long axis of the p→p oval was tilted 20° away from the dipping direction. However, in the 92 bilayer film shown in Figure 10, no asymmetry was observed. The difference in the results on the 24 and 92 bilayer films may be related to the high lateral defect density due to low transfer ratios of 7(A).

Our group initially reported that the average deposition ratio of **7**(A) (down stroke) was about 0.5 to 0.7, and the average transfer ratio of **8**(B) (up stroke) was about 0.9 to 1.5 (*10*). Recently, it was found that if the Langmuir films are replaced after only about 50% is consumed in the dipping process (compared to about 90% in the first case), transfer ratios were always between 0.95 and 1.0. Work is presently underway to characterize the optical properties of new bilayer films prepared with transfer ratios close to unity.

Recent results on the first bilayer films of **7**(A) and **8**(B) with transfer ratios close to one have shown that a 20-bilayer film comprised gave SHG signal intensity unchanged after aging in nitrogen for 28 days. A 30% loss of SHG signal intensity was observed in an identical experiment carried out in air. Thus, oxidative bleaching of the chromophores in thin films continues to be a problem.

A related experiment demonstrated that SHG signal intensity was completely lost after heating a second 24-bilayer film for an hour at 90°C. Thus, the Tg's of these polymers are too low to be useful in electro-optic devices. However, these polymers were useful in demonstrating the concept of polar film fabrication by the LBK technique. Similar polymers with higher T_g's are now in progress.

Conclusions

For the original high temperature accordion **1** (T_g = 205°C), a novel folded mainchain polymer containing α-cyanocinnamoyl chromophores, solid state NMR indicates no significant increases chromophore movement at temperatures up to the instrumentation heating limit, 110°C. Comparing NMR T_1 relaxation data between the *trans* **1** and *cis* **2** polymers suggests that the central region of the chromophore is more rigid in the case of the *trans* **1** isomer.

A photo-oxidative reaction in room lighting at elevated temperatures has been identified as the cause of variability in our previous NLO studies on polymer **1**. Processing in the dark (heating and poling) resulted in a d_{33} of 30 pm/V for **1** which is three times the best previous measurements obtained from poling in ambient light. An achieved alignment factor $\langle \cos^3\theta \rangle$ of 0.35, together with retention of 80% of NLO activity after 120 hrs at 125°C in the dark, indicate that the syndioregic architecture can be as well ordered as side chain architecture and is a promising stable structure for NLO polymers. The electro-optic coefficient (r_{33}) of **1** was found to be 8.5 pm/V, as measured in a Mach-Zehnder intensity modulator.

New mainchain syndioregic (accordion) polymers were synthesized, demonstrating the feasibility of incorporating different kinds of chromophores into polymers with this topology. With the appropriate chemistry, we believe it will be possible to make polymers with higher d_{33} coefficients, glass transition temperatures and thermal stabilities.

In an alternative approach, polar multilayer $-(AB)_n-$ films were fabricated by the Langmuir-Blodgett-Kuhn (LBK) technique. Two complementary amphiphilic polymers were alternatively deposited by Y-type deposition. In Polymer **7**(A), the chromophore's electron accepting end was connected to a relatively hydrophobic bridging unit, and its electron donating end was connected to a relatively hydrophilic bridging unit. The converse was true for Polymer **8**(B). New results on the fabrication of a 92-bilayer LBK NLO film from mainchain syndioregic chromophoric polymers are reported. These multilayer thin films have second order nonlinear optical properties which are stable at ambient temperature in the absence of oxygen. Although these polymers have rather low glass transition temperatures (75 to 95°C), they serve to demonstrate that this approach is potentially viable for the mass production of electro-optic waveguides without the need for electric-field poling.

Acknowledgments

The authors are grateful to Dr. Paul R. Ashley and his group at the US Army Missile Command, Redstone Arsenal, AL for fabricating the Mach-Zehnder modulators and providing those data; to Prof. L. Michael Hayden and his group at the U. of Maryland (Baltimore) for providing the thermal aging data; to Dr. Rena Yee for preparing spin-cast films and thermal analysis; to Dr. Robin Nissan for providing assistance in the NMR analysis; and to Jerrold Cline for assistance in optical measurements.

Literature Cited

1. (a) *Polymers for Second-Order Nonlinear Optics,* G. A. Lindsay and K. D. Singer, Eds., Am. Chem. Soc. Advances in Chemistry Series 601, Washington, D.C., 1995; (b) P. N. Prasad and D. J. Williams, *Introduction to Nonlinear Optical Effects in Molecules and Polymers*, John Wiley & Sons, Inc: New York, 1991.
2. (a) Singer, K.D.; Kuzyk, M.G.; Holland, W.R.; Sohn, J.E.; Lalama, S.L.;. Comizolli, R.B; Katz, H.E.; and Schilling, M.L. *Appl. Phys. Lett.* **1988,** *53*, 1800; (b) Lindsay, G.A.; Henry, R.A.; Hoover, J.M.; Knoeson, A.; and Mortazavi, M.A. *Macromol.* **1992,** *25*, 4888; (c) Herman, W.N.; Rosen, W.A.; Sperling, L.H. ; Murphy, C. J.; and Jain, H. *Proceedings of SPIE*, **1992**, *1560*, 206.
3. (a) Lindsay, G.A.; Stenger-Smith, J. D.; Henry, R.A.; Hoover, J.M.; Nissan, R.A.; and Wynne, K. J. *Macromol.* **1992,** *25*, 6075; (b) Ranon, P.M.; Shi, Y.; Steier, W.H.; Xu, C.; Wu, B.; and Dalton, L.R. *Appl. Phys. Lett.* **1993**, *62*, 2605; (c) Wright, M. E.; Mullick, S.; Lackritz, H. S.; Liu, L.-Y. *Macromolecules* **1994**, *27*, 3009..
4. (a) Mandal, B.K.; Chen, Y.M.; Lee, J.Y.; Kumar, J.; and Tripathy, S. *Appl. Phys. Lett.* **1991,** *58*, 2459; (b) Aramaki, S., Okamoto, Y., and Murayama, T. *Jpn. J. Appl. Phys* , Part 1, **1994,** *33*(10), 5759.
5. Verbiest, T.; Burland, D.M.; Jurich, M.C.; Lee, V.Y.; Miller, R.D.; Volksen, W. *Science*, **1995**, *286*, 1604.

6. Kakkar, A.; Yitzchaik, S.; Roscoe, S.; Marks, T.; Lin, W.; Wong, G. *Thin Solid Films* **1994,** *242*, 142.
7. Hall, R. C.; Lindsay, G. A.; Anderson, B.; Kowel, S. T.; Higgins, B.G.; Stroeve, P. *Mat. Res. Soc. Symp. Proc.* **1988**, *109*, 351.
8. Ashwell, G. J. in *Organic Materials for Nonlinear Optics III*, G. Ashwell and D. Bloor, Eds.; the Royal Society of Chem., 1993; Special Publication 137, pp. 31.
9. (a) Penner, T. L.; et al. *Nature* **1994,** *367*, 49; (b) Mötschmann, H.; Penner, T.; Armstrong, N.; Ezenyilimba *J. Phys. Chem.* **1993,** *97*, 3933.
10. Lindsay, G.; Wynne, K.; Herman, W.; Chafin, A.; Hollins, R.; Stenger-Smith, J.; Hoover, J.; Cline, J.; Roberts, J. *Nonlinear Optics, Sec. B*, **1996,** *15*(1-4), 139.
11. Stenger-Smith, J. D.; Fischer, J. W.; Henry, R. A.; Hoover, J. M.; Nadler, M. P.; Nissan, R. A.; Lindsay, G. A. *J. Poly. Sci.: Part A: Polym. Chem.* **1991**, *29*, 1623.
12. (a) Stenger-Smith, J. D.; Henry, R. A.; Hoover, J. M.; Lindsay, G. A.; Nadler, M. P.; and Nissan, R. A. *J. Poly. Sci. Part A: Chem. Ed.*, **1993**, *31*, 2899; (b) Stenger-Smith, J. D, et al., Chapter 14 in reference 1a above.
13. Henry, R. A.; Merwin, L. H.; Nissan, R. A., and Stenger-Smith, J. D. *Magnetic Resonance in Chemistry*, in press.
14. Solid-state NMR spectra were run on a Bruker MSL-200 NMR spectrometer operating at a field of 4.7 Tesla with a corresponding ^{13}C frequency of 50.3 MHz. All spectra were obtained using magic-angle spinning at speeds of ca. 3 to 4 kHz. 90° pulse widths of 5 μs were used throughout. For the cross-polarization (CP) experiments, contact times of 1 ms and recycle delays from 5 to 10 seconds were required. ^{13}C shift values are reported with respect to TMS and were measured by replacing the sample with the secondary reference, adamantane (high-frequency line = 38.5 ppm with respect to TMS).
15. (a) Lindsay, G. A.; Stenger-Smith, J. D.; Henry, R. A.; Nissan, R.A.; Merwin, L.H.; Chafin, A.P.; Yee, R.Y.; Herman, W.N.; and Hayden, L.M. *Organic Thin Films for Photonic Applications*, ACS/OSA Technical Digest Series, **1993**, *17*, 14; (b) Herman, W.N; Hayden, L.M.; Brower, S.; Lindsay, G. A.; Stenger-Smith, J. D.; and Henry, R. A. *Organic Thin Films for Photonic Applications*, ACS/OSA Technical Digest Series, **1993**, *17*, 18.
16. Ulrich, R.; and Torge, R. *Applied Optics* **1973**, 12, 2901.
17. The sample was rotated on a computer controlled Oriel rotation stage at the beam waist of a focused (100 mm lens) fundamental beam (1064 nm) produced by a Q-switched Nd:YAG laser. The pulse width was 150 ns and the repetition rate 1 kHz. A half wave plate was used to control the polarization of the fundamental beam to produce s- and p-polarized light incident on the sample. The second harmonic signal was detected by a Hamamatsu R928 photomultiplier tube in conjunction with a Stanford Research SR250 boxcar averager, which was operated using active baseline subtraction. Data was collected by computer as a function of the angle of incidence of the fundamental optical beam. The reference quartz crystal was X-cut and 0.381 mm thick. To account for laser power fluctuations, the fundamental was monitored along with the second harmonic in order to normalize the second harmonic signal.

18. Hayden, L. M.; et al. *J.Appl. Phys.* **1990**, *68*, 456. See also Chapter 20 in reference 1a.
19. Lindsay, G. A.; Stenger-Smith, J. D.; Henry, R. A.; Nissan, R.A.; Chafin, A.P.; Herman, W.N.; and Hayden, L.M. *ACS Polymer Preprints* **1993**, *34*, 796.
20. Oudar, J.L.; and Chemla, D.S. *J. Chem. Phys.* **1977**, *66*, 2664.
21. Page, R.H.; Jurich, M.C.; Reck, B.; Sen, A.; Twieg, R.J.; Swalen, J.D.; Bjorklund, G.J.; and Wilson, C.G. *J. Opt. Soc. Am.* **1990**, *B7*, 1239.
22. Singer, K.D.; Kuzyk, M.G.; and Sohn, J.E. *J. Opt. Soc. Am.* **1987**, *B4*, 968.
23. Singer, K.D.; Kuzyk, M.G.; Sohn, S.E. *J. Opt. Soc. Am.* **1987**,*B4*, 968.
24. Chafin, A. P.; et al., submitted to *Macromolecules.*
25. Stenger-Smith, J. D.; Henry, R. A.; Chafin, A.P.; Lindsay, G. A. *ACS Polymer Preprints* **1994**, *35*(2), 140.
26. Ashley, P. R., U. S. Army, AMSMI-RD-WS-CM, Redstone Arsenal, personal communication, 1995.
27. Hoover, J. M.; Henry, R. A.; Lindsay, G. A.; Nee, S. F.; Stenger-Smith, J. D. in *Organic Materials for Nonlinear Optics III*, G. J. Ashwell and D. Bloor, Eds.; the Royal Society of Chemistry: London, 1993; Special Publication 137.
28. Associating the designations A and B with the polymer composition number is meant to remind the reader of the polymer architecture shown in Figure 6.
29. Herman, W. N.; Hayden, L. M. *J. Opt. Soc. Am.*. **1995**, *B12*, 416.

Chapter 11

Novel Nonlinear Optical Polymers

Naoto Tsutsumi, Osamu Matsumoto, and Wataru Sakai

Department of Polymer Science and Engineering, Kyoto Institute of Technology, Matsugasaki, Sakyo, Kyoto 606, Japan

A novel class of nonlinear optical (NLO) urethane polymers for second harmonic generation is presented. These NLO polymers consist of NLO units which are embedded in the polymer backbone while the dipole moments are perpendicular to the main chain. The first is a linear polyurethane, T-polymer synthesized from 2,4-tolylene diisocyanate (TDI) and 4-[(2-hydroxyethyl)amino]-2-(hydroxymethyl)-4'-nitroazobenzene (T-AZODIOL) and the second is T-polymer 2 prepared from 4,4'-diphenylmethane diisocyanate (PDI) and T-AZODIOL. For comparison, a linear polyurethane, L-polymer, whose NLO chromophore is incorporated into the main chain, was synthesized from TDI and 4-[N-(2-hydroxyethyl)-N-methylamino]-3'-(hydroxymethyl)azobenzene (AZODIOL) and L-polymer 2 from PDI and AZODIOL. These polymers are amorphous with a high density of NLO chromophore moiety and optically transparent thin films can be processed by spin-casting. T-polymer poled for 60 min at an optimum condition of corona poling voltage of 8.0 kV and 95°C shows a large second order nonlinearity of $d_{33}=1.6 \times 10^{-7}$ esu (67 pm/V). Good thermal stability of nonlinearity for T-polymer and T-polymer 2 was observed at ambient condition. The oriented NLO dipole moments of T-polymer does not show the significant relaxation at ambient condition in 60 days except for a small initial decrease over a few days after poling. In comparison, the SHG activities of the L-polymer and L-polymer 2 were largely decayed at room temperature. The better thermal stability of this new class of T-polymer and T-polymer 2 can be related to the inherent smaller free volume.

Recent development of organic nonlinear optical (NLO) polymers has focused on the fabrication of thermally stable NLO materials in which the second order nonlinearity is stable at higher temperatures. Several strategies of molecular design have been presented for realizing such materials. Cross-linking is one way to suppress the reorientation of an NLO chromophore due to segmental molecular motion *(1-5)*. Utilization of linear polymers with high glass transition temperature (T_g) such as polyimides is another possibility *(6,7)*.

Recently, we synthesized a new type of NLO chromophore, based on an azobenzene dye, whose dipole moment is aligned transverse to the main chain. The resultant poled film displayed a large second harmonic generation efficiency with good thermal stability at ambient condition *(8)*. Present NLO chromophore is based on an azobenzene dye. As pointed out by the previous work *(9)*, an NLO chromophore in this arrangement can be easier to orient by an external electric field than in structures where their dipole moments are pointing along the polymer backbone. Namely, the system whose NLO dipole moments are aligned transverse to the polymer backbone requires less deformation of the main chain backbone on orienting the dipole moment to the poling field direction than in the system whose NLO dipole moments are incorporated along the polymer backbone. In this article, we present the synthesis and the second harmonic generation (SHG) properties of this novel type of NLO polymer material whose dipole moments are transversely aligned to the main chain, and compare these properties with those for the polymer whose NLO chromophore is aligned along the main chain.

Experimental Section

Materials. 4-[(2-Hydroxyethyl)amino]-2-(hydroxymethyl)-4'-nitroazobenzene (T-AZODIOL) *(10)* was used as the NLO chromophore whose dipole moment is aligned transverse to the main chain. T-AZODIOL was synthesized via two step reactions as shown in Schemes 1 and 2. As shown, 3-(2-hydroxylamino)-benzylalcohol (DIOL) prepared from *m*-aminobenzyl alcohol with 2-chloroethanol was coupled with diazotized *p*-nitroaniline to give rise to T-AZODIOL. The details of the synthetic procedures were previously reported *(8)*. 4-[N-(2-Hydroxyethyl)-N-methylamino]-3'-(hydroxymethyl)-azobenzene (AZODIOL) dye was the NLO chromophore monomer for preparing the polymer with NLO chromophore incorporated in the main chain. Detailed synthetic procedures of AZODIOL were previously reported *(5)*. Commercially available 2,4-tolylene diisocyanate (TDI) and 4,4'-diphenylmethane diisocyanate (PDI) were used without further purification. The chemical structures of these monomers are shown in Figure 1.

Polymer Synthesis and Film Processing. T-AZODIOL and TDI or PDI in equivalent molar ratios were reacted in dimethylacetamide solution at 85°C for 20 - 30 min under nitrogen atmosphere to prepare T-polymer or T-polymer 2, respectively. Similarly, L-polymer and L-polymer 2 were prepared from AZODIOL with TDI and PDI, respectively. Molecular structures of these polymers are shown in Figure 2. A spin-casting technique was employed to prepare thin films for SHG measurements.

Corona Poling. Spun-cast films were corona-poled at an elevated temperature up to 105°C to orient NLO chromophore to the poling direction. The distance between

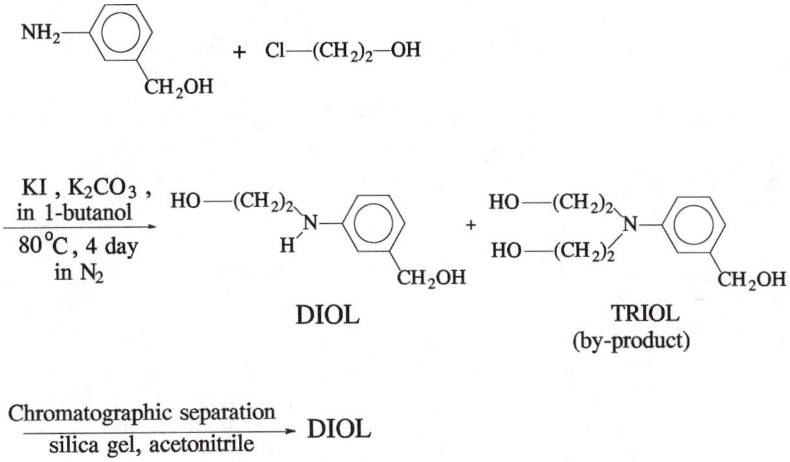

NH_2–C_6H_4–CH_2OH + Cl—$(CH_2)_2$—OH

KI, K_2CO_3, in 1-butanol / 80 °C, 4 day in N_2 → HO—$(CH_2)_2$–N(H)–C_6H_4–CH_2OH (DIOL) + [HO—$(CH_2)_2$]$_2$N–C_6H_4–CH_2OH (TRIOL (by-product))

Chromatographic separation / silica gel, acetonitrile → DIOL

Scheme 1

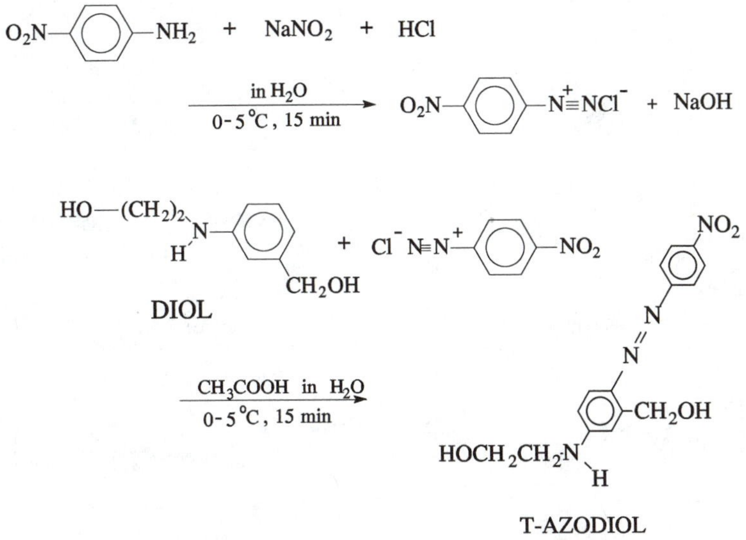

O_2N–C_6H_4–NH_2 + $NaNO_2$ + HCl

in H_2O / 0-5 °C, 15 min → O_2N–C_6H_4–$\overset{+}{N}\equiv NCl^-$ + NaOH

HO—$(CH_2)_2$–N(H)–C_6H_4–CH_2OH (DIOL) + $Cl^-\,N\equiv\overset{+}{N}$–$C_6H_4$–$NO_2$

CH_3COOH in H_2O / 0-5 °C, 15 min → $HOCH_2CH_2N(H)$–$C_6H_3(CH_2OH)$–N=N–C_6H_4–NO_2

T-AZODIOL

Scheme 2

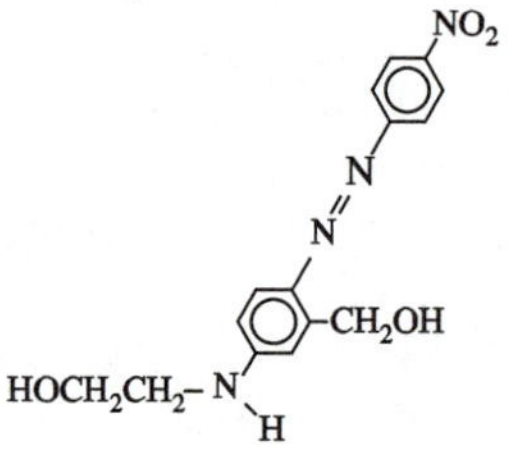

T-AZODIOL

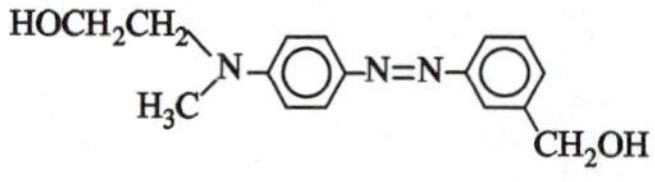

AZODIOL

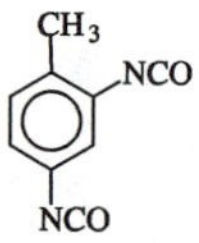

TDI

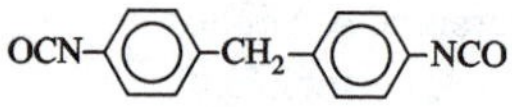

PDI

Figure 1. Chemical structures and codes of monomers.

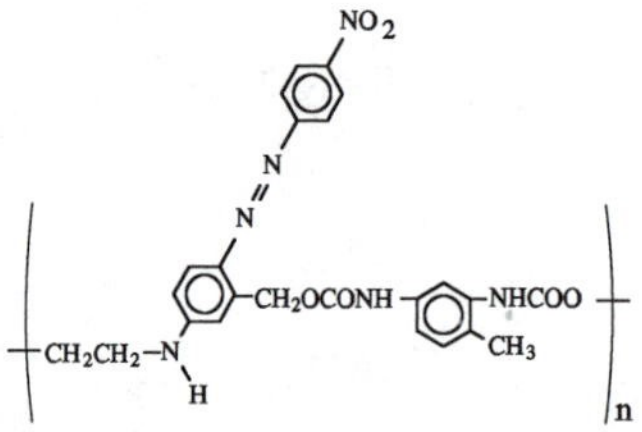

T-polymer

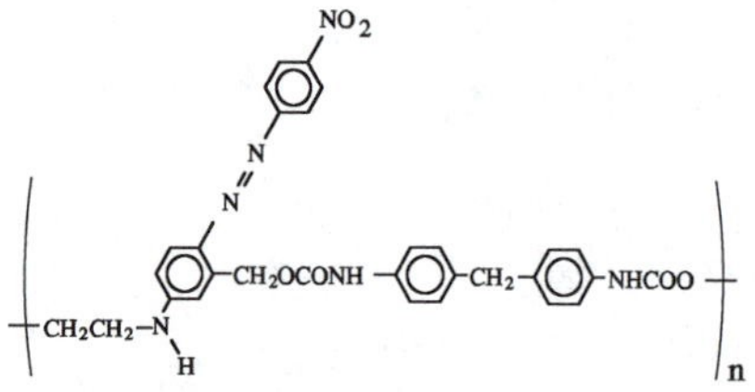

T-polymer 2

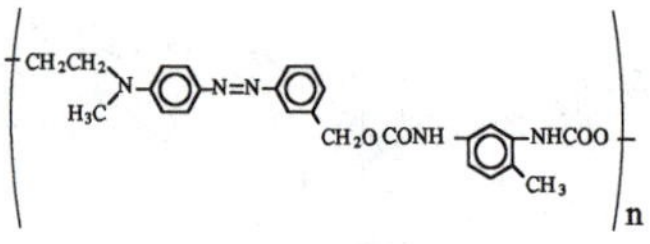

L-polymer

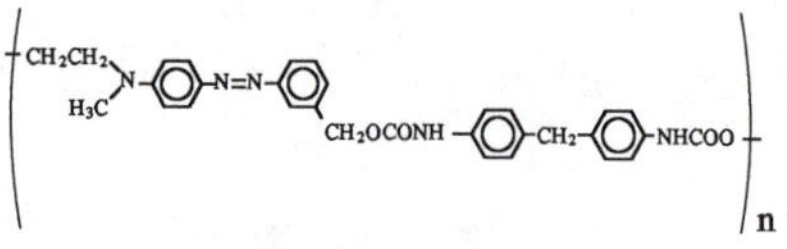

L-polymer 2

Figure 2. Molecular structures of polymers.

the sample and 0.1 ϕ tungsten wire for corona poling was kept constant at 1.4 cm. Figure 3 illustrates the schematic pictures of aligned NLO chromophores induced by poling in T- and L-polymers which indicate decreased deformation of the main chain in the T-polymer as compared to the L-polymer configuration.

SHG Measurements. The Maker fringe method *(11,12)* was employed to measure the SHG intensity of poled spun-cast films. The source was a Continuum model Surelite-10 Q-switched Nd:YAG pulse laser with 1064 nm *p*-polarized fundamental beam (320 mJ maximum energy, 7 ns pulse width and 10 Hz repeating rate). The second harmonic (SH) wave generated was detected by using a Hamamatsu model R928 photomultiplier. The SH signal averaged on a Stanford Research Systems (SRS) model SR-250 gated integrator and boxcar averager module was transferred to a microcomputer through a SRS model SR-245 computer interface module. The details of the experimental procedures were previously described *(13, 14)*.

Characterization. Reduced specific viscosity was measured in *N,N*-dimethylacetamide solution. The concentration of polymer solution was 1.0 g/dL. Ultraviolet-visible spectra of the films were measured on a Shimadzu model UV-2101PC spectrophotometer. Refractive indices of materials were measured by the m-Line method using a prism coupling apparatus. Laser sources were a polarized He-Ne laser (632.8 nm) and a laser diode (830 nm). The prism of TaFD21 (HOYA Glass) with a high refractive index (1.92588 at 632.8 nm) and a spin-coated or cast film was coupled with an air gap. Differential scanning calorimetry (DSC) was carried out at a heating rate of 10°C/min in a nitrogen atmosphere, using a Perkin Elmer DSC7 controlled by 1020 TA workstation. Thermally stimulated discharge current (TSC) was measured by the current flow in the external circuit under a constant temperature increase of 4°C/min, using the sandwich type cell with the positively corona-poled surface grounded. Thermogravimetry (TG) was performed with a Shimadzu model DT-30 thermogravimetric analyzer at a heating rate of 10°C/min under a nitrogen atmosphere. Density of the polymer film was measured in potassium iodide solution at 30°C using a sink and float test.

Results and Discussion

Polymer Synthesis. All polymers were prepared in dimethylacetamide at the temperature of 85°C for 20 - 30 min in nitrogen atmosphere. Table I summarizes the reduced specific viscosity (η_{sp}/C) , density (ρ), and glass transition temperature (T_g) for the polymers synthesized. T_g was determined by DSC measurements. All polymers can be provided as the colored transparent thin films by spin-casting.

Linear Optical Properties. Poling caused a decrease in absorption intensity and a shift of absorption peak to shorter wavelength. Thermal annealing under the same conditions of temperature and time does not cause the observed change of intensity and spectral shift. The absorption intensity change is ascribed to poling induced orientation of the azobenzene dye in the direction of the film thickness. Spectral blue shift has been reported for other cross-linked main-chain polymers *(3)*, which is in contrast to the red shift observed in most side-chain polymers *(15)*.

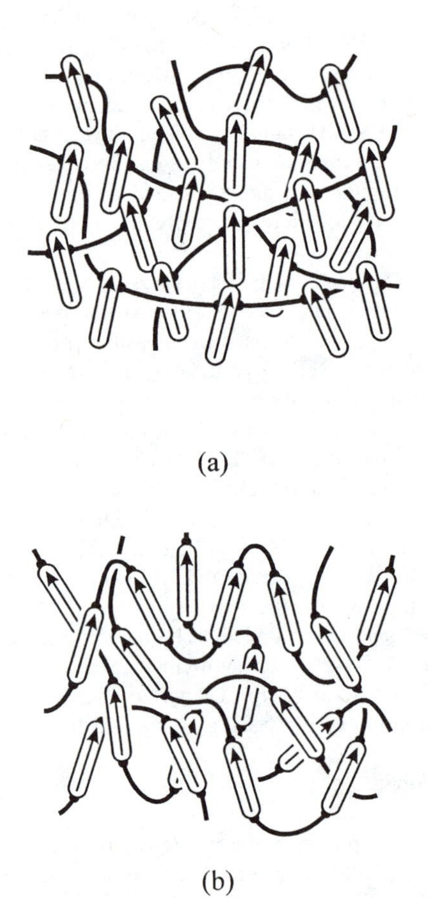

Figure 3. Schematic pictures of aligned NLO chromophores in (a) T-polymer and in (b) L-polymer.

The refractive indices (RI) for transverse electric field (TE) mode were measured using the m-line method at the wavelengths of 632.8 and 830 nm. Wavelength dispersion of RI, $n_f(\lambda)$, was fitted to a one-oscillator Sellmeier-dispersion formula,

$$n_f^2(\lambda) - 1 = \frac{q}{1/\lambda_0^2 - 1/\lambda^2} + A \qquad (1)$$

where λ_0 is the absorption wavelength of the dominant oscillator, q is a measure of the oscillator strength, and A is a constant containing the sum of all the other oscillators. Table II lists RI values measured at the wavelengths of 632.8 and 830 nm. Wavelength dispersion of RI was predicted using eq. (1) from the RI data obtained at 632.8 and 830 nm with λ_0 = 470 and 465 nm for T-polymer and T-polymer 2, respectively, and with λ_0 = 414 and 415 nm for L-polymer and L-polymer 2, respectively. RI values at 532 and 1064 nm obtained from the predicted plots using eq. (1) are also listed in Table II.

Table I. Reduced Specific Viscosity (η_{sp}/C) , Density (ρ), and Glass Transition Temperature (T_g)

Polymer	η_{sp}/C (dL/g)	ρ (g/cm^3)	T_g (°C)
T-polymer	0.14	1.246	57
L-polymer	0.30	1.147	91
T-polymer 2	0.19	1.218	82
L-polymer 2	0.42	1.133	62

Table II. Wavelength Dispersion of RI Values. RI Values at 632.8 and 830 nm Were Measured and Those at 532 and 1064 nm Were Predicted Using Eq. (1)

	Wavelength (nm)			
Polymer	632.8	830	532	1064
T-polymer	1.696	1.669	1.779	1.661
L-polymer	1.683	1.650	1.745	1.638
T-polymer 2	1.695	1.671	1.760	1.664
L-polymer 2	1.683	1.646	1.751	1.632

SHG Coefficients and Optimum Poling Condition. The SHG coefficients of the polymers are made relative to a Y-cut quartz plate (d_{11}=1.2 × 10^{-9} esu (0.5pm/V)). The typical Maker fringe pattern could be observed for both the *p*-polarized and *s*-polarized fundamental beams. It is important to optimize the poling condition (poling voltage, temperature, and time) to obtain maximum SHG activities. The SHG coefficient increased with increasing applied voltage and leveled out at a voltage above 8 kV for T-polymer and above 6 kV for T-polymer 2. Increasing poling temperature caused the SHG coefficient to increase, with the largest increase observed between 90 and 95°C, as shown in Figures 4 (a) and (b) for T-polymer and T-polymer 2, respectively. For L-polymer and L-polymer 2, d_{33} increased with increasing applied voltage and leveled out above 6 kV with a decrease above 8 kV. Figures 4 (c) and (d) show the d_{33} profile with increasing poling temperature for L-polymer and for L-polymer, respectively. Poling at the temperature above 70°C for L-polymer caused the film surface to become opaque. For L-polymer 2, d_{33} increased with increasing poling temperature and leveled out at temperatures above 53°C. For all polymers tested, increasing the poling time caused d_{33} to increase until 1 hour after which time d_{33} leveled out.

It is noted that T-polymer and T-polymer 2 show a two-step increase of d_{33} when the poling temperature is increased, whereas L-polymer and L-polymer 2 show a single-step increase of d_{33}. One possible explanation for this two-step increase of d_{33} is the existence of two different orientation states of NLO chromophores in T-polymer and T-polymer 2. A metastable oriented state of NLO chromophore may be achieved at the temperature around 60 - 70°C and then a more stable oriented state of NLO chromophore may be achieved at the temperature above 95°C, i.e., a higher degree of orientation of NLO chromophore is established at the higher temperature. When T-polymer was poled at the optimum conditions of poling voltage of 8 kV, temperature of 95°C, and time of 60 min, a d_{33} value of 1.6 × 10^{-7} esu (67 pm/V) was obtained. This value is larger than the SHG coefficient of lithium niobate ($LiNbO_3$) (d_{33} = 30 pm/V) *(16)*.

Theoretical Estimation of d_{33} Value for Polymer with T-AZODIOL. In the theoretical expression of SHG coefficient, d_{33} can be written as

$$d_{33} = \frac{N_d f_\omega^2 f_{2\omega} \beta \mu_g E_p}{10kT} \quad (2)$$

where N_d is the number density of noncentrosymmetric NLO molecules, β is the hyperpolarizability of the NLO guest, f_ω and $f_{2\omega}$ are Lorentz-Lorenz local field factors of the form (ε + 2)/3, μ_g is the dipole moment of NLO chromophore in the ground state, E_P is the poling electric field, *k* is the Boltzmann's constant, and *T* is the poling temperature. The value of ε has been taken as the square of the refractive index of the sample at either the fundamental or second harmonic frequency. The number density N_d is calculated using the film density. To evaluate d_{33} values from eq. (2), β, μ_g and E_p values must be estimated from the absorption spectra data.

E_p was determined from the intensity change in absorbance caused by the orientation of NLO chromophore, using electrochromic theory *(17,18)*. The orientation-induced intensity change in absorbance can be related to the electric field, using the following equation,

$$\frac{A(p)}{A(0)} = 1 - G(u) \tag{3}$$

where $A(p)$ and $A(0)$ are the absorbance with and without electric field, respectively, and

$$G(u) = 1 - \frac{3\coth(u)}{u} + \frac{3}{u^2} \tag{4}$$

and

$$u = \frac{\mu_g E_p}{kT} \tag{5}$$

β can be calculated using the two-level model *(19,20)*,

$$\beta = \frac{9e^2}{4m}\left(\frac{h}{2\pi}\right)^2 \frac{W}{[W^2-(2h\nu)^2][W^2-(h\nu)^2]} f\,\Delta\mu \tag{6}$$

where e is the elementary electric charge in esu, m is the rest mass of an electron, h is Planck's constant, W is the energy at the absorption wavelength of the dominant oscillator, and $h\nu$ and $2h\nu$ are the energies of the fundamental and second harmonic light. The oscillator strength of the dominant oscillator, f, and can be evaluated from the absorption spectrum of the dominant oscillator *(21,22)*;

$$f = \frac{2303mc^2}{\pi e^2 Nn}\int \epsilon_{\tilde{v}}\, d\tilde{v} = 4.38\times10^{-9}\int \epsilon_{\tilde{v}}\, d\tilde{v} \tag{7}$$

where c is the speed of light, N is Avogadro's number, n is the refractive index which is commonly omitted from the above expression, ε_v is the molar extinction coefficient and the integration is carried out over the absorption band of the dominant oscillator. The dipole moment difference between ground and excited states $\Delta\mu$ (=μ_e - μ_g) can be estimated from the well-known azobenzene dyes *(17)*. For T-polymer, the theoretical d_{33} value can be calculated as 1.0×10^{-7} esu using E_p = 2.7 MV/cm, μ_g = 8 D *(17)* and $\beta = 119 \times 10^{-30}$ esu. The experimentally obtained d_{33} value of 0.91×10^{-7} esu is in good agreement with the theoretical value.

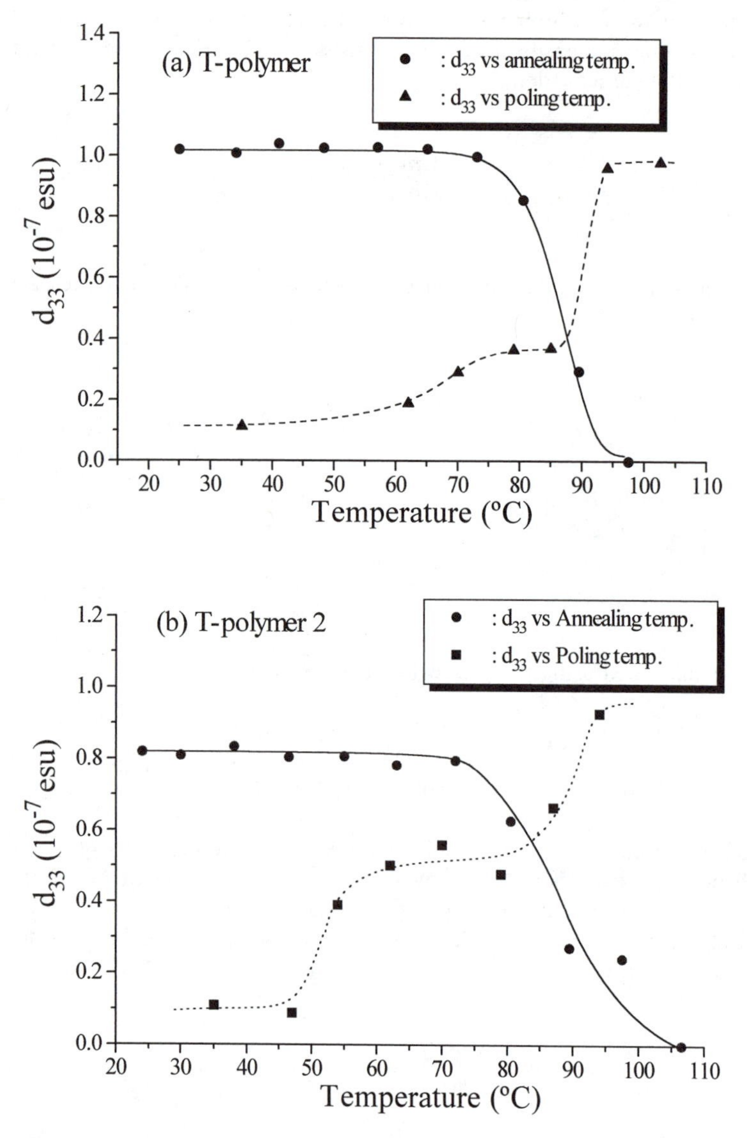

Figure 4. Poling and annealing temperature profiles for d_{33} of (a) T-polymer, (b) T-polymer 2, (c) L-polymer and (d) L-polymer 2. The SHG measurement was carried out at the fixed temperature shown on the horizontal axis in the figure. The dotted curve is the d_{33} profile on poling, and the temperature on the horizontal axis in this curve is the poling temperature.

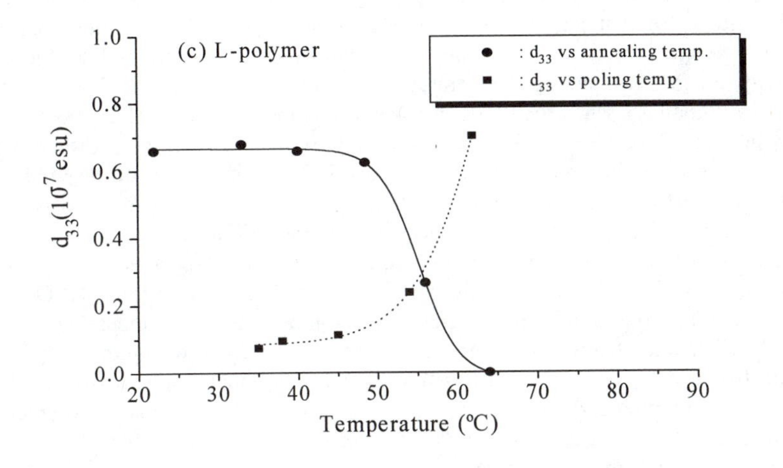

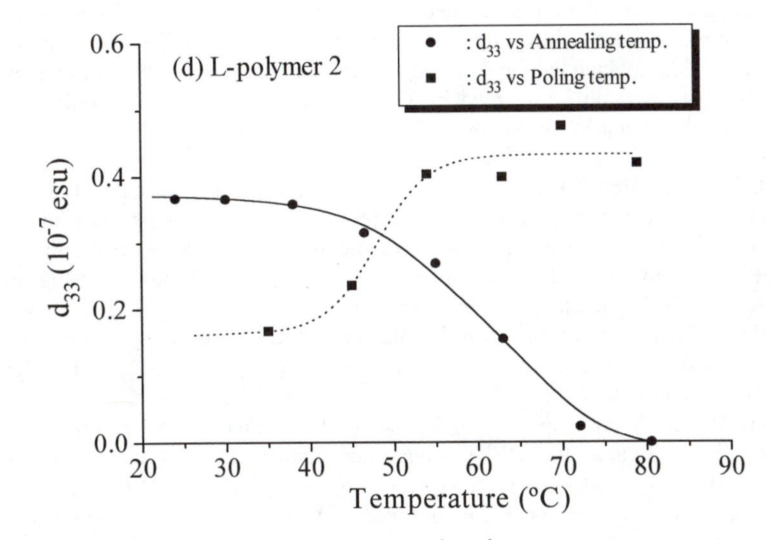

Figure 4. Continued.

Thermal Stability of SHG Activities upon Annealing. The SHG activity profiles vs. annealing temperature after corona-poling are also shown in Figure 4 for (a) T-polymer, (b) T-polymer 2, (c) L-polymer and (d) L-polymer 2. T-polymer and T-polymer 2 do not show significant loss of SHG activity until temperatures above 80°C. However, for L-polymer and L-polymer 2 the initial SHG loss starts at around 40°C with significant activity loss occurring around 50 - 55°C. It is noted that these SHG activity profiles vs. annealing temperatures can be compared to the SHG activity profiles vs. poling temperatures as shown in Figure 4. Values of d_{33} drastically increase when the poling temperature is increased from 85 to 95°C for T-polymer and T-polymer 2 and from 50 to 60°C for L-polymer. Either poling temperature range in which large increase of SHG activity occurs corresponds well to the fact that the SHG activity is significantly lost on annealing for each polymer. These trends imply that both processes of the alignment (orientation) and randomization (reorientation) of NLO chromophore should be subjected to the same thermal activation. In other words, the thermal mobility for aligning NLO dipole moments on poling is the same as that for randomizing them on annealing. Hence, the glass transition temperatures for the SHG activities may be within the temperature range between 80 and 85°C for T-polymer and T-polymer 2 and between 50 and 55°C for L-polymer and L-polymer 2. DSC measurements show that T-polymer and T-polymer 2 have T_g's of 57 and 82°C, respectively, and L-polymer and L-polymer 2 have T_g's of 91 and 62°C, respectively. These T_g values are not consistent with those estimated from the temporal thermal stability of SHG activity for all polymers shown above. These discrepancies may be related to the difference in mode of molecular motion which can be detected by each measurement. The reorientation of aligned dipole is mainly governed by the local motion of the surrounding dipole chromophore and not by the segmental molecular motion.

Thermally Stimulated Discharge Current. The thermally stimulated discharge current (TSC) technique is a useful tool for probing the depolarization profile, in this case the reorientation of aligned dipole moments and the release of mobile space charge created by injected charge in the polymeric matrix. When the NLO dipole moment and the segmental dipole moment randomize at favorable temperatures, the specific discharge current peak can be obtained in the TSC profile. Figure 5 shows the TSC profiles for the corona-poled (a) T-polymer, (b) T-polymer 2, (c) L-polymer, and (d) L-polymer 2. Two distinct peaks appear around 52 and 92°C for T-polymer, 60 and 95°C for T-polymer 2, 43 and 78°C for L-polymer, and 52 and 95°C for L-polymer 2. In figures, the TSC profiles after depolarization (dashed curve in the figures) are illustrated. After depolarization, no significant TSC peak can be detected. Thus the TSC peak for the polarized sample by corona-poling can be related to the relaxation of the oriented dipole moments and the molecular motion of the polymer matrix. T-polymer and T-polymer 2 have larger TSC peak around 90 to 95°C, whereas L-polymer and L-polymer 2 have larger one around 40 to 50°C. These temperature regions correspond well to those when significant SHG depression occurs as shown in Figure 4. Thus, the larger TSC peaks are considered to be due to the reorientation of aligned NLO dipole moments.

Long-term Thermal Stability of SHG Activity. Figure 6 shows the long-term thermal stability of d_{33} when the sample films were stored at room temperature for up to 60 days. The plot in the figure is of SHG coefficients d_{33} normalized by that measured at time zero. It is noted that T- polymer and T-polymer 2 display long-term thermal stability of d_{33} (no significant relaxation at the ambient condition in 60 days for T-polymer) except for the small activity loss within the first few days after poling. This is contrasted to the SHG coefficient d_{33} of L-polymer and L-polymer 2 which shows a significant daily decreases, reaching half the initial value after 2 weeks storage. These storage time profiles of SHG activity support the conclusion that T-polymer and T-polymer 2 have better thermal stability than L-polymer and L-polymer 2. These differences in SHG activity profiles against storage time between T-polymer and L-polymer groups are due to the differences of the mobility of the dipole moments between the two polymer groups.

The question arises as to what is the origin of stabilizing NLO dipole moments in T-polymer and T-polymer 2. One possibility is that the free volume of the polymer matrix provides the free space where the aligned dipole moment can be thermally reoriented. The densities of T-polymer and T-polymer 2 are larger than those of L-polymer and L-polymer 2 as shown in Table III. The larger densities of T-polymer and T-polymer 2 imply smaller free volumes. The free volume V_f can be estimated from the experimentally obtained specific volume V_t (the reciprocal value of measured density) and the zero point molar volume V_0,

$$V_f = V_t - V_0 \tag{8}$$

where V_0 can be calculated from the van der Waals volume V_w of the polymer, $V_0 = 1.3V_w$ *(23)*. V_w of polymers was determined by the summation of the van der Waals volume of group contributions *(24)*. Table III shows the calculated free volumes of T- and L-polymer groups. As expected, the free volumes of T-polymer and T-polymer 2 are smaller than those of L-polymer and L-polymer 2. Thus, the smaller free volume of T-polymer and T-polymer 2 significantly contributes to the restriction of molecular motion in the glassy state of the matrix. That is, the orientation of NLO chromophore in T-polymer and T-polymer 2 is sustained by the smaller free volume of matrix in them.

Table III. Density and Free Volume (V_f) Calculated by Eq. (8)

Polymer	Density (g/cm^3)	V_t (cm^3/g)	V_0 (cm^3/g)	V_f (cm^3/g)
T-polymer	1.246	0.803	0.632	0.171
L-polymer	1.147	0.872	0.662	0.210
T-polymer 2	1.218	0.821	0.644	0.177
L-polymer 2	1.133	0.883	0.672	0.211

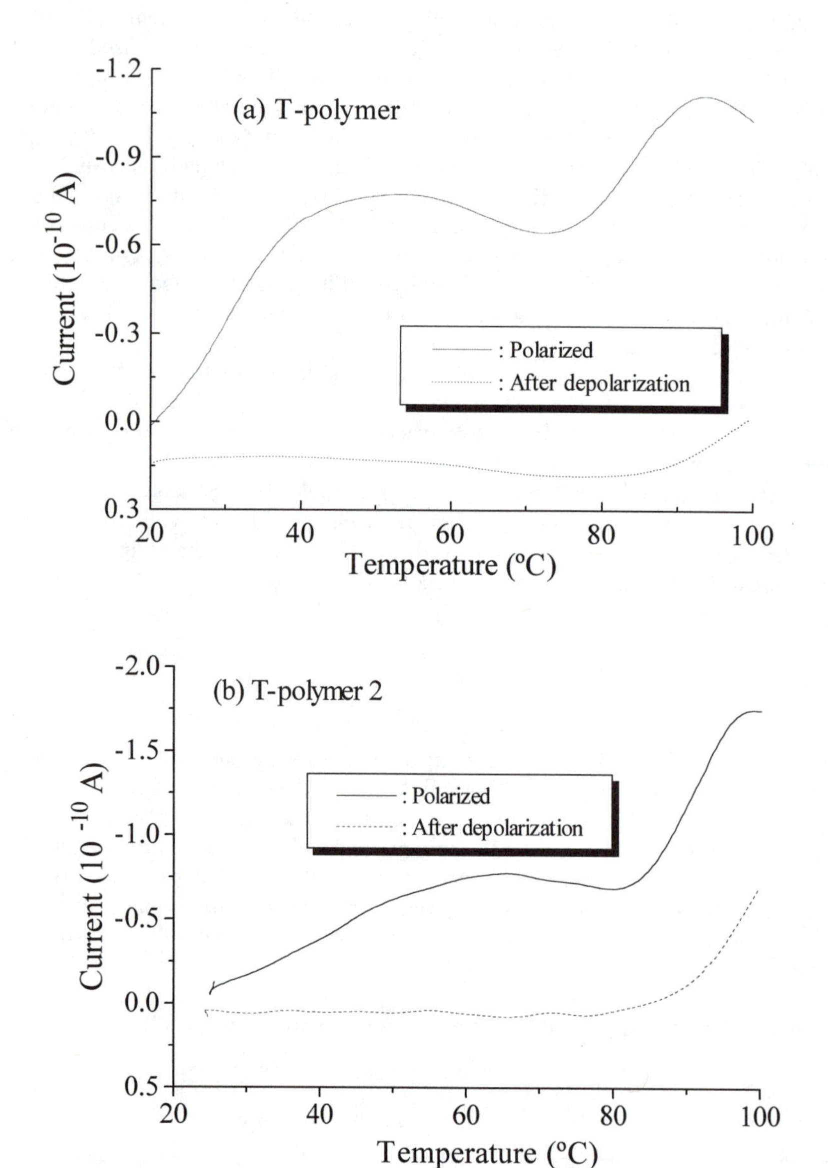

Figure 5. TSC profiles of corona-poled (a) T-polymer, (b) T-polymer 2, (c) L-polymer and (d) L-polymer 2. Dashed TSC curves in both figures are those for the depolarized samples.

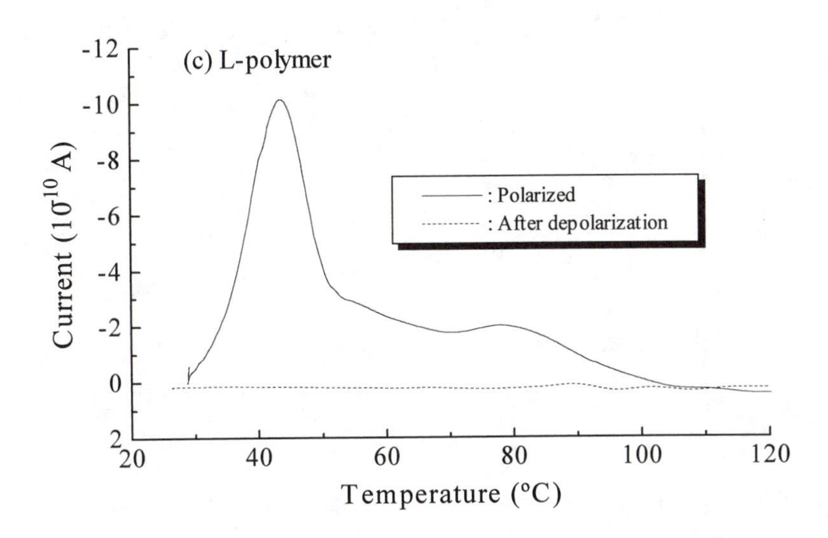

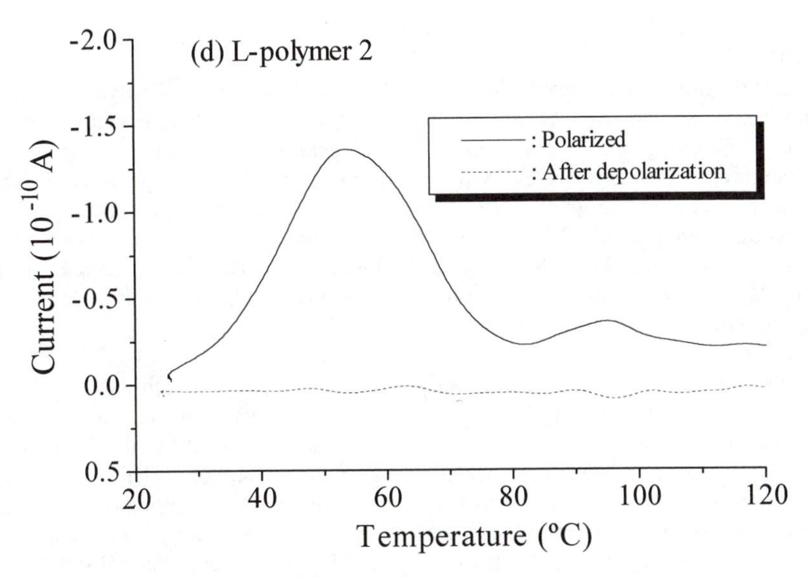

Figure 5. Continued.

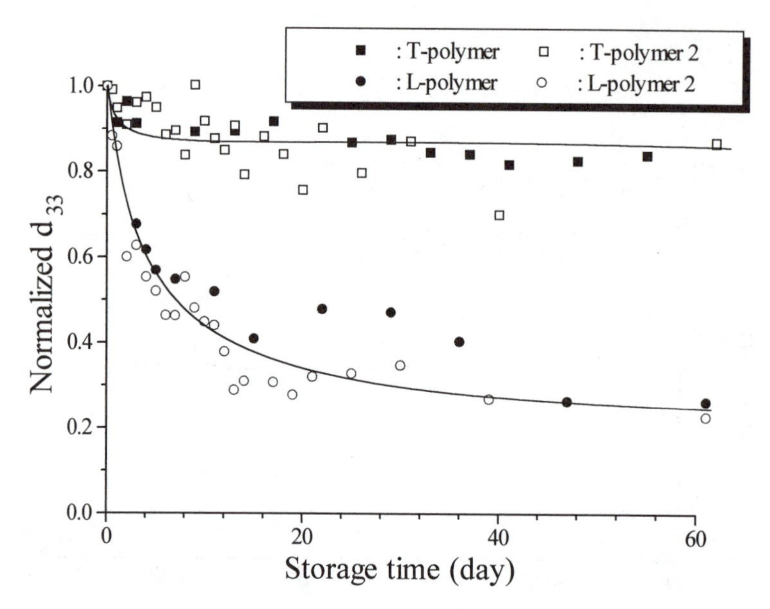

Figure 6. Long-term thermal stability of d_{33} when all polymers are stored at room temperature.

Chemical Stability. Chemical stability of NLO chromophores and polymers at high temperatures are the critical characteristics of organic NLO materials that determine device processing and device operation temperatures. Figure 7 shows the thermogravimetric behavior of T- and L-polymers. It is clear that L-polymer is stable up to temperatures around 300°C whereas T-polymer starts to decompose at temperatures around 160°C. This low chemical stability may be due to the secondary amine group in the T-AZODIOL unit in T-polymer. DSC measurement of T-AZODIOL showed that the large exotherms occurring just after the endothermic transition due to crystalline melting.

Conclusions. A new approach of molecular design for stabilizing NLO dipole moments in polymeric materials is presented. Advantage in using an NLO unit which is partly embedded in the polymer backbone while the NLO dipole moment extends transverse to the main chain has been demonstrated. This polymer is amorphous with a high density of NLO chromophore moiety and optically transparent thin films can be prepared by spin-casting. Poled T-polymer displayed a large second order nonlinearity of d_{33}=1.6 × 10^{-7} esu (67 pm/V). Good thermal stability of nonlinearity for T-polymer and T-polymer 2 was observed at ambient conditions, whereas L-polymer and L-polymer 2 showed inferior thermal stability of nonlinearity. The better thermal stability of T-polymer and T-polymer 2 can be related to the smaller free volume in them. This approach should provide a mechanism for fabricating electro-optic devices using this material which exhibits large second order nonlinearity with good thermal stability. The next step is the creation of a more rigid

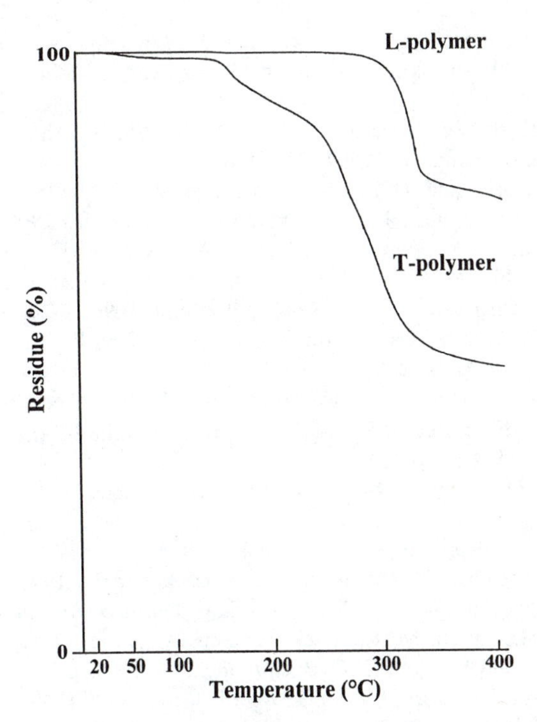

Figure 7. Thermogravimetric profiles of T-polymer and L-polymer.

main chain polymers embedded the same type of NLO chromophore to achieve a thermally stable NLO active polymer system at elevated temperatures.

Literature Cited

(1) Jungbauer, D; Reck, B; Twieg, R; Yoon, D. Y.; Willson, C. G.; Swalen, J. D. *Appl. Phys. Lett.* **1990**, *56*, 2610.
(2) Ranon P. M.; Shi, Y, Steier, W. H.; Xu, C; Wu, B; Dalton, L. R. *Appl. Phys. Lett.* **1993**, *62*, 2605.
(3) Boogers, J. A. F.; Klaase, P. Th. A.; de Vliger, J. J.; Tinnemans, A. A. *Macromolecules* **1994**, *27*, 205.
(4) White, K. M.; Francis, C. V.; Isackson, A. J. *Macromolecules*, **1994**, *27*, 3619.
(5) Tsutsumi, N.; Yoshizaki, S.; Sakai, W.; Kiyotsukuri, T. *Macromolecules*, **1995**, *28*, 6437.
(6) Prêtre, P.; Kaats, P.; Bohren, A.; Günter, P.; Zysset, B.; Ahlheim, M.; Stähelin M.; Lehr, F. *Macromolecules,* **1994**, *27*, 5476.
(7) Miller, R. D.; Burland, D. M.; Jurich, M.; Lee, V. Y.; Moylan, C. R.; Thackara, J. I.; Twieg, R. J.; Verbiest, T.; Volksen, W. *Macromolecules* **1995,** *28*, 4970.
(8) (a) Tsutsumi, N.; Matsumoto, O.; Sakai, W.; Kiyotsukuri, T. *Appl. Phys. Lett.* **1995**, *67,* 2272: ibid. **1996**, *68*, 2023. (b) Tsutsumi, N.; Matsumoto, O.; Sakai, W.; Kiyotsukuri, T. *Macromolecules*, **1996**, *29*, 592: ibid. **1996**, *29*, 3338.
(9) Wender, C.; Neuenschwander, P.; Suter, U. W.; Prêtre, P.; Kaatz, P.; Günter, P. *Macromolecules* **1994**, *27*, 2181.
(10) 3-[(2-Hydroxyethyl)amino]-5-(hydroxymethyl)-4'-nitroazobenzene (T-AZODIOL) in ref 8 should be corrected as 4-[(2-hydroxyethyl)amino]-2-(hydroxymethyl)-4'-nitroazobenzene (T-AZODIOL).
(11) Maker, P. D.; Terhune R. W.; Nisenoff, M.; Savage, C. M. *Phys. Rev. Lett.* **1962**, *8*, 21.
(12) Jerphagnon J.; Kurtz, S. K. *J. Appl. Phys.* **1970**, *40*, 1667.
(13) Tsutsumi, N.; Ono, T.; Kiyotsukuri, T. *Macromolecules* **1993**, *26*, 5447.
(14) Tsutsumi, N.; Fujii, I.; Ueda, Y.; Kiyotsukuri, T. *Macromolecules* **1995**, *28*, 950.
(15) Page, R. H.; Jurich, M. C.; Reck, B.; Sen, A.; Twieg, R. J.; Swalen, J. D.; Bjorklund G. C. and Willson, C. G. *J. Opt. Soc. Am.* **1990**, *B* 7, 1239
(16) Uematsu, Y. *Jpn. J. Appl. Phys.* **1974**, *13*, 1362.
(17) Liptay, W. in *Exited States Vol I*, edited by Lim, E. C.; Academic Press: New York and London, 1974; pp. 129-229.
(18) Havinga, E. E.; van Pelt, P. *Ber. Bunsen-Ges. Phys. Chem.* **1979**, *83*, 816.
(19) Oudar, J. L.; Chemla, D. S. *J. Chem. Phys.* **1977**, *66*, 2664.
(20) Levine, B. F.; Bethea, C. G. *J. Chem. Phys.* **1978**, *69*, 5240.
(21) Turro, N. J. in *Modern Molecular Photochemistry*; The Benjamin/Cummings Publishers: Menlo Park, 1978; pp.86-87.
(22) Birks, J. B. in *Photophysics of Aromatic Molecules*; John Wiley & Sons: London and New York, 1970; pp.51-52.
(23) Bondi, A. in *Physical Properties of Molecular Crystals, Liquids and Glasses;* Wiley: New York, 1968; Chapters 3 and 4.
(24) van Krevelen, D. W. in *Properties of Polymers*; Elsevier: Amsterdam, 1990; Chapter 4.

Chapter 12

Thermal Stability of Nonlinear Optical Chromophores

Ivan J. Goldfarb[1] and Hongtu Feng[2]

[1]Department of Chemistry, Wright State University, Dayton, OH 45435
[2]Canadian Explosives Research Laboratory, 555 Booth Street, Ottawa, Ontario K1A 0G1, Canada

A method utilizing differential scanning calorimetry has been developed for quick and reproducible estimation of the thermal stability of nonlinear optical chromophores under consideration for incorporation into polymers for use in second-order nonlinear optical devices. The method which uses sealed glass ampoules has been used to compare a large number of chromophores and has been compared to electric field-induced second harmonic generation methods.

In developing new organic nonlinear optical (NLO) materials, most of the attention in the literature has been directed at increasing the NLO activity of chromophores with some concern for their processing and temporal stability in a polymer matrix (*1*). Attempts to improve temporal stability in poled second-order NLO materials have been primarily by attaching the chromophore to a polymer chain and locking in the orientation of the chromophores by the glassy structure of the polymer or even by crosslinking. Little effort has been described for designing chromophores with adequate thermal stability to survive processing or long-term use conditions (2). Processing conditions could include solder bath immersion as well as polymer processing temperatures (melt or curing temperatures). Long-term use conditions could be 100°C or greater for periods of up to 10 years. Methods are needed to be able to estimate the behavior of NLO chromophores under these conditions.

Thermal Analysis techniques [differential scanning calorimetry (DSC) and thermogravimetric analysis (TGA)], have been used widely to provide a measure of the thermal stability of materials. These methods are complicated when it is desired to measure the stability of volatile organic compounds like many of the proposed NLO chromophores. Methods are needed that can distinguish between volatilization and decomposition.

The method that is used herein is a variation on high pressure DSC . Samples were sealed in glass capillaries to prevent volatilization and subjected to isothermal aging at several temperatures. DSC curves were run on each aged sample to determine the loss of either melting endotherm or decomposition exotherm. Using a thermal stability criterion defined as the temperature at which the chromophore (in neat form) loses 10% of its original NLO activity after 30 minutes of isothermal aging, we have determined either the melting point endotherm or the decomposition exotherm remaining after 1/2 hour and interpolated those results to find the temperature where 90% remained.

EXPERIMENTAL SECTION

Samples. All samples of chromophore materials were obtained from other researchers and used without further purification. Sources included Dr. W. A. Feld, Wright State University, Dr. Susan Ermer, Lockheed Martin Co., Dr. Robert Miller, IBM and Dr. Alex Jen, ROI Technology. The sources of each sample will be given when the results are presented.

Instrumental. All Differential Scanning Calorimetry (DSC) runs were conducted on a Perkin-Elmer DSC-7 attached through a TAC-7 Thermal Analysis Controller to a DEC computer station 325c. All runs were conducted at 10°C/min. in N_2 unless otherwise stated. Thermogravimetry was run on a Perkin Elmer TGA-7 attached through the same system as the DSC. Isothermal aging of samples in sealed capillaries were conducted in the DSC cell of a DuPont 900 Thermal Analyzer after temperature calibration.

Sealed Ampoules. A method for handling these volatile compounds has been developed(*3*) and is briefly described below. Kimax-51 melting point capillary tubes with diameter of 1.5-1.8 mm and length of 90 mm were used to seal samples for DSC runs. The sealing system consists of a jig containing a pin vise which doubles as a holder for the capillary and a heat sink to protect the sample from the heat of sealing. The end of the pin vise is placed in liquid nitrogen during the sealing operation and it thereby transfers heat from the capillary keeping the sample cool. The capillaries are sealed off with a miniature oxygen-gas torch to a length of approximately 6 mm which is just less than the inside diameter of the DSC sample cell. This sealed capillary is used as the sample holder for the DSC instead of the usual crimped aluminum pans. An empty sealed capillary was used on the reference side of the DSC and a baseline consisting of two empty sealed capillaries was subtracted from the raw data during the analysis.

Data Treatment. Inasmuch as the thermal stability criterion was to be the temperature at which 10% loss occurs in 30 minutes, we developed a strategy of isothermal aging at three different temperatures for 20, 40 and 60 minutes. This provided nine samples to be analyzed by DSC for the residual area of either the melting or decomposition peaks. The fractional residual area for the three different times for each temperature were fitted to a least squares straight line and interpolated for the 30 minute value. This was perceived to be more reproducible than just aging the sample for 30 minutes. The three 30 minute points were subsequently plotted versus aging temperature and interpolated for the 0.9 fractional residual area point (10% loss). This whole procedure was programmed on an Excel™ spreadsheet including the plots described above.

RESULTS AND DISCUSSION

Sealed Ampoule Technique Development. In order to overcome the difficulty with vaporization interfering with measurements of decomposition, a method was developed using samples sealed in melting point capillaries for DSC analysis. First it was necessary to compare results of conventional DSC runs in crimped aluminum pans with those of the new technique. An example of results of this technique compared with a run in an aluminum pan is shown in Figure 1. Also included is a

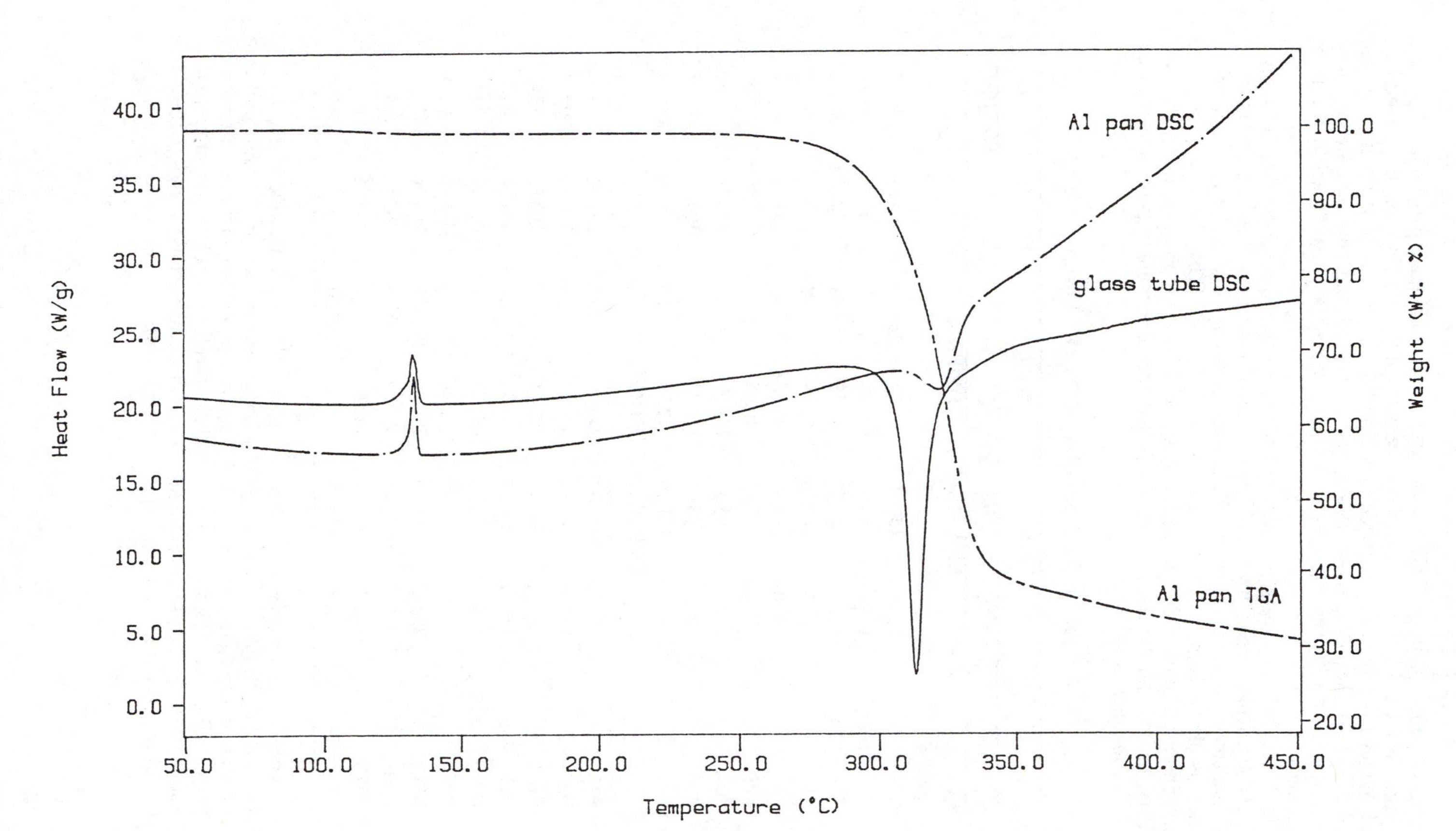

Figure 1. Comparison of conventional Al pan and glass capillary DSC.

thermogravimetric analysis (TGA) curve run in an aluminum pan under the same conditions as the DSC. It can be seen that, because of the vaporization of the sample, a broad endotherm almost obscures the decomposition exotherm of the aluminum pan sample. The capillary DSC shows a clear and distinct decomposition exotherm which displays evidence of a two-step decomposition, which is often observed for these compounds.

Melting points and heats of fusion and decomposition were compared for the two methods. Particular attention was paid to the fusion endotherm to ascertain whether the glass capillary method accurately measured the transition. Results of this comparison as well as the decomposition exotherms are given in Table I for several NLO chromophores. Compounds were included just to show the reproducibility of the glass capillary method and the comparison with conventional aluminum pans at least in the case of the melting endotherm.

Table I DSC Results of NLO Compounds

	Al pan		capil.(I)		capil(II)	
sample	peak, °C	ΔH, J/g	peak, °C	ΔH, J/g	peak, °C	ΔH, J/g
			MELTING ENDOTHERM			
HNPP	-	-	-	-	-	-
N.N	162.56	152.50	160.76	149.05	159.51	150.17
Dis. Red 1	164.12	80.62	162.38	81.27	163.15	81.21
NPP - TSB	100.75	64.53	97.11	64.54	90.05	62.68
NPP -S	113.63	108.28	107.44	110.38	115.65	109.80
VNL -S	80.07	144.49	80.48	135.02	81.19	136.62
NPP	13.96	115.93	110.05	109.28	112.49	110.85
PDA-NBS1	130.59	93.59	129.99	93.66	130.29	94.68
BNPT	105.75	66.63	104.23	66.31	104.29	65.98
NPP - 1B	115.09	111.78	112.89	111.55	111.70	111.54
			DECOMPOSITION EXOTHERM			
HNPP	273.90	-511.22	272.76	-792.39	271.76	-798.62
N.N	304.96	-705.86	309.99	-1959.58	310.49	-2000.44
Dis. Red 1	298.86	-837.44	292.41	-1428.47	293.85	-1452.87
NPP - TSB	155.62	-224.33	151.99	-187.58	144.98	-173.71
NPP -S	296.21	-801.58	296.27	-1419.13	295.77	-1439.56
VNL -S	-	-	338.94	-671.42	371.28	-662.40
NPP	295.11	-769.06	298.68	-1435.27	299.50	-1419.65
PDA-NBS1	311.33	-427.22	307.39	1413.38	303.35	-1763.53
BNPT	152.34	-965.85	149.59	-977.50	148.75	-1000.72
NPP - 1B	292.87	-925.19	298.71	-1506.51	294.91	-1463.72

The glass capillary runs were run in duplicate to be able to ascertain the reproducibility of this technique. Although the melting endotherms were broader in the capillaries than in the aluminum pans due to slower heat diffusion, the heats of fusion compared very favorably in all cases. This conclusion and the excellent baselines on most runs in glass capillaries satisfied us concerning the accuracy of this technique and no need was seen for using a holder in the DSC cell for the capillaries.

Thermal Aging Technique Development. The next task was to see what changes thermal aging produced on the DSC runs as a way to follow the aging process. For this purpose temperature of a DSC cell on an old thermal analyzer was calibrated and it was used as an isothermal aging oven. An illustration of this aging for Disperse Red-1(DR-1) is shown in Figure 2 for isothermal aging at 200°C for up to 6 hours. It can be seen that the melting endotherm decreases in both peak maximum and heat of fusion, consistent with the disappearance of the starting material with time at a constant temperature. The changes in the heat of fusion and heat of decomposition with time for this compound are shown in Figure 3. It should be noted that the heat of fusion decreases to zero within 6 hours while the decomposition heat has not even decreased to half if its initial value. This is consistent with the observation of a two-step decomposition exotherm with the first exotherm disappearing along with the heat of fusion while the second exotherm apparently represents the exotherm associated with the decomposition of the product of the first reaction and does not decrease with the decrease in the fusion endotherm. The product of this first reaction now decomposes at a higher temperature as shown by this second exotherm. Apparently this second reaction does not occur at 200°C.

This study indicates that following the heat of fusion is apparently a good method of following the disappearance of starting material although it may not agree with the decomposition exotherm if the decomposition reaction is complex (e.g., multi-step). Since the objective of this program was to ascertain the loss of NLO activity, it remained to be determined which one of these measures of thermal stability would agree most closely with the loss of NLO activity as determined by EFISH methods.

Thermal Decomposition Index. Since the temporal stability of a poled polymeric chromophore, either in a guest-host system or where the chromophore is chemically incorporated into the polymer chain, has been shown to increase with an increase in the difference between the glass transition temperature (T_g) of the polymer and the operating temperature of the NLO system, it is desirable to use polymeric systems with Tg's as high as practical. But poling of polymeric NLO materials must be accomplished at or near the T_g, so increasing the temporal stability of the system requires higher temperature poling and therefore more thermally stable chromophores. In order to compare various chromophores that may be utilized in a polymeric matrix, it is desirable to utilize a figure-of-merit for stability of the chromophore in the matrix during poling (or other high temperature processing).

It has been suggested (*4*) that an appropriate measure of chromophore stability would be the temperature at which the NLO activity dropped to 90% of its initial value after 1/2 hour exposure. We have therefore defined a $T_d(1/2)$ as the temperature where 10% decomposition takes place in 30 minutes. The procedure followed was to thermally age the samples in sealed capillaries for 20, 40 and 60 minutes at each of several temperatures and run a DSC temperature scan of each sample. Decomposition was followed by the gradual disappearance of the melting endotherm or decomposition exotherm. The data were entered into a preprogrammed spreadsheet to perform the appropriate calculations as described above. The plots as well as the data are placed on the spreadsheets, a sample of which is shown on Figure 4.

Comparison with EFISH and UV Data. Inasmuch as the chosen measure of chromophore stability is based on the loss of NLO activity, it was considered critical to compare our results on loss of melting point or decomposition point peaks with a more direct measurement of NLO activity. Fortunately, we were able to obtain a number of samples from Dr. Robert Miller, IBM Almaden Laboratories, some of which they had aged isothermally for one half hour at several temperatures and measured the fractional residual electric field induced second harmonic generation (EFISH). They also measured the residual UV peak absorption intensity which they felt was also a

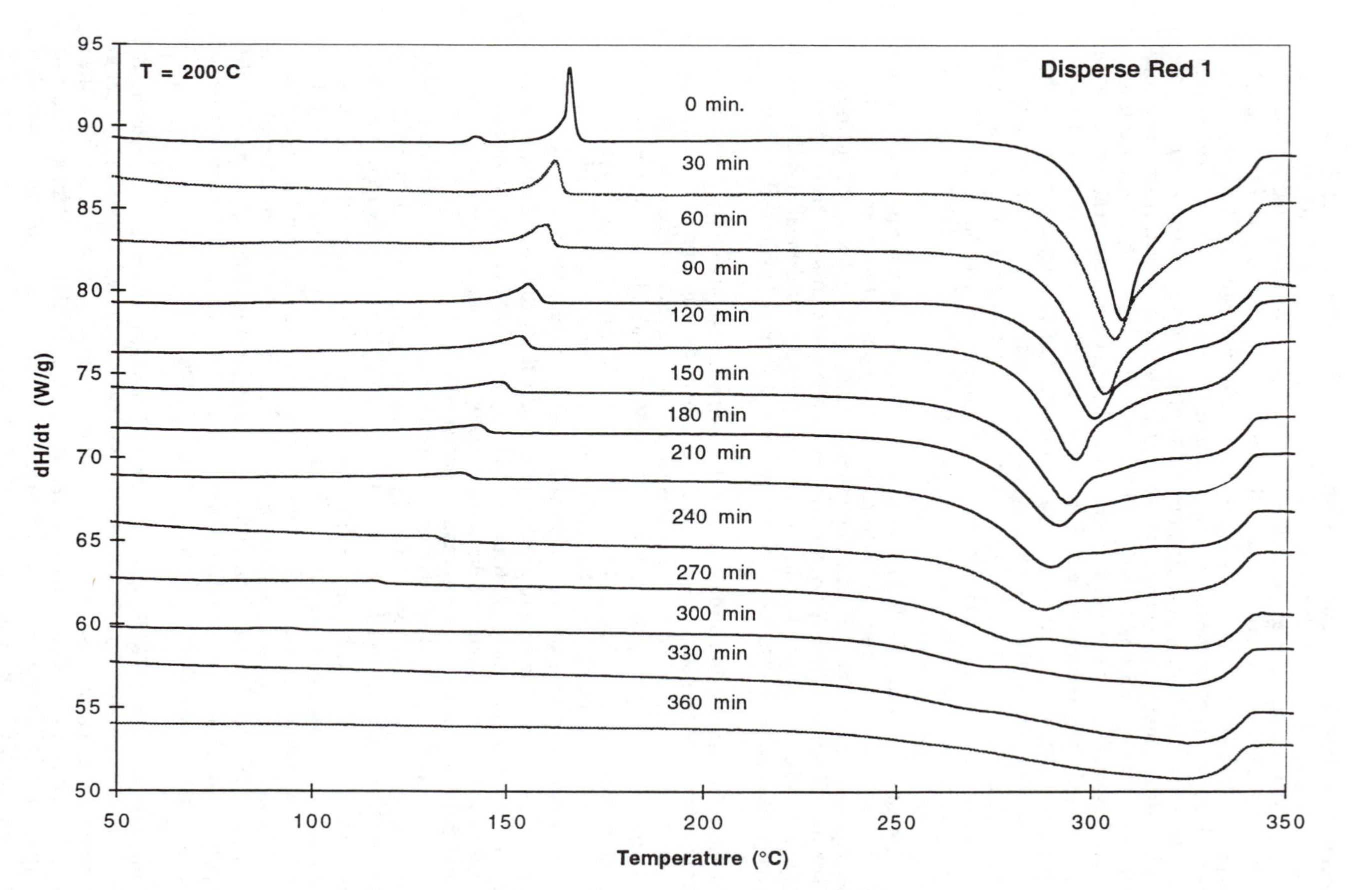

Figure 2. DSC of Isothermally aged Disperse Red-1.

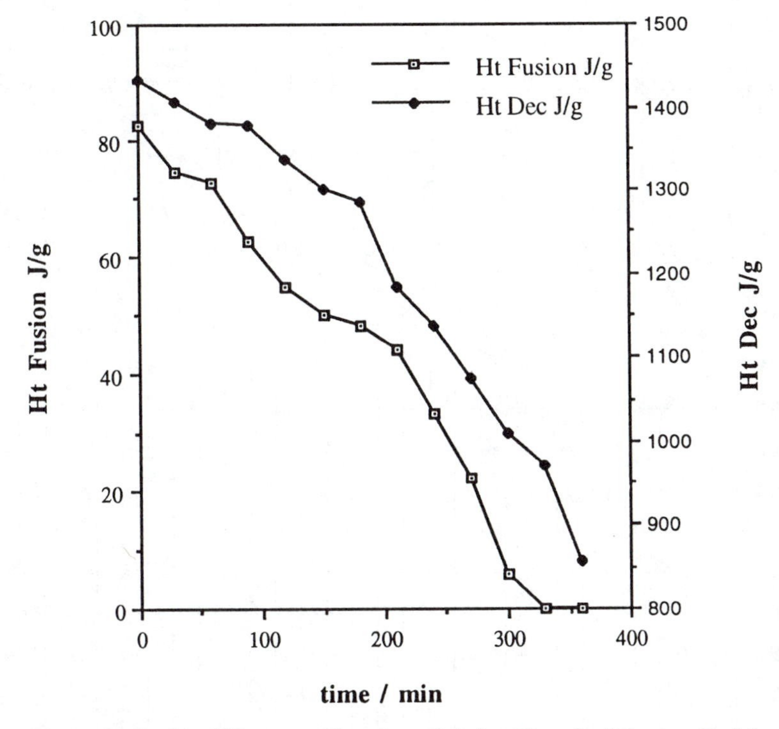

Figure 3. Residual Heats as a function of Aging Time for Disperse Red-1

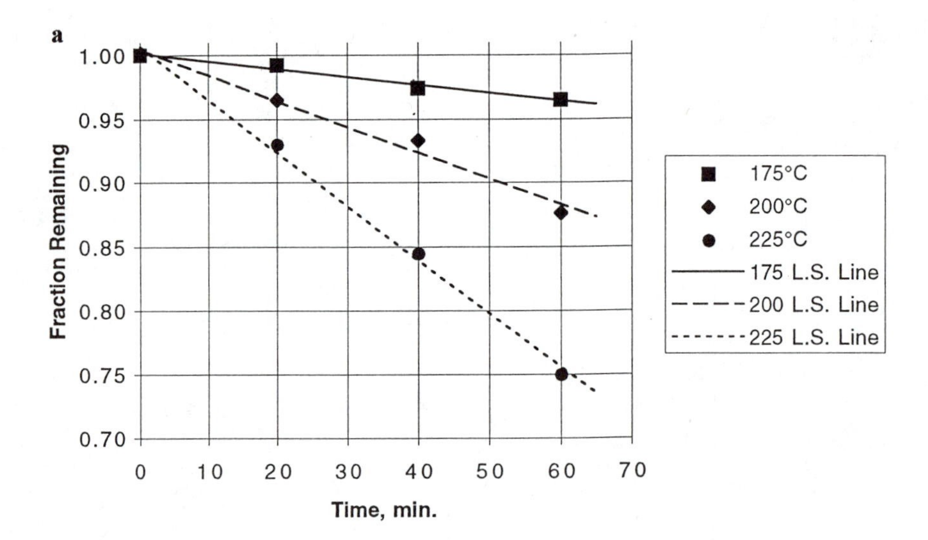

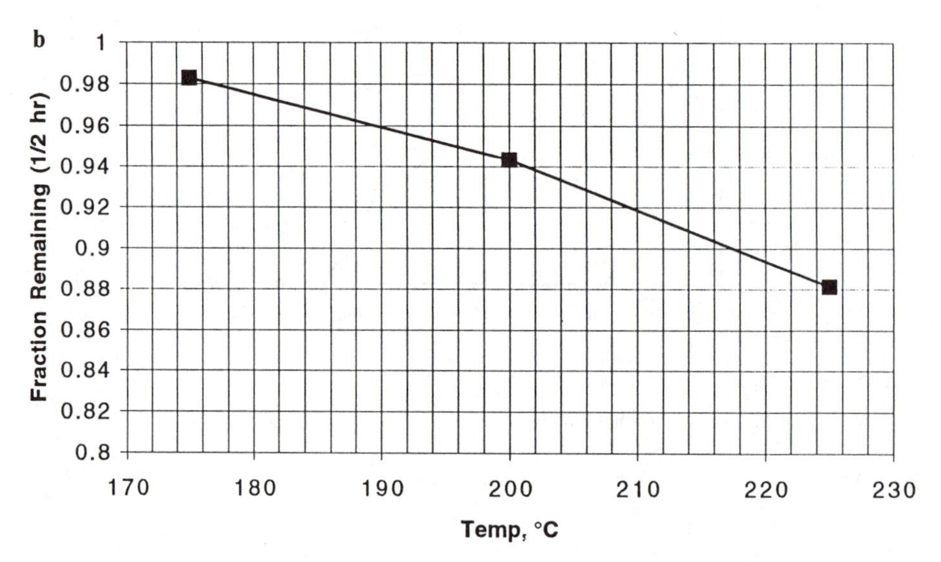

Figure 4. Determination of $T_d(1/2)$ for Disperse Red-1. a) Fraction remaining by residual heat at different aging temperatures and times. b) interpolation to obtain $T_d(1/2)$

reasonable measure of residual NLO activity. Their results and the comparison with ours are shown in Table II.
In each case, the IBM data represents the fractional EFISH intensity or peak UV absorbance after 30 minutes aging at the indicated temperature. From these data , we interpolated for the temperature where the fraction would be 0.9 and used that as the $T_d(1/2)$. It should be noted that, in the two cases where we were able to obtain melting point endotherm decreases, the agreement with EFISH was not nearly as good as the decomposition exotherm decreases and indeed all four decomposition exotherm data were in good agreement with the EFISH studies.

Table II. Comparison of $T_d(1/2)$ By DSC, EFISH and UV

Compound	Temp	Rel. EFISH	Rel. UV	melt endo	decomp. exo
IBM #1	25	1	1		
	200	0.88	0.85		
	250	0.037	0.01		
	$T_d(1/2)$	171	142	-	168
IBM#3	25	1	1		
	250	0.93	0.97		
	275	0.014	0.02		
	Td(1/2)	251	252	210	251
IBM#4	25	1	1		
	250	0.93	0.97		
	300	0.83	0.85		
	$T_d(1/2)$	265	279	-	252
IBM#5	25	1	1		
	250	1	0.98		
	300	1	0.96		
	350	0.88	0.85		
	$T_d(1/2)$	308	327	281	312

The UV data roughly agreed with the EFISH data but were not always as close as the DSC decomposition exotherm data. It was therefore decided to use the decomposition exotherm data for our measure of $T_d(1/2)$. We reasoned that the agreement was better with the decomposition exotherm data than the melting point endotherm data because the decrease in melting point peak signals the loss of the original compound, whereas the decrease in decomposition exotherm is a measure of the loss of the original compound and any subsequent products which are formed. It was clear from our earlier data for DR-1 that this material appears to decompose by a two-step mechanism. If the product of the first decomposition still had NLO activity, the EFISH data would reflect the disappearance of both products. The decomposition exotherm likewise measures whatever decomposition occurs and if they have similar heats of decomposition it would respond to the sum of the two. Thus, although the original NLO chromophore may have decomposed giving rise to a loss in the melting endotherm for that compound, a new compound is formed that is still a chromophore and can exhibit its own decomposition exotherm.

Sample Test Data. Thirty different NLO chromophores were tested to provide a thermal stability index. Of these, seven were studied using both melting endotherm and decomposition exotherm disappearance whereas the other twenty-three compounds were studied using only the decomposition exotherm decreases. This makes a total of thirty-seven samples whose data are given in Table III. The chemical structures corresponding to the sample names are shown in Figures 5, 6 and 7.

Table III. NLO Activity And Thermal Stability Of Chromophores

Chromophore	Source	$T_m(1/2)$ °C	$T_d(1/2)$ °C	β_o*10^{30} esu	M.W.	β_o/M.W.
5RM 47A	1		193			
DAD	2		264	70	434	0.161
DADB	2		269	108	602	0.179
DADC	2		321	69	582	0.119
DADCH	2		293			
DADI*	2		299	52	454	0.115
DADIH	2		269	52		
DADOH	2		247			
DADT	2		229			
DADTB	2		229			
DCIOH	2		265			
DCIP	2		301			
DCIT*	2		219	136	349	0.390
DCM	2	231	249	58	303	0.191
DR-1	2	202	218	50	314	0.159
IBM #1	3		168	66.6	357	0.187
IBM #3	3	210	251	49	298	0.164
IBM #4	3		252	18.9	401	0.047
IBM #5	3	281	312	54	394	0.137
IBM #6	3	248	254	71.8	439	0.164
IBM #7	3		261			
IBM #8	3		227			
H-NNP	1	207	252	38	166	0.229
NPP	1	164	196	18.3	222	0.082
PTA	4		219	12	185	0.065
PTE	4	231				

*β_o from an analogous structure

Sources: 1. W. A. Feld, Department of Chemistry, Wright State University; 2. Dr. S. Ermer, Lockheed Martin Missiles & Space Co., Palo Alto, CA; 3. Dr. R. D. Miller, IBM Almaden Research Center, San Jose, CA; 4. B. Reinhardt, AF Wright Laboratory, WPAFB, OH

In order to provide some graphical presentation of the potential usefulness of these NLO chromophores it is necessary to provide some measure of their relative NLO activity as well as their thermal stability. An appropriate measure of NLO activity would be β (2ω,ω,0) the second-order molecular hyperpolarizability, usually obtained by EFISH measurements. So that this value would not be influenced by any spectral resonance contributions, β_o, the value of β extrapolated to infinite wavelength, commonly obtained using the two-level model approximation (*5*), is actually used.

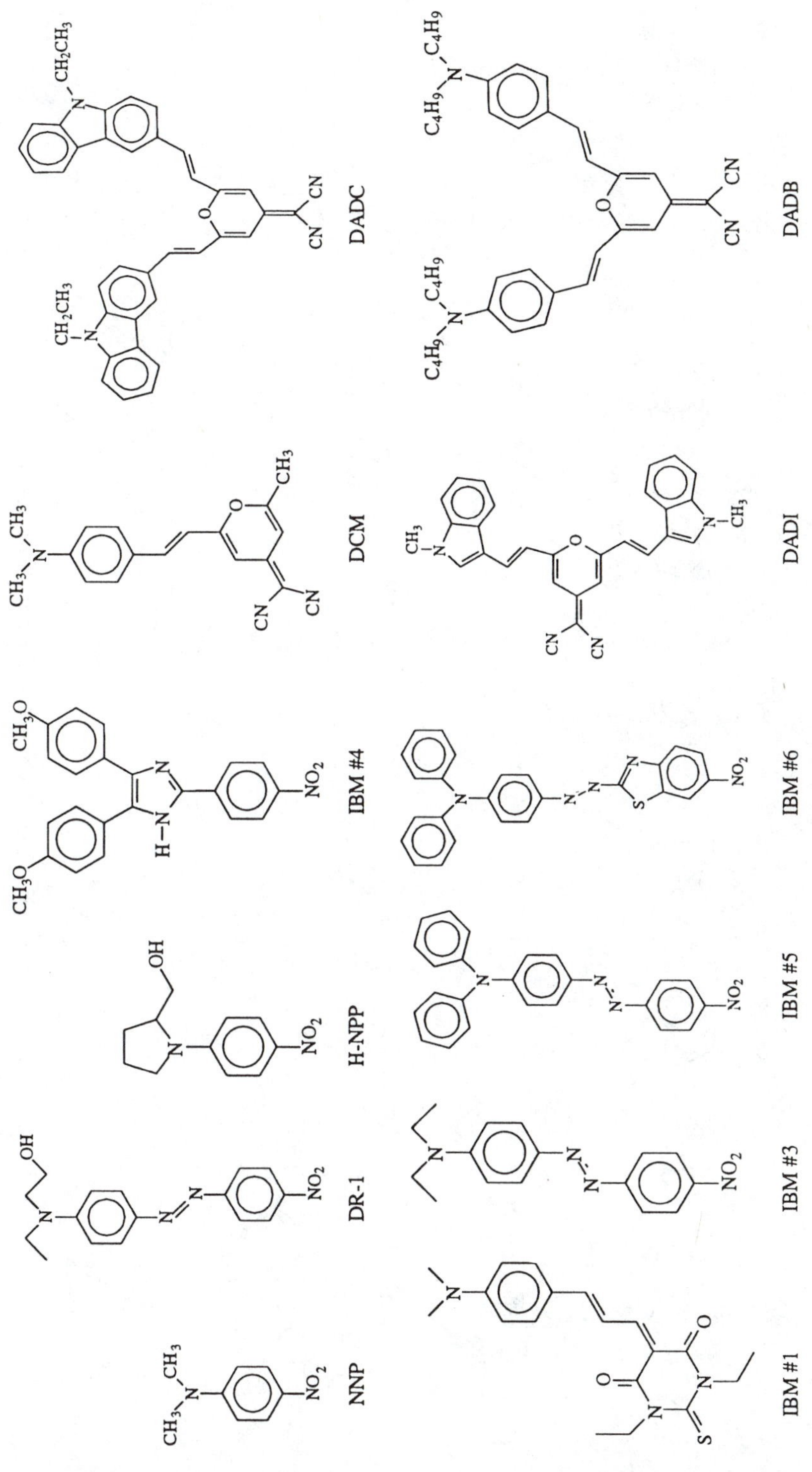

NNP DR-1 H-NPP IBM #4 DCM DADC

IBM #1 IBM #3 IBM #5 IBM #6 DADI DADB

Figure 5. NLO Chromophore Chemical Structures

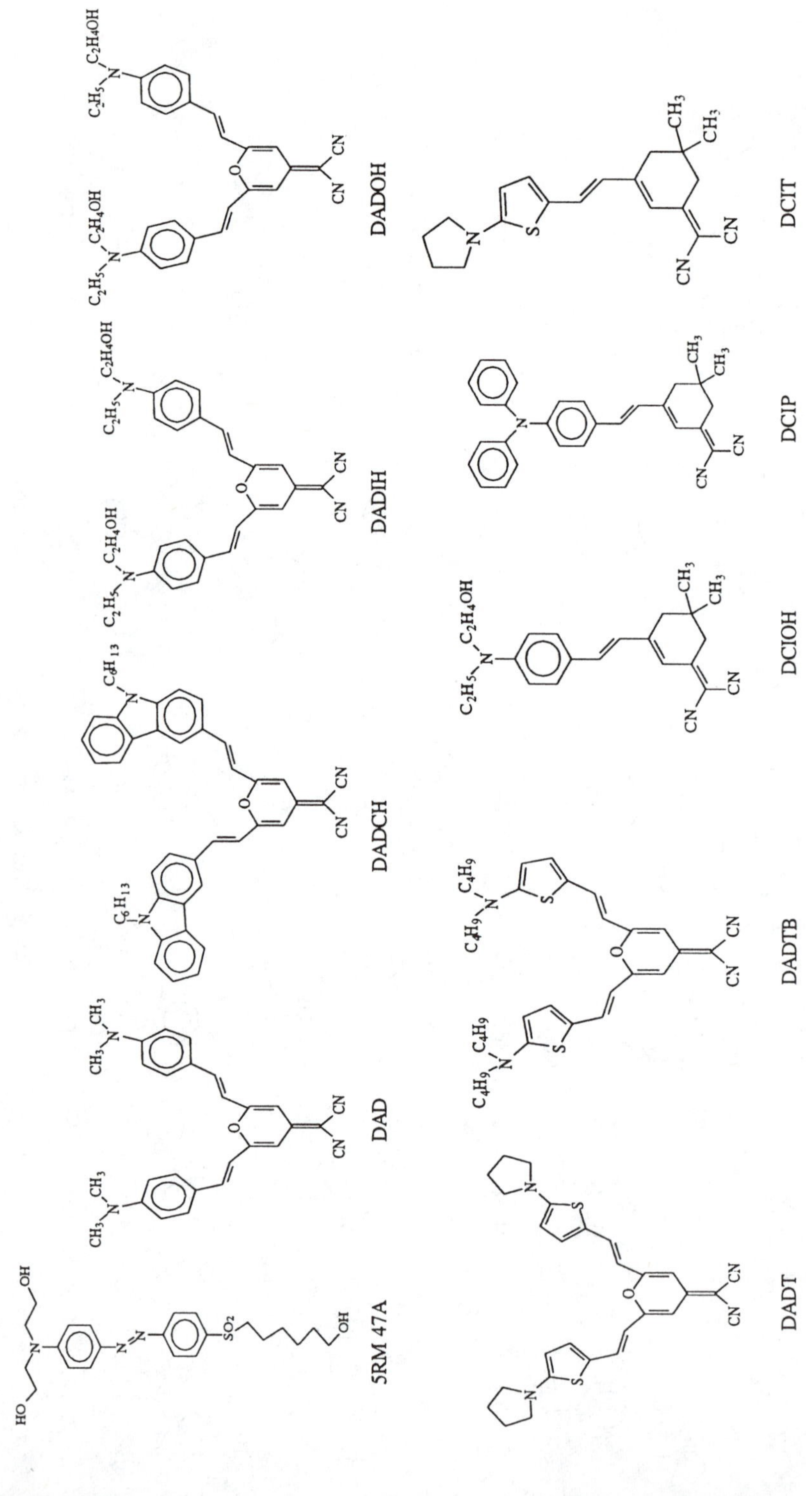

5RM 47A

DAD

DADCH

DADIH

DADOH

DADT

DADTB

DCIOH

DCIP

DCIT

Figure 6. NLO Chromophore Chemical Structures, II

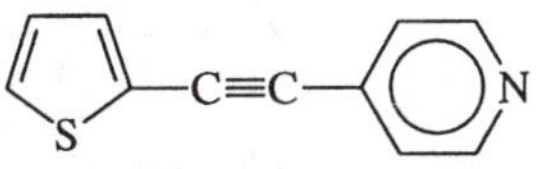

PTA

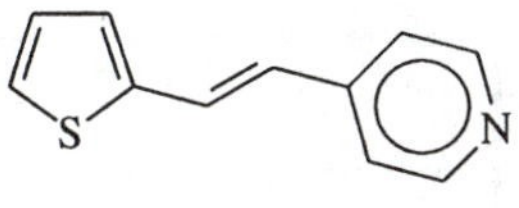

PTE

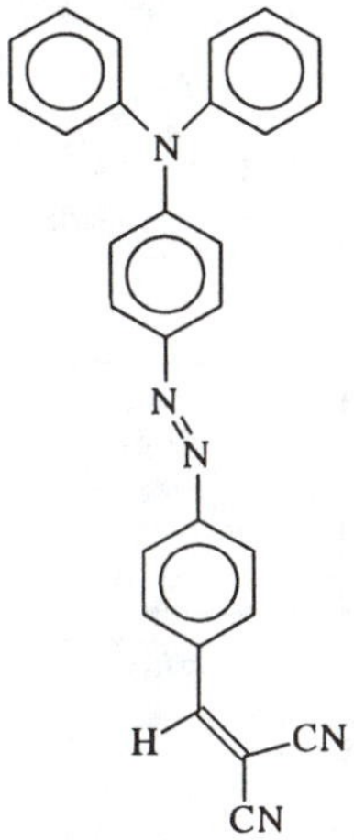

IBM#7

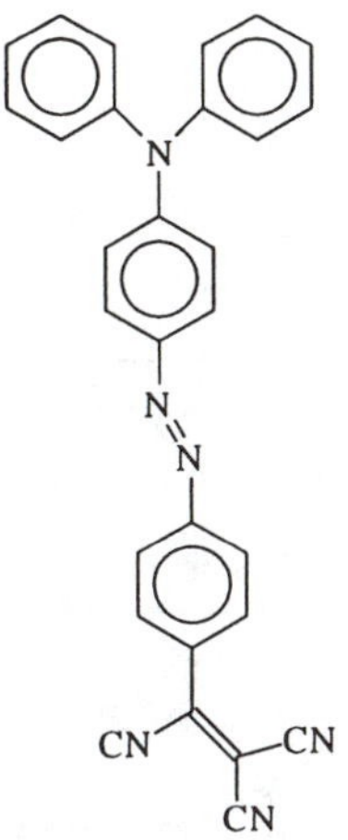

IBM#8

Figure 7. NLO Chromophore Chemical Structures, III

Furthermore, since these chromophores are expected to be used by incorporating them into a polymer matrix, either as a guest-host mixture or by chemical bonding, the relative NLO activity per unit weight is given by β_o /MW. It is this figure of merit which is plotted versus $T_d(1/2)$ to give a potential usefulness curve. These data from Table III are plotted in Figure 8.

It should be noted that we were unable to obtain the NLO activity data or, in some cases, even the molecular structures for all the compounds that were investigated, so not all of our thirty compounds are shown on the graph of Figure 8. All samples are listed in Table III, however, with whatever data we were able to gather.

The highest decomposition value we measured of a sample for which we have structural information is DADC at 321°C. This compound is a donor-acceptor-donor molecule that has two N-ethyl carbazole groups as electron donors. Note that changing the ethyl groups to n-hexyl groups in DADCH reduces $T_d(1/2)$ to 293°C. The next highest stability is shown by IBM #5 where the alkyl substituents on the nitrogen in disperse red I have both been replaced by phenyl groups. As has been pointed out previously this is a very powerful way of improving the thermal stability of chromophores (*2*). The highest β_o/MW value was achieved by DCIT although the value of β was actually obtained for an analogous alkyl substituted compound in order to obtain sufficient solubility to obtain EFISH data (*6*).

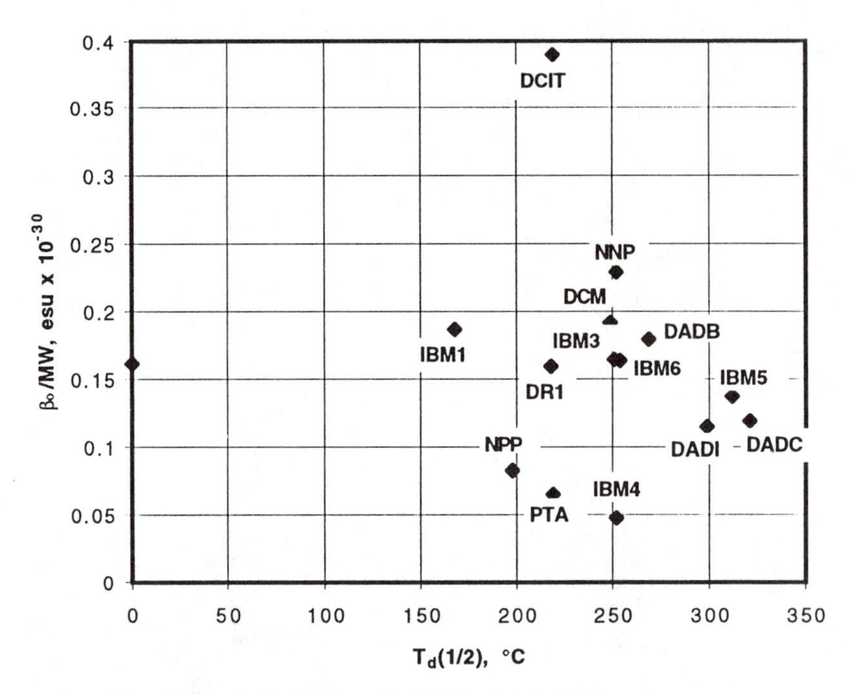

Figure 8. β_o/MW vs. $T_d(1/2)$ for NLO Chromophores

CONCLUSIONS

From the results presented, it has been shown that the sealed capillary DSC technique for estimating thermal stability can be used to determine the comparative stabilities of a variety of NLO chromophores. Furthermore, the data compare favorably with actual nonlinear optical measurements, namely the decrease in EFISH values with time at temperature.

Another conclusion that could be drawn from these data is that it is important for proper evaluation of NLO chromophores to use materials representative of the chromophore alone without the destabilizing effect of reactive functional groups that might be added to attach it chemically to a polymer host. For example, the structure of IBM #3 is similar to that of disperse red-1 without the hydroxyl group on one of the ethyl groups of the diethylamino donor. This hydroxyl group appears to reduce the stability by over 30 degrees, from 251°C to 218°C, leading to a prediction of lower stability for the basic chromophore unit.

LITERATURE CITED

(1) Prasad, P. N.; Williams, D. J. Introduction to Nonlinear Optical Effects in Molecules and Polymers; Wiley: New York, NY, 1991.
(2) Moylan, C. R.; Miller, R. D.; Twieg, R. J.; Betterton, K. M.; Lee, V. Y.; Matray, T. J. Nguyen, C. Chem. Mater.,**1993,** 5, 1499.
(3) Goldfarb, I. J.; Feng, H; Oxley, J. C. Thermochimica Acta, **1996**, 275,139).
(4) Lee, C. Y-C. Private communication.
(5) Oudar, J.L., and Chemla, D.S., J. Chem. Phys. **1977**, 66, 2664.
(6) Ermer, S. private communication.

Chapter 13

Polydiacetylene Microcrystals for Third-Order Nonlinear Optics

Hachiro Nakanishi and Hitoshi Kasai

Institute for Chemical Reaction Science, Tohoku University, Sendai 980–77, Japan

Microcrystals of some diacetylenes, prepared by the reprecipitation method, have been studied as dispersions in liquid media. Interesting behavior has been observed in the solid-state polymerization of diacetylene monomers and with the optical properties of polydiacetylene (PDA) microcrystals. First, the polymerization perfectly proceeded from one end to the other end of the diacetylene microcrystals. Next, the excitonic absorption peak position was found to shift to higher energy side with decreasing size of the PDA microcrystals. The size effect was observed even for crystals as large as 100 nm or more in contrast to conventional quantum effect of inorganic semiconductors where size effect is observed only for microcrystals of less than about 10 nm size. In addition, since the microcrystal dispersions in water have low optical loss, the optical Kerr shutter response of PDA microcrystals could be measured, and the non-resonant $\chi^{(3)}$ value was estimated to be on the order of 10^{-13} to 10^{-12} esu in very low concentrations (*ca.* 10^{-3} M).

Organic compounds with delocalized π-electron systems, *e.g.*, π-conjugated polymers, are considered to be candidates for third-order nonlinear optical materials. Among them, polydiacetylenes (PDAs) are an important class of conjugated polymers that has attracted investigators from many different fields (*1,2*). PDAs, which can be obtained as single crystals by topochemical solid-state polymerization (*3*), have been extensively studied since 1976 (*4*). PDAs show large third-order nonlinear optical susceptibilities (*5*) and ultrafast optical responses; therefore, they have been considered to be very promising materials for future photonic technologies (*6-8*).

However, it is difficult to prepare large PDA crystals with good optical quality. This is a serious problem as far as the application of PDA bulk crystals for photonic technologies are concerned. On the other hand, organic materials can readily make glasses or liquids with high optical quality. Using liquid media, the optical Kerr shutter (OKS) is considered to be one of the most promising devices for ultrafast all-optical signal processing based on third-order nonlinearity (*9-11*). OKS materials are required to satisfy the following factors: (A) high optical transparency, (B) large susceptibilities, and (C) ultrafast response. OKS experiments with organic materials have so far been mainly performed with solutions.

Recently, many researchers have studied fine particles or microcrystals to clarify the intermediate state between bulk crystals and isolated atoms and molecules (*12-16*). From these studies in the field of nonlinear optics, Hanamura predicted excitonic and surface-state enhancements of third-order optical nonlinearity in microcrystals because of quantum confinement effects, which is one of several size effects theoretically established by Ekimov *et al.* and Brus (*17-19*). Following these theories, enhancements of $\chi^{(3)}$ have been reported in semiconductor-microcrystallite-doped glasses and polymers (*20,21*). Organic microcrystals, however, have attracted very little attention so far (*22,23*) owing to the difficulties of preparing them.

Since 1992, we have succeeded in obtaining some organic microcrystals by the conventional reprecipitation method (*24*). For PDA microcrystals, we have been able to find interesting optical properties as well as characteristic polymerization behavior (*25,26*). Moreover, among the many compounds investigated, some of the microcrystal-dispersed liquids were found to show little light scattering because the crystal sizes were smaller than the relevant wavelengths. They are stable without cohesion in liquids or matrices for more than half a year in spite of high concentration. Therefore, if these microcrystal dispersions are applied for OKS, it will be possible to achieve low optical-loss and large $\chi^{(3)}$ materials. Many promising crystalline materials could thereby be used for OKS.

In this article, we will describe in detail the preparation of diacetylene (DA) microcrystals and their solid-state polymerization behavior and optical properties as a function of crystal size. In addition, we also report OKS response of organic microcrystal dispersions.

EXPERIMENTAL SECTION

Preparation of DA Microcrystals. The DA derivatives used in this work were 5,7-dodecadiynylene bis(*N*-(butoxycarbonylmethyl)carbamate) (4BCMU), 10,12-heptacosadiynoic acid (14-8ADA), and 1,6-di(*N*-carbazolyl)-

2,4-hexadiyne (DCHD), as shown in Figure 1. These compounds are known to have large third-order nonlinear optical properties(*4*).

Their microcrystal dispersions in water were prepared by the simple and easy reprecipitation method developed and established by our group(*24*). For example, in the case of 14-8ADA, 250 μl of 1.0×10^{-2} M acetone solution was injected into 10 ml of water with 20 mg of poly(vinly alcohol) at room temperature. PDA microcrystals were obtained by UV irradiation of the microcrystal dispersions of the corresponding monomers.

Characterization of PDA Microcrystals. Scanning electron microscope (SEM; Hitachi S-900) and dynamic light scattering (DLS; Otsuka electronics DLS-7000) were used to evaluate the size of microcrystals. The molecular weight of poly(4BCMU) microcrystals was estimated by gel permeation chromatography (GPC; Shimadzu LC-8A, Shodex GPC K-80M). Monodispersed polystyrene (Toso) was used as a standard sample. Solid-state polymerization of DA microcrystals was monitored by observing the changes in UV absorption spectrum and by determining the conversion using Differential Scanning Calorimetry (DSC; Rigaku 8240B).

Experimental System for OKS. The optical system for OKS measurements is shown in Figure 2. All the dispersions were put in a 10-mm quartz cell between a polarizer and an analyzer in cross Nicol configurations. OKS experiment was performed using a Nd:YAG laser at 1064 nm with 6 ns pulse width, 6.3 mW power and 10 Hz repetition as a gate beam, and a laser diode at 813 nm with 100 ns pulse width as a probe beam. A Photomultiplier tube with a 2 ns response was employed as a detector. The refractive index (n) and $\chi^{(3)}$ value of carbon disulfide used as a reference are 1.62 and 3.7×10^{-12} esu, respectively (*25*). Details of the experimental set-up are described in previous papers (*26*).

RESULTS AND DISCUSSION

Solid-State Polymerization of 14-8ADA Microcrystals. Solid-state polymerization is naturally influenced by the lattice mismatch between monomer and polymer crystals as well as the lattice rigidity. Tieke *et al.* (*27*) reported that strain in the bulk crystal is relaxed by a phase transition to some extent, in which the continuous phase is transformed from monomer rich region to polymer rich one in the course of the polymerization. Such a phase transition is considered to occur easily in a microcrystal, compared with a bulk crystal, because of loosening of the crystal lattice with decreasing crystal size. This must especially be the case with 14-8ADA which has a rather weak crystalline lattice due to its long alkyl chains.

DSC measurements of DA crystals show that the area of the endothermic peak decreases upon UV or γ-ray irradiation due to the formation of

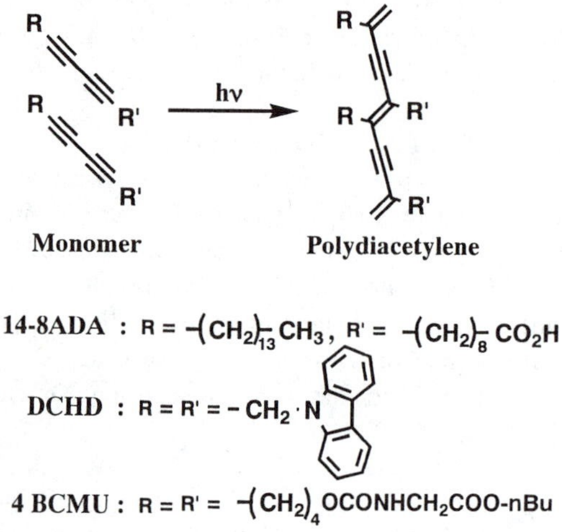

R
R'
hν
Monomer
Polydiacetylene
14-8ADA : R = —(CH2)13 CH3 , R' = —(CH2)8 CO2H
DCHD : R = R' = - CH2 · N
4 BCMU : R = R' = —(CH2)4 OCONHCH2COO-nBu

Figure 1. Diacetyrene compounds used for the preparation of microcrystals.

Pump beam
45°
Sample
Probe beam
y
Polarizer
Analyzer
x

Pump beam :	wave length	1064 nm
	duration	6 ns
	power	6.3 mW
	cycle	10 Hz
Probe beam :	wave length	813 nm
	duration	100 ns
	cycle	10 Hz
Pass length :		10 mm

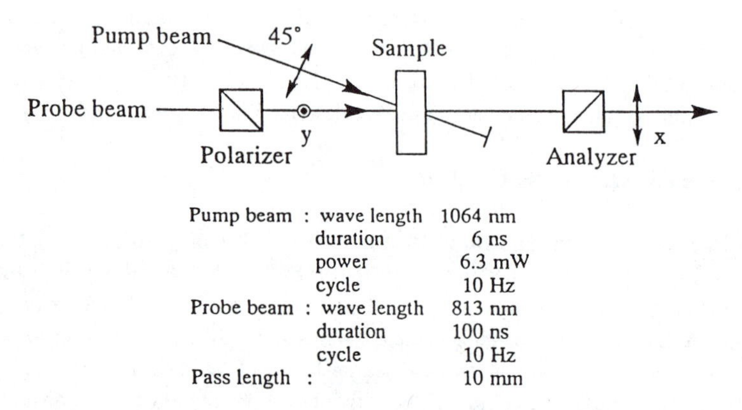

Figure 2. Experimental set-up for the measurements of Kerr effect.

polydiacetylene(*28*). The amount of residual monomer in a crystal could be estimated from the heat of fusion. The final conversion of 14-8ADA thus determined is summarized in Table I. The results with either UV or γ-ray irradiation were almost the same. Interestingly, the final polymer conversions with microcrystals were larger than with bulk crystals. It turns out that the loose crystalline lattice of microcrystals allows the strain to be easily relaxed. In microcrystals of several nanometer size, however, the conversion was 75% which is lower than that of submicrometer crystals, in spite of the smaller crystal size. This is due to excessive enhancement of thermal vibrations of the crystal lattice in this type of long-alkyl-substituted diacetylene. This is considered to be a kind of dynamic defect in the crystal lattice that disturbs the solid-state polymerization. Such enhanced thermal vibration is also estimated to be one of the main reasons for the size-dependent shift of absorption peaks as mentioned later.

Table I. The Final Polymer Conversion of 14-8ADA Crystals with Different Sizes

Methods for polymerization	Bulk crystals	Microcrystals (Sub μm)	Microcrystals (A few nm)
UV irradiation	35%	97%	75%
X-ray irradiation	49%	~100%	76%

Degree of Polymerization of Poly(4BCMU) Microcrystals. Another point that we want to emphasize is the possibility of synthesizing PDAs with narrow molecular weight distribution; this may lead to sharper absorption due to regular and uniform electronic states in the PDA backbones. The principle idea to achieve this is as follows: when the polymerization proceeds in microcrystals, whose size is smaller than the propagation length, every single polymer chain must reach from one end to the other of the microcrystal. Upper views in Figure 3 represent microcrystals with different sizes and shapes. Horizontal lines in these figures are for conjugated backbones of PDA and outlines show the edges of the microcrystals. Lower views in Figure 3 indicate molecular weight distributions monitored by GPC. The horizontal axes are retention time; longer retention time corresponds to lower molecular weight. Considering that solid-state polymerization is stopped by defects, polymerization in microcrystals is expected to proceed continuously until the polymerization reaches both edges of the microcrystals as shown in (a) to (c) because the number of defects should be reduced owing to the small number of molecules in the microcrystals. If we obtain size-controlled microcrystals with the rectangular shapes (a) and (b), molecular weight distribution becomes more mono-dispersed and the molecular weight of the polymer depends on the size of the microcrystals. Even in the case of (c), the molecular weight of the polymer must be roughly controlled by crystal size though its distribution is broader than (a) and (b).

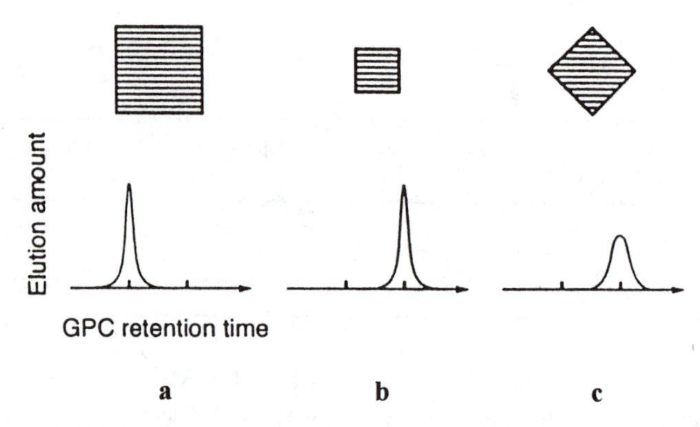

Figure 3. Schematic diagrams on the polymerization of diacetylene microcrystals. In the upper figure, size of microcrystals and polymer backbones are indicated by thick outlines and thin lines, respectively. The lower figures show GPC data expected for the polymers obtained from each upper microcrystal.

As a preliminary experiment to prove this hypothesis, microcrystal fabrication of poly(4BCMU) was investigated. Since this polymer is soluble in chloroform, molecular weight distribution was evaluated using GPC. Two kinds of poly(4BCMU) microcrystals with different crystal sizes (300 nm and 1 μm) were obtained by changing water temperature in the reprecipitation method, as shown in Figure 4. It was expected that high and low molecular weight poly(4BCMU) would be obtained from the sample with crystals of large (sample A) and small (sample B) size, respectively. Indeed, the polymer prepared from larger microcrystals showed higher molecular weight as summarized in Table II *(28,29)* Surprisingly, these values coincide well with those calculated from their crystal sizes on the assumption that the repeat length of poly(4BCMU) is *ca.* 0.5 nm. Thus, solid-state polymerization of this PDA monomer is estimated to proceed from one end to the other of the microcrystals. The molecular weight distribution of poly(4BCMU) bulk crystals was much broader than that of microcrystals, and the molecular weight was lower, indicating that polymer lengths in bulk crystals are regulated by the distance between randomly distributed defects. Though the size control of the microcrystals was not exact, the present results clearly demonstrate the possibility of synthesis of monodisperse polymers using topochemical polymerization of microcrystals. In order to decrease Mw/Mn value we are making efforts to prepare the microcrystals with monodispersed size and polymerize them from one end to the other.

Table II. Molecular Weight of Poly(4BCMU) Microcrystals

Size of microcrystals (μm)	Molecule weight[a] Mw	Mn	Mw / Mn
0.5 - 1.5	1.6×10^6	5.2×10^5	3.1
0.2 - 0.4	5.5×10^5	1.6×10^5	3.4

[a] Determined by GPC using polystylene as a standard.

Optical Properties of PDA Microcrystals of Different Sizes. The effect of average molecular weight on the functions of PDA microcrystals is expected to be important from the point of view that the electronic and nonlinear optical properties of quantum-wire and quantum-dot structures differ significantly from those of bulk crystals. An enhancement of several orders of magnitude in third-order optical nonlinearities has been predicted in the case of quantized microcrystallites.

To examine such size effects, different sizes of DCHD microcrystals were

a

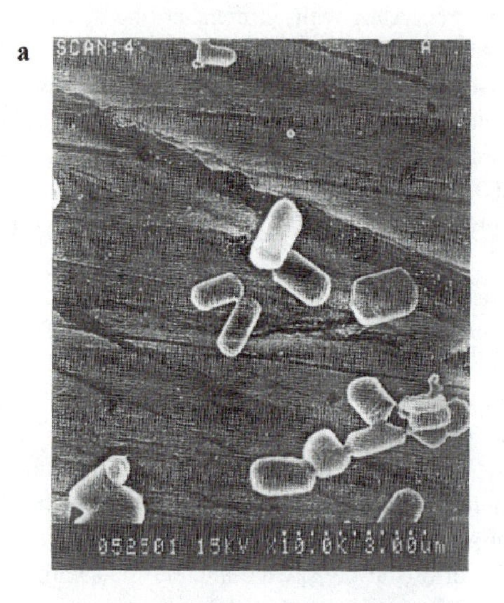

b

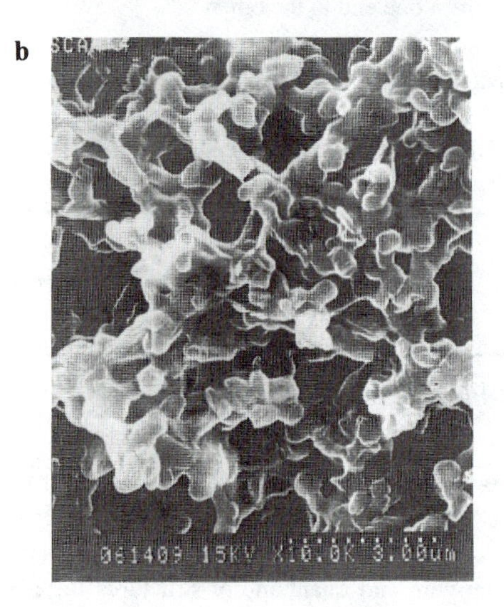

Figure 4. SEM photographs of poly(4BCMU) microcrystals with average size of (a) 1 μm and (b) 300 nm.

prepared using the reprecipitation method *(28,30,33)*. DCHD was selected because of its rigid and ideal packing and the peak position of the excitonic absorption of the resulting polymers for each microcrystal size never moved in the course of polymerization. Therefore, the size effect could be unambiguously examined.

There are many factors which are believed to play a role in controlling the size of microcrystals. One of the most effective factors is the water temperature. For example, in the case of DCHD, when 250 μl of 7.0×10^{-3} M acetone solution was injected into 10 ml of water at 343, 298, and 273 K, the crystal size was 70, 100, and 150 nm, respectively (Figure 5). The shape of the DCHD microcrystals was found to be intact by SEM observation before and after solid-state polymerization. Figure 6 shows the optical absorption spectra of poly(DCHD) microcrystals in water with different crystal sizes. The excitonic absorption peak positions of poly(DCHD) microcrystals of 70, 100 and 150 nm in size were 652 nm, 646 nm, and 640 nm, respectively. That is, the absorption energy of exciton was found to shift to the high energy side with decreasing crystal size.

Reasons for Size Effects in Organic Microcrystals. Table III summarizes the size-dependence of optical absorption properties among organic superlattice, PDA, perylene, and semiconductor microcrystals (*20,31-35*). Size effects in organic superlattice as well as semiconductor microcrystals appear upon reducing the size to less than 10 nm, and the cause of these effects has been explained in terms of quantum confinement. However, interestingly, in the case of organic microcrystals such as PDA and perylene, the high energy shifts in the excitonic absorption were observed, even though the crystal size was much larger than about 100 nm. The cause for this peculiar phenomenon is not clear. Two factors are presumed to be the cause of the size effect in organic microcrystals. One is the change of lattice state. Because the lattice would soften by microcrystallization, the coulombic interaction energies between molecules get smaller. As a result, optical absorption properties of molecules would change in microcrystals. The other is the electric field effect of the surrounding media with varying dielectric constants. This is currently under investigation. In the near future, this size effect of organic microcrystals might be explained by a new quantum confinement model. At present, however, it is important to prepare organic microcrystals with sizes less than 10 nm in the free-standing state, and to clarify their optical properties.

OKS Measurements of PDA Microcrystals. In OKS experiments, the probe beam through two polarizers in a cross Nicol state was detected because of the induced birefringence resulting from high intensity gate light incident upon the sample. The $\chi^{(3)}$ value was evaluated by probe transmittance (T). Equations (1) and (2) show the relationship between T and the nonlinear optical coefficient (n_2) as follow;

a

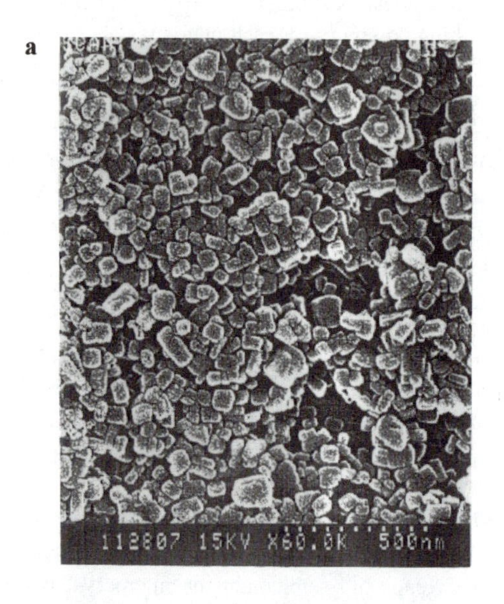

b

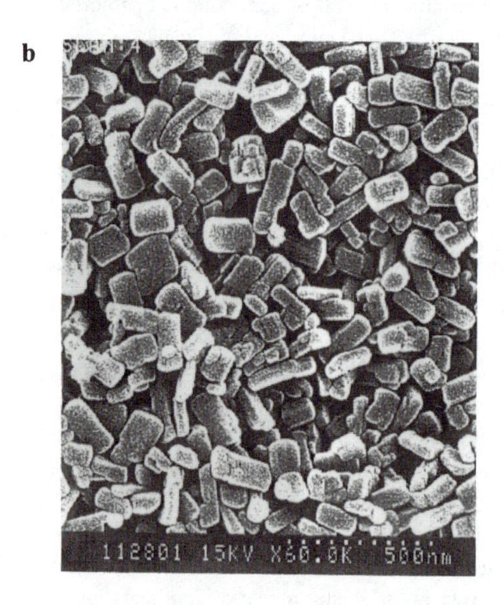

Figure 5. SEM photographs of poly(DCHD) microcrystals with average size of (a) 70 nm (b) 100 nm and (c) 150 nm.

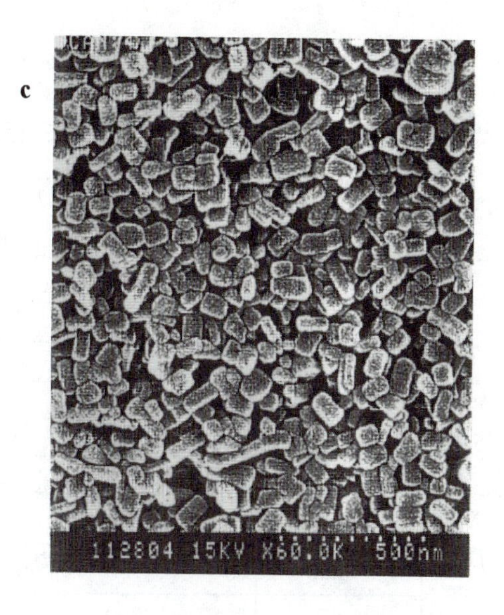

Figure 5. Continued.

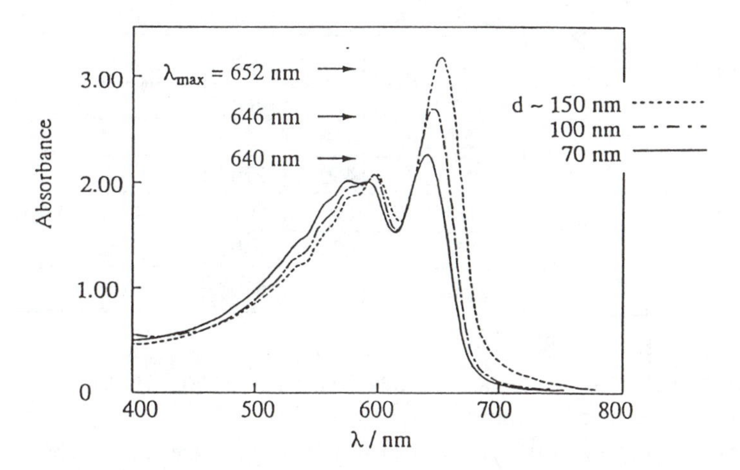

Figure 6. Relationship between visible absorption spectra of poly(DCHD) microcrystals dispersed in water and their crystal size.

Table III. The Size-Dependence of Peak Position of Excitonic Absorption Peak Positions (λ max) and These Energy Shift from Bulk Crystal (Δ E) of Various Microcrystals

	Compound	Crystal Size	λ max nm	λ max (cm^{-1})	Δ E /cm^{-1}	Meas. Temp.
Inorganics	CuCl[a]	Bulk	386	(25900)	0	77 K
		3.3 nm	384	(26000)	100	
		2.1 nm	382	(26200)	300	
	CdSe[a]	Bulk	674	(14800)	0	2 K
		12 nm	664	(15100)	300	
		1.8 nm	507	(19700)	4900	
Organics	NTCDA/ PTCDA[b] (Super lattice)	Bulk	560	(17860)	0	R.T.
		4 nm	558	(17900)	40	
		1 nm	555	(19000)	160	
	PPy/PBT[c] (Polymer Super lattice)	Bulk		(16130)	0	R.T.
		9 nm		(16210)	80	
		3 nm		(16390)	260	
	Perylene[d]	Bulk	480	(20800)	0	R.T.
		200 nm	470	(21300)	500	
		50 nm	450	(22200)	1400	
	PDA (DCHD)	Bulk	665	(15000)	0	R.T.
		150 nm	652	(15300)	300	
		50 nm	635	(15700)	700	

[a]Ref. 20, 32.

[b] NTCDA (3, 4, 9, 10-perylenetetracarboxylic dianhydride), PTCDA (3, 4, 7, 8-naphthalenetetracarboxylic dianhydride)(Ref. 34).

[c] PPy (polypyrrole), PBT (polybithiophene) These data were measured by photoluminescense.(Ref. 31).

[d]Ref. 33, 35.

$$T = T_0 \sin^2(\Delta\phi/2) \tag{1}$$

$$\Delta\phi = 2\pi n_2 L_{eff.} I/\lambda \tag{2}$$

where, T_0 is the maximum possible transmittance, $\Delta\phi$ is the phase shift, I is the gate power intensity, and λ is the probe light wavelength. $L_{eff.}$ is the effective medium length and is given by equation (3):

$$L_{eff.} = (1-e^{-\alpha L})/\alpha \tag{3}$$

where, α is the absorption loss, L is the medium length. The n_2 value for the microcrystal dispersions can be calculated using $n_{2(CS2)}$ as a reference sample from equation (4).

$$n2 = n2_{(CS2)} \times \frac{Leff._{(CS2)}}{Leff.} \times \sqrt{\frac{(T/T0)}{(T/T0)_{(CS2)}}} \tag{4}$$

The relationship between n_2 and $\chi^{(3)}$ is expressed by equation (5):

$$n_2 = (8\pi\chi^{(3)} \times 10^7)/Cn^2 \tag{5}$$

where C and n are light velocity and refractive index, respectively. $\chi^{(3)}$ is estimated from equation (6) which combines equations (4) and (5):

$$\chi^{(3)} = \left(\frac{n}{n_{(CS2)}}\right)^2 \times \frac{Leff._{(CS2)}}{Leff.} \times \sqrt{\frac{(T/T0)}{(T/T0)_{(CS2)}}} \tag{6}$$

$\chi^{(3)}$ of PDA Microcrystals in OKS Response. Table IV summarizes concentration *(c)*, crystal size, and optical data of PDA microcrystal dispersions *(36,37)*. Dispersion concentrations were of the order 10^{-3} M for the polydiacetylenes. Crystal size was diverse to some extent and Table IV lists representative values for each sample. The absorption maximum of dispersions of poly(14-8ADA) is 645 nm. Those of poly(DCHD) varied depending on microcrystal size, i. e. absorption maxima of the crystals with 70, 100 and 150 nm in size are 640, 646 and 652 nm, respectively. Since there are no apparent absorption at both 813 and 1064 nm for these microcrystals, decrease of transmittance *(T)* is mainly due to scattering loss in the dispersion state. The transparency of poly(14-8ADA) dispersions at the wavelengths used is larger than that of poly(DCHD) dispersions. Enhancement by two-photon absorption can also be ignored because the transmittance did not change with increasing gate beam intensity in the range used in this study. Thus, for a pulsed laser with a low repetition rate, thermal effect of the laser beam should be negligible under our experimental conditions.

Table IV. Concentration(c), Crystal Size and Optical Data of PDA Microcrystals Dispersed in Water

Microcrystals	c / M	Crystal size / nm	T at 813 nm / %	T at 1064 nm / %	$\chi^{(3)}$ / esu	$\chi^{(3)} c^{-1}$ / esuM^{-1}
Poly(14-8ADA)	1.0×10^{-3}	300	43	>90	0.86×10^{-12}	0.86×10^{-9}
Poly(DCHD)	2.3×10^{-3}	70	1.6	17	1.4×10^{-12}	0.61×10^{-9}
Poly(DCHD)	2.3×10^{-3}	100	1.6	17	2.2×10^{-12}	0.93×10^{-9}
Poly(DCHD)	1.6×10^{-3}	150	0.03	16	1.6×10^{-12}	1.0×10^{-9}

Measured $\chi^{(3)}$ values range from 10^{-13} to 10^{-12} esu and the largest $\chi^{(3)}$ is 2.2×10^{-12} esu for poly(DCHD) (*37*). This is slightly less than that of the maximum $\chi^{(3)}$ so far reported for organic dyes in solution. However, the $\chi^{(3)}$ values normalized by the concentration for the dispersion system are 10^{-9} esuM^{-1}, about two orders of magnitude larger than those for any other solution systems. These results suggest that microcrystal dispersion systems of large $\chi^{(3)}$ compounds are applicable for OKS materials when the concentration increases without increasing scattering loss. Though a size effect on absorption maximum has been observed for poly(DCHD) microcrystals, clear differences between $\chi^{(3)}$ values on these samples were not observed in this study. If the obtainable microcrystal size reaches several nanometers, quantum confinement effects which have been observed in inorganic semiconductors, may appear in these organic materials. Another virtue expected for microcrystal dispersion systems is ultrafast response since molecular orientation effects seem to be negligible due to the larger mass of microcrystals compared to molecules. However, in this study, it was impossible to measure the ultrafast response, because the pulse width of the laser used was of the order of 10^{-9} sec. This is currently under investigation.

CONCLUSIONS

We have established an effective and simple method for preparing a variety of organic microcrystals in water. The solid-state polymerization of 4BCMU microcrystals was estimated to proceed from one end to the other end. The possibility of preparing PDAs with controlled molecular weight was qualitatively demonstrated. In the case of microcrystals of 14-8ADA, size-dependent conversion is found and can be explained by the looseness or thermal vibrations of the crystal lattice.

Regarding the optical properties size was clearly found to affect the absorption maxima of PDA microcrystals. If crystal sizes down to several nanometers can be achieved, they are expected to show interesting nonlinear optical properties. Moreover, we found that organic microcrystal dispersions are applicable for OKS devices, with actual $\chi^{(3)}$ values and those normalized by concentration for PDA dispersions are $10^{-13} \sim 10^{-12}$ esu and 10^{-9} esuM^{-1}, respectively. Further efforts to increase dispersion concentration will establish microcrystallization as a new technique to incorprate crystalline compounds into large isotropic media. In addition, microcrystals composed of single chains of conjugated polymers from one end to the other could be useful for making molecular devices as well.

ACKNOWLEDGEMENTS

We thank Drs. H. Kanbara and T. Kaino for OKS measurements of PDA microcrystals and for useful discussions. We express our sincere gratitude to Dr. H. Matsuda, of National Institute of Materials and Chemical Reasearch; Drs. H. S. Nalwa and A. Kakuta, of Hitachi Labolatory; Drs. S. Okada, H. Oikawa, and R. Iida, of Institute for Chemical Reaction Science, Tohoku University, for collaborative research. These works were mostly supported by the Center for Interdisciplinary Research, Tokoku University and the New Energy Development Organization, MITI.

LITERATURE CITED

(1) *Polydiacetylenes: Synthesis, Structure and Electronic Properties;* Bloor, D.; Chance, R. R., Eds.; Martinus Nijhoff: Dordrecht, 1985.
(2) Enkelmann, V. *Adv. Polym. Sci.* **1984,** *63,* 91.
(3) Wegner, G. *Z. Naturforsch* **1969,** *24b,* 824.
(4) Sauteret, C.; Hermann, J. P.; Frey, R.; Pradere F.; Ducuing, J.; Baughman, R. H.; Chance, R. R. *Phys. Rev. Lett.* **1976,** *36,* 956.
(5) Etemad, S.; Baker, G. L.; Soos, Z. G. *Molecular Nonlinear Optics;* Zyss, J., Ed.; Academic Press: San Diego, 1994, p. 433.
(6) Molyneux, S.; Matsuda, H; Kar, A. K.; Wherret, B. S.; Okada, S.; Nakanishi, H. *Nonlinear Opt.* **1993,** *4,* 299.
(7) Kajzar F.; Messier, J. *Polym. J.* **1987,** *19,* 275.
(8) Nakanishi, H.; Matsuda, H.; Okada, S.; Kato, M. *Polym. Adv. Technol.* **1990,** *1,* 75.
(9) Kanbara, H.; Asobe, M.; Kubodera, K.; Kaino, T.; Kurihara, T. *Appl. Phys. Lett.* **1992,** *61,* 2290.
(10) Townsend, P. D.; Jackel, J. L.; Baker, G. L.; Shelburne, J. A.; Etemad, S. *Appl. Phys. Lett.* **1989,** *55,* 1829.
(11) Duguay, M. A.; Hansen, J. W. *Appl. Phys. Lett.* **1969,** *15,* 192.

(12) Depasse, J.; Watilon, A. *J. Collid Interface Sci.* **1970,** *33,* 430.
(13) van de Hulst, H. C. *Light Scattering by Small Particles*; Wiley: 1957.
(14) Iijima, S.; Ichihashi, T. *Jpn. J. Appl. Phys.* **1985,** *24,* L125.
(15) Iwama, S.; Hayakawa, K.; Arizumi, T. *J.Cryst.Growth* **1984,** *66,* 189.
(16) Buffat, P.; Borel, J. P. *Phys. Rev. A* **1976,** *13,* 2287.
(17) Hanamura, E. *Phys. Rev. B* **1988,** *37,* 1273.
(18) Brus, L. E. *J. Chem. Phys.* **1984,** *80,* 4403.
(19) Ekimov, A. I.; Efros, Al. L.; Onushchenko, A. A. *Solid State Commun.* **1985,** *56,* 921.
(20) Nakamura, A.; Yamada, H.; Tokizaki, T. *Phys. Rev. B* **1989,** *40,* 8585.
(21) Wang, Y. *J. Opt. Soc. Am. B* **1991,** *8,* 981.
(22) Toyotama, H. *Kinouzairyo* **1987,** *6,* 44 (in Japanese).
(23) Yase, K.; Inoue, T.; Okada, M.; Funada, T.; Hirano, J. *Hyomen Kagaku* **1989,** *8,* 434 (in Japanese).
(24) Kasai, H.; Nalwa, H. S.; Oikawa, H.; Okada, S.; Matsuda, H.; Minami, N.; Kakuta, A.; Ono, K.; Mukoh, A.; Nakanishi, H. *Jpn. J. Appl. Phys.,* **1992,** *31,* L1132.
(25) Chang, T. Y. *Opt. Eng.* **1981,** *20,* 220.
(26) Kanbara, H.; Kobayashi, H.; Kaino, T.; Kurihara T.; Ooba, N. *J. Opt. Soc. Am. B.,* in press.
(27) Tieke, B.; Bloor, D.; Young, R. J. *J. Mater. Sci.* **1982,** *17,* 1156.
(28) Iida, R.; Kamatani, H.; Kasai, H.; Okada, S.; Oikawa, H.; Matsuda, H.; Kakuta, A.; Nakanishi, H. *Mol. Cryst. Liq. Cryst.* **1995,** *267,* 95.
(29) Nalwa, H. S.; Kasai, H.; Okada, S.; Matsuda, H.; Oikawa, H.; Minami, N.; Kakuta, A.; Ono, K.; Mukoh, A.; Nakanishi, H. *Polym. Adv. Technol.* **1995,** *6,* 69.
(30) Iida R., Master thesis, Tohoku Univ. (1994), to be published.
(31) Fujitsuka, M.; Nakahara, R.; Iyoda, T.; Shimidzu, T.; Tsuchiya, H. *J. Appl. Phys.* **1993,** *74,* 1283.
(32) Tokizaki, T.; Akiyama, H.; Tanaka, M.; Nakamura, A. *J. Cryst. Growth* **1992,** *117,* 603.
(33) Nakanishi, H.; Kasai, H. *Koubunshikakou* **1996,** *45,* 15 (in Japanese).
(34) So, F. F.; Forrest, S. R.; Shi, Y. Q.; Steier, W. H. *Appl. Phys. Lett.* **1990,** *56,* 674.
(35) Kasai, H.; Kamatani, H.; Okada, S.; Oikawa, H.; Matsuda, H.; Nakanishi, H. *Jpn. J. Appl. Phys.* **1996,** *35,* L221.
(36) Kasai, H.; Kanbara, H.; Iida, R.; Okada, S.; Matsuda, H.; Oikawa, H.; Nakanishi, H. *Jpn. J. Appl. Phys.* **1995,** *34,* L1208.
(37) Kasai, H.; Iida, R.; Kanbara, H.; Okada, S.; Matsuda, H.; Oikawa, H.; Kaino, T.; Nakanishi, H. *Nonlinear Opt.* **1996,** *15,* 263.

Chapter 14

Third-Order Nonlinear Susceptibility of Polydiacetylene-Containing Polymeric Systems

M. P. Carreón[1], L. Fomina[2], S. Fomine[2], D. V. G. L. N. Rao[3], F. J. Aranda[3], and T. Ogawa[2,4]

[1]Instituto de Ciencias Nucleares and [2]Instituto de Investigaciones en Materiales, Universidad Nacional Autónoma de México, Apartado Postal 70–360, Ciudad Universitaria, Coyoacán, Mexico DF 04510, Mexico
[3]Department of Physics, University of Massachusetts, Boston, MA 07160

Unconventional polydiacetylenes which include polydiacetylene-containing polymers, amorphous polydiacetylenes and polydiacetylene microcrystal-host polymer composites, are described. These methods give films of high optical quality by simple casting or spin coating. Various new diacetylene molecules were synthesized and their polymerization is discussed. The third-order nonlinear optical susceptibility of these polydiacetylenes, determined by degenerated four wave mixing using 532 nm laser, lies in the range of 10^{-11} - 10^{-9} esu.

Since Wegner (*1*) reported in 1969 the unique polymerization of some diacetylenes in the solid state, much attention has been paid to the subject, and many studies have been reported in the literatures (*2*). The polydiacetylenes (PDAs) thus obtained are completely crystalline materials having conjugated chains consisting of a triple-single-double-single bond linkage sequence, and they possess third order nonlinear optical (NLO) susceptibility (*3*). However, the crystalline PDAs are a group of polymers for which processing into films is not simple, and hence various techniques have been applied to obtain thin films of PDAs, including: (1) Single crystals; (2) Langmuir-Blodgett (LB) membranes of aliphatic amphiphilic DAs; (3) Vacuum evaporation epitaxy; (4) Solution casting; and (5) Photodeposition from solution (*4*). The third order susceptibility, $\chi^{(3)}$ of the PDA films thus prepared have been reported in literature (*5*). However, the above mentioned techniques have serious limitations. In general, it is not easy to obtain defectless large single crystals of diacetylenes which undergo topochemical polymerization, although a few have been reported in the literature (*6,7*). The LB membrane technique has been investigated by many (*8,9*) and shown to be a useful method to obtain ultra thinfilms of PDAs, and $\chi^{(3)}$ values of such PDA films have been reported to be in the range of 10^{-12} - 10^{-11} esu (*5*). These values are not satisfactorily high for NLO applications, and the method is rather too

[4]Corresponding author

laborious for preparation of thick films. The vacuum evaporation-deposition technique is limited as not many diacetylenes are volatile. On the other hand, the solution casting of soluble PDAs tends to give opaque films due to crystallization into polycrystalline films.

Alternative methods to obtain PDA films, which are technically much simpler than the above mentioned, include: (6) Synthesis of processable polymers which contain diacetylenes, and developing PDA networks in the polymer films; (7) Polymer-PDA composites: PDA micro crystals dispersed in amorphous polymers; (8) Polymerization of diacetylenes in the molten state. Studies on the synthesis, characterization, and properties of diacetylene (DA)-containing polymers have been recently reviewed (*10, 11*). Some of these DA-containing polymers can be cast into films, and the DA groups in the polymer films are cross-polymerized to develop PDA networks as shown in Scheme 1. However, similar to the case of DAs, not every DA group incorporated in polymers is topochemically polymerizable, some being highly radiation-sensitive and others being inert to radiation.

It is interesting to attempt to disperse microcrystals of light sensitive DAs in transparent amorphous polymers to obtain PDA-containing transparent films. It is possible to prepare films with excellent optical quality by this method (*12*). If the crystals or particles of PDAs dispersed in the media, are smaller than the wavelength of the light, the materials are transparent. Prasad (*13*) has attempted to fill the micropores in transparent silica with organic functional materials, and Nakanishi *et al* have reported a system consisting of microemulsion of PDA crystals in water (*14*).

Some DAs can be readily polymerized by heating them at temperatures above their melting points and the polymerized products are highly transparent, red, glassy materials. It is noteworthy that many light insensitive DAs are readily polymerized in the molten state, whereas the majority of light sensitive DAs are not. An important advantage of polymerization in the molten state is that one can obtain transparent materials with extremely high optical quality and with any desired dimension. In this article, our laboratory's recent results on synthesis, characterization, and third-order NLO susceptibility of PDA-containing systems, are described.

EXPERIMENTAL SECTION

Synthesis.

DA-containing polymers. The polymers which contain DA groups in their main chains can be prepared by (1) direct condensation or addition polymerization of bifunctional monomers which contain DA groups, or (2) oxidative coupling polymerization of terminal bisacetylenic monomers. Because of the reactive DA groups, the first method can be performed only under mild conditions, principally in solutions, and the molecular weights of polymers obtained are generally not very high. Using the second method, polymers with high molecular weights can be obtained with quantitative yields under mild conditions (below 80°C) in solvents in which the polymers are soluble, provided that the monomers and catalysts are highly pure.

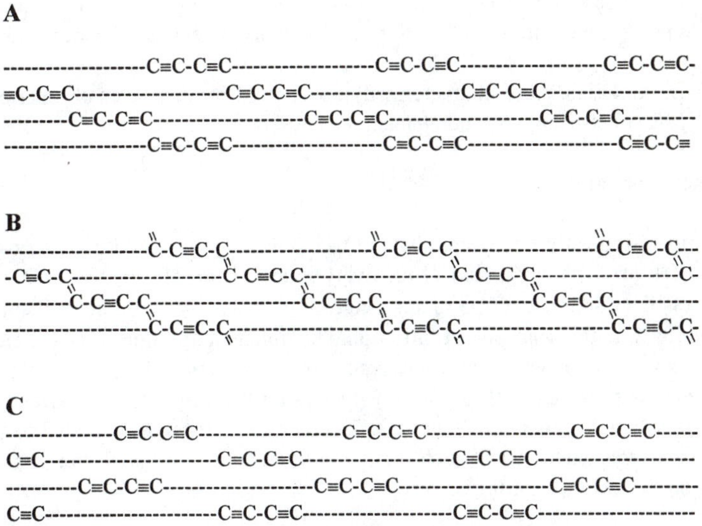

A

```
-------------------C≡C-C≡C------------------C≡C-C≡C------------------C≡C-C≡C-
≡C-C≡C-------------------C≡C-C≡C-------------------C≡C-C≡C------------------
---------C≡C-C≡C-------------------C≡C-C≡C-------------------C≡C-C≡C---------
-------------------C≡C-C≡C-------------------C≡C-C≡C-------------------C≡C-C≡
```

B

```
-------------------C-C≡C-C------------------C-C≡C-C------------------C-C≡C-C---
-C≡C-C-------------------C-C≡C-C-------------------C-C≡C-C------------------C-
---------C-C≡C-C-------------------C-C≡C-C-------------------C-C≡C-C-----------
-------------------C-C≡C-C-------------------C-C≡C-C-------------------C-C≡C-C-
```

C

```
---------------C≡C-C≡C------------------C≡C-C≡C------------------C≡C-C≡C------
C≡C-----------------------C≡C-C≡C-------------------C≡C-C≡C--------------------
----------C≡C-C≡C-------------------C≡C-C≡C-------------------C≡C-C≡C---------
C≡C-----------------------C≡C-C≡C-------------------C≡C-C≡C--------------------
```

Scheme 1. Schematic models of DA-containing polymer chains.
A: Before cross-polymerization, **B:** PDA networks. **C:** Light insensitive model.

However, it is often difficult to remove trace amounts of copper adsorbed by the polymer. The synthesis, characterization, and some properties of the light-sensitive DA-containing polymers, **1** and **2**, employed for this study have been reported previously by the authors (*15,16,17,18,19*). A few novel aromatic DA-containing, light-insensitive polyesters, **3** and **4**, were also synthesized. Their structures are shown in Chart 1.

A few methacrylates (**5**) and dimethacrylate (**6**) which contain aliphatic DA groups were synthesized (*20*) and are shown in Chart 2. In order to compare the NLO properties of PDA-containing polymers with those of the PDAs obtained by the molten state polymerization, several novel diacetylenes were synthesized, and their chemical structures are shown in Chart 2 for aromatic diesters (**7**, **8**) and aromatic diamide (**9**). These compounds were synthesized by the Heck reaction of the corresponding bromo compounds with trimethylsilylacetylene, as shown in Chart 3.

All of these materials were characterized and verified by thin layer chromatography, IR and NMR spectroscopies, and elemental analysis.

Preparation of samples.

PDA-Containing polymer films. The DA-containing polymer films were cast from solutions (chloroform for **1** and N-methylpyrrolidone for **2**). In the case of **2** the solvent was evaporated at 60°C under reduced pressure. When a heated spin coater was used for **2**, highly transparent films could be obtained, but due to their amorphous nature, no crosspolymerization took place on irradiation. In the case of **1**, the transparency (crystallinity) of the films depended on the number of methylene units (x and y). For example, when x is 2 and y is 5, the film is always completely opaque, and in other cases reasonably transparent films can be obtained, although the transparency depends on the casting conditions.

PDA-Polymer mixed systems. N,N'-di-*n*-alkylocta-3,5-diynylenediurethanes were synthesized by the oxidative coupling reaction of N-n-alkyl-3-butynylurethanes obtained by the reaction of corresponding isocyanates with 3-butyn-1-ol. n-Octyl, n-butyl and ethyl diurethanes were synthesized, but only the n-octyl diurethane was used because of its higher miscibility with the host polymer. Various polymers such as poly(methylmethacrylate) and poly(vinyl acetate) were tested, but it was found that poly(N,N-dimethylaminoethyl methacrylate) gave the best results. Required amounts of the diacetylenic diurethane and the host polymer were dissolved in chloroform, and films were cast or spin coated on quartz panes. The films were then irradiated with UV light or electron beam (*12*).

Amorphous PDAs. The cast films of **1** were heated between two quartz panes (1" x 1", 1 mm thick) at above their melting points, upon which thermal cross-polymerization of DA groups took place and orange to red brown transparent materials were obtained. It seems that simultaneous irradiation with UV light (from a medium pressure Hg lamp) helps the thermal polymerization. Some of the polymers **2** only undergo thermal polymerization in the amorphous state (*19*).

Chart 1

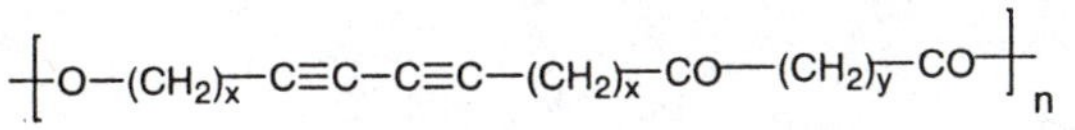

$$\left[O-(CH_2)_x-C{\equiv}C-C{\equiv}C-(CH_2)_x-CO-(CH_2)_y-CO \right]_n$$

1

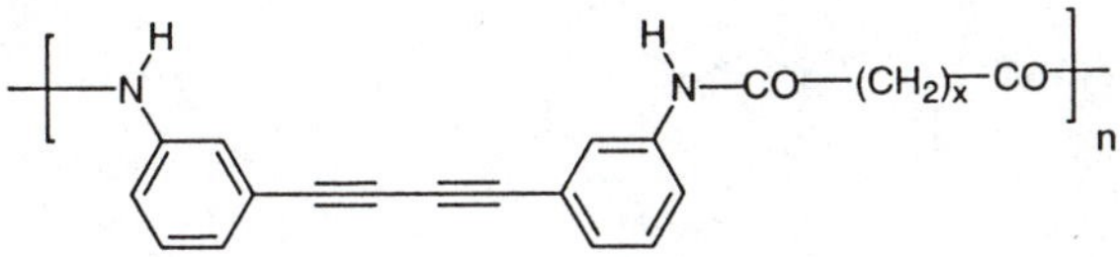

2

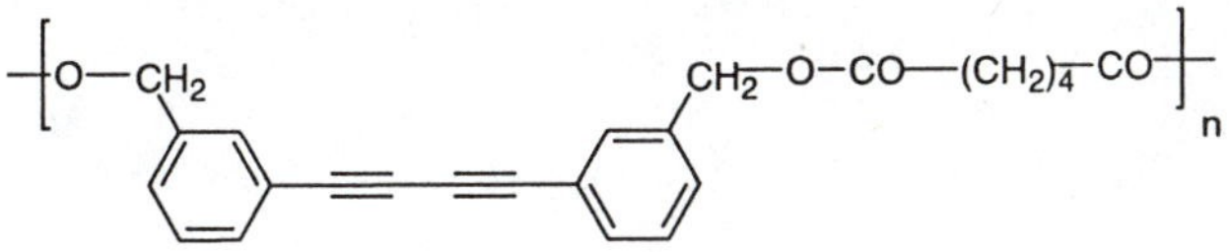

3

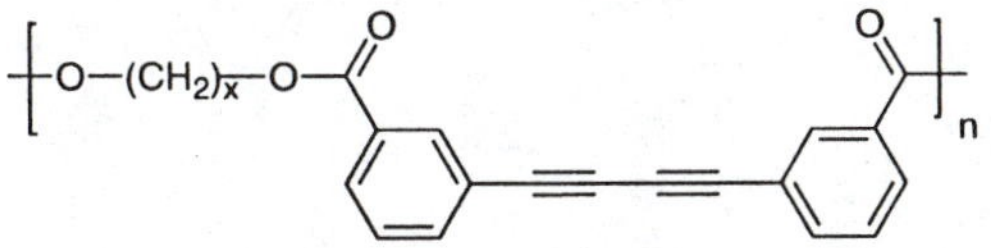

4

Chart 2

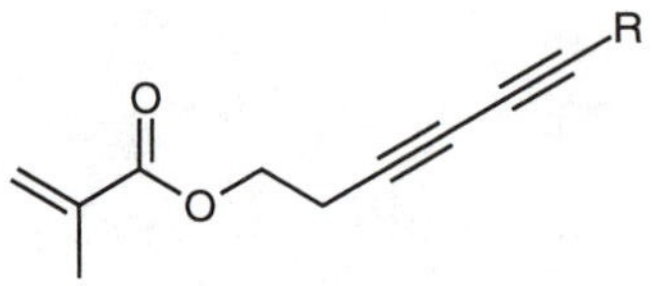

5

R = ^{n}Bu, ^{n}Hex, and ^{n}Oct

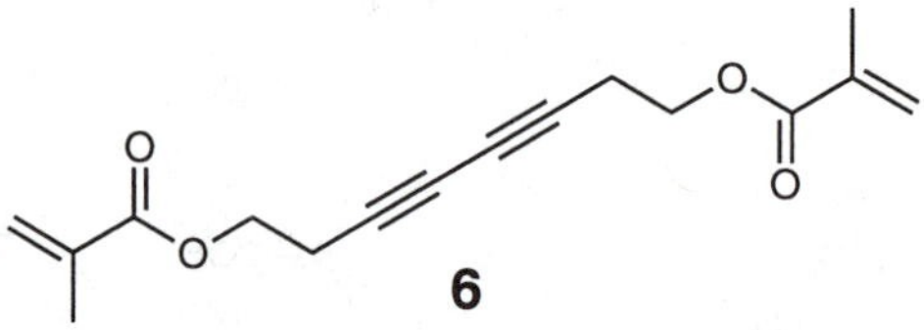

6

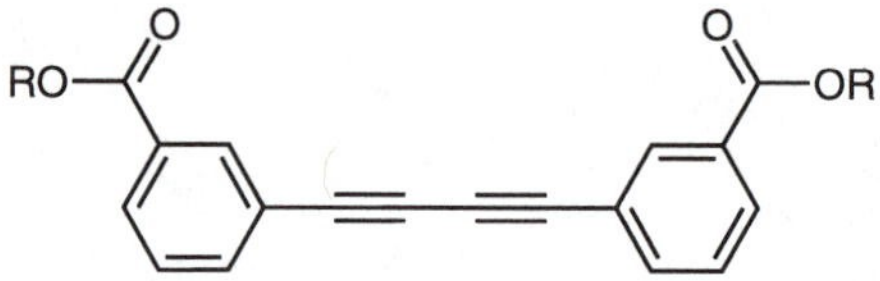

7

8

9

Chart 3

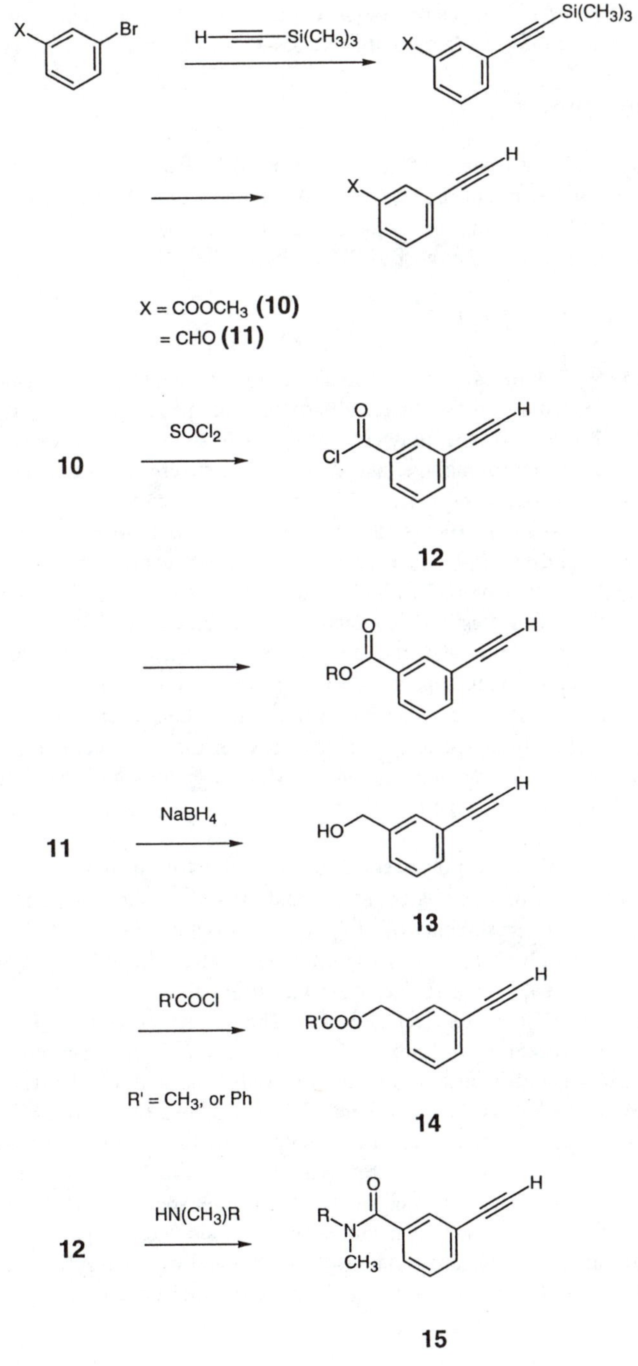

X $-C_6H_4-$Br $\xrightarrow{H-C\equiv C-Si(CH_3)_3}$ X $-C_6H_4-C\equiv C-Si(CH_3)_3$

$\longrightarrow$ X $-C_6H_4-C\equiv C-H$

X = $COOCH_3$ **(10)**
= CHO **(11)**

10 $\xrightarrow{SOCl_2}$ $ClC(=O)-C_6H_4-C\equiv C-H$

12

$\longrightarrow$ $ROC(=O)-C_6H_4-C\equiv C-H$

11 $\xrightarrow{NaBH_4}$ $HOCH_2-C_6H_4-C\equiv C-H$

13

$\xrightarrow{R'COCl}$ $R'COOCH_2-C_6H_4-C\equiv C-H$

R' = CH_3, or Ph

14

12 $\xrightarrow{HN(CH_3)R}$ $R-N(CH_3)-C(=O)-C_6H_4-C\equiv C-H$

15

The DA groups of polymers **3** and **4** are not light sensitive and they polymerize only in the molten state. They were melted between two quartz windows and irradiated with UV light. The DAs **3-9** were polymerized between two quartz windows at temperatures above their melting points with simultaneous irradiation of UV light.

$\chi^{(3)}$ Measurements.

The third order NLO susceptibility was determined by degenerate four wave mixing using a pico second laser consisting of a mode-locked Quantel Nd:Yag laser that was frequency doubled to 532 nm. It should be noted that some $\chi^{(3)}$ values thus observed are resonance enhanced as the most of PDAs absorb the 532 nm light.

RESULTS AND DISCUSSION

DA-Containing Polymers. The crystallinity of DA-containing polymers differs tremendously depending on the chemical structure of spacer groups. In the cases of polydianilides **2** (*18, 19*), the polymer with $x = 3$ and a copolymer were totally amorphous and not photosensitive, and they underwent crosspolymerization only by heating at 180°C, while the polymers with $x = 2$, 4 and 8, and some copolymers gave crystalline films in which microcrystallites of about 30 Å were dispersed, and the films become blue or bluish purple on irradiation. The $\chi^{(3)}$ values of these films were determined and they were found to lie between 10^{-10} - 10^{-9} esu (*21*). The irradiated microcrystalline films of these DA-containing polyamides have absorption maxima at around 560 and 610 nm, and the amorphous thermally treated films have no absorption maximum, but the absorption tails from shorter wavelength region up to 650nm. This means that the measured $\chi^{(3)}$ values are resonant enhanced. It is worth mentioning that the difference in $\chi^{(3)}$ values between the crystalline and amorphous films is about an order of magnitude, and that the amorphous films have much better transparency than the crystalline films.

In the case of the aliphatic polyesters **1**, the cast films irradiated by electron beam develop yellow to red orange colors (*15-17*) and their $\chi^{(3)}$ values range from 0.5 to 7 x 10^{-10} esu (*21*). The crystallinity of the polyester **1** with x=1 is less (32%) than that of the polymer with x=2 (64%). This difference causes the difference in the $\chi^{(3)}$ values, the former being 0.5 x 10^{-10} esu and the latter being 1.2 x 10^{-10} esu, with both films irradiated with 50 Mrads electron beam. The polyesters with $y = 7$ and $x = 1$, 2, 3, 4 and 9 were recently synthesized (*22*). When x is 2 the cast polymer film is completely opaque and it became orange on irradiation, and when x is 3, the films became blue on irradiation with UV light. Figure 1 shows the absorption spectra of polyester films irradiated or heated. The absorption in the region of the wave length shorter than 400 nm is mainly due to scattering, as the films are not completely transparent but contain large crystallites. The absorption spectra A, B, C, and D are of a copolyester obtained from the following two monomers; dipropargyldecamethylenedicarboxylate (40 mol %) and dipropargylterephthalate (60 mol %). The copolyester forms a liquid crystal phase and the DA groups of the

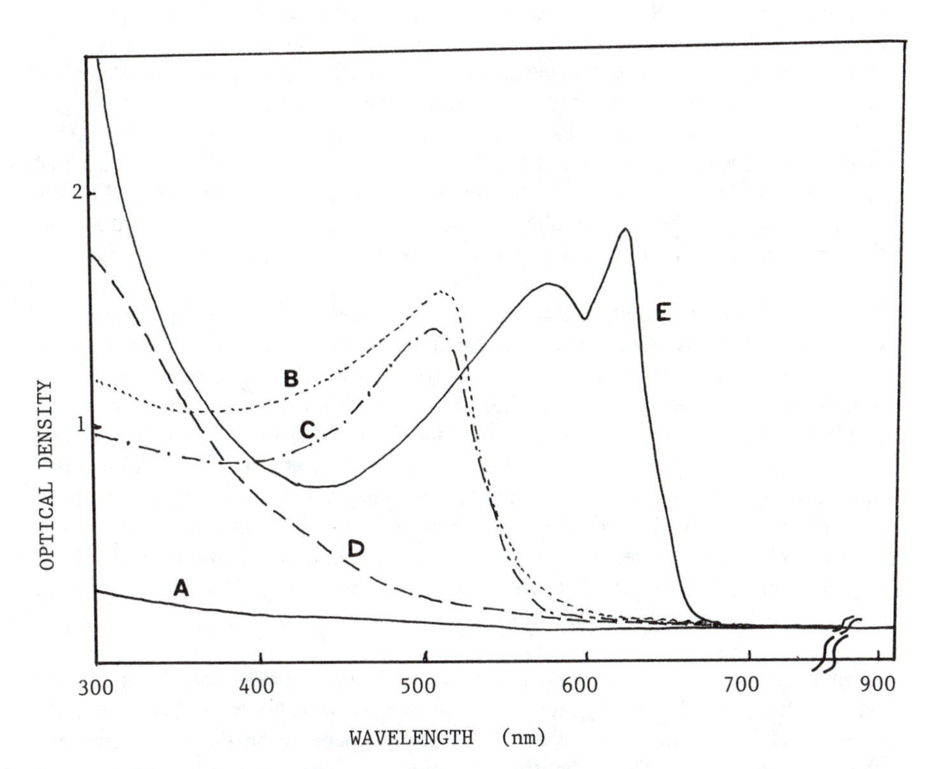

Figure 1. UV-Visible absorption spectra of some cross-polymerized DA-containing polyesters.
A: Copolymer before cross-polymerization. **B:** Copolymer irradiated with electron beam (50 Mrads) at 20°C. **C:** Copolymer irradiated at 60°C (Liquid crystalline state) with UV light. **D:** Copolymer heated at 100°C (molten state). **E:** Polyester 1 (x=3, y=7) irradiated with UV light.

copolymer are photosensitive and the film becomes orange by irradiation. The spectra show that the PDA structures formed in the solid state and liquid crystalline state polymerizations are same.

The blue film of polyester from 1-pentyn-3-ol and azelaic acid had a $\chi^{(3)}$ value of 3 x 10^{-10} esu, which is somewhat higher than that of the red orange films (*22*). IR spectra of these irradiated polyester films indicate that almost all the DA groups in the polymers had been consumed for cross-polymerization by the irradiation with 50 Mrads electron beam. The topochemical polymerization is supposed to take place only in the crystalline state, and therefore it seems that both 1,4- and 1,2-cross-polymerizations could take place also in the amorphous region for these DA-containing polymers when irradiated with high energy radiation or heated at elevated temperatures. The IR spectra of the amorphous state cross-polymerized films and those of solid state polymerized films are often identical indicating the structure of the former consists mainly of the 1,4-polydiacetylene structure (Figure 2).

The crystallinity and transparency of these films depend on the film preparation conditions. In the case of polymers **2**, rapid evaporation of the solvent on a heated spin coater gives completely amorphous, transparent films, but no topochemical polymerization of the DA groups takes place. In the case of polymers **1** it is also possible to obtain completely transparent films by rapid evaporation of the solvent, however, such amorphous films do not form appreciable amounts of PDA networks on irradiation and thus their $\chi^{(3)}$ are negligible. This is a significant disadvantage of the polymer films which contain topochemically polymerizable DA groups, when optical applications are desired. Recently, it was found in our laboratory that solvent impregnation into the cross-polymerized, semi-crystalline PDA films, improve considerably optical quality of the films.

From these results, it can be concluded that the $\chi^{(3)}$ values of PDA containing polymer thin films measured by the degenerate four wave mixing method at 532 nm, are in the order of 10^{-10} esu and do not vary significantly depending on the crystallinity and chemical structures. These factors can influence only within the same order of magnitude, and the values of 10^{-10} esu seem to be the characteristic value for the PDA networks.

Amorphous PDAs. Yu et al.(*23*) prepared poly(hexa-2,4-diynylene terephthalate), which is not photosensitive, but does polymerize by heating at 150°C. A $\chi^{(3)}$ value of 3.2 x 10^{-10} esu (determined the degenerate four wave mixing technique at 532 nm.) has been reported for this material. The polymers **3** and **4** (Chart 4) are not photosensitive, but underwent cross-polymerization when heated at 180°C (in the molten state) for 2.5 hours with simultaneous UV irradiation, giving red transparent materials. The $\chi^{(3)}$ values for these materials were found to be 1.9 - 3.5 x 10^{-10} esu for polymers **3** and 2.7 -2.9 x 10^{-10} esu for polymers **4**. Absorption spectra of one of the polymers **4** are shown in Figure 3. The films have an absorption maximum at 400 nm and a trough at 340-350 nm, but absorption tails down towards 700 nm due to their amorphous nature.

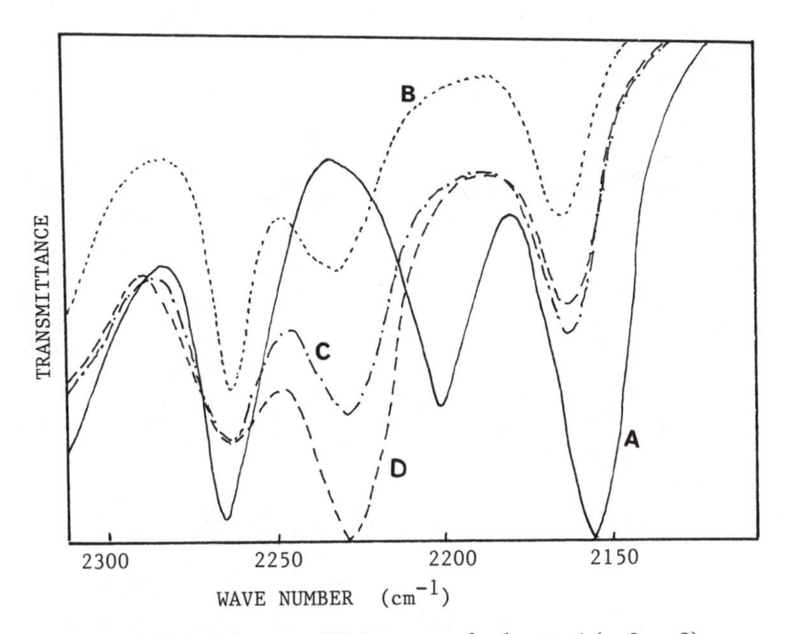

Figure 2. FT-IR Spectra of DA groups of polyester **1** (x=2, y=8). **A:** Before cross-polymerization. **B:** Irradiated with UV light for 30 minutes at room temperature. **C:** Heated at 90°C for 96 hours (solid state thermal cross-polymerization). **D:** Heated at 180°C for 4 hours (amorphous state cross-polymerization).

Chart 4

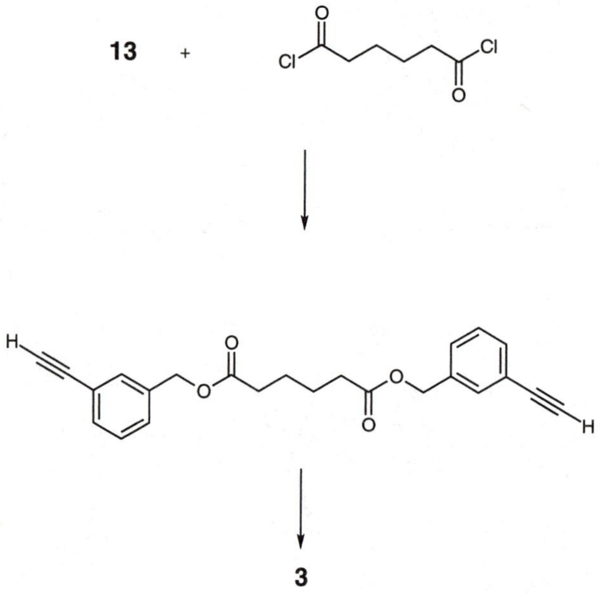

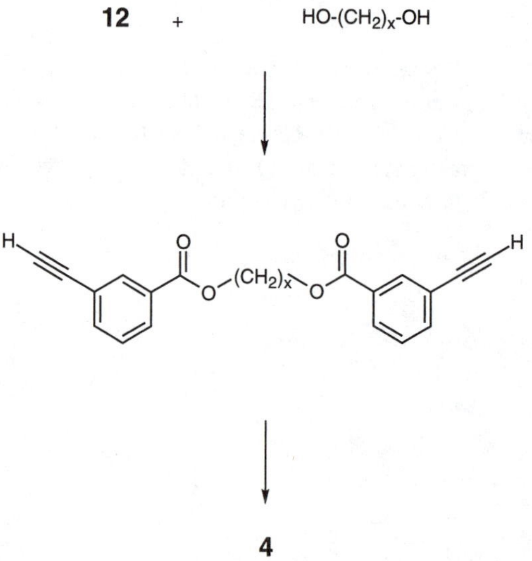

13 + Cl O O Cl

3

12 + $HO-(CH_2)_x-OH$

4

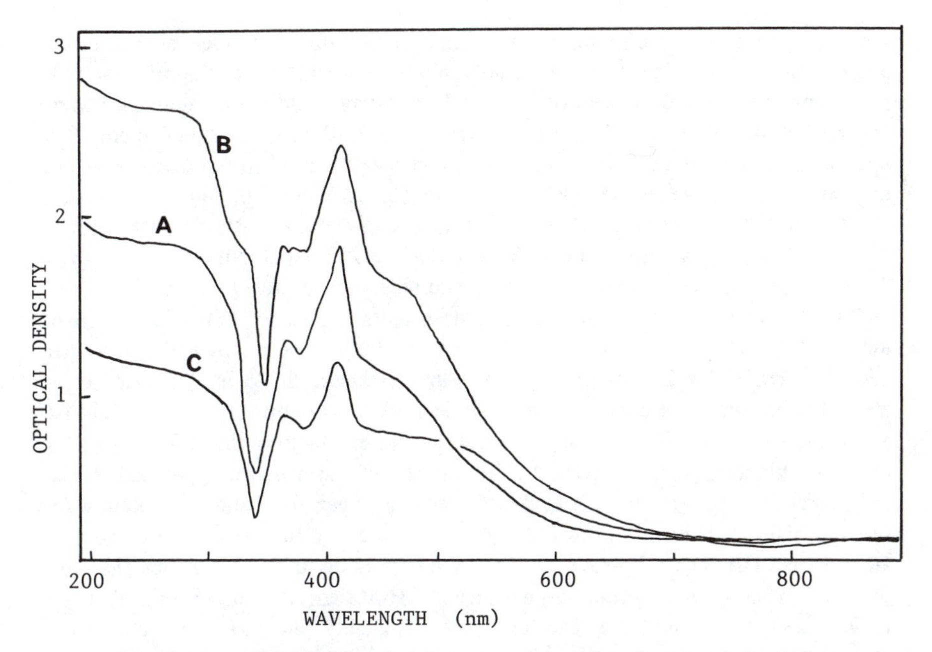

Figure 3. UV-Visible absorption spectra of polyester **4** (x=10) heated at 180°C.
A: 1 hr, **B:** 2 hrs, **C:** 3 hrs. Film thickness: 0.011 mm.

The aliphatic DAs linked to methacrylate groups **5** and **6** undergo rapid polymerization; the vinyl polymerization is followed by the diacetylenic polymerization, resulting in yellow, highly transparent, glassy materials. The $\chi^{(3)}$ values for these materials obtained by heating at 140°C for 60 minutes (almost total consumption of DA groups in the molecules) (*20*) were found to be on the order of 5 -6.5 x 10^{-12} esu, independently of the types: Polymer from **5** R= n-Bu: 6.5, R= n-Hex: 6.2, R= n-Oct: 5.7; Polymer from **6** : 5.2 x 10^{-12} esu. The visible absorption spectra of these materials do not have maxima, but the spectra decrease and show almost no absorption at the wavelengths greater than 400 nm. It seems that the materials consist of mixtures of short conjugated units.

Since the DA groups incorporated in polymer chains are not topochemically cross-polymerizable, it is worth studying the bulk, molten state polymerization of these DAs for comparison. Therefore, aromatic diacetylenic esters **7** and **8** and an amide **9** were prepared, and they were polymerized at 180°C for 2.5 hours with simultaneous UV irradiation. These amorphous materials are highly transparent to the naked eyes and they are deep red when the thickness of the film is about 0.05 mm. They have absorption maxima around 420 nm but the spectra continuously decease towards 600 nm. The $\chi^{(3)}$ values of films of **8** with thickness of 0.040 mm were found to be 4.2 x 10^{-10} and 3.5 x 10^{-10} esu for R'= CH_3 and phenyl, respectively. The films from **7** with thicknesses of 0.027-0.030 mm were deep red and their $\chi^{(3)}$ values could not be measured due to excessive absorption. The DA **9** underwent polymerization when heated at 180°C for 2.5 hrs giving a red film (thickness: 0.038 mm) which had an absorption maximum around 420 nm, with the peak tailing down to 700 nm. It has a $\chi^{(3)}$ value of 1.1 x 10^{-9} esu and an excellent transparency to the naked eyes. An aromatic, highly conjugated DA **14** (Chart 4), was synthesized (*24*) and it was polymerized in the molten state at 225°C under nitrogen for 3 min. to obtain a red film on glass. This material showed a $\chi^{(3)}$ value of 5.6 x 10^{-10} esu. These aromatic amorphous PDAs generally have absorption in the region 600-700 nm, and therefore their $\chi^{(3)}$ values are resonance enhanced. However, it can be said that the polymerized materials have excellent optical quality due to their completely amorphous nature, and that their values are comparable with those of PDA-containing polymer films.

Z-scan experiment was carried out for some amorphous PDAs using wavelength of 1064 nm. A typical example is shown in Figure 4. Two photon absorption coefficient, β, was found to be 2.7 cm/GW.

PDA-Polymer Composite Systems. If microcrystals of PDAs can be homogeneously dispersed in an amorphous, transparent polymer, and if the crystal size is smaller than the wavelength of the applied laser, the materials can probably have applications. An example of this was previously mentioned (*12*) and the $\chi^{(3)}$ values were measured (*25*) for the system consisting of N,N'-dioctylocta-3,5-diynylenediurethane dispersed in poly(N,N-dimethylaminoethyl methacrylate). Up to

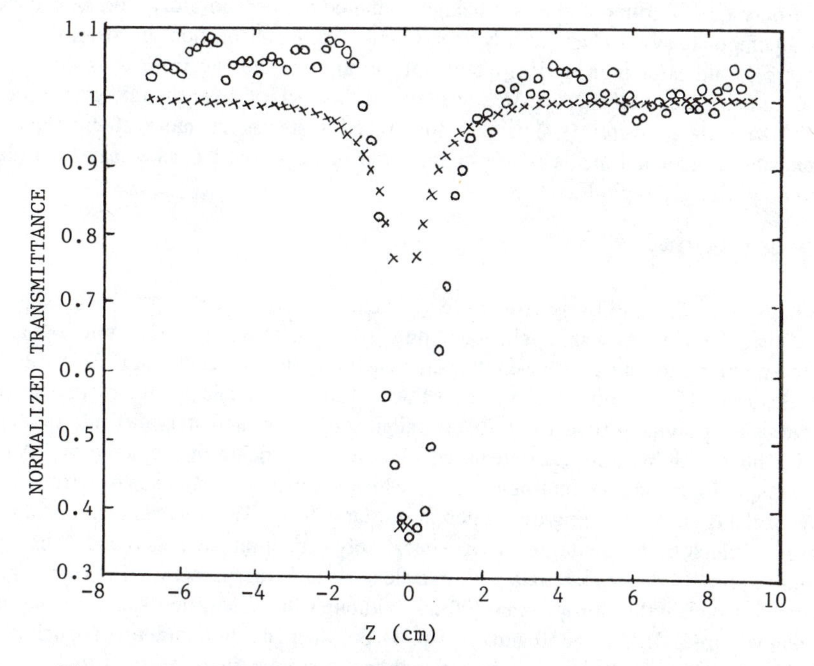

Figure 4. Z-Scan experimemt of 4,4'-butadiynylenedi-*n*-nonyldibenzoate polymerized at 190°C for 1 hr between 2 quartz windows.
x : calculated. O : observed.

values of 2.0 - 2.9 x 10^{-10} esu, almost independent of the applied dose (5 - 100 Mrads), indicating that about 20 Mrads are sufficient to polymerize the DA crystals dispersed in the host polymer (*25*). Amounts of the DA higher than 20% causes formation of larger crystals and the films have poor transparency. The irradiated films have an absorption maximum around 550 nm. The optical quality of these films, however, was not satisfactory and their absorption spectra were rather broad over 400 - 700 nm with a maxima at 550 nm. This broadness seems to be due to wide distribution of crystal size. By selecting suitable combinations of DAs and host polymers, it is possible to obtain high optical quality films which contain PDA microcrystals. Further studies are being conducted in our laboratory, and an excellent nanocomposite system has recently been found (*26*). Its absorption spectrum shown in Figure 5 indicates high optical quality of the film, as the spectrum is as defined as those of single crystal PDAs. An advantage of this method over the microemulsion of PDA crystals in water (*14*) is that the mixed systems are more stable than the emulsion in aqueous media where coagulation of dispersed particles may take place over a prolonged period.

CONCLUSIONS

As described these PDA systems have $\chi^{(3)}$ values on the order of 10^{-10} esu. Some of the values are resonant enhanced due to absorption of 532 nm radiation. Measurements at longer wavelengths such as 1064 nm, are needed in order to further evaluate the NLO properties of these PDAs. The topochemically polymerizable DA-containing polymers tend to form crystalline polymers, and it is difficult to obtain thick films with high optical transparency without sacrificing the crystallinity. A new technique for obtaining amorphous films which involves topochemically formed PDA networks is currently being developed in our laboratory. The polymers containing DA groups which do not undergo topochemical polymerization, such as **3** and **4**, can give orange to red, glassy materials with excellent optical transparency. Their absorption covers wavelengths shorter than 700 nm, although the absorption spectra have local minima around 330 - 350 nm. The DAs which do not undergo topochemical polymerization, but polymerize in the molten state, give similar results to the above materials. It should be mentioned that the DAs which polymerize at high temperatures are not useful because thermal decomposition often accompanies polymerization, thus decreasing the optical quality of materials. Therefore, it is desirable to polymerize the DAs rapidly below 200°C.

These systems, which consists of microcrystals of photosensitive DAs in host polymers, are interesting materials. Very little has yet been studied on such systems. DA microcrystals dispersed in an amorphous polymer can be converted to PDA microcrystals by irradiation, and when the crystal size is small compared with the wavelength of the laser, the materials may be useful for NLO applications. Systems with new combinations of DAs and host polymers are expected to appear in the future.

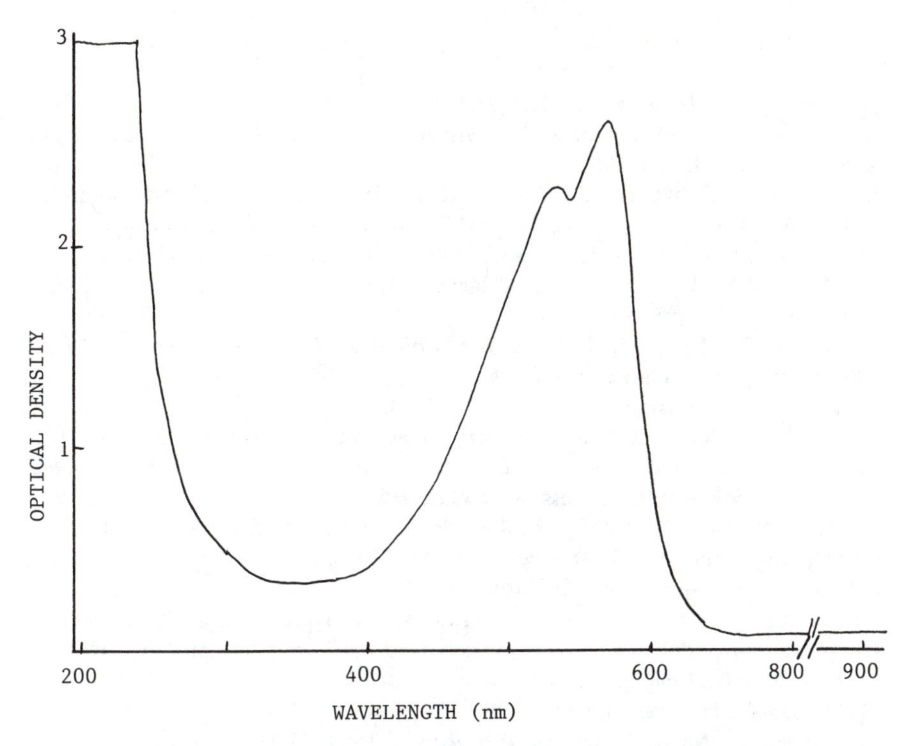

Figure 5. UV-Visible absorption spectra of a composite film of N,N'-di-*n*-butylocta-3,5-diynylene diurethane dispersed in poly(vinyl acetate-*co*-vinylpyrrolidone). 20% by weight DA/host polymer. Film thickness: 10 μm. UV-Irradiated for 2.5 hrs.

Safety considerations

The DAs are high energy compounds and their polymerization is highly exothermic. Therefore, heating a large amount at elevated temperatures may cause an explosive reaction. The topochemically polymerizable DAs are also sensitive to pressure. We tried to prepare filaments of polyester **1** (x=2, y=8) in an extruder, and when about 4 g of the polymer was melted in the extruder it suddenly exploded. Temperature (probably overheating) and pressure caused an explosion.

Acknowledgement

This study was supported by Grants from CONACyT (Consejo Nacional de Ciencia y Tecnologia) of Mexican Government, Grant Nos. 3595-0540E and 4264-E9406, and from DGAPA (Dirección General de Asuntos del Personal Académico) of our university, Grant Nos. IN-101492 and IN-101793.

Literature Cited

(1) Wegner, G. *Z.Naturforsch.* **1969**, *24B*, 824.
(2) Cantow, H.-J. Ed. *Advances in Polymer Science*, vol. 63, "Polydiacetylenes", Springer-Verlag: Berlin, 1984.
(3) (a) Bloor, D; Chance, R.R.. Eds. *Polydiacetylenes*, Martinus Nijihoff: Dordrecht 1985. (b) Chamla, D.S; Zyss, J., Eds. *Nonlinear Optical Properties of Organic Molecules and Crystals*," vol.2, Academic Press: Orlando, Florida, 1987.
(4) Paley, M.S.; Frazier, D.O.; Abdeldeyem, H.; Armstrong, S.; McManus, S.P. *J.Am.Chem.Soc.*, **1995**, *117*, 4775.
(5) Kajzar, F.; Messier, J.; in *Conjugated Polymers*, Brédas, J.L.; Silbey, R., Eds. Kluwer Academic Publishers: Dordrecht, 1991; pp. 517-541.
(6) Wegner, G., *Macromol. Chem.*, **1971**, *145,* 85.
(7) Yee, K.C.; Chance, R.R., *J.Polym.Sci., Polym.Phys.Ed.,* **1978**, *16*, 431.
(8) Ulman, A., *An Introduction to Ultrathin Organic Films from Langmuir-Blodgett to Self-Assembly;* Academic Press: New York, 1991.
(9) Bader, H.; Dorn, K.; Hupfer, B.; Ringdorf, H., *Adv.Polym.Sci.,* **1985**, *61*, 1.
(10) Ogawa, T.; Fomine, S., *Trrends Polym. Sci.,* **1994**, *2*, 308.
(11) Ogawa, T., *Prog. Polym. Sci.,* **1995**, *20,* 943.
(12) Alexandrova, L.; Chavarin, C.; Ogawa, T., Paper presented at POLYMEX'93 (An Intnat. Symp. Polym.), Cancún, Mexico. Nov. 1993. Preprints, pp 215 (1993).
(13) Prasad, P.N., Chapter in this Book.
(14) Nakanishi, H., Chapter in this Book.
(15) Fomine, S.; Neyra, R.; Ogawa, T., *Polym. J.,* **1994**, *26*, 845.
(16) Fomine, S.; Maciel, A.; Ogawa, T., *Polym. J.,* **1994**, *26*, 1270.
(17) Fomine, S.; Sabchez, S.; Ogawa, T., *Polym. J.,* **1995**, *27*, 165.
(18) Fomine, S.; Ogawa, T., *Polym. J.,* **1994**, *26*, 93.
(19) Fomine, S.; Marin, M.; Ogawa, T., *Macromol. Symp.,* **1994**, *84*, 91.
(20) Fomina, L.; Fomine, S.; Ogawa, T., *Polym. Bull.,* **1995**, *34*, 547.
(21) Fomine, S.; Adem, E.; Rao, D.; Ogawa, T., *Polym. Bull.,* **1995**, *34*, 169.
(22) Pedrosa, G.; Fomine, S.; Ogawa, T., unpublished results.
(23) Yu, L.; Chen, M.; Dalton, L.R., *J.Polym.Sci., Polym.Chem.Ed.,* **1991**, *29*, 127.
(24) Fomine, S.; Fomina, L.; Quiroz-Florentino, H.; Mendez, J.M.; Ogawa, T., *Polym. J.,* **1995**, *27*, 1085.
(25) Valverde, C.; Alexandrova, L.; Adem, E.; Rao, D.; Ogawa, T., *Polym. Adv. Tech.,* **1996**, *7*, 27.
(26) Ogawa, T., Unpublished results.

Optical Information Storage and Processing

Chapter 15

Organic Polymers for Photorefractive Applications

B. Kippelen, Sandalphon, B. L. Volodin, K. Meerholz[1], and N. Peyghambarian

Optical Sciences Center, University of Arizona, Tucson, AZ 85721–0094

We review the physics of highly efficient low glass transition temperature photorefractive polymers. Different design approaches to photorefractivity in organic polymers are discussed. Frequency-dependent ellipsometry and dielectric experiments are presented in poly(N-vinylcarbazole)-based materials. Orientational enhancement effects that lead to high efficiencies are clearly identified. Observation of non-Bragg diffraction orders is reported and their properties including phase-conjugation and phase-doubling are demonstrated

Organic synthetic polymers, whose applications have strongly influenced most aspects of our daily lives, are emerging as key materials for advanced information and communication technology. Owing to their high performance, good quality, manufacturing flexibility and low cost, polymers are expected to play a major role in optical technology especially where large volume manufacturing of devices at low cost is crucial. Polymers are being intensively investigated for the fabrication of passive optical waveguide devices such as splitters, combiners, polarizers, interconnects and wavelength division multiplexers (WDM) systems (*1,2*). Simultaneously, polymers have been developed that exhibit electrical (*3*) and nonlinear optical properties (*4-6*) that were formerly found only in inorganic materials. Piezoelectric, photoconductive, light-emitting, second- and third-order nonlinear optical (NLO) polymers are being studied. Owing to their structural flexibility, organic polymers can combine several of the above mentioned functionalities.

Among multifunctional polymers, amorphous organic photorefractive (PR) polymers have emerged recently (*7-10*). These new materials look promising as recording medium for holographic storage and for real-time optical information processing applications. In traditional photorefractive materials, the optical encoding of the hologram is based on the recording of a volume phase grating in second-order

[1]Current address: Department of Physical Chemistry, University of Munich, Sophienstrasse 11, 80333 Munich, Germany

NLO materials (*11,12*). The photogeneration of carriers, their transport over macroscopic distances, and their trapping result in the build-up of an internal electric field that modulates the refractive index of the material through the linear electro-optic effect (Pockels effect). The photorefractive effect is very light-sensitive since high refractive index changes can be achieved with low power light sources but is rather slow compared to purely electronic third-order NLO processes for instance. The speed is limited by the photogeneration rate of carriers and by their migration over distances in the range of several times the wavelength of the laser source. However, since the processing of optical information is mainly achieved in parallel, the data bit transfer rates can be very high. For holographic storage applications, the recording or retrieval of a 1024x1024 bits image in 10 ms for instance, corresponds to a transfer time of 10 ns per bit.

For several decades, the fields of photoconducting (*13*) and purely electro-optic polymers (*14*) have been very active but had almost no direct overlap. With the development of photorefractive polymers in the early nineties, the knowledge of these two research areas could be combined and has led to a rapid improvement of the performance of existing photorefractive polymers. The photorefractive polymer composite DMNPAA:PVK:ECZ:TNF (DMNPAA: 2,5 -dimethyl-4-(p-nitrophenyl-azo)anisole; PVK: poly(N-vinylcarbazole); ECZ: N-ethylcarbazole; TNF: 2,4,7-trinitrofluorenone) we developed recently (*9*) has reached a level of performance that competes with that of the best inorganic photorefractive crystals (*11,12*). With the recent progress achieved in the development of new chromophores for electro-optic applications (*15*), the efficiency of these new materials is expected to be significantly further improved.

Design of Photorefractive Polymers

In order to be photorefractive, a material must combine optical absorption, charge separation through transport, trapping and electro-optic properties. Owing to the structural flexibility of organic polymers, these functionalities can be incorporated in a given material in different ways (*16*). Photorefractive polymers can be fabricated by mixing several molecules in a polymer binder (the so-called guest/host approach) or by synthesis of fully functionalized structures. In the latter approach, the different functional groups can be side-chains of a linear polymer chain, incorporated in the main chain of the polymer, or consist of interpenetrated three-dimensional cross-linked functional networks. A few examples of different polymer designs are illustrated in Figure 1. Generally, guest/host materials have a lower glass transition temperature than fully functionalized structures. The glass transition temperature (T_g) of the polymer is an important parameter since the physics of low and high T_g polymers is quite different.

The first photorefractive polymer (*17*) was a guest/host system based on a partially cross-linked form of the electro-optic side-chain polymer bisA-NPDA (bis-phenol-A-diglycidylether (bis-A) and 4-nitro-1,2-phenylenediamine (NPDA)) doped with the transport agent DEH (diethylamino-benzaldehyde diphenylhydrazone). NPDA also provided photosensitivity. Simultaneously, we developed at the University of Arizona a fully functionalized side-chain polymer that showed electro-optic response, photosensitivity and photoconductivity intrinsically (*18-20*). Similar side-chain approaches were also developed at the University of Chicago (*21*). The efficiencies of these first generation polymers were rather small. Significant improvement in the performance of PR polymers was observed in polymers based on the photoconductor matrix PVK doped with TNF for charge generation and nonlinear molecules for the electro-optic properties. This approach was followed simultaneously in Prasad's group at Suny Buffalo (*22*) and in Moerner's group at IBM Almaden (*23,24*). In the polymer doped with F-DEANST (3-fluoro-4-N,N-diethylamino-β-nitrostyrene) developed at IBM Almaden (*23*), a diffraction efficiency of 1% could be measured in 125-μm thick films at an applied field of 40 V/μm and a grating growth-time of 100 ms for writing intensity of 1 W/cm^2. The improvement in diffraction efficiency compared to previous materials was two orders of magnitude. The photorefractive origin of these efficient gratings was confirmed by asymmetric two-beam coupling experiments and for the first time, PR gain exceeding the absorption of the sample could be observed in a polymer (*23*). Such high efficiencies were indeed unexpected. Calculations based on the standard models of photorefractivity and that took into account the independently measured electro-optic properties of these materials, led to efficiencies that were smaller than the experimental efficiencies. The orientational enhancement mechanism was proposed by Moerner and coworkers (*25*) as a possible explanation. This effect is based on the reorientation of the nonlinear optical molecules by the internal electric field leading to enhanced electro-optic properties and to a new contribution to the total refractive index change due to the anisotropy of the rod-like molecules. This orientational enhancement effect turned out to be a way to tremendously enhance the efficiency of photorefractive polymers by using NLO molecules that have a strong polarizability anisotropy and by adjusting the glass transition temperature of the composite close to room temperature by adding a plasticizer (*26*). The photorefractive polymer composite DMNPAA:PVK:ECZ:TNF we developed recently (*9*) shows a very strong orientational enhancement effect that leads to fully reversible photorefractive refractive index changes as high as $\Delta n = 0.007$ at an applied field of 90 V/μm, for a grating spacing value of 3 μm, and at 1 W/cm^2 writing intensity. Polymers containing dual-function dopants have also been fabricated but the diffraction efficiency that could be measured so far is only 7% for polymers doped with DTNBI (1,3-dimethyl-2,2-tetramethylene-5-nitrobenzimidazoline) (*27*) and less than one percent for those with DPANST (4-(N, N'-diphenylamino) - (β) - nitrostyrene) (*28*). Net gain coefficients comparable to those measured in DMNPAA:PVK:ECZ:TNF, have been obtained recently in a polysiloxane polymer with carbazole side groups and DMNPAA as a dopant (*29*). This functionalized polysiloxane polymer has a much lower T_g than PVK and does not need any additional plasticizer to show orientational mobility at room temperature. The chemical structures of some of the molecules and polymers described in this section are shown in Table I.

Photogeneration, Transport, Trapping, and Electro-Optic Properties

The space-charge build-up process can be divided in two steps: the electron-hole generation process followed by the transport of generally one carrier species.

Table I. Chemical structure and functionality of some of the compounds used in photorefractive polymers

Chemical structure	*Name and functionality*
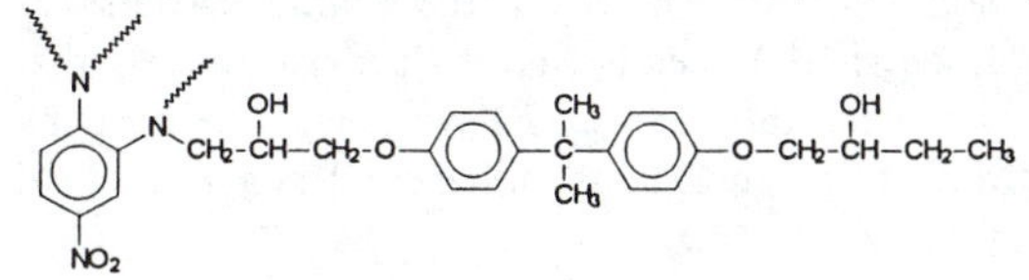	**bisA-NPDA** (electro-optic and sensitizer)
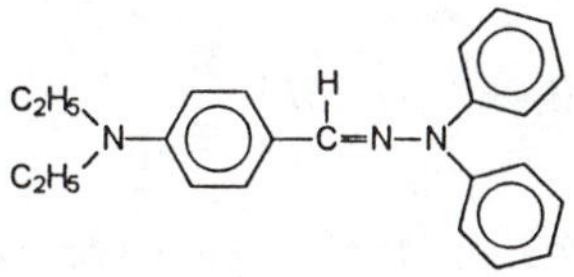	**DEH** (hole transport)
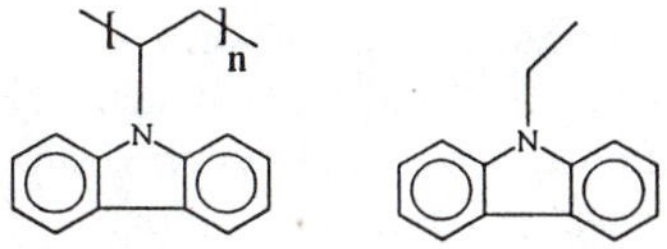	**PVK - ECZ** (charge-transporting polymer and charge transporting plasticizer)
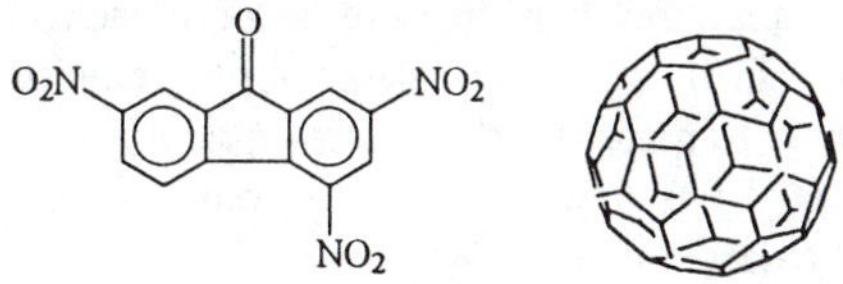	**TNF - C_{60}** (both form a charge transfer complex with carbazole and are used as sensitizers)
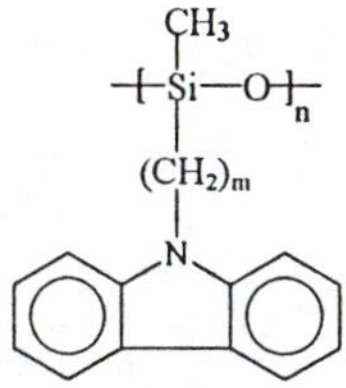	**Polysiloxane with carbazole side groups** (low T_g charge transporting polymer)
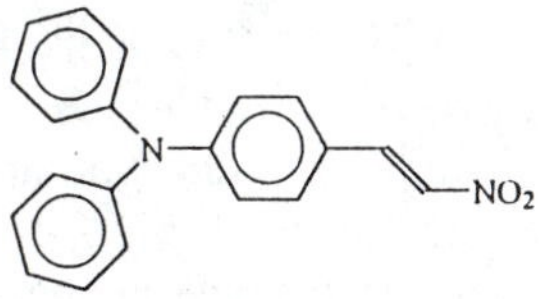	**DPANST** (dual function dopant, electro-optic and transport)

The quantum efficiency for carrier generation in organics is strongly field-dependent and increases with the applied field. A theory developed by Onsager (*30*) for the dissociation of ion pairs in weak electrolytes under an applied field has been found to describe reasonably well the temperature and field dependence of the photogeneration efficiency in most of the organic photoconductors (*31*).

After photogeneration and charge separation, charge transport can occur by a sequence of electron transfer steps between neighboring neutral and charged moieties. Pioneering work by Scher and Montroll (*32*) explained this hopping-type transport by a continuous time random walk model (CTRW) in which the charge carriers are hopping between equally spaced lattice sites. Recently, more elaborate models that include the effects of disorder between the different sites have been developed by Bässler and coworkers (*33*). Transport in amorphous polymers leads to a field dependence of the mobility $\mu(E)$ of the type (*34,35*):

$$\mu(E) \propto \exp\left[-\left(E_A(E) - \beta_T\, E^{1/2}\right) / \, k_B T\right] \tag{1}$$

where E_A is the activation energy and β_T is a constant which is positive below the glass transition temperature of the polymer but can become negative above.

As for purely electro-optic polymers the electro-optic functionality can be achieved in a variety of different ways including guest/host systems, side-chain and main-chain polymers, crosslinked polymers and self-assembly approaches (*36-38*). In amorphous polymers, the NLO chromophores which have a permanent dipole moment are oriented with an electric field to induce electro-optic effects (*39*). Orientation of these dipoles leads not only to macroscopic electro-optic properties but also to birefringence (*40*). In the oriented gas model and for a poling field applied along the Z axis these two effects can be described by (*39*):

$$\Delta n_Z^{(1)}(\omega) = \frac{2\pi}{n} N\, F^{(1)} (\alpha_{//} - \alpha_{\perp}) \left(\left\langle \cos^2 \theta \right\rangle - 1/3\right) \tag{2}$$

$$\chi_{ZZZ}^{(2)}(-\omega; \omega, 0) = N\, F^{(2)}\, \beta \left\langle \cos^3 \theta \right\rangle \tag{3}$$

where $F^{(1)}$ and $F^{(2)}$ are local field correction factors, $(\alpha_{//} - \alpha_{\perp})$ is the polarizability anisotropy of the NLO molecule, β the first hyperpolarizability, N the density of molecules and θ the polar angle between the poling field direction and the dipole moment of the molecule. The dispersion free values of the first hyperpolarizability β_0 deduced from EFISH experiments, the permanent dipole moment μ, and the wavelength of maximum absorption of some of the chromophores used in photorefractive polymers (*41,42*) are summarized in Table II together with a state-of-the-art new chromophore with high nonlinearity developed recently for purely electro-optic applications (*15*).

In high glass transition temperature (T_g) photorefractive polymers, the molecules have an averaged spatially uniform orientation that leads to a constant and uniform electro-optic coefficient. The spatially modulated photorefractive space-

Table II. Properties of nonlinear optical chromophores: absorption maximum λ_{max}, dipole moment μ, and dispersion free first hyperpolarizability β_0

Compound	λ_{max} *(nm)*	μ *(10^{-18} esu)*	β_0 *(10^{-30} esu)*	$\mu\beta_0$ *(10^{-48} esu)*
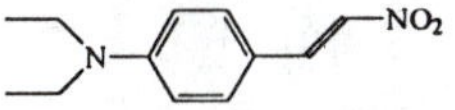 DEANST	282	8	60	480
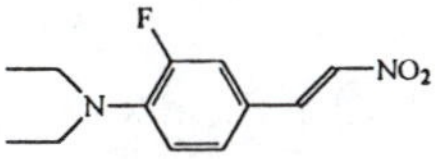 F-DEANST	274	6.8	58	394
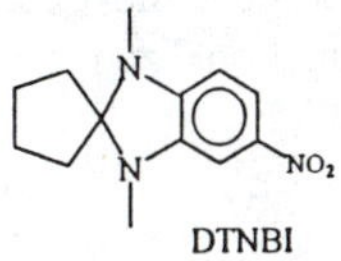 DTNBI	480	8.2	18	148
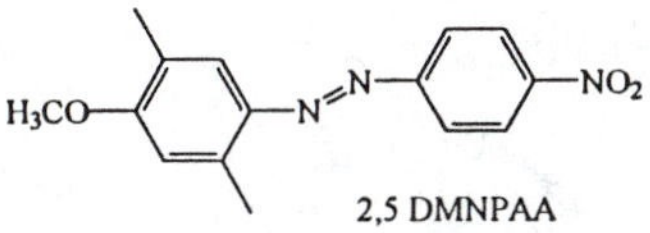 2,5 DMNPAA	398	5.5	38	209
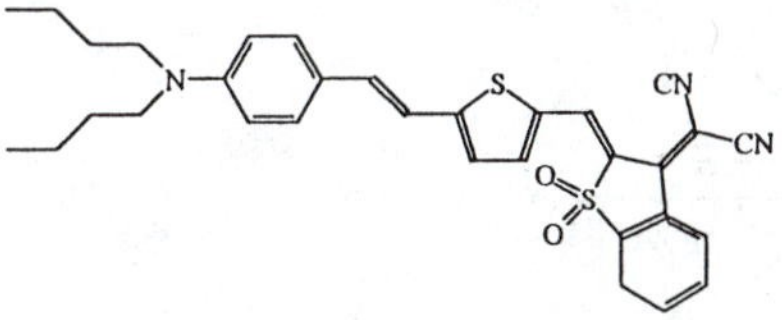 (structure: N, S, CN, CN, O=S=O)	744	-	-	5000

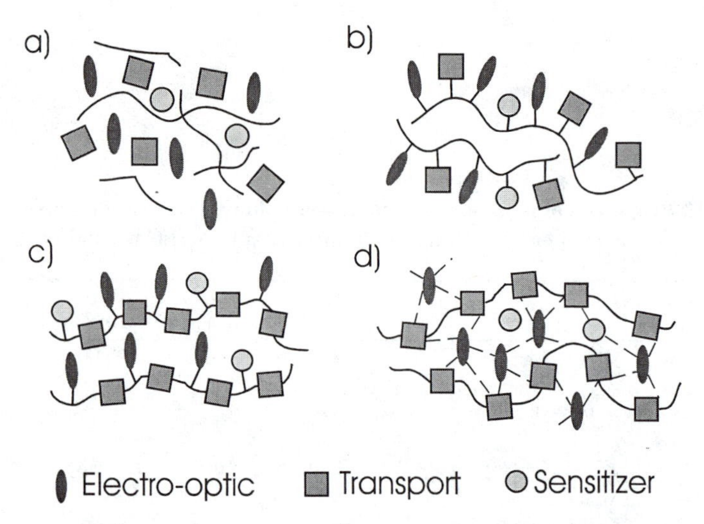

Figure 1. Examples of photorefractive polymer design: a) guest/host composite; b) fully functionalized side-chain polymer; c) fully functionalized main chain polymer; d) interpenetrated cross-linked polymer network.

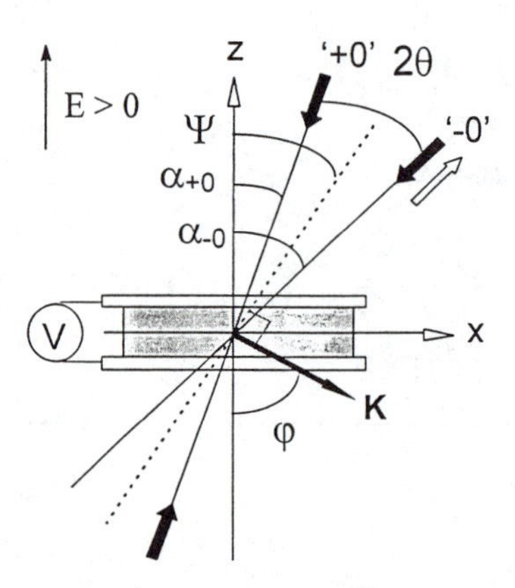

Figure 2. Geometry for the four-wave mixing and two-beam coupling experiments.

charge field modulates the refractive index of the material through the well-known Pockels effect:

$$\Delta n(K) = -\frac{1}{2} n^3 r_{eff} E_{sc}(K) \quad (4)$$

where E_{sc} is the amplitude of the space-charge field, and r_{eff} the effective electro-optic coefficient for a given configuration. ***K*** is the grating vector. The electro-optic tensor elements r_{IJ} are related to the second-order susceptibility tensor elements $\chi^{(2)}_{IJ}$ through (in CGS units) (*5*):

$$r_{IJ} = -8\pi \chi^{(2)}_{IJ}(-\omega; 0, \omega) / n^4 \quad (5)$$

where n is the average refractive index of the polymer.

Identification and Characterization of Photorefractive Properties

Two standard wave-mixing techniques are commonly used to characterize PR polymers: the four-wave mixing and the two-beam coupling technique (*43*). The samples consist of the polymer sandwiched between two transparent ITO (Indium-Tin-Oxide) electrodes. The external field is applied to the electrodes and is therefore perpendicular to the polymer film. To have a component of the electric field along the grating vector, the sample is tilted as shown in Figure 2.

In four-wave mixing experiments, index gratings recorded in the PR material are probed by a weak beam, counterpropagating with one of the writing beams as shown in Figure 2 and the intensities of the transmitted and the diffracted light are monitored. The diffraction efficiency is defined as the ratio of the intensity of the diffracted beam measured after the sample and the intensity of the reading beam before the sample.

Whereas in a four-wave mixing experiment the total grating amplitude is probed independently of the phase shift between the light fringe pattern and the index modulation, the beam coupling experiment is sensitive only to the component of the grating which is 90° phase shifted with respect to the original light pattern (*44*). Therefore, two-beam coupling experiments should be carried out together with four-wave mixing experiments in order to verify the nonlocal nature of the index grating and to prove its photorefractive origin. In the two-beam coupling experiments, the intensities of the two transmitted beams are measured after the sample. In a typical photorefractive material, the two interfering beams that propagate in the sample couple through the photorefractive grating that they generate and exchange energy. Due to the phase shift between the refractive index grating and the interference pattern, one beam gains energy at the expense of the other one. The direction of energy exchange is determined by the symmetry of the sample and by the nature of the carriers that participate in the transport process. When the polarity of the applied field is reversed or when the sample is rotated by 180° with the same field polarity, the direction of the energy transfer is reversed, i.e., the beam that was gaining energy previously is now transferring its energy to the other beam. This effect is called asymmetric beam-coupling and can be considered as a proof of photorefractivity if the experiments are performed in samples that are thick enough to treat the gratings as thick phase volume holograms. In thick samples and for small refractive index modulation amplitudes the Bragg condition is strong enough not to allow Raman-Nath diffraction orders. As has been shown in liquid crystals (*45*), for instance, energy transfer can be observed in the Raman-Nath regime even with local thermal gratings. It is due to the grating written by the interference of the first-order diffracted beam and the zero order beam, and the $\pi/2$ phase shift between these two diffraction orders. Moreover, energy transfer can be observed with local gratings in a transient regime (*44*). Asymmetric energy transfer can be considered as a proof of photorefractivity only when measured in steady-state conditions in thick samples.

The best steady-state performance of a PR polymer to date was observed in the composite DMNPAA:PVK:ECZ:TNF (50:33:16:1 %wt) (*9*). These experiments were performed with beam intensities of 1 W/cm^2 at 675 nm in 105 μm-thick samples. The geometry was $\psi = 60°$ and $2\theta = 22°$ leading to a grating spacing of 3 μm. The sample has a refractive index of n = 1.75 (measured by ellipsometry). In four-wave mixing experiments, a diffraction efficiency of 86% was measured. At 90 V/μm, a total index change of Δn = 0.007 was deduced and a two-beam coupling net gain coefficient of 207 cm^{-1} was measured (*9*). Independent determination of the hyperpolarizability of the DMNPAA molecule by EFISH measurements (*42*) indicated that the high refractive index modulation amplitudes that were measured in this material could not be explained solely by electro-optic properties and that the orientational enhancement mechanism of low T_g photorefractive polymers was playing a major role.

Origin of The Refractive Index Changes in Low T_g Polymers

In low T_g polymers, the orientation of the NLO molecules is not temporally stable after the poling field is switched-off. The necessity for an applied field can be seen as a drawback but it has a major advantage because it enables the orientation of the chromophores at room temperature. As a result, the total poling field that orients the molecules is the superposition of the uniform external field and the spatially modulated internal space-charge field. This property leads to a spatially modulated birefringence in the polymer film and therefore, to a linear contribution to the total refractive index modulation at the origin of photorefractive grating formation. This effect was originally referred to as *orientational enhancement* mechanism (*25*). In the low poling field limit of the oriented gas model ($\mu E/kT < 1$), the first and second-order contributions to the refractive index modulation with spatial frequency K can be written as (*25*):

$$\Delta n^{(1)}(K) = \frac{4\pi}{n} B\, G^{(1)}_{s,p} E_{ext}\, E_{sc}(K) \tag{6}$$

$$\Delta n^{(2)}(K) = \frac{8\pi}{n} C\, G^{(2)}_{s,p}\, E_{ext}\, E_{sc}(K) \tag{7}$$

with

$$B = \frac{2}{45} N F^{(1)} (\alpha_{//} - \alpha_{\perp}) \left(\frac{\mu}{kT} \right)^2 \tag{8}$$

$$C = N F^{(2)} \beta \mu / 5kT \tag{9}$$

where $G_{s,p}^{(1)}$ and $G_{s,p}^{(2)}$ are factors that depend on the configuration of the four-wave mixing experiment and the polarization of the beam that is testing the photorefractive grating. $G_s^{(1)}$ and $G_p^{(1)}$ have an opposite sign. While four-wave mixing experiments are insensitive to the sign of the refractive index changes, two beam coupling experiments are. A change in the sign of the refractive index modulation leads to a change of the direction of the energy transfer and consequently to a change in sign of the gain coefficient. This change in sign was observed experimentally when the polarization of the beams was changed from "p" to "s" (*9*).

To further investigate these several contributions to the total refractive index changes in these materials, we developed a new technique that is based on frequency-dependent ellipsometry experiments (*46, 47*). The experimental set-up shown in the inset of Figure 3 is based on the ellipsometric experiments developed previously by Schildkraut (*48*), and by Teng and Man (*49*). Refractive index changes induced in the sample by the applied field, result in a change in transmitted intensity through the crossed polarizer and analyzer. The field applied to the sample is the superposition of a dc component E_B and a modulated component $E(\omega) = E_m \sin\omega t$. Depending on the ability of the chromophores to orient with a time constant that is higher or lower compared with the inverse of the frequency of the modulated voltage, the poling field E_p will be different when the frequency of the modulated field is varied. At high frequencies, the chromophores cannot orient under the influence of the modulated field and the poling field is given solely by the bias field $E_p = E_B$. At low frequencies, when the changes of the modulated field are slow enough to reorient the chromophores in the sample, the poling field will be given by the total field $E_p = E_B + E(\omega)$. Thus, the refractive index changes induced in the sample and detected after the analyzer will be different in the two frequency regimes. The study of the modulated intensity $I_m(\omega)$ at the frequency ω and its second harmonic $I_m(2\omega)$ in the low and high frequency limits enables the determination of the first-, second-, and third-order contributions to the total refractive index changes (*46*). By analogy with permanently poled polymers, the nonlinear response of the sample can be characterized by the following response functions :

$$R_{eff}^{(\omega)} = \frac{3 \lambda G}{\pi n^3} \frac{I_m(\omega) d}{I_i V_B V_m} \tag{10}$$

$$R_{eff}^{(2\omega)} = \frac{3\lambda G}{\pi n^3} \frac{I_m(2\omega) d}{I_i V_m^2} \quad (11)$$

where $G = (n (n^2 - sin^2\alpha_i)^{1/2}) / (sin^2\alpha_i)$ is a geometrical factor which has a value of $G = 5.6$ for an incidence angle $\alpha_i = 45°$ and for an average refractive index of the sample of $n = 1.75$, λ is the wavelength, V_B and V_m the applied bias and modulated voltage amplitude, respectively, and I_i is the maximum intensity change measured after the analyzer when the relative phase is changed from 0 to π with the Soleil-Babinet compensator. The response functions as a function of frequency for DMNPAA:PVK:ECZ:TNF samples are shown in Figure 3. The response measured at ω at the lowest frequencies is more than 15 times higher than the one measured at ω at the highest frequencies indicating clearly that the birefringence contribution is dominating the overall refractive index modulation in the low frequency regime. The non-vanishing response detected at 2ω at the highest frequencies indicates that a Kerr contribution is also present.

Dielectric Properties

In order to analyze carefully the frequency-dependent ellipsometric measurements described in the previous section, a precise determination of the frequency dependence of the dielectric constant ε is needed. While, the dielectric constant of nonpolar polymers is nearly constant over a wide range of frequencies, that of polar materials decreases with increasing frequency (*50*). In the optical range, ε generally increases with the frequency and this behavior is known as normal dispersion. At these high frequencies, the origin of the polarizability is mainly electronic. However, at moderate and low frequencies the dielectric constant is enhanced compared with its optical frequency value due to the motion of the molecular dipoles. This regime is called anomalous dispersion. The orientational and electronic contributions are found in the well-known Clausius-Mossotti formula for instance. In the simplest model, the frequency dependence of the dielectric constant can be described by the Debye formula (*50*):

$$\varepsilon(\omega) = \varepsilon_\infty + (\varepsilon_{dc} - \varepsilon_\infty)/(1 + \omega^2 \tau^2) \quad (12)$$

where ε_∞ is the purely electronic contribution at optical frequencies, and ε_{dc} is the static dielectric constant. τ is the dielectric relaxation time and is approximately the mean time interval during which an orientation of a polar molecule takes place under the influence of an applied field in a glassy polymer. In the Debye model, the time constant for dipole orientation is given by (*50*):

$$\tau = (4\pi\eta_0 R^3)/kT \quad (13)$$

where η_0 denotes the viscosity coefficient of the matrix and R the radius of the dipole molecule that, for simplicity, is assumed spherical. Equation 13 shows that the size of the chromophore in a glassy polymer matrix will strongly influence its orientation time. Due to the structural complexity of polymers, the dielectric constant of most of the polar polymers does not follow the simple Debye equation (equation 12) but can be represented by a sum of several Debye functions with a distribution of relaxation times (*50*):

$$\varepsilon(\omega) = \varepsilon_\infty + \sum_i A_i\left((\varepsilon_{dc} - \varepsilon_\infty)/\ (1+\omega^2\ \tau_i^2)\right) \quad \text{with} \quad \sum_i A_i = 1 \tag{14}$$

For polymers with a glass transition temperature well above room temperature, the dipole contribution to the dielectric constant will be weak. However, low T_g polymers exhibit a strong contribution as shown in Figure 4 for the composite DMNPAA:PVK:ECZ:TNF with $T_g = 16°$. The frequency-dependence of the dielectric constant has been deduced for this material from frequency-dependent impedance measurements and the sample was approximated to a capacitor and a resistor in parallel. In the range of frequencies $f = \omega / 2\pi = 0$ to 1000 Hz, a good fit to the experimental data is found with the superposition of just two Debye functions with the following parameters: $\varepsilon_\infty = 3.55$, $\varepsilon_{dc} = 6.4$, $A_1 = 0.8$, $A_2 = 0.2$, $\tau_1 = 0.004$ s and $\tau_2 = 3.2 \times 10^{-4}$ s.

Non-Bragg Diffraction Orders

Due to the high refractive index modulation amplitudes that can be achieved in these highly efficient photorefractive polymers ($\Delta n = 0.007$), the recording of thick phase grating in these materials can also lead to non-Bragg diffraction orders (*51*). This effect is different from higher diffraction orders observed in the Raman-Nath diffraction on thin gratings (*43*). For instance, diffraction is not obtained for beams incident at arbitrary angles as in thin gratings and the higher diffraction orders can not be described by Bessel functions.

The geometry of the experiment is similar to that of a two-beam coupling experiment. Two beams with wave-vectors $\boldsymbol{k}_{-0}$ and $\boldsymbol{k}_{+0}$ are interacting in the photorefractive polymer during their propagation in the sample and they create a grating with wave-vector $\boldsymbol{K} = \boldsymbol{k}_{+0} - \boldsymbol{k}_{-0}$. According to Bragg's condition, energy can be diffracted in the directions ($\boldsymbol{k}_{+0} \pm \boldsymbol{K}$) and ($\boldsymbol{k}_{-0} \pm \boldsymbol{K}$) leading to energy transfer between the beams with wave-vector $\boldsymbol{k}_{-0}$ and $\boldsymbol{k}_{+0}$, and to the emergence of higher diffraction order beams (*52*) with wave-vector $\boldsymbol{k}_{-1}$ and $\boldsymbol{k}_{+1}$ as shown in the inset of Figure 5. The diffraction on a thick grating is treated by solving the time-dependent wave equation for the total optical field:

$$\nabla(\nabla \vec{E}) - \nabla^2 \vec{E} = k_0^2\ (\hat{\varepsilon}{:}\vec{E}) \tag{15}$$

where $k_0 = \omega / c$ and $\hat{\varepsilon} = n^2 + \hat{\varepsilon}^{(1)}$ is the dielectric tensor of the medium. $\hat{\varepsilon}^{(1)}$ is the change in the dielectric function induced by the combined action of the photorefractive grating and the static electric field and it contains a contribution that is spatially homogeneous and a contribution that is modulated at the spatial frequency

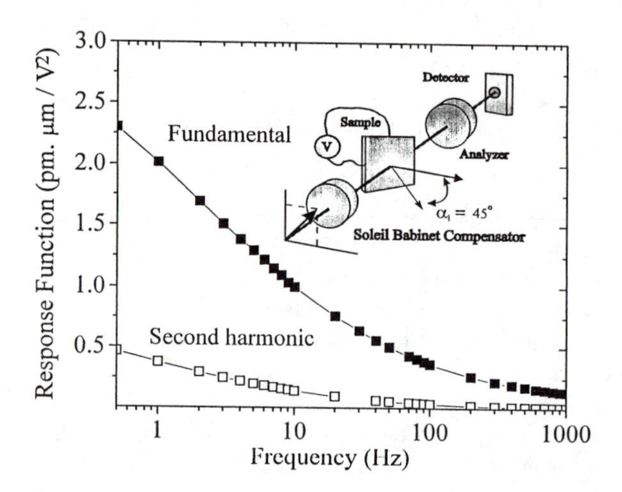

Figure 3. Effective electro-optic response function measured at the frequency of the applied ac field (solid squares) and at the second harmonic (open squares), for different frequencies of the applied voltage. The inset shows the configuration of the set-up.

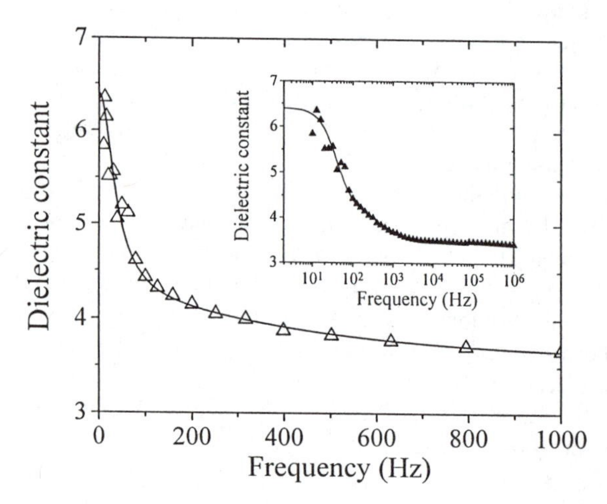

Figure 4. Frequency dependence of the dielectric constant of DMNPAA:PVK:ECZ:TNF photorefractive polymers.

K. When the angle between the incident beams is large, the phase mismatch of the higher diffraction order beams is significant (denoted by Δk_1 in the inset of Figure 5) and the intensity of these beams vanishes after propagation in thick samples. In this case, solving equation 15 leads to the well-known coupled-wave equations that describe two-beam coupling. However, for intermediate spatial frequencies and for large refractive index modulation amplitudes, these higher order beams, called non-Bragg orders do appear behind the sample and they can have a significant intensity (*53,54*). In the undepleted pump beams approximation, analytical solutions can be obtained from equation 15 for the diffraction efficiency of these non-Bragg orders (*51*). Due to the phase shift of the photorefractive grating, the field dependence of these two non-Bragg orders is highly asymmetric with respect to the applied field polarity. This behavior is due to the asymmetric energy coupling of the interacting beams and is therefore a signature of the photorefractive effect. The field dependence of the diffraction efficiency of the non-Bragg beams with wave-vector k_{+1} and k_{-1} measured after propagation in a 105 μm-thick DMNPAA:PVK:ECZ:TNF sample is shown in Figure 5. The electric field asymmetry with respect to its polarity can be clearly seen. The diffraction efficiency of these non-Bragg orders can reach ≈ 10% for a grating spacing of 7 μm and at an applied field > 90 V/μm. For a grating spacing ≤ 3 μm, the diffraction efficiency of these higher orders is < 1 %.

The analytical expressions for the non-Bragg order intensities show that one beam represents the phase-conjugated replica of one of the input waves and the other the phase-doubled replica of the second input wave. To verify this property we performed a beam-coupling experiment with a collimated (-0) and a slightly divergent input beams (+0). As shown in Figure 6, after the sample, one of the non-Bragg beam (-1) was convergent and the other non-Bragg order (+1) was diverging twice as much as the divergent input beam (+0). This effect can be used for phase conjugation applications in a forward configuration and uses only two beams instead of three as in a classical phase conjugation configuration. Moreover, other mathematical operations such as spatial correlation of two images can be performed. We also demonstrated incoherent to coherent image conversion using this process.

Towards Applications

The demonstration of high diffraction efficiency and high net gain coefficients in these new materials show that organic polymers can compete or sometimes even outperform inorganic materials that have been studied for a much longer time. The recent demonstration (*15*) of an electro-optic coefficient of 55 pm/V at 1.3 μm in an electro-optic polymer (to be compared with 30 pm/V for $LiNbO_3$) confirms the constant progress of organic materials and their potential. In the case of photorefractive polymers, the performance has been improved by four orders of magnitude since the first demonstration material. The performance level of current materials makes them promising for a variety of applications including holographic storage, and real-time optical processing. We have demonstrated (*55*) their use in holographic storage, non-destructive testing and in optical correlation applications (*56*). All these optical systems were operated with low power semiconductor laser diodes. Recently, 64 kbit data images could be recorded and retrieved holographically

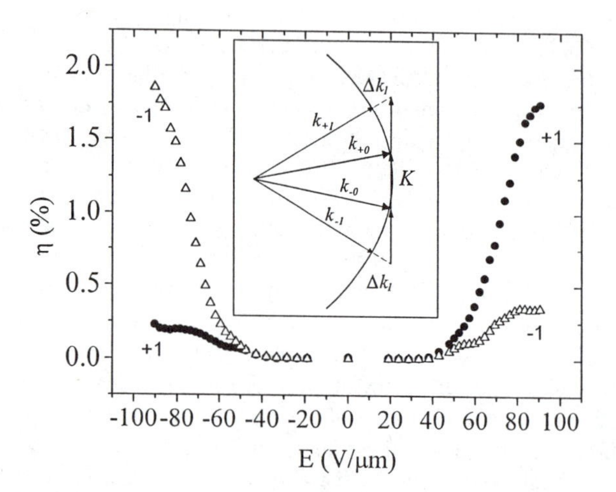

Figure 5. Electric field dependence of the diffraction efficiency of the non-Bragg orders. The field polarity convention is illustrated in Figure 2.

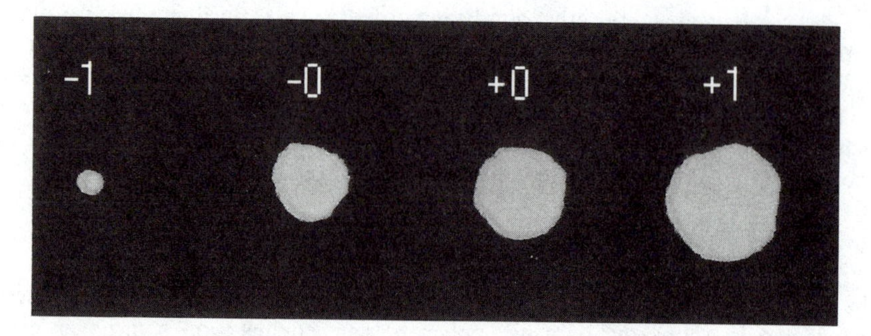

Figure 6. Photograph of the transmitted beams and the non-Bragg diffraction orders. Beam (-0) is collimated, beam (+0) is diverging, beam (-1) is converging demonstrating the phase-conjugate relation to beam (+0), and beam (+1) is diverging two times faster than beam (+0), demonstrating the phase-doubling property.

in a PMMA-based dual dopant polymer composite (*54*). Further improvements are required including long dark storage, low background scattering, non-destructive readout. For real applications, polymers must show good long term shelf stability (chemical and thermal), low fatigue and no photobleaching during multiple write/read cycles. Due to the structural flexibility of organics, there is good hope that these problems can be solved in future materials.

Acknowledgments

This work has been supported by AFOSR, NSF, a NSF/CNRS international program, and by the Center for Advanced Multifunctional Nonlinear Optical Polymers and Molecular Assemblies (CAMP) funded by ONR. The authors would like to thank CAMP collaborators for fruitful discussion, Dr. N. V. Kukhtarev from Alabama A&M University for his contribution to the study of the non-Bragg orders, Dr. P. M. Allemand from Donnelly Corporation, Tucson, for his help during the impedance measurements, and Drs. A. Fort, M. Barzoukas, and C. Runser from IPCMS France for the EFISH characterization experiments.

Literature Cited

1. Wong, C. P. *Polymers for Electronic and Photonic Applications*; Academic Press, Inc.: San Diego, 1993.
2. *Polymers for Lightwave and Integrated Optics, Technology and Applications*; Hornak, L. A., Ed.; Marcel Dekker, Inc.: New York, 1992.
3. Chilton, J. A.; Goosey, M. T. *Special Polymers for Electronics and Optoelectronics*; Chapman & Hall: London, 1995.
4. *Molecular Nonlinear Optics, Materials, Physics, and Devices*; Zyss, J., Ed.; Academic Press, Inc.: San Diego, 1994.
5. Prasad, P. N.; Williams, D. J. *Nonlinear Optical Effects in Molecules and Polymers*; John Wiley & Sons, Inc.: New York, 1991.
6. Bosshard, C.; Sutter, K.; Prêtre, P.; Hulliger, J.; Flörsheimer, M.; Kaatz, P.; Günter, P. *Organic Nonlinear Optical Materials*; Gordon and Breach Publishers: Basel, 1995.
7. Moerner, W. E.; Silence, S. M. *Chem. Rev.* **1994**, *94*, 127.
8. Liphard, M.; Goonesekera, A.; Jones, B. E.; Ducharme, S.; Takacs, J. M.; Zhang, L. *Science* **1994**, *263*, 367.
9. Meerholz, K.; Volodin, B. L.; Sandalphon; Kippelen, B.; Peyghambarian, N. *Nature* **1994**, *371*, 497.
10. Moerner, W. E.; Peyghambarian, N. In *Optics & Photonics News*, 1995; Vol. 6., 24
11. Günter, P.; Huignard, J.-P. *Photorefractive Materials and Their Applications I*; Springer-Verlag: Berlin, 1988.
12. Günter, P.; Huignard, J.-P. *Photorefractive Materials and Their Applications II*; Springer-Verlag: Berlin, 1989.
13. Mort, J. *Adv. Phys.* **1980**, *29*, 367.

14. *Nonlinear Optical Properties of Organic Molecules and Crystals*; Chemla, D. S.; Zyss, J., Eds.; Academic Press, Inc.: Orlando, 1987.
15. Ahlheim, M.; Barzoukas, M.; Bedworth, P. V.; Blanchard-Desce, M.; Fort, A.; Hu, Z.-Y.; Marder, S. R.; Perry, J. W.; Runser, C.; Staehelin, M.; Zysset, B. *Science* **1996**, *271*, 335.
16. Kippelen, B.; Peyghambarian, N. In *Chemistry & Industry*, **1995**, 22, 917.
17. Ducharme, S.; Scott, J. C.; Twieg, R. J.; Moerner, W. E. *Phys. Rev. Lett.* **1991**, *66*, 1846.
18. Tamura, K.; Padias, A. B.; Hall, H. K. J.; Peyghambarian, N. *Appl. Phys. Lett.* **1992**, *60*, 1803.
19. Kippelen, B.; Tamura, K.; Peyghambarian, N.; Padias, A. B.; Hall, H. K. Jr. *Phys. Rev. B* **1993**, *48*, 10710.
20. Kippelen, B.; Tamura, K.; Peyghambarian, N.; Padias, A. B.; Hall, H. K. Jr. *J. Appl. Phys.* **1993**, *74*, 3617.
21. Yu, L.; Chan, W.; Bao, Z.; Cao, S. X. F. *Macromolecules* **1993**, *26*, 2216.
22. Zhang, Y.; Cui, Y.; Prasad, P. N. *Phys. Rev. B* **1992**, *46*, 9900.
23. Donckers, M. C. J. M.; Silence, S. M.; Walsh, C. A.; Hache, F.; Burland, D. M.; Moerner, W. E. *Opt. Lett.* **1993**, *18*, 1044.
24. Silence, S. M.; Donckers, M. C. J. M.; Walsh, C. A.; Burland, D. M.; Twieg, R. J.; Moerner, W. E. *Appl. Opt.* **1994**, *33*, 2218.
25. Moerner, W. E.; Silence, S. M.; Hache, F.; Bjorklund, G. C. *J. Opt. Soc. Am. B* **1994**, *11*, 320.
26. Kippelen, B.; Sandalphon; Peyghambarian, N.; Lyon, S. R.; Padias, A. B.; Hall, H. K. J. *Electronics Lett.* **1993**, *29*, 1873.
27. Silence, S. M.; Scott, J. C.; Stankus, J. J.; Moerner, W. E.; Moylan, C. R.; Bjorklund, G. C.; Twieg, R. J. *J. Phys. Chem.* **1995**, *99*, 4096.
28. Zhang, Y.; Ghosal, S.; Casstevens, M. K.; Burzynski, R. *Appl. Phys. Lett.* **1995**, *66*, 256.
29. Zobel, O.; Eckl, M.; Strohriegl, P.; Haarer, D. *Adv. Mater.* **1995**, *7*, 911.
30. Onsager, L. *Phys. Rev.* **1938**, *54*, 554.
31. Law, K. Y. *Chem. Rev.* **1993**, *93*, 449.
32. Scher, H.; Montroll, E. W. *Phys. Rev. B* **1975**, *12*, 2455.
33. Van der Auweraer, M.; De Schryver, F. C.; Borsenberger, P. M.; Bässler, H. *Adv. Mater.* **1994**, *6*, 199.
34. Gill, W. D. *J. Appl. Phys.* **1972**, *43*, 5033.
35. Fujino, M.; Mikawa, H.; Yokoyama, M. *J. Noncryst. Solids* **1984**, *64*, 163-172.
36. Burland, D. M.; Miller, R. D.; Walsh, C. A., *Chem. Rev.* **1994**, *94*, 31.
37. *Polymers for second-order nonlinear optics*; Lindsay, G. A.; Singer, K. D., Eds.; American Chemical Society: Washington, 1995; Vol. 601.
38. Dalton, L. R.; Harper, A. W.; Ghosn, R.; Steier, W. H.; Ziari, M.; Fetterman, H.; Shi, Y.; Mustacich, R. V.; Jen, A. K.-Y.; Shea, K. J. *Chem. Mater.* **1995**, *7*, 1060.
39. Singer, K. D.; Kuzyk, M. G.; Sohn, J. E. *J. Opt. Soc. Am. B* **1987**, *4*, 968.
40. Wu, J. W. *J. Opt. Soc. Am. B* **1991**, *8*, 142.

41. Moylan, C. R.; Miller, R. D.; Twieg, R. J.; Lee, V. Y. In *Polymers for second-order nonlinear optics*; Lindsay, G. A., Singer, K. D., Eds.; ACS Symposium Series: Washington, 1995; Vol. 601.
42. Kippelen, B.; Runser, C.; Meerholz, K.; Sandalphon; Volodin, B. L.; Peyghambarian, N. 8th International Symposium on Electrets, Paris, France, 1994; p 781.
43. Eichler, H. J.; Günter, P.; Pohl, D. W. *Laser-induced dynamic gratings*; Springer-Verlag: Berlin, 1986.
44. Vinetskii, V. L.; Kukhtarev, N. V.; Odulov, S. G.; Soskin, M. S. *Sov. Phys. Usp.* **1979**, *22*, 742.
45. Sanchez, F.; Kayoun, P. H.; Huignard, J. P. *J. Appl. Phys.* **1988**, *64*, 26.
46. Kippelen, B.; Sandalphon; Meerholz, K.; Peyghambarian, N. *Appl. Phys. Lett.* **1996**, *68*, 1748.
47. Sandalphon; Kippelen, B.; Meerholz, K.; Peyghambarian, N. *Appl. Opt.* **1996**, *35*, 2346.
48. Schildkraut, J. S. *Appl. Opt.* **1990**, *29*, 2839.
49. Teng, C. C.; Man, H. T. *Appl. Phys. Lett.* **1990**, *56*, 1734.
50. Prock, A.; McConkey, G. *Topics in Chemical Physics*; Elsevier: Amsterdam, 1962.
51. Volodin, B. L. ; Kippelen, B.; Meerholz, K.; Kukhtarev, N. V.; Peyghambarian, N. *Opt. Lett.* **1996**, *21*, 519.
52. Au, L.-B.; Solymar, L. *IEEE J. Quant. Electron.* **1988**, *24*, 162.
53. Roy, A.; Singh, K. *J. Appl. Phys.* **1992**, *71*, 5332.
54. Khoo, I. C.; Liu, T. H. *Phys. Rev. A* **1989**, *39*, 4036.
55. Volodin, B. L.; Sandalphon; Meerholz, K.; Kippelen, B.; Kukhtarev, N.; Peyghambarian, N. *Opt. Eng.* **1995**, *34*, 2213.
56. Halvorson, C.; Kraabel, B.; Heeger, A. J.; Volodin, B. L.; Meerholz, K.; Sandalphon; Peyghambarian, N. *Opt. Lett.* **1995**, *20*, 76.
57. Poga, C.; Burland, D. M.; Hanemann, T.; Jia, Y.; Moylan, C. R.; Stankus, J. J.; Twieg, R. J.; Moerner, W. E. *SPIE proceeedings* **1995**, Vol. *2526*.

Chapter 16

Azobenzene-Containing Polymers: Digital and Holographic Storage

Almeria Natansohn[1] and Paul Rochon[2]

[1]Department of Chemistry, Queen's University, Kingston, Ontario K7L 3N6, Canada
[2]Department of Physics, Royal Military College, Kingston, Ontario K7K 5L0, Canada

Films of amorphous high-Tg azo-containing polymers can be used as reversible optical storage materials. Information can be "written" by inducing orientation with a linearly polarized laser beam and "erased" by restoring disorder with a circularly polarized laser beam. The higher the Tg of the polymer, the higher the stability of the "written" information. One important factor affecting the rate of writing and the level of induced birefringence is cooperative motion of neighboring azo groups, due to dipolar coupling taking place between these groups. Volume holograms can be inscribed by this orientation process, but the efficiency is below 1%. Long exposure times to circularly polarized light produce **surface** gratings of high efficiency (30% is typical), by a mechanism which involves photoisomerization. Such gratings can be used as coupling elements into a film waveguide. These two phenomena are reported on a carbazole-containing polymer of high Tg (160^0 C). Asymmetrical two-beam gain coupling in the presence and absence of an electric field is demonstrated, allowing for design of optical switches. In principle, a whole photonic device can be designed on a simple polymer film.

Azobenzene-containing polymers are reported in the literature at a rate of about 150 papers/year for the last five years. There are two main directions of interest: the first relates to the donor-acceptor substituted azobenzenes as the structure generating second-order polarizability and - by noncentrosymmetrical alignment - second order nonlinear optical properties in "poled" films. The second direction of interest is related to the photoinduced isomerization between the trans (more stable) and cis configurations of the azobenzene. This photoisomerization has a wealth of unexpected and useful consequences which were noted only in the last decade. Both these directions appeared because of the availability of lasers, which on the one hand allowed the observation of the nonlinear optical properties and on the other hand enhanced the rate of the photochemical isomerization. A review of the recent literature is not within the scope of the present chapter, but some recently published reviews are recommended for general reading on azo polymers[1], nonlinear optical applications[2], photorefractive properties[3,4] and liquid crystal alignment using azo compounds[5, 6].

Research in our laboratory concentrates mainly on the photoisomerization of various azobenzenes bound as side groups in amorphous high-Tg polymers and copolymers. We are also studying the effects of photoisomerization on the orientation of the photoactive groups and on the refractive indices of the random and oriented parts of the polymer film. This paper presents first an overview of the relevant structural factors affecting the orientation process, and then concentrates on two novel findings related to orientation: surface gratings formation and two-beam gain. Their potential applications in photonics are also briefly discussed.

Photoinduced Birefringence in Amorphous Azo Polymer Films: Digital Storage

Linearly polarized light activates the trans-cis photoisomerization of azobenzene groups bound into polymer chain in a selective manner. Since only azobenzene groups having a dipole component along the electric field vector of the light are activated, groups perpendicular to the polarization direction will remain inert and all other groups which happen to fall along the perpendicular direction will become inert. As long as the light is on, the number of azobenzene groups perpendicular to the polarization direction will continuously grow and eventually reach a saturation concentration. This excess number of azobenzene groups perpendicular to the laser polarization direction can be observed as dichroism or as birefringence. When illumination is terminated, there is some relaxation of the photoinduced orientation towards randomization, but the extent of randomization depends significantly on some thermal properties of the polymer film. If the film is amorphous, for example, and has a glass transition temperature (Tg) higher than the operating temperature, a certain amount of oriented azobenzene groups will remain oriented (frozen-in) for an infinitely long time, unless the film is heated to its Tg. This long term photoinduced birefringence is very stable, but can be selectively destroyed by optical means, i.e. by using circularly polarized light. Circularly polarized light will activate **all** azobenzene groups in the plane of the film, with no exclusion for any single direction, which will eventually restore the original randomness of orientation, thus "erasing" the dichroism and birefringence. A typical experiment of inducing, letting it relax and erasing birefringence is shown in Figure 1.

This phenomenon is not new. It had been reported (the photoinduced birefringence part) in amorphous polymer films having azo dyes dissolved in the material[7]. The main disadvantages of using a doped polymer system are the limited loading due to the limited solubility and the intrinsic decrease of the Tg of the material when a small molecule is doped into a polymer. As Tg decreases and comes closer to the operating temperature (usually room temperature), the stability of the photoinduced birefringence starts to vanish. More impetus was generated into this type of research with the advent of liquid crystalline polymers. On a preoriented liquid crystal polymer film, the orientation could be changed by 90° using polarized laser light and generating quite a difference in the refractive indices of the initial and final orientations[8]. The most exciting aspect of the photoinduced birefringence was the possibility to use it in hologram applications. Interference patterns could be inscribed onto polymer films with excellent contrasts and stability. A significant amount of publications exploiting various aspects of photoinduced birefringence on liquid crystalline polymer films followed, but will not be detailed here.

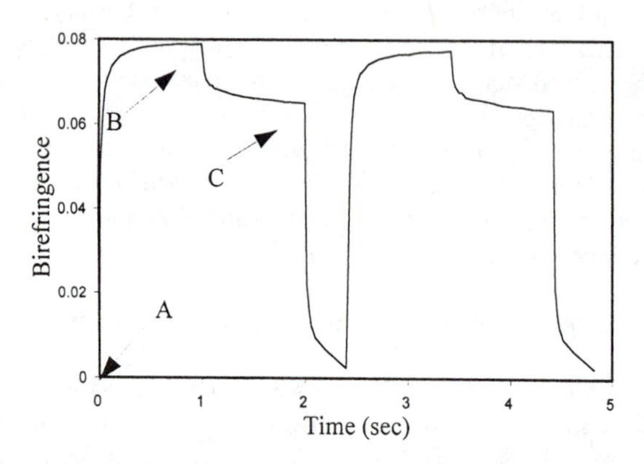

Figure 1. Birefringence is induced (A), it relaxes (B) and is erased (C).

Although most papers investigating such liquid crystalline polymer films had one or two examples of polymers which did not exhibit liquid crystalline phases, but were active in the photoinduced orientation process, the main idea in the literature was that a spacer between the azo group and the main chain was necessary in order to allow the azobenzene group enough freedom to be able to move and achieve the desired orientation. The amorphous polymers investigated in such papers had such spacers, even if they were not long enough to generate liquid crystalline phases.

In fact, contrary to the accepted mechanism at the time, this spacer is not necessary, and the motion of such side groups with short or no spacers was becoming obvious from the study of polymer materials for nonlinear optics. Polymers with higher Tg were proposed and actually used in the poling process, where a polar alignment is achieved with the help of an electric field.

We reported photoinduced orientation on a polymer containing a Disperse Red 1 residue and a very short (2 methylene units) spacer between the azobenzene group and the main chain[9, 10]. The structure of this first polymer is shown - together with other polymers investigated in our laboratory in Scheme 1.

We regard the photoinduced birefringence as a procedure for digital optical storage. At point A in Figure 1, one "writes" information on the polymer film. Subsequent to point B, the "written" part is conserved for years and can be "read" without affecting it at any time during storage. At point C in Figure 1, "erasing" is being performed, and that specific point on the polymer film is ready for a new cycle of storage. Tens of thousand of "writing/erasing" cycles can be performed on the same point on the polymer film without distortion. After a very long exposure time to the laser, the polymer can eventually degrade. "Writing" a line, instead of a point, on the polymer film, provides an optically-inscribed waveguide, which can be optically "erased" or "re-written". A variety of structural factors affecting the photoinduced orientation phenomenon have been identified.

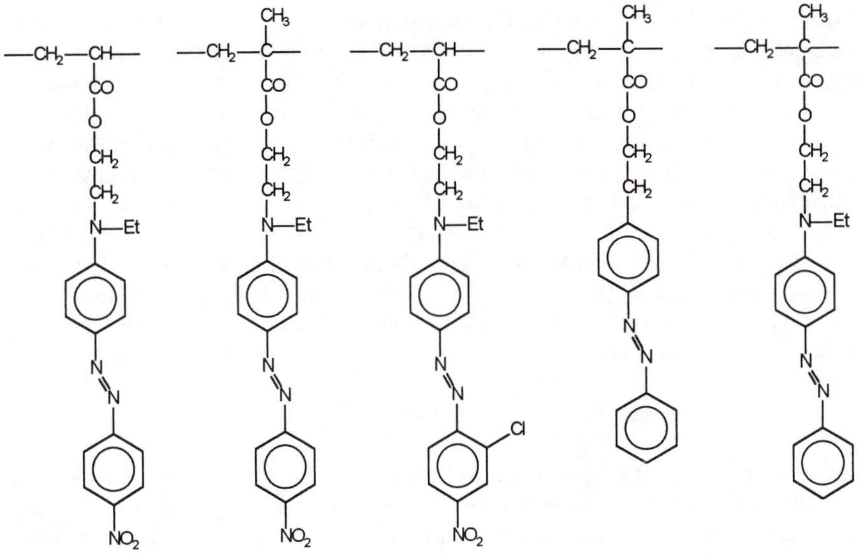

pDR1A pDR1M pDR13A pMEA pMAEA

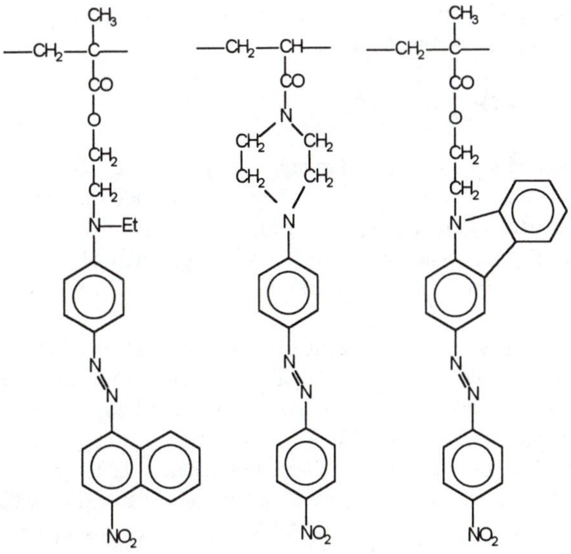

pNDR1M pANPP pCARBA

Scheme 1. Polymers studied in our laboratory (obtained by polymerization)

Structural factors affecting the photoinduced orientation. One obvious important parameter determining the rate and stability of the photoinduced orientation is the glass transition temperature of the polymer film. The higher the Tg, the lower the propensity of the azobenzene groups to move at room temperature, the usual "working" temperature. Thus, for polymers with relatively high Tg one expects a slower rate of the photoinduced birefringence, but also a better stability of the orientation after the light has been turned off. The two structural parameters related to the Tg are the main chain structure and the structure of the spacer or link between the azo group and the main chain. We have not undertaken yet a separate study of these two parameters, so only the analysis of the spacer structure will be shown below.

To describe the growth of the photoinduced birefringence, a biexponential equation can be used as follows[11]:

$$\Delta n = A[1 - \exp(-k_a t)] + B[1 - \exp(-k_b t)] \quad (1)$$

where k_a and k_b are the rate constants of the two exponential processes and A and B are the weighting coefficients representing the relative contribution of the two processes to the overall birefringence (Δn) growth. In general, the "fast" process is associated with the azobenzene group isomerization and reorientation, while the "slow" process is supposed to involve motional coupling between the azobenzene group and the main chain, as well as any motion of the main chain.

For the relaxation of the photoinduced orientation when the "writing" beam is turned off, a similar biexponential equation can describe the process[11]:

$$\Delta n = C\exp(-k_c t) + D\exp(-k_d t) + E \quad (2)$$

where k_c and k_d are the rate constants of the two exponential processes, C and D are their weighting coefficients and E is the residual long term birefringence (independent on time). Again, two main processes are assumed to occur and they are associated, as before, with motions of the azobenzene groups (including thermal cis-trans isomerization) and with coupled motions involving the main chain.

The polarity of the azo group. Rau[12] classified azobenzene groups into three categories, according to their isomerization characteristics. The first category, called "azobenzenes" has a relatively slow cis-trans thermal isomerization and consists mainly of groups containing no polar substituents. pMEA, in Scheme 1, belongs to this category. The second category, called "aminoazobenzenes", has a faster cis-trans thermal isomerization rate and, as the name suggests, has amino substituents, which generate larger dipole moments. In Scheme 1, pMAEA belongs to this category. Finally, the third category is called "pseudostilbenes", due to the similarity of isomerization with stilbene molecules, and it typically contains donor-acceptor substituted azobenzene, with large dipole moments. The thermal cis-trans isomerization is very fast in this category, but it is also noteworthy that the trans and cis absorbances are typically superimposed and the maximum wavelengths lie in the visible range. Most of the other polymers in Scheme 1 belong to this category.

We have studied pMEA, pMAEA and pDR1M in comparison, in order to determine what is the importance of Rau's classification on the photoinduced birefringence[13]. These three polymers have increasing dipole moments, and their comparison clearly indicate that the "pseudostilbene"-type azobenzenes are the best candidates for photoinduced orientation. Their absorbance in the visible range of the spectrum allows the use of lower power lasers (514 nm), the coincidental absorbances of the cis and trans isomers allows photoexcitation of both trans-cis and cis-trans isomerization processes. Both are necessary for orientation, and the lower the polarity of the azobenzene, the slower the cis-trans thermal isomerization process. The levels, rates and stabilities of the photoinduced birefringence, all are higher for pDR1M in comparison with the other two, as is the efficiency of the process. Almost all our research is concentrated on the donor-acceptor substituted azobenzenes.

The bulkiness of the azo group. Since isomerization is essential for a photoinduced birefringence, a bulkier azobenzene group, which would require more free volume to isomerize, should be detrimental to the orientation process. This is not the case when one compares the levels of photoinduced birefringence for the bulkier azobenzene investigated by us: pDR13A[14], pNDR1M[11] and pCARBA[15], see Scheme 1. The saturation levels of the birefringence are similar or even greater than those achieved in comparable homopolymers (pDR1A compared with pDR13A), due to differences in λ_{max}, but the main differences appear only in the **rate** by which the birefringence is achieved. A bulkier azobenzene group would take longer to achieve the orientation.

The structure of the spacer between the azo group and the main chain. This parameter affects directly the motion of the azobenzene group, since a rigid link may prevent free motion necessary in achieving orientation. As discussed above, a rigid link would have a double effect: it would require more energy to achieve orientation, but it would also confer a better stability of the orientation, after the light has been switched off. pANPP in Scheme 1 has been compared with pDR1A[16]. The two polymers are extremely similar, the main difference in their structures is the spacer (link) between the azobenzene group and the main chain. This spacer is more rigid in pANPP, and although the level of the photoinduced birefringence is similar, the rate of achieving it is much slower for pANPP. As expected, the birefringence is much more stable in pANPP[16]. A similar result is obtained for a polymer obtained by polycondensation (pMNAP, shown in Scheme 2[17]).

In the case of pMNAP, the azobenzene group is tethered to the main chain by two bonds, and the spacer allowing some motion is much shorter than in all the previously analyzed polymers. Again, due to this structural feature, the Tg of the polymer is much higher, and so are the levels of photoinduced birefringence and its stability in the absence of light.

The azo group concentration. The homopolymers shown in Scheme 1 contain one azobenzene group per structural unit. In principle, one would expect a better orientation if more azobenzene groups are present. To check this assumption, we have investigated a number of copolymers containing azobenzene and "inert" groups, and blends of azobenzene-containing polymers with, for example, poly(methyl methacrylate) (pMMA). This investigation revealed a very intriguing neighboring group effect, thus it was expanded on other copolymers. Blends were not investigated in comparison, since most of the homopolymers were not compatible. Scheme 3 shows the structures of copolymers investigated in our laboratory.

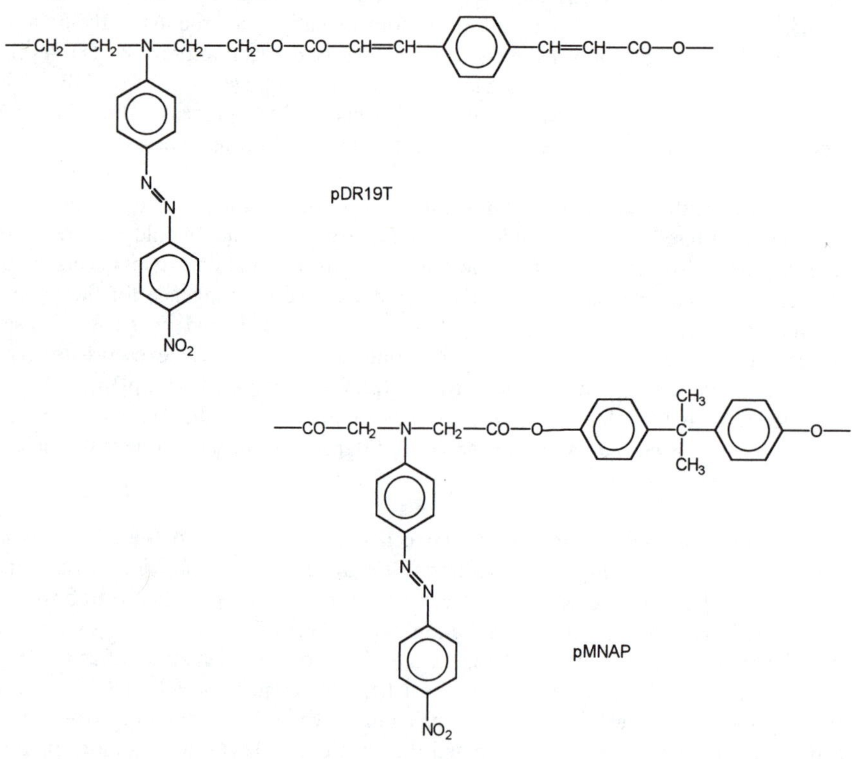

—CH_2—CH_2—N—CH_2—CH_2—O—CO—CH=CH—C$_6$H$_4$—CH=CH—CO—O—

N

N

NO_2

pDR19T

CH_3

—CO—CH_2—N—CH_2—CO—O—C$_6$H$_4$—C—C$_6$H$_4$—O—

CH_3

N

N

NO_2

pMNAP

Scheme 2. Polymers studied in our laboratory (obtained by polycondensation)

As expected, the photoinduced birefringence level does depend on the azo concentration, the more azobenzene groups are present in the copolymer, the higher the birefringence level. This is observed for all copolymers[18-21] but some of the copolymers show a clear deviation from a linear dependence, which has been assigned to a neighboring group effect.

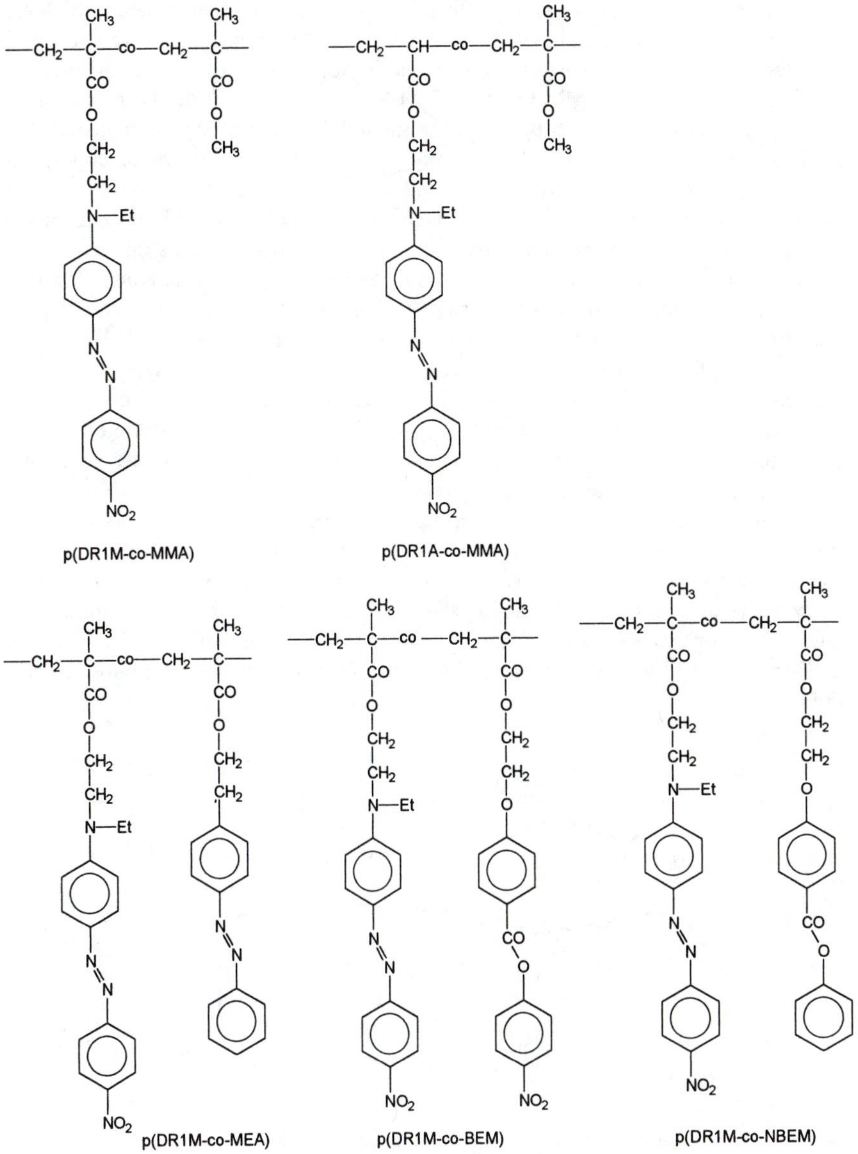

p(DR1M-co-MMA)

p(DR1A-co-MMA)

p(DR1M-co-MEA)

p(DR1M-co-BEM)

p(DR1M-co-NBEM)

Scheme 3. Copolymers studied in our laboratory

The neighboring group effect. This is an unexpected phenomenon, which is usually common in crystalline and liquid crystalline polymers. If two azobenzene groups are situated close to each other they tend to move in concert. This means that neighboring groups will be more difficult to reorient, but it also means that after achieving the new orientation they will have a higher tendency to remain in the oriented position than isolated azobenzene groups. This phenomenon was first observed in p(DR1M-co-MMA) and p(DR1A-co-MMA) (Scheme 3) in comparison with a blend of pDR1A with pMMA. The photoinduced birefringence calculated per azobenzene group is much higher at very low azobenzene content in copolymers, suggesting that the isolated azobenzenes are much easier to move than the azobenzenes neighboring each other. In the blend, the photoinduced birefringence does not appear to depend on the azobenzene concentration, which means that neighboring groups influence each other mainly if they belong to the same polymer chain; the intrachain neighboring effect is negligible[18]. The thermal cis-trans isomerization rate appears also to depend on the presence of neighboring azobenzene groups[22].

In liquid crystalline polymers (and in semicrystalline polymers) cooperative motion of this kind is very common and its cause is the steric effect of forming organized domains. To find out if in these amorphous polymers the steric effect is the dominant one, or if other effects are more important, a few copolymer series were synthesized, some of which are presented in Scheme 3.

Apart from this steric effect, a dipolar interaction is possible between these polar azobenzene groups. The methacrylates are mostly syndiotactic, as demonstrated by the proton and carbon NMR spectra[23]. Electronic spectra show that the azobenzene groups tend to be antiparallel to each other in the random state, which means that orientation will reinforce this tendency. Thus, one would expect an antiparallel more stable arrangement of two neighboring azobenzene groups as shown in Figure 2.

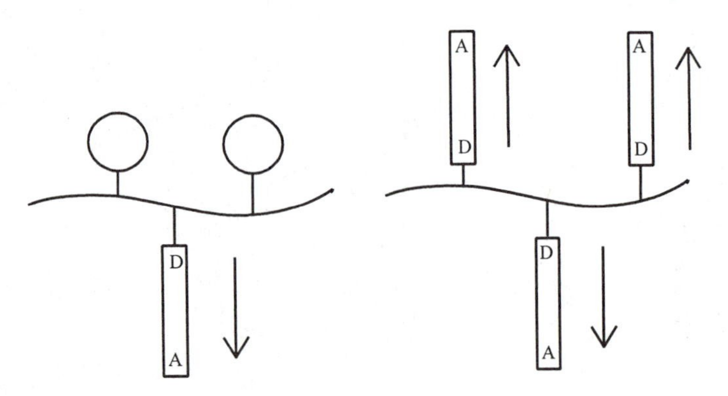

Figure 2. Arrangement of neighboring azobenzene groups.

If the dipolar factor is the dominant one in the cooperative motion of the azobenzene groups, then less polar groups should show a lesser degree of influence between neighbors. To test this hypothesis, p(DR1M-co-MEA) copolymers covering the whole range of compositions have been synthesized and their optical behavior investigated[19]. The results were not too conclusive, since here there are two types of azobenzene groups, reacting differently to the same laser wavelength. Overall, there seemed to be much less of cooperative motion in these copolymers than in p(DR1M-co-MMA), but one cannot assign this to the lesser polarity of the MEA structural unit, since the same neighboring effect in DR1M units should be present here as well.

The answer to this question (dipolar or steric effect) can be given by analyzing the behavior of p(DR1M-co-BEM) and p(DR1M-co-NBEM) copolymer series (Scheme 3). These copolymers contain DR1M structural units, which generate photoinduced birefringence by repeated isomerization cycles, and "inert" structural units of the same shape, but very different polarities. The dipole moment of BEM is about 7D, similar to the dipole moment of DR1M[20], while the dipole moment of NBEM is close to 1D[21]. These two copolymer series behave entirely differently. In p(DR1M-co-BEM), there is an enhanced birefringence at low azobenzene contents, much more than in p(DR1M-co-MMA). This is due to the cooperative motion of the BEM structural units in concert with the DR1M units. Results obtained using polarized IR difference spectroscopy in Prof. M. Pézolet's laboratory at Laval University (Quebec, Canada) clearly show that the DR1M groups and the BEM groups move at different rates, and that BEM is being moved by the DR1M motion[24]. Calculations using birefringence data[20] indicate that the BEM groups can produce as much as 80% of the birefringence obtained by the DR1M groups.

A very similar copolymer series, p(DR1M-co-NBEM) produces a linear dependence of the photoinduced birefringence on the azobenzene concentration[21]. This means that a sterically equivalent group (NBEM) does **not** move in concert with DR1M, and the only possible explanation is the absence of dipolar interactions between the two groups. Hence, it appears that the steric interaction which dominates liquid crystalline self-organization, plays a minor (if any) role in the cooperative motion of azobenzene and "inert" groups. The polarity of such groups is the dominant factor.

The thermodynamic tendency towards order. Most of the publications on photoinduced birefringence in azobenzene-containing polymers used liquid crystalline polymers. Since the liquid crystalline polymers have an intrinsic tendency to organize into oriented domains, the order parameter is much higher in such films, and so is the photoinduced birefringence. The main disadvantage of using a liquid crystalline (or semicrystalline) polymer is the propensity of these materials to conserve the order created (or facilitated) by light. Destroying this photoinduced orientation is not a trivial matter, hence "erasing" it (point C in Figure 1) is extremely difficult. Total erasure is actually impossible. We believe that the mechanism of erasure involves randomization of the orientation of various oriented domains, so that overall no order is perceived, while within each domain the liquid crystalline order is preserved.

Our studies have only included one semicrystalline polymer depicted in Scheme 2 (pDR19T)[25]. Its degree of crystallinity is 16%, as determined by X-ray diffraction. The polymer forms amorphous films in which birefringence can be photoinduced at much higher levels than in the intrinsically amorphous polymer films (about three times higher). At the same time, however, the "written" spot forms its crystalline domains, and "erasing" becomes very difficult as explained above. One very interesting phenomenon is that the rigid groups present into the main chain participate in the alignment of the azobenzene groups, probably as part of the crystalline domains. This cooperative motion was also observed by infrared difference spectroscopy[25].

We plan to extend our studies to a variety of semicrystalline and liquid crystalline polymers.

Surface Gratings Inscription: Holographic Storage

If, instead of addressing one point on the surface of the polymer film with a laser beam, one creates an interference pattern, a holographic grating will be created in the polymer film, with regions of different refractive indices where the beams interact constructively or destructively. This is the basis of holographic storage, which makes extensive use of the whole film surface, thus affording a high density of data. There are quite a few reports of holographic storage especially in liquid crystalline polymer films (where the birefringence is higher than in amorphous films)[26-28], but - at least in some cases - the diffraction efficiencies of such holographic gratings clearly exceed any efficiency which could be calculated with birefringence values of around 0.3 (typical for a liquid crystalline polymer).

While inscribing such holographic gratings on our amorphous polymers, it has become obvious that, with increasing the exposure time, the surface of the polymer was changing[29]. A similar observation was reported by Tripathy et al[30]. The surface grating depth depends on the laser exposure time and on the light polarization[31], the best diffraction efficiency being obtained with circularly polarized light[32]. An atomic force microscopic profile of a surface after exposure to the laser is shown in Figure 3.

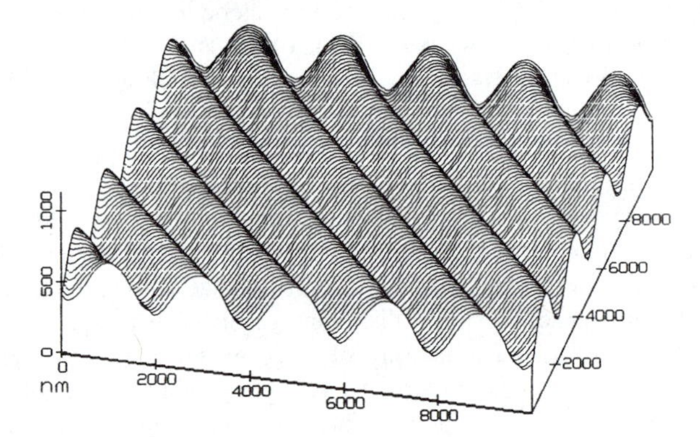

Figure 3. Surface of a polymer film after being subjected to a laser interfering pattern.

The mechanism of producing such a huge change in the surface of a polymer film below Tg (the heating induced by the laser is minimal, it has been calculated to be of the order of magnitude of a few degrees centigrade) is not yet clearly understood, although the presence of photoisomerizing azobenzene is essential. Such isomerization required free volume and may induce a pressure in the parts of the polymer film where there is high isomerization (circularly polarized light) and this may create a flow of the polymer towards the regions of low isomerization (where the interfering light is linearly polarized)[33]. Clearly, more work is necessary in order to explain this completely unexpected and counterintuitive phenomenon. However, creation of such gratings and control of their depth and diffraction efficiencies are relatively easy to achieve, thus one can envisage their use in photonics. We believe that at least some of the papers mentioned above as having reported volume gratings were actually dealing with surface gratings, based on the high diffraction efficiencies.

Surface holographic storage. The main difference between orienting the azobenzene groups and moving whole polymer molecules away from light (photoinduced birefringence and surface gratings inscription) is that the first process is optically reversible, i.e. the orientation can be locally randomized to its initial state by addressing with a circularly polarized light, while the surface gratings are "permanent", i.e. they cannot be optically erased. The only way to erase the grating is to heat the polymer film above its Tg, thus flattening the surface. New gratings can subsequently be inscribed on the surface. This is another argument in favor of the surface gratings interpretation in the previously published articles[26-28]. If these gratings were volume gratings (based on photoinduced birefringence), they would have been optically erasable, but the authors only mention erasing by heating above Tg.

The fact that inscribed surface gratings are not optically erasable can actually be exploited to increase the density of the holographic storage. Different gratings (i.e. having different orientations and/or different spacings) can be inscribed on the same film surface. We have inscribed, for example, eight gratings using only two directions at 90^0 to each other. The four gratings inscribed using the same direction differ by the spacing between the lines. These eight holograms can be identified separately, by looking at the each diffraction spot (Figure 4). Even more interesting, since the whole spectrum of an argon laser was used to inscribe these gratings, separate diffraction peaks can be read for the green and the blue wavelengths, which actually makes sixteen holograms inscribed in the same polymer film. Obviously, this is not an upper limit for storage, and it makes very little use of directional possibilities.

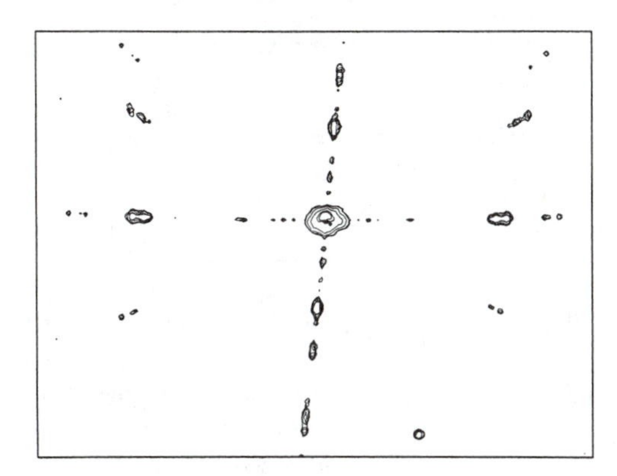

Figure 4. Diffractions of multiple holograms inscribed on a polymer film.

Coupling of light into and out of a film. Since the surface gratings are extremely efficient at directing the light into a diffracted direction, one can think of their use as points of coupling of lightwaves into and out of a film (typically into and out of a waveguide inscribed on a film). Experiments are being done in our laboratory to test the efficiency of coupling and the losses within a waveguide laser-inscribed on a film of the azobenzene-containing polymer.

Two-beam coupling. Optical switching.

Using pCARBA[15], and relying on previous reports of photorefractivity, such as poly(vinyl carbazole) films appropriately doped[34], or a stilbene-based NLO polymer[35], we wanted to test photorefractivity in our polymer films. The idea was that if the amino substituent on a stilbene group gives enough photoconductivity, then the carbazole group would be better. It is well known that all azobenzene-containing polymers can show electrooptic activity upon poling [2], and this property, coupled with the intrinsic photoconductivity of a carbazole-based polymer, should fulfill the conditions for the photorefractive effect[3].

Photorefractive properties in a polymer generate a lot of possible uses in photonics, but from our point of view, the most interesting was the possibility to introduce an optical switch onto the polymer film. Having waveguides optically inscribed and erased (thus allowing changes after the patterns has been established) and having points of entry end exit created a basis for generating a whole photonic device on a polymer film by optical means. A switch would complete the possibilities, and a scheme of such a device is shown in Figure 5.

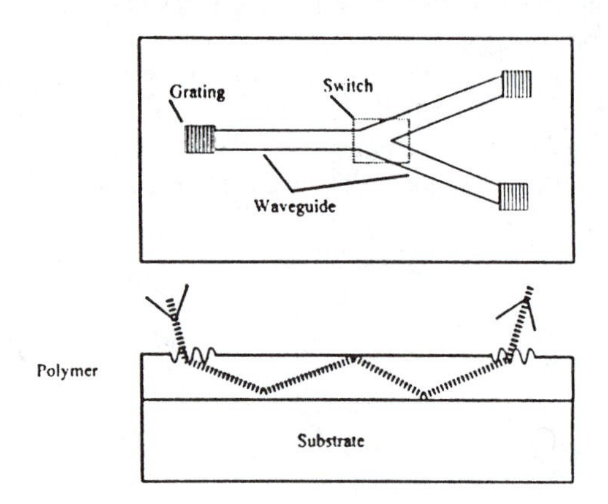

Figure 5. Photonic device optically inscribed on a polymer film.

On pCARBA, we have successfully obtained both asymmetric and symmetric two beam coupling, but the explanation is more complex than just the presence of the photorefractive effect. Typically, we use polarized light in the presence of an electric field to photo-pole the polymer film. High intensity interfering beams used for photoinduced poling also create volume and surface gratings, as described above. Such gratings diffract the light of the beams used for coupling, and symmetric two-beam coupling can actually be obtained in the absence of an electric field. A lot more work is needed to elucidate the phenomena contributing to these coupling effects. However, since asymmetric two beam coupling can be obtained, as demonstrated in Figure 6, in principle optical switching is possible, and a photonic "device" can be designed and operated.

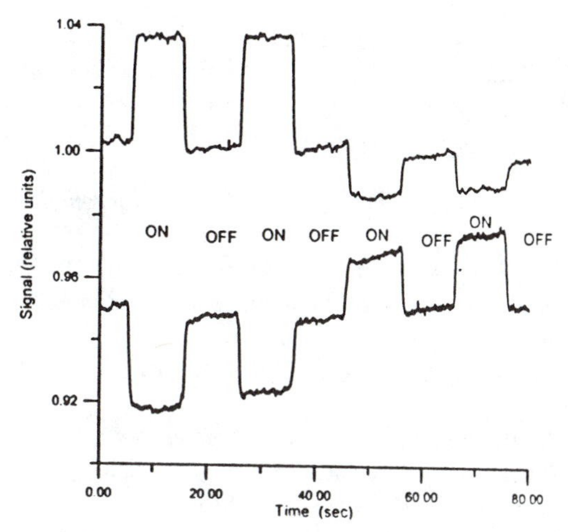

Figure 6. An asymmetric two-beam coupling experiment.

Conclusions

The correlation between photoinduced birefringence levels, rates and efficiencies and the azobenzene polymer structure is well understood for amorphous polymers. The most intriguing structural aspect is the existence and the magnitude of a neighboring group effect. The arguments in favor of a dipolar mechanism governing this neighboring group effect are fairly convincing.

A more serious investigation is required in establishing the role of intrinsic order, or thermodynamic tendency to order, in the process of photoinducing birefringence.

The most exciting aspect of this research is the possibility to use these properties (photoinduced birefringence, surface gratings and two-beam coupling) in photonic devices design, given the simplicity of the material and of the necessary preliminary manipulation in order to obtain the desired effects.

Acknowledgments

Funding from the Office of Naval Research is gratefully acknowledged. A. N. thanks Canada Council for a Killam Research Fellowship. Contributions from our students and postdoctoral associates were essential: Dr. Shuang Xie, Dr. Mei Sing Ho, Dr. Sean Meng, Christopher Barrett, Soi To, Darryl Brown and Jon Paterson.

Literature Cited

1. Xie, S.; Natansohn, A.; Rochon, P. *Chem. Mater.* **1995,** *5,* 403.
2. Burland, D.M.; Miller, R.D.; Walsh, C.A. *Chem. Rev.* **1994,** *94,* 31.
3. Moerner, W.E.; Silence, S.M. *Chem. Rev.* **1994,** *94,* 127.
4. Yu, L.; Chan, W.K.; Peng, Z.; Gharavi, A. *Acc. Chem. Res.* **1996,** *29,* 13.
5. Ichimura, K., In *Polymers as Electrooptical Components*, Shibaev, Ed., Springer Verlag, 1996, p. 138.

6. Ikeda, T.; Tsmtsumi, O. *Science* **1995,** *268,* 1873.
7. Todorov, T.; Nikolova, L.; Tomova, N. *Appl. Opt.* **1984,** *23,* 4309.
8. Eich, M.; Wendorff, J. *Makromol. Chem.* **1987,** *8,* 59.
9. Rochon, P.; Gosselin, J.; Natansohn, A.; Xie, S. *Appl. Phys. Lett.* **1992,** *60,* 4.
10. Natansohn, A.; Rochon, P.; Gosselin, J.; Xie, S. *Macromolecules* **1992,** *25,* 2268.
11. Ho, M.S.; Natansohn, A.; Rochon, P. *Macromolecules* **1995,** *28,* 6124.
12. Rau, H., In *Photochemistry and Photophysics* J.K. Rabek, Ed., CRC Press, Inc: Boca Raton, Fl.,. 1990,. Vol. 2. p. 119.
13. Ho, M.S.; Natansohn, A.; Barrett, C.; Rochon, P. *Can. J. Chem.* **1995,** *73,* 1773.
14. Natansohn, A.; Rochon, P.; Xie, S. *Macromolecules* **1992,** *25,* 5531.
15. Ho, M.S.; Barrett, C.; Paterson, J.; Esteghamatian, M.; Natansohn, A.; Rochon, P. *Macromolecules* **1996,** *29,* 4613.
16. Meng, X.; Natansohn, A.; Rochon, P. *Supramolecular Science* submitted.
17. Meng, X.; Natansohn, A.; Rochon, P. *J. Polym. Sci., Part B. Polym. Phys.* **1996,** *34,* 1461.
18. Brown, D.; Natansohn, A.; Rochon, P. *Macromolecules* **1995,** *28,* 6116.
19. Ho, M.S.; Natansohn, A.; Rochon, P. *Macromolecules* **1996,** *29,* 44.
20. Meng, X.; Natansohn, A.; Barrett, C.; Rochon, P. *Macromolecules* **1996,** *29,* 946.
21. Meng, X.; Natansohn, A.; Barrett, C.; Rochon, P. *Polym. Prepr. (ACS)* **1996,** *37,* 127.
22. Barrett, C.; Natansohn, A.; Rochon, P. *Macromolecules* **1994,** *27,* 4781.
23. Xie, S.; Natansohn, A.; Rochon, P. *Macromolecules* **1994,** *27,* 1885.
24. Buffeteau, T.; Natansohn, A.; Rochon, P.; Pezolet, M. *Macromolecules* submitted.
25. Natansohn, A.; Rochon, P.; Pezolet, M.; Audet, P.; Brown, D.; To, S. *Macromolecules* **1994,** *27,* 2580.
26. Eich, M.; Wendoff, J. *Makromol. Chem.* **1987,** *8,* 467.
27. Eich, M.; Wendorff, J. *J Opt. Soc. Am. B.* **1990,** *7,* 1428.
28. Hvilsted, S.; Andruzzi, F.; Kulinna, C.; Siesler, H.W.; Ramanujam, P. *Macromolecules* **1995,** *28,* 2172.
29. Rochon, P.; Batalla, E.; Natansohn, A. *Appl. Phys. Lett.* **1995,** *66,* 136.
30. Kim, D.Y.; Tripathy, S.K.; Li, L.; Kumar, J. *Appl. Phys. Lett.* **1995,** *66,* 1166.
31. Kim, D.Y.; Li, L.; Jiang, X.L.; Shivshankar, V.; J.Kumar; Tripathy, S.K. *Macromolecules* **1995,** *28,* 8835.
32. Natansohn, A.; Rochon, P.; Ho, M.S.; Barrett, C. *Macromolecules* **1995,** *28,* 4179.
33. Barrett, C.J.; Natansohn, A.L.; Rochon, P.L. *J. Phys. Chem.* **1996,** *100,* 8836.
34. Meerholz, K.; Volodin, B.; Sandalphon; Kippelen, B.; Peyghambarian, N. *Nature* **1994,** *371,* 497.
35. Sansone, M.J.; Teng, C.C.; East, A.J.; Kwiatek, M.S. *Opt. Lett.* **1993,** *18,* 1400.

Chapter 17

Photofabrication of Surface Relief Gratings

D. Y. Kim[1], X. L. Jiang[2], L. Li[2], J. Kumar[2], and S. K. Tripathy[1,3]

[1]Department of Chemistry and [2]Department of Physics, Center for Advanced Materials, University of Massachusetts, Lowell, MA 01854

Surface relief gratings were optically produced on a series of azobenzene-based polymer films. The surface grating formation was investigated by monitoring the diffraction efficiency and using atomic force microscopy. The effects of structure of the chromophores and polymer backbones on the surface grating formation were investigated. The surface deformation process depended on the polarization state of the writing beams. The localized variations of the light intensity and alteration of the resulting electric field polarization were essential writing conditions to the formation of the surface relief gratings.

It has been known over a decade that azobenzene groups in polymer matrices give rise to optical birefringence when excited by polarized light (*1*). This process involves repeated *trans-cis* photo-isomerization and thermal *cis-trans* relaxation of azobenzene groups, which result in orientation of the azobenzene groups perpendicular to the polarization direction of the incident beam. This photoinduced orientation of azobenzene groups has been studied in various polymer matrices and also employed to produce birefringence gratings by a number of research groups (*2-6*). However, surface relief grating formation in these polymers has only been reported very recently (*7-10*).

We have recently reported direct formation of large amplitude (>1000 Å) holographic surface relief gratings on epoxy-based nonlinear optical (NLO) polymer films containing azobenzene groups (*7, 8*). The formation of surface relief gratings on an acrylate polymer with azobenzene groups was also recently reported by Rochon *et al* (*9, 10*). These surface relief gratings were produced upon exposure to an interference pattern of Ar^+ laser beams at modest intensities without any subsequent processing steps. This type of surface relief grating formation has not been observed before in the orientation birefringence studies of azo dye incorporated polymers. The gratings were very stable when kept below the glass transition temperature (T_g) of the polymer. The gratings could be erased by heating the polymer above T_g. Since the amplitude of the surface variation is large and the relief gratings can be conveniently recorded on the polymer films, such polymers have significant potential applications for various optical devices and optical elements.

The driving force and the mechanism of this process is not well understood at

[3]Corresponding author

the present time. Large scale molecular motion and volume change due to reorientation seem to be occurring simultaneously. To understand this process, the role of the structural elements, chromophore side groups and polymer backbone, should be investigated. Especially, the structure of the chromophore side groups appears to be critical to the process. We have also observed that p-polarized recording beams induced much larger surface relief structures than s-polarized recording beams (*8*). It demonstrates that this large mass transfer process is not simply a thermal process. It also implies that the direction of orientation of the azobenzene groups has a very important role to this process. In this paper we report the effects of the structures of various chromophores and backbones on the writing process. A detailed investigation of the polarization dependent recording process was also carried out.

Experimental Section

Polymers were synthesized using the procedure described previously (*11, 12*). The chemical structures of the polymers are presented in Figure 1. The polymer, PBDO3, was synthesized from 1,4-butanediglycidyl ether and Disperse Orange 3 by the same procedure. T_gs of the polymers were measured from DSC (Dupont Thermal Analysis). Molecular weights (Mw) of the polymers were in the range of 5000-10000. Optical quality polymer films were prepared by spin coating of filtered polymer solutions on microscope glass slides. The typical thickness of the polymer films was approximately 0.8 μm. The spin-coated polymer films were optically isotropic.

The experimental setup for the grating recording is shown in Figure 2. A linearly polarized laser beam at 488 nm from an Ar^+ laser was used. The polarized laser beam passed through a halfwave plate, and was then expanded and collimated. Half of the collimated beam passed through another halfwave plate and was incident on the sample directly. The other portion of the beam was reflected from an aluminum coated mirror. Two sets of experiments were carried out. In the first set of experiments, the second halfwave plate was removed. Laser beams with different polarizations (defined by an angle α, with respect to s-polarization) was achieved by rotating the first halfwave plate. By replacing this halfwave plate with a quarter wave plate or a depolarizer, circularly polarized and unpolarized recording beams were obtained respectively. Due to the complex refractive index of aluminum, the polarization of the beam reflected from the mirror became elliptically polarized except for $\alpha=0^\circ$ (s-polarization) and $\alpha=90^\circ$ (p-polarization). In the second set of experiments, by selecting either $\alpha=0^\circ$ or $\alpha=90^\circ$ and positioning the second halfwave plate in one of the recording beams as mentioned earlier, two orthogonally polarized recording beams (polarization recording) could be obtained. In the case of polarization dependent study, the typical intensity of the recording beam after the collimating lens was 55 mW/cm^2 and the recording time was about 45 minutes. The incident angle θ of the recording beams was selected to be 14°, resulting in grating spacing of about 1 μm. The diffraction efficiency of the first order diffracted beam from the gratings in transmission mode was probed with an unpolarized low power He-Ne laser beam at 633 nm. The probe beam was incident at normal to the surface of the polymer film. Surface relief gratings on polymer films were investigated by atomic force microscopy (AFM).

Results and Discussion

The surface relief grating formation on the epoxy polymer, PDO3, was previously reported (*8*). The PDO3 has the azobenzene chromophores with strong electron

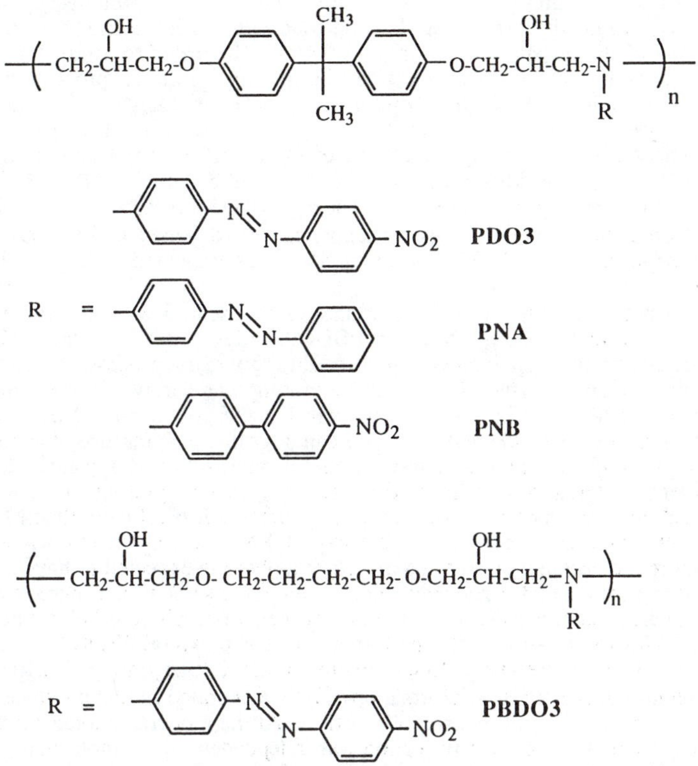

$$-\!\left(\mathrm{CH_2\text{-}CH(OH)\text{-}CH_2\text{-}O}\text{-}C_6H_4\text{-}C(CH_3)_2\text{-}C_6H_4\text{-}\mathrm{O\text{-}CH_2\text{-}CH(OH)\text{-}CH_2\text{-}N(R)}\right)_n\!-$$

R = $-C_6H_4-N{=}N-C_6H_4-NO_2$ **PDO3**

$-C_6H_4-N{=}N-C_6H_4$ **PNA**

$-C_6H_4-C_6H_4-NO_2$ **PNB**

$$-\!\left(\mathrm{CH_2\text{-}CH(OH)\text{-}CH_2\text{-}O\text{-}CH_2\text{-}CH_2\text{-}CH_2\text{-}CH_2\text{-}O\text{-}CH_2\text{-}CH(OH)\text{-}CH_2\text{-}N(R)}\right)_n\!-$$

R = $-C_6H_4-N{=}N-C_6H_4-NO_2$ **PBDO3**

Figure 1. The chemical structures of the polymers.

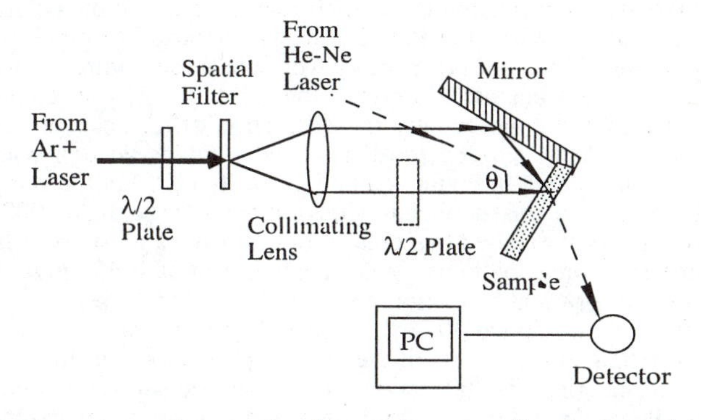

Figure 2. Experimental setup for the grating recording by laser beams with different polarizations.

donor-acceptor substituents (Disperse Orange 3). The design rationale to incorporate 4-aminoazobenzene group in the polymer PNA which has no strong acceptor group was to investigate the effects of the donor-acceptor electronic structure of the chromophore on the surface grating formation. In order to study the role of photoisomerizable azobenzene groups to the grating formation process, biphenyl groups with the same electron donor-acceptor substituents were chosen as chromophores in the polymer PNB, which are not able to undergo photoisomerization. To study the effects of the mobility of the polymer, low T_g flexible alkyl chain backbone was incorporated in the polymer PBDO3. All the synthesized polymers were amorphous. The T_gs of the polymers, PDO3, PNA, and PNB, which have Bisphenol A-based backbones were detected at near 100 °C. The T_g of the polymer, PBDO3, was about 35 °C as expected given the flexible backbone.

Formation of surface relief gratings were observed on the polymers with azobenzene groups, PDO3, PNA and PBDO3. Figure 3 shows the diffraction efficiency curves of the polymers monitored during writing process. In case of the biphenyl side chain polymer, PNB, surface grating was barely observed under the same exposure level. When it is considered that the polymer PNB has the same backbone structure as PDO3 and PNA, we can infer that the presence of azobenzene side groups which can undergo *trans-cis* photoisomerization is a critical structural requirement for the surface deformation process. Surface grating formation from PNA which has no nitro groups (electron acceptor) was less efficient than PDO3. It may be attributed to lower optical density of PNA at the writing wavelength. Furthermore the acceptor substituted azo chromophore is expected to have a shorter excited state lifetime and hence cycle more often in the same time frame. It appears that strong electron donor-acceptor structure of the chromophore is helpful but is not a critical factor for the surface grating formation. The polymer, PBDO3, which has a low T_g backbone, showed very slow formation of the surface grating. It appears that even if photoinduced surface deformation may be generated in this polymer, it may dynamically relax to result in a smooth surface. At the moderate molecular weights employed in our work, grating formation was independent of the molecular weights. Strong molecular weight dependence has been shown at high molecular weights by Natansohn *et al* (*10*).

As we have already reported, polarization states of the writing beams significantly influenced the grating formation process (*8*). Our recent results indicate that the grating formation is a complicated function of the polarization state of the writing beams. When the PDO3 film was exposed to p-polarized beams, significantly larger diffraction efficiencies were observed compared with the film exposed to s-polarized beams. Although the light intensity is comparable, the surface modulation and consequently the diffraction efficiencies are considerably different. Typical AFM surface profiles of the samples written with s- and p-polarized beams are shown in Figure 4 and 5, respectively. For the sample exposed to p-polarized writing beams, surface modulation depth larger than 3000 Å could be obtained, while the film exposed to s-polarization showed almost no surface gratings under the same exposure conditions. When the polarization is 45° with respect to the s-polarization, the grating formation was most efficient. We could achieve a diffraction efficiency as high as 30 % with this polarization state.

We investigated the effects of the various polarization of the writing beams on the grating formation. The diffraction efficiencies and surface modulations of the gratings recorded under different recording conditions are summarized in Table I. All the diffraction efficiency values in the table were measured at least one day after the gratings were recorded to ensure that there are no transient effects involved. The surface modulations of the gratings were studied by AFM.

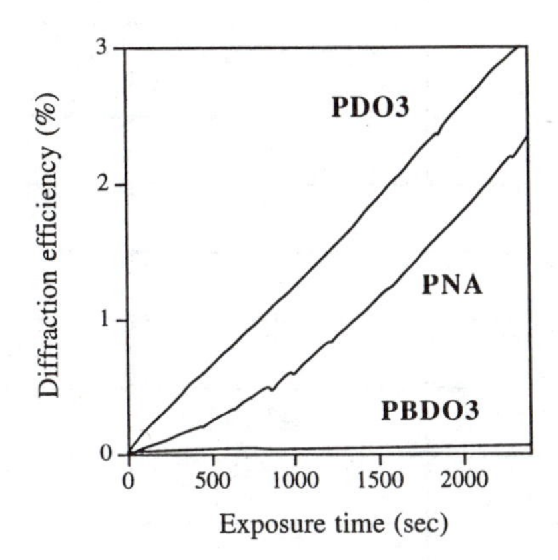

Figure 3. The diffraction efficiency curves for the polymers, PDO3, PNA, and PBDO3.

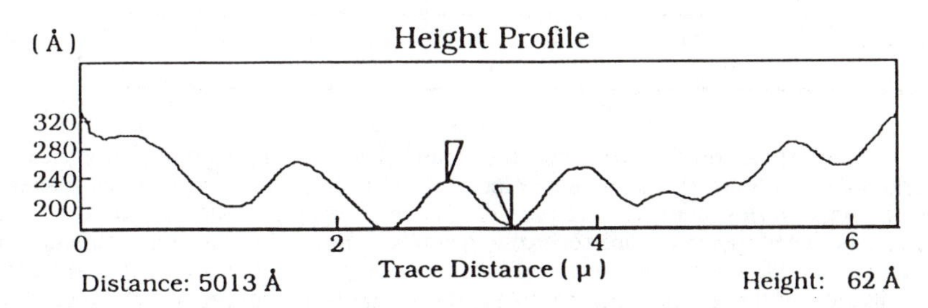

Figure 4. AFM surface profile of the grating written with s-polarized beams.

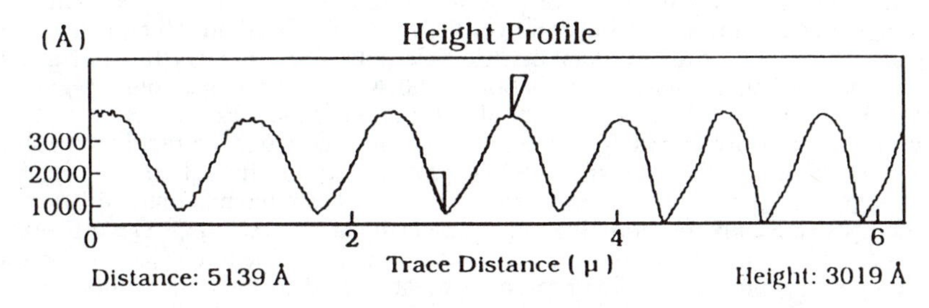

Figure 5. AFM surface profile of the grating written with p-polarized beams.

Table I. The diffraction efficiencies and surface modulations under different recording conditions

Recording conditions	Diffraction efficiency (%)	Surface modulation (Å)
$\alpha = 0°$	<0.01	<100
$\alpha = 8°$	0.4	250
$\alpha = 16°$	5.5	1470
$\alpha = 24°$	15	2140
$\alpha = 45°$	27	3600
$\alpha = 65°$	17	2770
$\alpha = 90°$	15.2	2540
Unpolarized	16.5	2560
Circularly polarized	30	3500
Polarization recording	<0.05	<100

Superposition of two recording beams with different polarizations cause certain distributions of the resultant electric field in the film. The resultant electric field vector varies spatially and periodically in both magnitude and direction. It should be noted that since one recording beam is reflected from an aluminum coated mirror, the reflectivity of electric field is a complex number and it makes the distribution of the resultant electric filed more complicated except for the recording conditions with either s- or p-polarized beams.

In the case of intensity recording (α=0°, two s-polarized beams), interference of the two recording beams with parallel polarization will give rise to the largest light intensity variation. However, the resulting electric field is always linearly polarized and in the same direction over the entire irradiated area (i.e., there is no spatial alternation of the resultant electric field polarization). Very low diffraction efficiency and small surface modulation (<100 Å) were obtained from the grating recorded as shown in Figure 4. The diffraction efficiency as a function of time during the intensity recording process is shown in Figure 6. The rapid increase of the diffraction efficiency at the initial stage is due to the formation of refractive index grating created by photoinduced orientation. From the maximum diffraction efficiency, a change in refractive index was estimated to be 0.02. The efficiency then dropped to a very small value and remained at that value throughout the rest of the recording process. This drop could be attributed to the cancellation of the refractive index grating by small surface modulation.

Under the polarization recording condition, the greatest alternation of the resultant electric field polarization occurs on the film, which results from the superposition of the field of the two recording beams with orthogonal polarization. However, the light intensity on the film is uniform over the entire irradiated area.

Very small surface modulation and diffraction efficiency were also obtained under this recording condition. Figure 7 shows the diffraction efficiency during the polarization recording process. As the orientation grating formed, the diffraction efficiency increased, saturated and remained at a constant value throughout the rest of the recording process. After the laser beams were switched off, the diffraction efficiency decayed nearly to zero, indicating that most of the orientation grating had been erased.

Under the other recording conditions, which could be termed mixed recording conditions, variations of both light intensity and the resultant electric field polarization on the film exist. Surface relief gratings could be formed, leading to much greater values of diffraction efficiency and surface modulation than those from intensity recording or from polarization recording. This seems to indicate that the existence of both light intensity and resultant electric field polarization variations is essential to the formation of surface relief gratings. It is expected that by choosing certain polarization of the recording beams, maximum diffraction efficiency and surface modulation could be achieved. Under the recording condition of $\alpha=45°$, a very large diffraction efficiency of 27% and surface modulation as large as 3600 Å were obtained. A typical grating formation process of mixed recording as probed by the diffraction efficiency is shown in Figure 8. The photoinduced orientation effects leading to refractive index and orientation gratings contributed to the initial increase of the diffraction efficiency. It is experimentally verified by the observation of AFM that in the first minute of recording, the surface relief gratings are not formed. The following stage might involve the saturation of orientation grating and partial cancellation of the refractive index grating by the surface modulation. In the later stage, the diffraction efficiency increased almost linearly until saturation. We therefore infer that the increase of diffraction efficiency after the first minute or so of recording indicates the formation process of the surface relief grating. The gratings recorded with circularly polarized and unpolarized laser beams also revealed large surface modulations with high diffraction efficiencies.

Conclusions

We have established that polymers containing chromophores which can not cycle *trans-cis-trans* isomerization do not give rise to appreciable surface relief gratings. In case of a polymer with low Tg, the surface grating formation was not efficient. The formation of surface relief gratings on the azobenzene-containing polymer films strongly depends on the polarization of the recording beams. Very large surface modulation (> 3000 Å) or virtually no surface change could be achieved by controlling the polarization of recording beams. Although the exact driving force of the migration of the polymer chains is not well understood, we have established that the existence of the spatial variations of both magnitude and direction of the resultant electric field vector in the films is an essential condition to induce surface modulation. The results of polarization dependent experiments clearly imply that spatially varied thermal effects due to light absorption is not sufficient to record the large surface relief gratings. We therefore speculate that the co-operative motions of chromophore dipoles attached to the polymer chains result in migration under a certain driving force (such as due to a space charge field), as they are continuously cycled through the *trans-cis-trans* isomerization process.

Acknowledgments

Financial support from ONR and NSF-DMR is gratefully acknowledged.

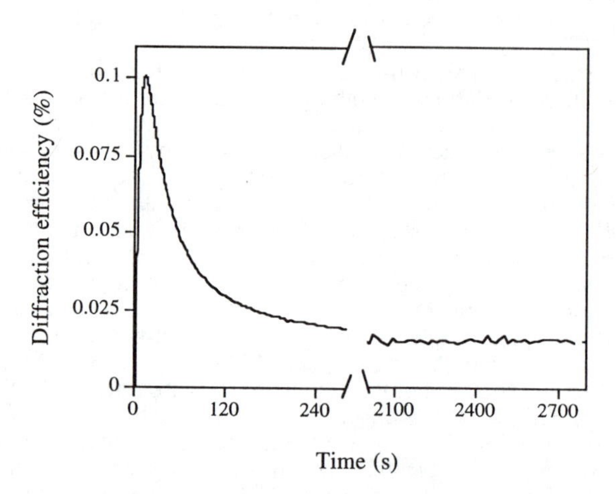

Figure 6. Diffraction efficiency under the intensity recording condition.

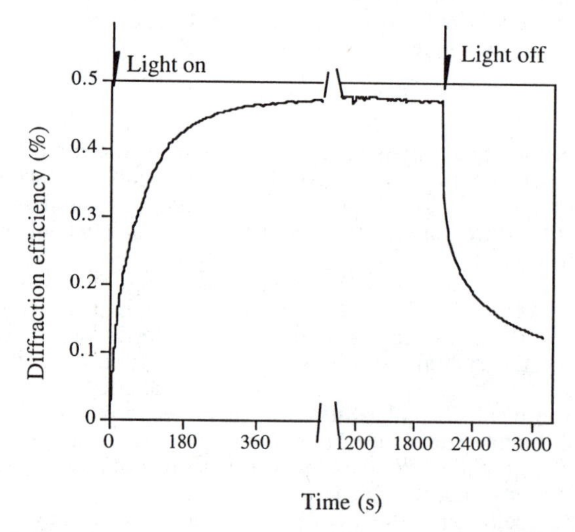

Figure 7. Diffraction efficiency under the polarization recording condition.

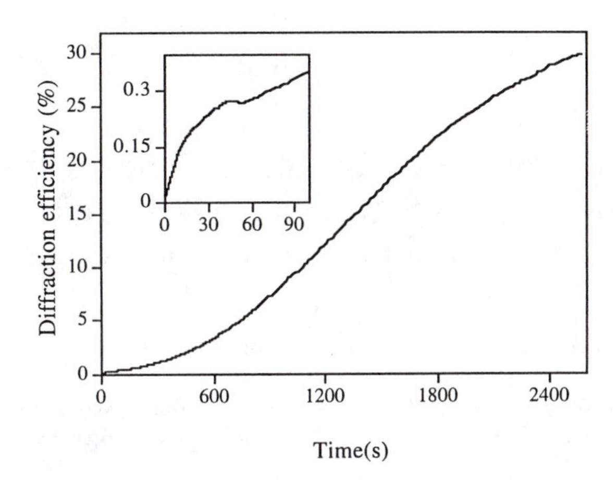

Figure 8. Diffraction efficiency under a mixed recording condition (α=45°). The inset shows the initial stage of the grating formation (about 90 seconds).

Literature Cited

1. Todorov, T.; Nikolova, L.; Tomova, N. *Appl. Opt.* **1984**, *23*, 4309.
2. Eich, M.; Wendorff, J. H.; Reck, B.; Ringsdorf, H. *Makromol. Chem., Rapid Commun.* **1987**, *8*, 59.
3. Eich, M.; Wendorff, J., H. *Makromol. Chem., Rapid Commun.* **1987**, *8*, 467.
4. Eich, M.; Wendorff, J., H. *J. Opt. Soc. Am., B: Opt. Phys.* **1990**, *7*, 1428.
5. Wiesner, U.; Antonietti, M.; Boeffel, C.; Spiess, H. W. *Makromol. Chem.* **1990**, *191*, 2133.
6. Wiesner, U.; Reynolds, N.; Boeffel, C.; Spiess, H. W. *Makromol. Chem., Rapid Commun.* **1991**, *12*, 457.
7. Kim, D. Y.; Li, L.; Kumar, J.; Tripathy, S. K. *Appl. Phys. Lett.* **1995,** *66*, 1166.
8. Kim, D. Y.; Li, L.; Xiang, X. L.; Shivshankar, V.; Kumar, J.; Tripathy, S. K. *Macromoecules*, **1995**, *28*, 8835.
9. Rochon, P.; Batalla, E.; Natansohn, A. *Appl. Phys. Lett.* **1995,** *66*, 136.
10. Barrett, C.; Natansohn, A.; Rochon, P. *J. Phys. Chem.* **1996**, *100*, 8836.
11. Mandal, B. K.; Jeng, R. J.; Kumar, J.; Tripathy, S. K. *Makromol. Chem., Rapid Commun.* **1991**, *12*, 607.
12. Jeng, R. J.; Chen, Y. M.; Kumar, J.; Tripathy, S. K. *J. M. S.- Pure Appl. Chem.* **1992**, *A29*, 1115.

Chapter 18

Photoinduced Refractive Index Changes of Polymer Films Containing Photochromic Dyes and Evaluation of the Minimal Switching Energy

Shin'ya Morino[1] and Kazuyuki Horie

Department of Chemistry and Biotechnology, Graduate School of Engineering, University of Tokyo, Hongo 7–3–1, Bunkyo-ku, Tokyo 113, Japan

Refractive index spectra and their changes were measured for some polymers during photochromic reactions. A poly(methyl methacrylate) copolymer, P(MMA-*co*-GMA-PNCA), which includes norbornadiene groups shows large photoinduced refractive index changes of *ca.* 0.01 in a transparent wavelength region. Refractive index changes for PMMA films containing photochromic dyes are of the order of 10^{-3}. The properties of a Mach-Zehnder photo-optical switching devices using a fulgide derivative or *p*-methoxyazobenzene were evaluated by considering a relationship between absorption losses and wavelength dependence of refractive index changes.

Optical properties of organic compounds are determined by the polarization of electrons along the induced electric field. On the other hand, photochemical reactions change the chemical structures and electronic states of organic materials. Many researchers are interested in photoisomerization of azo dyes because it changes orientations of dye molecules and can be controlled with light irradiation. This effect is utilized for optically-induced alignment of liquid crystals(*1*, *2*) and

[1]Current address: Research Laboratory of Resources Utilization, Photofunctional Chemistry Division, Tokyo Institute of Technology, Nagatsuta-cho 4259, Midori-ku, Yokohama 226, Japan

liquid crystalline polymers(*3*, *4*), photo-assisted poling(*5*), all-optical poling(*6*, *7*), birefringence formation in polymer matrices(*8-10*) and so on.

We have estimated the efficiency of photoinduced reorientation of dye molecules dispersed in an amorphous polymer(*10*). It has been seen that 14% of *p*-dimethylaminoazobenzene (DAAB) molecules disappears from the direction parallel to the polarization of the light inducing its *trans*-to-*cis* photoisomerization in poly(methylmethacrylate) (PMMA). Furthermore, refractive index and absorbance could be controlled by photochromic reactions and utilized for photonic applications such as optical memories, optical switching devices(*11-13*), hologram recording(*9*), waveguide lithography and so on. Increasing number of researchers reporting for photoinduced changes in refractive indices of dye-doped polymer films and some applications have been demonstrated (*11-13*). Ebisawa et. al.(*11*) reported a self-supporting photochromic polymer Mach-Zehnder optical switch. The crosstalk of their switch is -12 dB at 1.55 μm.

In order to evaluate the properties of photochromic dyes for photonics applications, we have to measure absolute refractive index values, their changes and their wavelength dispersions. Absolute refractive index measurements are needed because design of photonic devices strongly depend on these values as long as they take advantage of changes in refraction, reflection and phase shift. Some papers have reported photoinduced refractive index changes of the order of 10^{-3} for polymer systems(*10*, *15-17*). In most cases, the reported results are at one wavelength, 632.8 nm of a He-Ne laser or 488 nm of an Ar^+ laser. Measurement of wavelength dependence of both refractive index and their changes due to photoisomerization are needed for better understanding and application of the materials.

In the present study, we report the results of measurements of refractive index, absorbance, their photoinduced changes and their wavelength dispersions. Based on these results, we will discuss the applicability of photochromic systems to photonic applications.

Experimental Section

Refractive Index Measurements. Refractive index measurements were carried out by an *m*-line method(*18*, *19*). Figure 1 depicts our measurement system. Films on fused silica substrates work as slab waveguides. The light sources are an Ar^+ laser

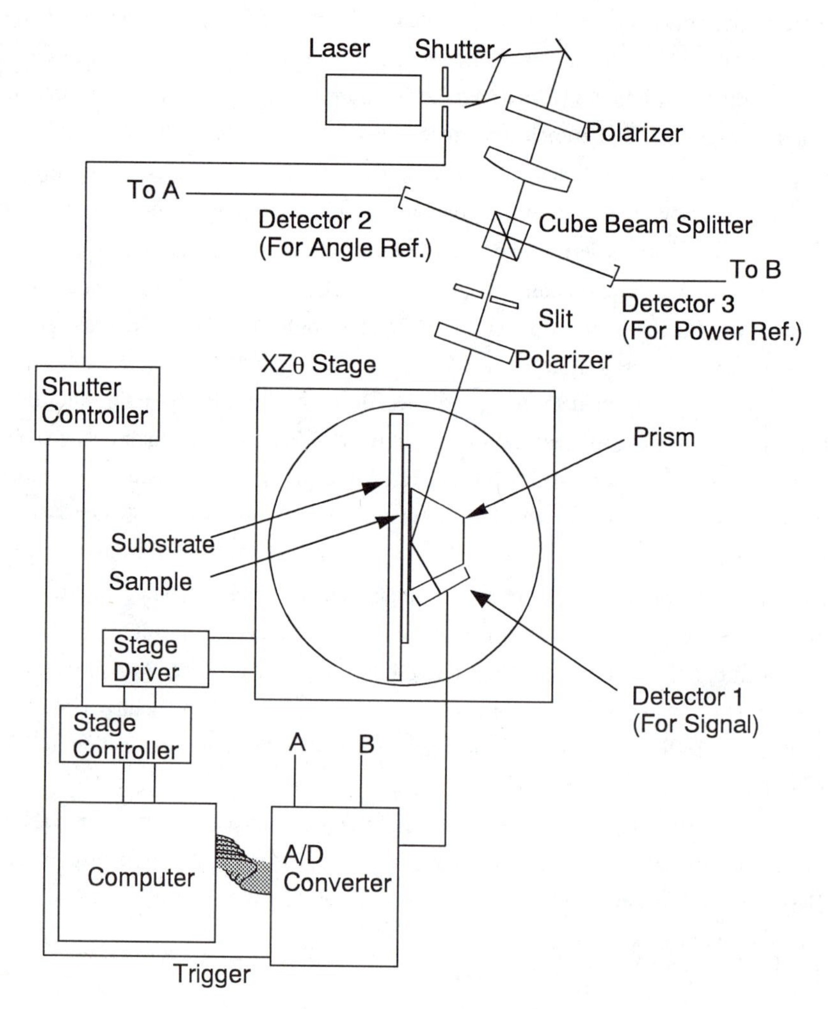

Figure 1. Experimental setup for *m*-line refractive index measurements.

(488 nm), a line-selectable tunable He-Ne laser (543 nm, 594 nm, 604 nm, 612 nm, 633 nm) and an Ar^+ laser-pumped ring dye laser(Dye : DCM, 633 nm ~ 700 nm). Waveguided modes have two orthogonal modes, TE and TM. TE modes have the electric component parallel to the film plane, and TM modes have the magnetic component parallel to the film plane. Both modes are selected by rotating the plane of polarization of incident laser beam. Refractive index of in-plane orientation, n_{TE}, and refractive index of out-of-plane orientation, n_{TM}, were measured for polymer films casted on substrates. The light source is a He-Ne laser (632.8 nm). Using a prism coupler, waveguided modes were excited. Effective refractive indices, n_{eff}, were measured for each mode by measuring incident angles to the prism plane when waveguided modes were excited. Intensity of reflected light from the bottom of the prism was measured. Effective refractive indices were calculated by using angles that gave minimum reflectance at the bottom of the prism as shown in eq. (1).

$$n_{eff} = n_p sin\left(\alpha + sin^{-1}\left[\frac{n_{air}}{n_p} sin\ \{\theta_{obs}\}\right]\right) \tag{1}$$

Here n_{eff} is the effective refractive index, n_p and n_{air} are the refractive indices of the prism and air, respectively, and α is the angle of the prism, and θ_{obs} is the incident angle to the prism. The relationship between the effective refractive index, the film thickness, W, and the refractive index of the film, n_{film}, are described by the following eigenvalue equation,

$$k_0 W \sqrt{n_i^2 - n_{eff}^2} - \Phi_{sub} - \Phi_{air} = m\pi \tag{2}$$

$$\Phi_j = arctan\left(\zeta\left(\frac{n_i}{n_j}\right)^a \sqrt{\frac{n_{eff}^2 - n_j^2}{n_i^2 - n_{eff}^2}}\right)$$

i = TE, TM

j= *air*, *sub*

where k_0 is the wavenumber of the incident laser beam in vacuum, m is the mode number, and n_{sub} is the refractive index of the substrate. For TE modes, $\zeta = 1$ and a = 0. For TM modes, $\zeta = (n_{TE} / n_{TM})$ and a = 2. The n_{TE}, n_{TM}, and W were calculated by a least square fitting using eq 1, 2 with appropriate values for n_p, n_{air}, n_{sub}, and α.

Refractive index obtained in this measurement is converted to one for an isotropic medium by using eq. (3).

$$n = \frac{2\, n_{TE} + n_{TM}}{3} \tag{3}$$

Kramers-Kronig Transformation. In order to obtain continuous refractive index spectra for visible-light region from several discrete observed values of refractive indices, we have carried out Kramers-Kronig transformation by using absorption spectra from 200 nm to 2.0 μm with some assumptions. Absorption bands located at wavelengths shorter than 200 nm are simplified to two absorption bands. One of them is assumed to be located at a short wavelength enough to be simplified to a constant. Another transition band is assumed to be located in the wavelength region near the wavelength region of our interest and gives wavelength dispersion. The width of the second absorption band was ignored. Finally we have obtained following equation,

$$n(\omega_0) = A + B\frac{\omega_1}{\omega_1{}^2 - \omega_0{}^2} + \frac{2}{\pi}\int_{\omega_{2\mu m}}^{\omega_{200\,nm}} \frac{\omega\, \kappa(\omega)}{\omega^2 - \omega_0{}^2}\, d\omega \qquad (3),$$

where ω_0 is the frequency where refractive index is calculated, $\kappa(\omega)$ is the extinction coefficient at a frequency, ω. The symbols $\omega_{200\ nm}$ and $\omega_{\ 2\mu m}$ are the frequencies of the wavelengths 200 nm and 2μm. The following equation is used to obtain the spectra of κ,

$$\kappa(\omega) = \frac{c\, OD(\omega)}{2\omega L} ln\,(10) \qquad (4),$$

where c [m/s] is the light velocity, L [m] is the film thickness, OD(ω) is the optical density at frequency ω. Unknown constants, A, B and ω_1, were determined by using refractive indices measured by an m-line method with a least square fitting.

Absorption Spectra Measurements. UVDEC-660(JASCO) was used for absorption spectra measurements from 200 nm to 800 nm. MPS-5000 (Shimadzu) was used for absorption spectra measurements from 800 nm to 2 μm.

Sample Preparations. Figure 2 shows the chemical structures of NBD (norbornadiene) derivatives and other dyes used in this study. NBDA (norbornadiene dicarboxylic acid) and P(MMA-*co*-GMA-PNCA)(*20*) were supplied by Prof. Nishikubo of Kanagawa University. P(NBDA/*m*XDA) was synthesized as previously reported(*21*). Dyes of DAAB (*p*-dimethylaminoazobenzene), FG540 ((E)-a-2,5-dimethyl-3-furylethylidene(isopropyl)succinic anhydride) and MAB (*p*-methoxyazobenzene) were of commercial purity. PMMA (poly(methylmethacrylate)) was twice repreciptated and dried under vacuum at 110°C for one day. The solution of the mixture of a dye and PMMA was casted on a fused silica substrate and dried for 12 hours under vacuum at room temperature.

Results and Discussion

Concentration Dependence of Refractive Indices. We have measured concentration dependence of refractive indices for some dye/polymer systems at 633 nm with a He-Ne laser. Figure 3 shows the concentration dependence of refractive index of PMMA films containing NBDA or QCA(*22*). This figure shows a linear relationship between dye content and refractive indices. Refractive indices of NBDA/PMMA films are higher than those of QCA/PMMA films. This shows that the linear polarizability of NBDA is larger than that of QCA, because NBDA has two double bonds.

The photochromic reaction of NBDA to QCA does not occur in PMMA, and we could not measure photoinduced refractive index changes. However refractive index changes between NBDA/PMMA and QCA/PMMA would have approximately corresponded to the photoinduced changes if this photochromic reaction had occurred in PMMA, which is ascertained by the following evidence. The NBDA and QCA are small compared to the free volume in PMMA, and densities of both films would be approximately the same. Changes in refractive indices agree with the changes in molar refraction estimated from molar refraction of functional groups and its additivity rule.

Figure 4(22) shows refractive index changes during photoirradiation for P(NBDA/*m*XDA) at 633 nm. The film thickness was 2.2 μm. Refractive index changes are proportional to absorbance changes. This confirms that these changes are induced by photochromic reaction.

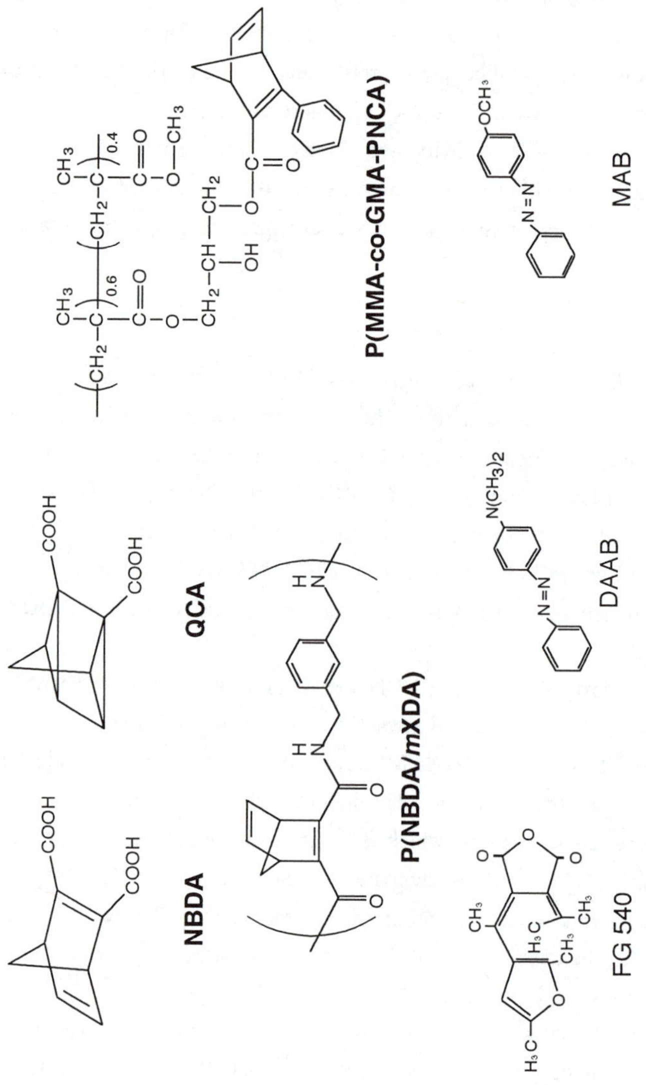

NBDA

QCA

P(NBDA/*m*XDA)

P(MMA-co-GMA-PNCA)

FG 540

DAAB

MAB

Figure 2. Chemical structures of photochromic dyes.

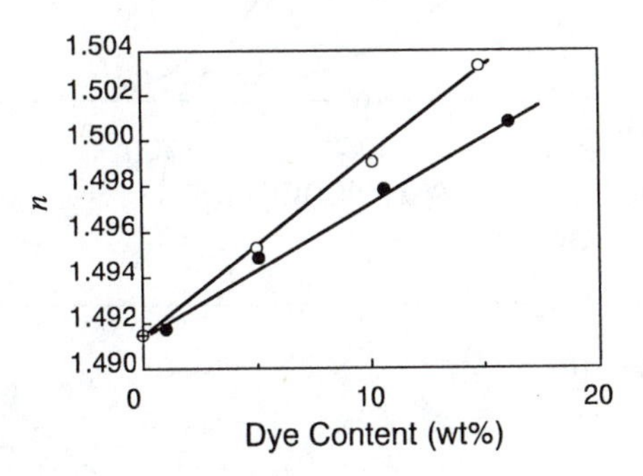

Figure 3. Dye content dependence of refractive index of PMMA films containing NBDA (○) or QCA(•) at 633 *nm*.

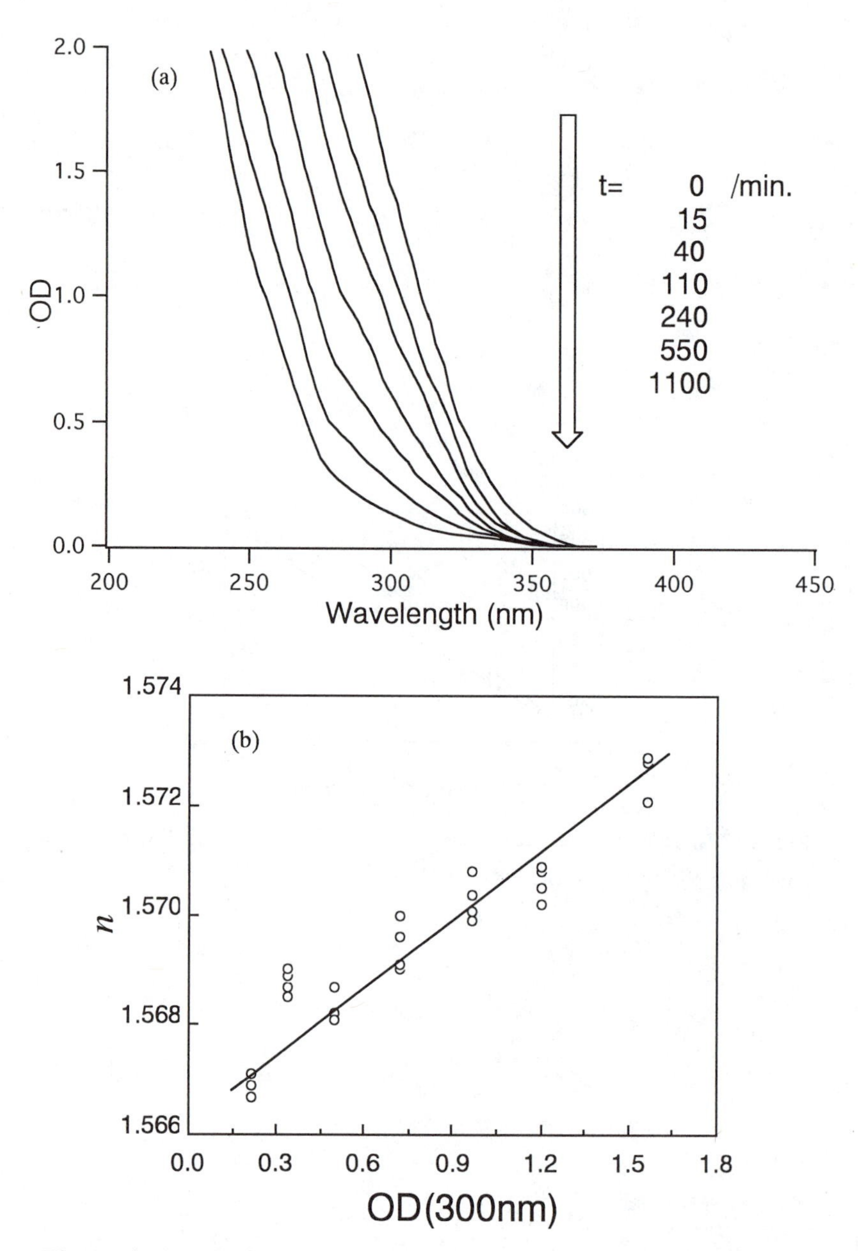

Figure 4. Changes in absorption spectra (a) and refractive index at 633 *nm* (b) of P(NBDA/*m*XDA) during photoirradiation. The arrow represents the direction of the change.

Figure 5 shows concentration dependence of refractive index for DAAB/PMMA films(*10*). Refractive indices were measured for unirradiated and irradiated samples. Unirradiated samples include the *trans* form alone. Photoirradiation was carried out by a Hg lamp with a filter of V-VIA (Toshiba). The conversion of *trans*-to-*cis* form in the photochromic reaction was 0.70 as calculated by using the molar extinction coefficients of both isomers. Similar to the results of measurements for NBD derivatives, refractive indices are higher in the region of higher any dye content, and refractive indices of unirradiated samples are higher than those of irradiated samples at any contents. It reflects the decrease in polarizability caused by disappearance of the π-π* transition band of the *trans* form.

Determination of Refractive Index Spectra. Continuous refractive index spectra were obtained for some polymer systems, PMMA films containing FG540 or MAB(*23*) and P(MMA-*co*-GMA-PNCA)(*24*) by using refractive indices at some wavelengths, absorption spectra and the Kramers-Kronig relation of eq. (3). For all polymer systems investigated, no absorbance change was observed at wavelength longer than 700 nm.

Figure 6(a) shows the absorption and refractive index spectra for FG540/PMMA samples. The dye concentration is 0.25 *mol/l*. An absorption band appeared around 500 nm after the photochromic reaction. The conversion of the photochromic reaction under the photostationary state condition was calculated to be 0.90 by using molar extinction coefficients(*25*). Refractive indices increased after the photochromic reaction, and the difference in refractive indices between the unirradiated and irradiated samples was larger at wavelength nearer the appearing absorption band due to a resonance effect. Figure 6(b) shows a difference in refractive indices before and after photochromic reaction in FG540/PMMA. It also shows a spectrum obtained from Kramers-Kronig transformation calculated by using the difference spectra of absorbance from 200 nm to 2 μm. Both are approximately in good agreement. It is concluded that the observed refractive index changes from 500 to 700 nm are induced by the changes in electronic states evidenced by changes in the UV-Visible absorption spectra.

Figure 7(a) shows the absorption and refractive index spectra for MAB/PMMA samples before and after the photochromic reaction. The dye concentration is 0.20 mol/l. The conversion of the photochromic reaction is assumed to be a unity, which is supported by the following facts. We could not observe the π-π* transition band of the *trans* isomer in the irradiated sample, and the

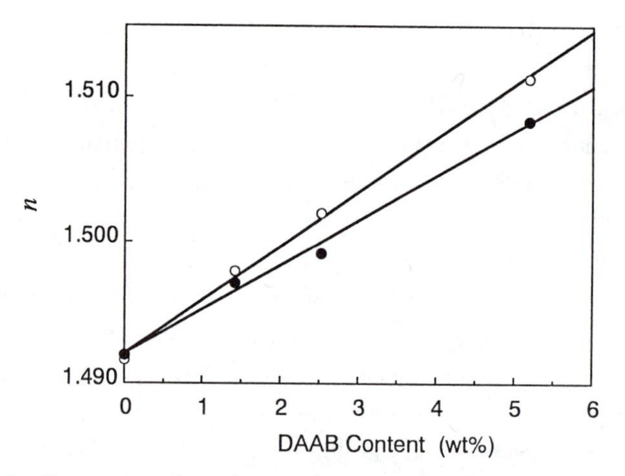

Figure 5. Dye content dependence of refractive index at 633 *nm* for PMMA films containing DAAB before irradiation (○) and after irradiation (•).

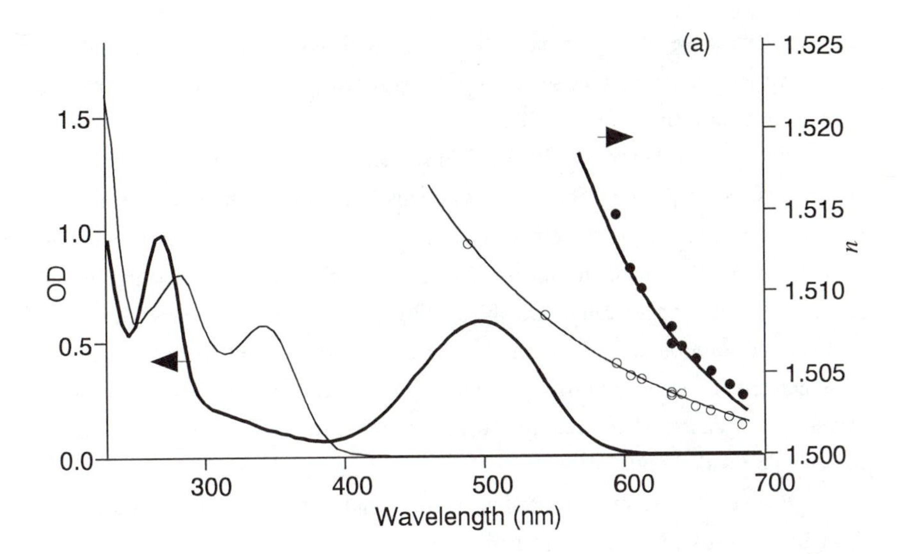

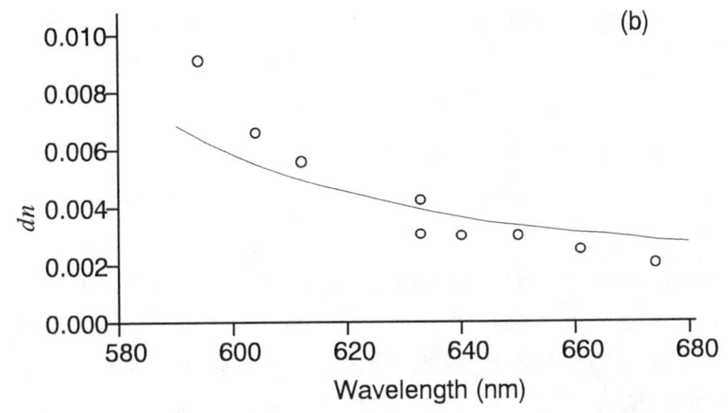

Figure 6. (a) Absorption spectra and refractive index spectra of FG540/PMMA. For absorption spectra, the thin line is for an unirradiated sample and the bold line is for an irradiated sample. For refractive index spectra, circles represent refractive indices measured by an *m*-line method for an unirradiated sample (○) and irradiated sample(●). Lines represent continuous spectra of refractive indices obtained by Kramers-Kronig transformation. (b) Spectra of refractive index changes. Circles represent refractive index changes measured by an m-line method and the line represents refractive index changes obtained from Kramers-Kronig transformation by using a difference absorption spectrum from 200 nm to 2 μm.

ratio of absorbance of an irradiated to the unirradiated sample is 0.1 at 360 nm. Refractive indices increase at shorter wavelengths but the wavelength dispersion is small compared with that of FG540/PMMA.

Figure 7(b) shows the difference in refractive indices and a spectrum of the integral part of eq. (3). The wavelength dispersion in both results is similar, however, the amount of refractive index changes obtained by measurements are larger than that calculated from the integral part of eq. (3). This suggests that refractive index changes cannot be obtained by the Kramers-Kronig transformation by using absorption spectra from 200 nm to 2 μm. An explanation of the disagreement is that absorption bands may appear during photochromic reaction at wavelengths shorter than 200 nm which could affect refractive index in the wavelength region of our interest.

Figure 8(a) shows absorption spectra and refractive index spectra for P(MMA-co-GMA-PNCA) before and after the photochromic reaction. The dye concentration is approximately 2.4 mol/l assuming that the density of this polymer is 1.0 g/cm^3. The actual concentration could be higher. Refractive indices were measured at wavelengths where they showed normal dispersion. Therefore, we simplified eq. (3) to the form which does not depend on ω_1 as shown below.

$$n = A' + B'\omega^2 + \frac{2}{\pi}\int_{\omega_{2\,\mu m}}^{\omega_{200\,nm}} \frac{\omega\,\kappa(\omega)}{\omega^2 - \omega_0{}^2}\,d\omega \qquad (5)$$

This polymer shows large refractive index changes of *ca*. 0.01 although there is no absorption at wavelengths longer than 400 nm. This material suggests the possibility of satisfying both requirements of large refractive index change and optical transparency in a wide wavelength region.

Figure 8(b) shows the difference in refractive indices and a spectrum of the integral part of eq. (5). The possibility of absorption at wavelength shorter than 200 nm was not considered in this calculation. It is clear that the Kramers-Kronig transformation calculated with difference absorption spectra alone does not give correct values in this polymer system.

Estimation of Minimal Switching Energy. The estimation of the minimal switching energy for devices utilizing refractive index changes induced by photochromic reactions was carried out in order to evaluate the properties of photo-

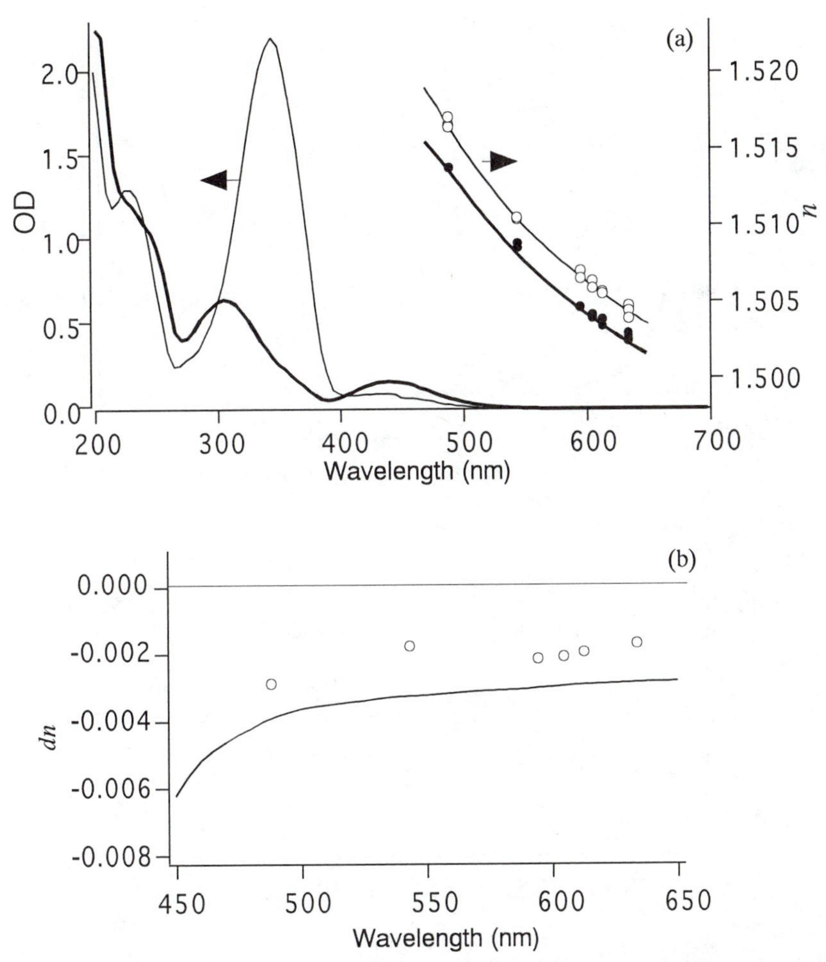

Figure 7. (a) Absorption spectra and refractive index spectra of MAB/PMMA. For absorption spectra, the thin line is for an unirradiated sample and the bold line is for an irradiated sample. For refractive index spectra, circles represent refractive indices measured by an m-line method for an unirradiated sample (○) and irradiated sample(•). Lines represent continuous spectra of refractive indices obtained by Kramers-Kronig transformation. (b) Spectra of refractive index changes. Circles represent refractive index changes measured by an m-line method, the line represents refractive index changes obtained from Kramers-Kronig transformation by using a difference absorption spectrum from 200 nm to 2 μm.

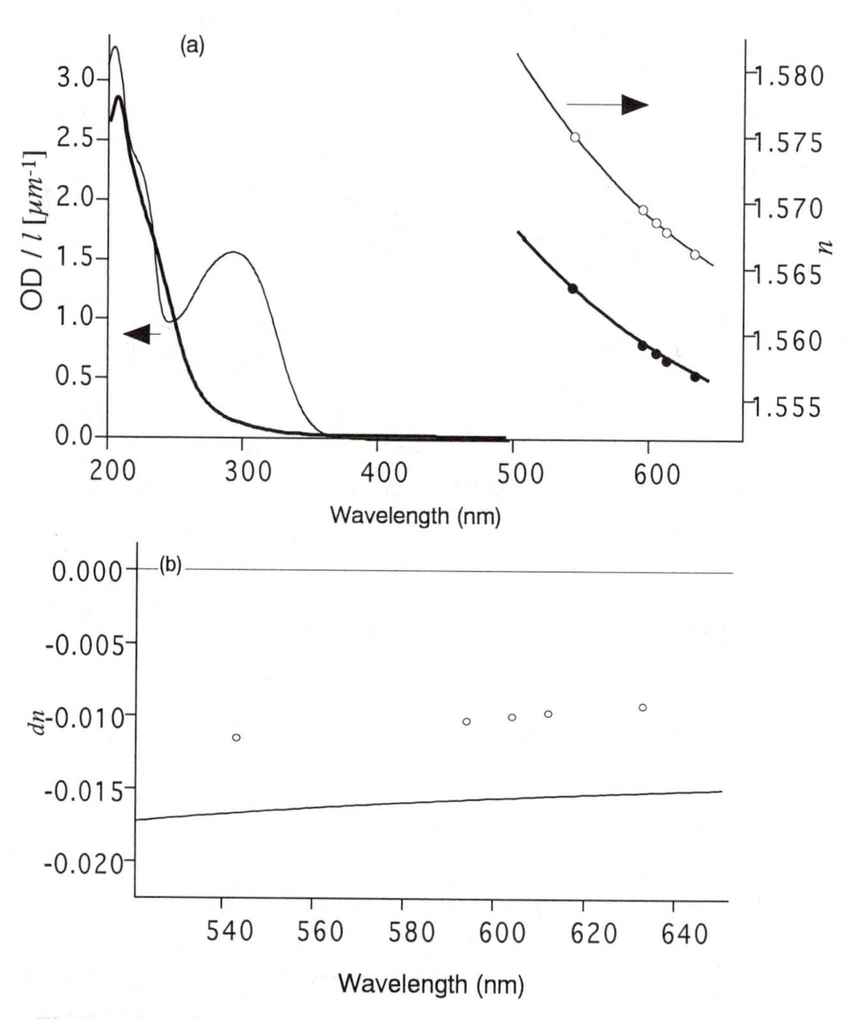

Figure 8. (a) Absorption spectra and refractive index spectra of P(MMA-*co*-GMA-PNCA) before and after irradiation. For absorption spectra, the thin line is for an unirradiated sample and the bold line is for an irradiated sample. For refractive index spectra, circles represent refractive indices measured by an *m*-line method for an unirradiated sample (○) and irradiated sample(●). Lines represent continuous spectra of refractive index obtained with eq. (5). (b) Spectra of refractive index changes. Circles represent refractive index changes measured by an m-line method, the line represents refractive index changes obtained from Kramers-Kronig transformation by using a difference absorption spectrum from 200 nm to 2 μm.

optical devices. We consider a Mach-Zehnder optical switch shown in Figure 9. A photochromic dye is included in the core of the waveguide. Optical output is controlled by the phase shift induced by the refractive index changes during a photochromic reaction. First, we will describe a method to estimate the wavelength of the light to be switched, because refractive index change depends on wavelength.

In order to act as an optical switch, light should pass through the waveguide containing a photochromic dye. This condition is described according to Lambert-Beer's law,

$$L_{max} = -\frac{1}{100\,\varepsilon(\lambda)\,C} \log T \quad (6).$$

Here, C [mol/l] is the dye concentration, L_{max} [m] is the acceptable longest optical path, $\varepsilon(\lambda)$ [l/mol•cm] is the molar extinction coefficient at the wavelength of the light to be switched, λ, and T is the transmittance of the waveguide containing a photochromic dye. To simplify calculations, we introduce the reduced refractive index change, dN [mol^{-1}], based on the fact that the refractive index change, dn, is proportional to dye concentration and the conversion of a photochromic reaction, x, as given in eq. (7),

$$dN = \frac{dn}{C\,x} \quad (7).$$

Finally, we obtain an equation which gives the maximal available phase shift, ϕ_{max} .

$$\phi_{max} = k_0\, dn(\lambda)\, L_{max} = -\frac{2\,\pi\, dN(\lambda)}{100\,\lambda\,\varepsilon(\lambda)} \log T \quad (8)$$

The Spectra of dN and ε have been obtained. By putting them into eq. (8), we have obtained the wavelength dependence of ϕ_{max} shown in Figure 10. As a phase shift of π is required for full switching, the optimal wavelength of the light to be switched which meets this requirement is 624 nm for FG540/PMMA and 577 nm for MAB/PMMA films.

The minimal energy of light required for full switching is the energy needed for all of the photochromic molecules to be isomerized. The molar number of a

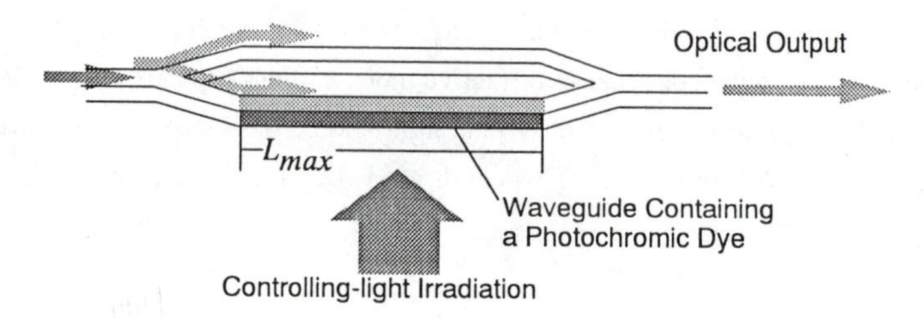

Figure 9. Configuration of a Mach-Zehnder photo-optical switch.

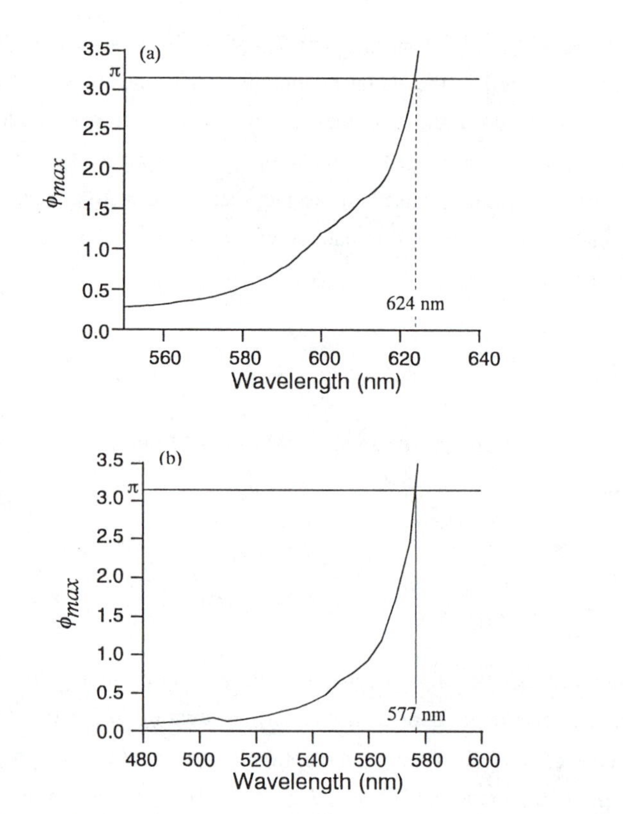

Figure 10. Wavelength dependence of available phase shifts: FG540/PMMA(a), MAB/PMMA(b)

photochromic dye in the waveguide is given by 1000 C L_{max} S in which the cross-section of the waveguide, S [m^2], is defined.

$$1000\, C\, L_{max}\, S = -\frac{10}{\varepsilon(\lambda)} \log T_{min} \tag{9}$$

Then the equation for the minimal energy required for full switching, E_{min}, is obtained as follows:

$$E_{min} = \frac{h\,c}{\lambda_{irr}\Phi}\, 1000\, C\, L_{max}\, S\, N_A \tag{10}$$

Here, h [J•s] is the Planck's constant, Φ is the quantum efficiency of the photochromic reaction, c is the light velocity in vacuum, N_A is the Avogadro's number, λ_{irr} is the wavelength of light inducing the photochromic reaction. Furthermore, we assumed that the value of S is 9×10^{-12} m^2 (= 3 μm × 3 μm), and T_{min} is 0.9. Table I shows the results of the estimation. A minimal energy of 180 *nJ* is obtained for FG540/PMMA and 840 *nJ* is obtained for MAB/PMMA. To further reduce these values requires the reduction of the size of this device, however, the cross-section of the waveguide cannot be smaller than the wavelength of light. Thus, it would not be possible to reduce the minimal energy of switching drastically below these values.

Conclusions

Refractive index spectra and their changes during photochromic reactions have been measured for several dye/polymer systems. The copolymer P(MMA-*co*-GMA-PNCA) shows remarkably large photoinduced refractive index changes of *ca*. 0.01 in a transparent wavelength region. FG540/PMMA and MAB/PMMA systems show larger wavelength dispersion of refractive index changes at wavelengths near resonant wavelength regions where absorbance changes during photochromic reactions. The properties of a Mach-Zehnder photo-optical switching device using FG540 and MAB were evaluated based on a trade-off relationship between the wavelength dependence of absorption loss and refractive index. The optimal wavelength and minimal switching energy were obtained as 624 nm and 170 nJ for

Table I. Estimated minimal switching energies

	FG540/PMMA	MAB/PMMA
Wavelength of switched light	624 nm	577 nm
Wavelength of Controlling light	345 nm	345 nm
Quantum efficiency	0.20	≈0.1 (26)
Amount of dye for full switching	9.6×10^{-14} *mol*	2.4×10^{-13} *mol*
Minimal switching energy	170 *nJ*	820 *nJ*

the FG540/PMMA system and 577 nm and 840 nJ for the MAB/PMMA system, respectively.

Literature Cited.

(1) Gibbons, W. M.; Shannon, P. J.; Sun, S.; and Swetlin, B. J. *Nature*, **1991**, *351*, 49.

(2) Ichimura, K.; Seki, T.; Kawanishi, Y.; Suzuki, Y.; Sakuragi, M.; Tamaki, T. In *Ultrahigh Density Optical Memory*; Irie, M. Ed.; Elsevier, Amusterdam, 1994, pp 130-158.

(3) Xie, S.; Nathanson, A.; Rochon, P.*Chem. Mater.*, **1993**, *5*, 201.

(4) Läsker, L.; Stumpe, J.; Fischer, T.; Rotloh, M.; Kostromin, S.; Ruhmann, R. *Mol. Cryst. Liq. Cryst.*, **1995**, *261*, 371.

(5) Hill, R. A.; Knoesen A.; Yankelevich, D. R.; Twieg, R. In *OSA Technical Digest Series 'Organic thin films for photonics applications'*, Portland, Oregon, September 11-14, 1995, pp 454-457.

(6) Charra, F.; Devaux, F.; Nunzi, J. M.; Raimond, P. *Phys. Rev. Lett.*, **1992**, *68*, 2440.

(7) Charra, F.; Kajzar, F.; Devaux, F.; Nunzi, J. M.; Raimond, P.; Idiart, F. *Opt. Lett.*, **1993**, *18*, 941.

(8) Todorov, T.; Nikolova, L.; Tomova, N. *Appl. Opt.*, **1984**, *23*, 4309.

(9) Todorov, T.; Nikolova, L.; Stoyanova, K.; Tomova, N. *Appl. Opt.*, **1985**, *24*, 785.

(10) Morino, S.; Machida S.; Yamashita T.; Horie K. *J. Phys. Chem.*, **1995**, *99*, 10280.

(11) Ebisawa, F.; Hoshino, M.; Sukegawa, K. *Appl. Phys. Lett*., **1994**, *65*, 2919.
(12) Kirkby, C. J. G.; Cush, R.; Bennion, I. *Opt. Commun.*, **1983**, *56*, 288.
(13) Lückemeyer, Th.; Franke, H. *Appl. Phys. Lett.*, **1988**, *53*, 2017.
(14) Lückemeyer, Th.; Franke, H.; *Appl. Phys. A*, **1992,** *55*, 41.
(15) Kardinahl, T.; Franke, H.; Appl. Phys. A, **1995**, *55*, 23.
(16) Tanio, N.; Irie, M.; *Jpn. J. Appl. Phys.*, **1994**, *33*, 1550.
(17) Tanio, N.; Irie, M.; *Jpn. J. Appl. Phys.*, **1994**, *33*, 3942.
(18) Tien, P. K. *Appl. Opt.*, **1971**, *10*, 2395.
(19) Prest, Jr. W. M.; Luca, D.J.; *J. Appl. Phys.*, **1979**, *50*, 6067.
(20) Nishikubo, T.; Kawashima, T.; Watanabe, S.; J. *Polym. Sci. A: Polym. Chem.*, **1993**, *31*, 1659.
(21) Nishikubo, T; Kameyama, A.; Nakajima, T.; Kishi, K.; *Polym. J.*, **1992**, *24*, 1165.
(22) Morino, S.; Watanabe, T.; Magaya, Y.; Yamashita, T.; Horie, K.; Nishikubo, T*J. Photopolym. Sci. Technol.* , **1994**, *7*, 121
(23) Morino, S.; Machida, S.; Yamashita, T.; Horie, K.; in *OSA Technical Digest Series 'Organic thin films for photonics applications'*, Portland, Oregon, September 11-14, 1995, pp 365-368
(24) Kinoshita, K.; Morino, S.; Horie, K.; Nishikubo, T.; to be published.
(25) Heller, H. G. in *Chemistry of Functional Dyes*, Yoshida, Z.; Kitao, T., Ed. Mita Press, Tokyo, 1989, pp. 267-287
(26) Rau, H. In *Photochromism, Molecules and Systems*, Dürr, H.; Bounas-Laurent, H., ed. Elsevier, Amsterdam, 1994, pp 165-192

Chapter 19

Molecular Dynamics of Liquid-Crystalline Side-Group Polymers with Fluorine-Containing Azobenzene Chromophores

A. Schönhals[1], R. Ruhmann[1], Th. Thiele[1], and D. Prescher[2]

[1]Institut für Angewandte Chemie Berlin-Adlershof, Rudower Chaussee 5, D–12484 Berlin, Germany
[2]Institut für Festkörperphysik, Universität Potsdam, Rudower Chaussee 5, D–12484 Berlin, Germany

Dielectric measurements on liquid-crystalline polymethacrylates with 4'-trifluoromethoxy-azobenzene mesogenic side groups and alkylene spacers of varying length (2-6) are reported. Several relaxation regions have been identified. The β-process (rotational fluctuation of the mesogenic unit around its long axis) and the δ-relaxation (rotational fluctuations of the mesogen around the short axis) show a behavior which is similar to that of the corresponding systems with phenylbenzoate as the mesogenic unit. The relaxation behavior of the intermediate α-process is quite unusually and has not been understood fully at present.

For several reasons there is still considerable interest in the synthesis and in the properties of comb-like polymers which have calamitic mesogens in the side group. From the pointof view of basic research these systems combine properties of anisotropic liquid-crystalline materials with those of polymeric systems (*1*). From a more practical point of view such compounds can be used as active materials for optical data storage, holographic applications, and electro-optical devices. This is especially true for systems having azobenzene-chromophores in the mesogenic unit because the properties (shape and dipole moment) can be changed by irradiation via a trans-cis isomerization process. Moreover, in recent years increasing attention is paid to fluorine- containing liquid crystals because enhancement of properties such as favorable mesophase behavior, low viscosity, high thermal stability, and high polarity is expected (*2-6*).

For an understanding of the phase behavior of liquid-crystalline systems in general knowledge of the dynamics is necessary. Moreover, in these materials molecular, dynamics is contemporary and important for their application in memory devices because the storage of information is directly connected with a reorientation of molecules or parts of it. Dielectric relaxation spectroscopy has proven to be a very suitable tool to study the molecular motion in these compounds for several reasons. Firstly, mesogens are usually polar. If different molecular reorientations involve different polar groups these motions can be detected as separate relaxation processes. Secondly, due to the broad frequency range which can be covered by dielectric spectroscopy (*7*) the change of these motions at phase transitions can be investigated. During the last years several reviews concerning this topic have appeared (*8-10*), where the most extensive one is Moscicki's work (*9*). The pioneering work in this field was done by Kresse and coworkers (*11*). Zentel et al. (*12*) have studied a variety of liquid-crystalline polyacrylates and polymethacrylates containing phenylbenzoate derivatives in the side group and they established a nomenclature for the different dielectric relaxation processes.(Investigations on fluorine-containing phenylbenzoate side groups have also been reported (*8*).) In unoriented samples in general, five relaxation processes can be detected. At the lowest temperatures, torsional fluctuations of the spacer and of the tail group of the mesogenic side group take place and cause the so called γ_1 and γ_2 relaxation peaks, respectively. With increasing temperature these processes are followed by a β-relaxation that was assigned to the rotational fluctuation of the mesogenic group around its long axis. This interpretation was further supported by a study on a set of combined main-chain side-group liquid-crystalline polymers (*13*) and NMR-investigations (*14,15*).

Whereas the temperature dependence of the relaxation rate of these processes is Arrhenius-like, the relaxation rate of the next (with increasing temperature) observed relaxation process, called α-relaxation, shows a curved dependence on 1/T. In many cases at low frequencies this temperature dependence agrees well with the glass transition temperature T_g (*12,16*). Therefore the α-relaxation is ascribed to the dynamic glass transition of the system very often (*12,16*). In contradiction to this interpretation it has also bee suggested that the α-relaxation is due to the transverse component of the molecular dipole moment of the mesogenic moiety (*17*). At higher temperatures than the α-relaxation (or equivalently at lower frequencies) a further relaxation process called δ-relaxation is found which seems characteristic for comb-like systems. Detailed investigations show that this relaxation process is caused by rotational fluctuations of the dipole component which is in parallel to the mesogenic side group about the local director axis (*12,16,17*). But however, the exact molecular motional mechanism of the δ-relaxation remains unclear up to now.

Compared to dielectric investigations on liquid-crystalline side-group polymers carrying derivatives of phenyl benzoate as the mesogenic unit studies on materials having azobenzene in the side group are rather rare (see for instance (*18*)). Therefore, we concentrate in this paper on dielectric experiments on liquid-crystalline side-group polymethacrylates with azobenzene in the mesogenic unit. Moreover, the azobenzene chromophores have fluorinated tail groups.

Experimental Part

As mesogenic groups, 4'-trifluoromethoxy-azobenzene units are used and linked to the methacrylate main chain by alkylene spacer, where the number n of the spacer groups was varied from 2 to 6. The chemical structure of the repeating unit of the polymers PM_n (n=2-6) is given in Figure 1. Details of the synthesis of the monomer, the condition for the radical polymerization, and the determination of the molecular masses were reported elsewhere (*19,20*). The phase transition temperatures of the polymers were determined by differential scanning calorimetry (DSC). Phase identification was done by both X-ray techniques and polarizing microscopy. More details can be found elsewhere (*20*). All characteristic sample parameters are collected in Table I. It should be noted that the polymers with spacer lengths n=5 and 6 have a smectic A phase at high and a nematic phase at low temperatures. This is a quite unusual behavior and has been reported for liquid-crystalline polymethacrylates for the first time (*20*). Moreover, the comparison of the X-ray data with the results of a molecular modeling study shows that overlapping of the mesogenic side groups is different in the smectic phases achieved by different spacer lengths. Whereas for the longest spacer lengths (PM_5 and PM_6) a bilayer structure was observed, the side group for the PM_4 sample overlaps totally in such a way that the F-C dipoles probably do not interact with the π-electron system of the phenyl ring of lateral neighboring side groups, but rather with phenyl rings of neighboring side groups. For the polymer PM_3, overlap of the phenyl rings of lateral neighboring side groups is possible without any steric hindrance. It is important to note that the influence of the main chain was not considered in the molecular modeling calculations.

For the dielectric measurements the samples were pressed between two gold-plated stainless steel electrodes (diameter 20 mm) at a temperature of 423 K. The spacing of the electrodes, 50±1 μm, was maintained by fused silica fibers. The complex dielectric permittivity ϵ^*,

$$\epsilon^*(f) = \epsilon'(f) - i\epsilon''(f) \tag{1}$$

(f-frequency, ϵ'-real part, ϵ''-imaginary or loss part, $i=\sqrt{-1}$), was measured in a frequency range from 10^{-2} Hz to 10^{6} Hz by a Schlumberger frequency-response analyzer model FRA 1260 interfaced to a buffer amplifier of variable gain (*21*) (Chelsea Dielectric Interface). A relaxation region is indicated by a step in ϵ' and a peak in ϵ''. The frequency of maximum loss f_p (f_p is also called the relaxation rate) is directly related to a characteristic relaxation time $\tau=1/(2\pi f_p)$ of the process under investigation.The temperature of the sample was varied from 170 K to 430 K by a custom made nitrogen gas jet heating system with a resolution of ±0.02 K. All samples were measured in nominal unaligned state. Measurements on aligned samples are in progress.

Figure 2 shows the dielectric loss for the polymer PM_5 versus temperature for different frequencies. Different relaxation regions indicated by peaks in ϵ'' are observed. At low temperatures a β-relaxation takes place. (For the highest frequency a further relaxation process can be seen which takes place at lower temperatures than the β-relaxation. This process has to be assigned to the γ-relaxation which is due to rotational fluctuation of the tail group.) This relaxation process is followed by the α-relaxation at higher and by the δ-process at the highest measuring temperatures. This relaxational behavior is quite the same for the other polymers (see Figure 3). However, it should to be noted that the α-relaxation is more suppressed for the even and shorter spacer lengths.

The temperature dependence of the relaxation rates of each process will be discussed in much more detail in the next sections for all polymers. But before doing

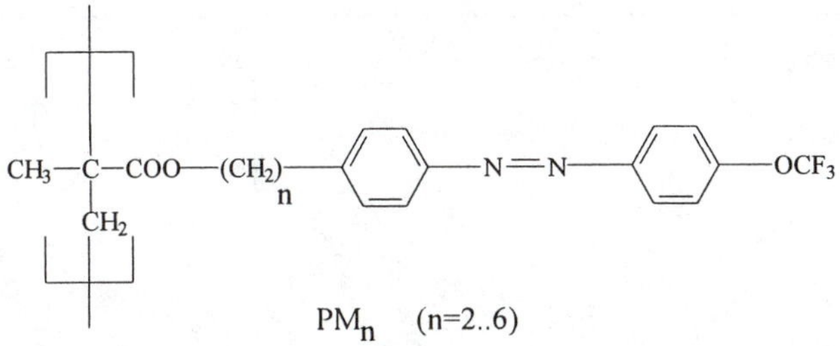

$CH_3-C(-CH_2-)-COO-(CH_2)_n-C_6H_4-N=N-C_6H_4-OCF_3$

PM_n (n=2..6)

Figure 1. Structure of the repeating unit of the polymers studied.

Table I. Chemical Structure and Transition Temperatures of the polymers

Polymer	M_w*10^4 g/mol	T_g K	Phase behavior[a]
PM_2	5.90	346	S_A 144 I[b]
PM_3	5.31	348	S_A 122 I[b]
PM_4	7.21	340	S_A 369 N 387 I
PM_5	3.93	339	N 393...398 S_A 405 I
PM_6	7.18	331	N 363 S_A 382 I

[a]S_A - Smectic A, N - Nematic, I - Isotropic; [b]On cooling only, one additional narrow smectic phase occurs just below the clearing temperature.

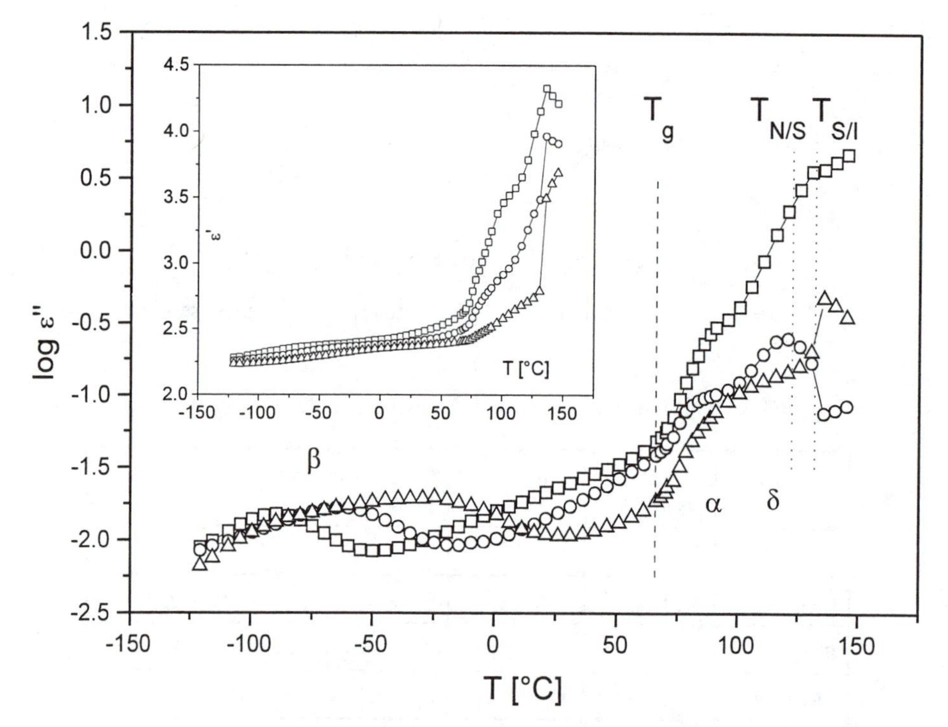

Figure 2. log ε'' vs. temperature at different frequencies for the sample PM5 : □ - 10 Hz, ○ - 10^3 Hz, Δ- 10^5 Hz. The inset shows ε′. The greek characters assign the different relaxation regions. Lines mark the transition temperatures.

that two additional points should be raised. Firstly, in the temperature range between the β- and the α-relaxation the dielectric loss increases relative strongly with increasing temperature but no clear relaxation peak can be detected. Moreover this effect seems to be more pronounced for shorter spacer lengths. Nevertheless according to the fluctuation dissipation theorem these losses are caused by real motional processes (*22*). Unfortunately they cannot be assigned to a specific mechanism at the moment. Such a behavior has not been observed for liquid-crystalline polymethacrylates with derivative of phenylbenzoate as mesogenic moiety (*16,23*). Secondly, the conductivity contribution is very high at low frequencies in the liquid crystalline phase (see Figure 2). This is due to a so-called Maxwell-Wagner-effect (*24*) which is caused by blocking of charge carriers at interfaces like electrodes or internal phase boundaries. This effect is accompanied by an upturne of ϵ' at low frequencies. If internal interfaces are formed for instance during the phase transition this effect should be enhanced. For this reason, in Figure 4 ϵ' is plotted versus frequencies for temperatures just above and below the mesophase formation. As one can judge from this plot for the lower temperature the upturn of ϵ' is more pronounced and starts at higher frequencies than for the higher temperature. This result can be taken as a direct proof of blocking of charge carriers at internal interfaces built up by the formation of liquid-crystalline structures. Clearly this has to be considered in the evaluation of the measurements.

Results and Discussion

β-Relaxation. To analyze the β-relaxation in more detail from the isothermal measured frequency scans temperature dependent charts at fixed frequency were constructed. From these plots the temperature of maximal loss was taken at the selected frequency. The results are shown in Figure 5 where the logarithm of the relaxation rate of the β-process log $f_{p\beta}$ is plotted versus the reciprocal temperature. For all samples the temperature dependence of the relaxation rate $f_{p\beta}$ can be described by an Arrhenius-equation

$$f_{p\beta} = f_{\beta\infty} \exp \left[- \frac{E_{A\beta}}{kT} \right] \quad , \tag{2}$$

where $f_{\beta\infty}$ is the preexponential factor and $E_{A\beta}$ is the activation energy. The quality of the fits is demonstrated by the lines in Figure 5. In Table II log $f_{\beta\infty}$ and $E_{A\beta}$ are summarized and compared to the low temperature mesophase structure which is frozen in at T_g. Both log $f_{\beta\infty}$ and $E_{A\beta}$ (see inset of Figure 5) increase with increasing spacer length up to n=4. For n=5 and n=6 both quantities decrease slightly. It also should be noted that especially log $f_{\beta\infty}$ is much too high for a truly activated process. From a formal point of view that could be understood in terms of an activation entropy which obviously becomes more important for longer spacers. In Figure 6 both parameters, log $f_{\beta\infty}$ and $E_{A\beta}$, are plotted vice versa in a so-called compensation (*16*) plot and it can be seen that both quantities correlate linearly.

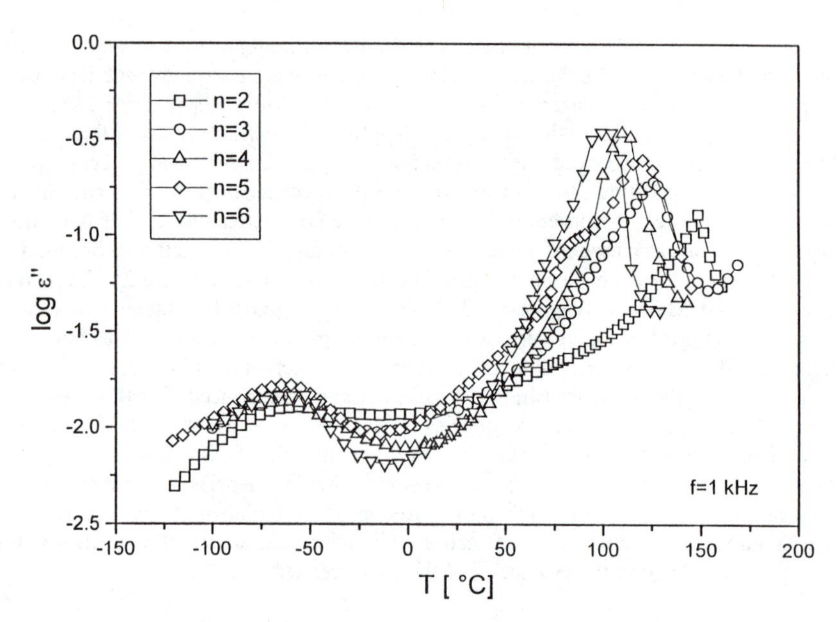

Figure 3. Log ϵ" vs. temperature at a frequency of 1000 Hz for the different liquid-crystalline polymers.

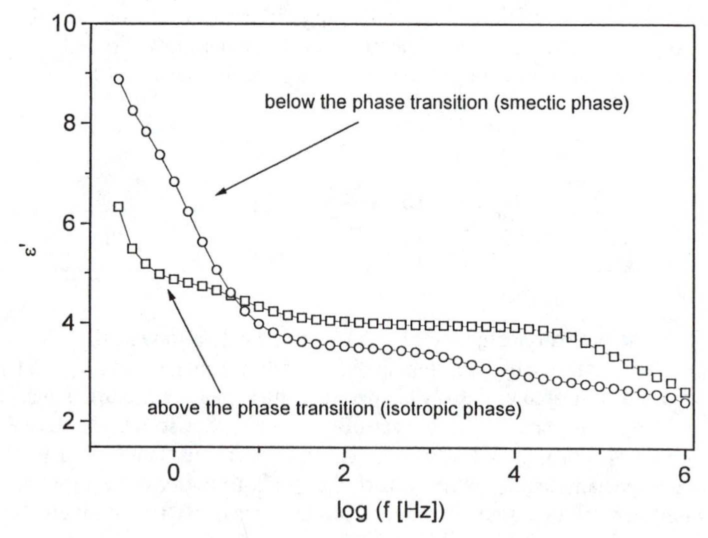

Figure 4. ϵ´ vs. logarithm of frequency for the polymer PM_5: □ - 408.8 K (isotropic phase); O - 398.85 K (smectic A phase). The steep increase of ϵ´ at low frequencies is due to a so-called Maxwell-Wagner polarization. This effcet is enhanced for the liquid-crystalline state because the charge carriers can be also blocked in this case on internal interfaces like the smectic layers.

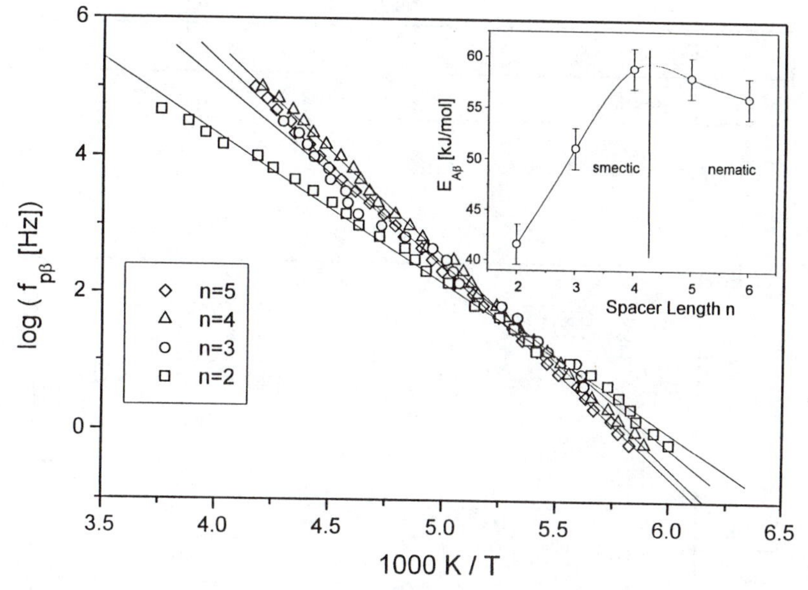

Figure 5. Relaxation rate of the β-relaxation versus 1/T. Lines are fits of the Arrhenius-equation to the data. The inset displays the dependence of the activation engery $E_{A\beta}$ on spacer lengths.

Table II. Activation Parameters of the β-Relaxation

Spacer lengths	Low temperature phase[1]	log ($f_{\beta\infty}$ [Hz])	$E_{A\beta}$ [kJ /mol]
n=2	smectic	13.6 ± 0.3	41.6 ± 1.4
n=3	smectic	15.7 ± 0.5	51.0 ± 2
n=4	smectic	17.9 ± 0,2	58.9 ± 1
n=5	nematic	17.5 ±0.3	57.9 ± 1
n=6	nematic	16.8 ±0.5	56.0 ± 2

[1]Which is frozen-in at the glass transition.

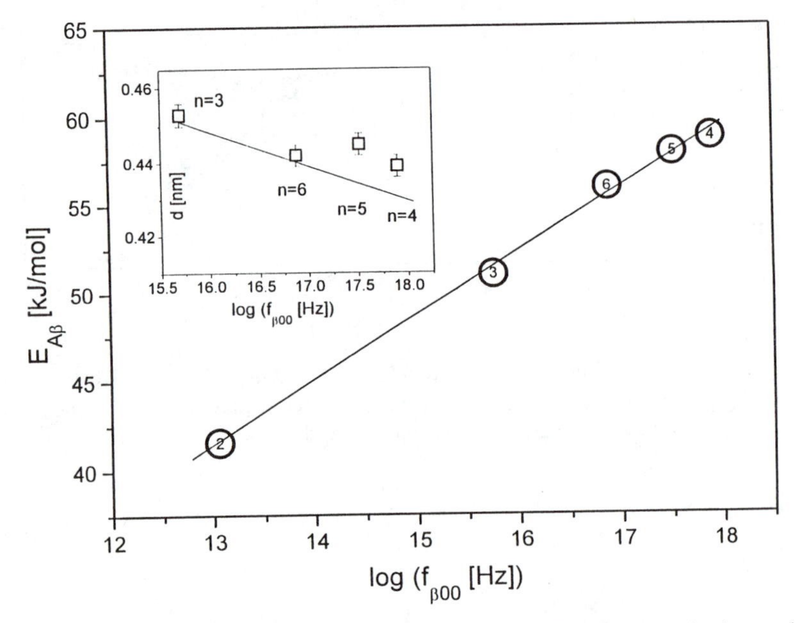

Figure 6. Log $f_{\beta\infty}$ versus $E_{A\beta}$ for the different samples. The numbers in the symbols characterizes the spacer lengths. The inset displays the mean lateral distance of mesogens d vs. log $f_{\beta\infty}$.

Recently, for polymethacrylates with derivatives of phenylbenzoate as the mesogenic unit it was found that both the preexponential factor and the activation energy increase with the order of the mesophases (*16*). This means that both parameters have smaller values for less ordered nematic phases and adopted higher values for well ordered smectic phases. Because the β-relaxation has been assigned to rotational fluctuation of the mesogenic unit around its long axis such a motional process should be more complicated in well organized phases than in poorly ordered ones. This is exactly reflected in the higher values for the preexponential factor as well as in the activation energy. Bearing these results in mind, the results for the fluorinated azobenzene containing polymers can be interpreted in the following way. Although the low temperature phases for PM_2 and PM_3 are smectic phases due to the short spacer lengths the decoupling of the mesogens from the backbone is restricted and so the molecular order of these smectic phases is rather low which yields the relative low values for the activation parameters. With increasing spacer lengths the decoupling of the mesogenic unit from the main chain is enhanced and the mesogenic moieties in the mesophases are better organized. Consequently the activation parameters for the β-relaxation increase. For the spacer lengths n=5 and n=6 the activation parameters decrease slightly. As our previous results imply this should be the case for nematic compared to smectic phases. However, it should be noted that both log $f_{\beta\infty}$ and $E_{A\beta}$ are larger for the nematic phases of the fluorinated azobenzene samples than the nematic phase achieved by shorter spacers in phenyl-benzoate containing polymethacrylates (*16*).

For phenylbenzoate-containing polymethacrylates with odd spacer lengths it was found (*16*) that the preexponential factor correlates linearly with the mean lateral distance of the mesogenic unit which can be estimated from X-ray experiments. A similar plot is given for the presented systems in the inset of Figure 6. Although the number of data points is limited, a correlation seems toalso exist for the investigated systems. A comparison of liquid-crystalline polymethacrylates with azobenzene and phenylbenzoate mesophase forming units in the side group is in preparation. That study will also include phenylbenzoate-containing liquid-crystalline polymethacrylates with odd spacer lengths. Also a detailed comparison of the β-relaxation of this two types of side group liquid-crystalline polymers is also in preparation.

α-and δ-Relaxation. In the most cases model functions such as the Havriliak-Negami-function (*25*) have been widely used (*8,23*) to analyze the relaxational behavior of liquid-crystalline side group polymers. This evaluation method can also be applied to separate overlapping relaxation processes if these processes have comparable relaxation strengths and do not overlap so closely. Unfortunately for the systems investigated, firstly, the δ-relaxation is much more intensive than the α-process (compare Figure 3). Secondly, α- and δ-processes overlap very strongly where the α-relaxation is generally very broad. Only for the polymer PM_5 both relaxation processes are clearly resolved in the temperature domain (see Figure 3). (Note that generally a plot in the temperature domain resolves overlapping relaxation processes more than a corresponding plot in the frequency domain.) For these reasons the number of free fit parameters must be reduced (mostly done by not so well founded arguments) or the unrestricted fit of the model functions to the data will lead to relative large errors in the estimated fit parameters.

Therefore to analyze the α- and the δ-relaxation of the polymers investigated another method has been employed. This is based on taking the derivative of ϵ' with respect to log f i.e. $\Delta = \partial\epsilon' / \partial \log f$. This way of analyzing dielectric data was introduced by van Turnhout and Wübbenhorst (van Turnhout, J.; Wübbenhorst, M., personal coomunication). It can be shown for Debye-like relaxation processes that

$$\Delta = \frac{\partial \epsilon'}{\partial \log f} \sim -\epsilon''^2 \tag{3}$$

holds. This means, firstly, that a relaxation peak in ϵ'' gives a minimum in Δ. Secondly, because of the square in relationship (3) the minimum is much more narrower than the peak in ϵ''. So strongly overlapped relaxation peaks are better resolved. Further, very broad dielectric loss peaks which are typical for crystalline and liquid-crystalline materials become narrower in such a representation and so the position of maximum loss can be better identified. This argument is supported by the fact that relationship (3) can be regarded as the simplest approximation of the dielectric relaxation spectra (*26*) which is known to be much smaller than the peak in the loss part itself. (Note that the dielectric relaxation spectrum for the Debye-function is the singular Dirac-function.) Furthermore conductivity contributions according to $\epsilon'' \sim 1/f$ do not play any role because for such a dependence the real part ϵ' is independent of frequency. Therefore electrode polarization and Maxwell-Wagner effects can be also analyzed in more detail. A theoretical study of the proposed method is in preparation (Schönhals, A. in preparation).
The power of the proposed method is demonstrated in Figure 7. Figure 7a displays the dielectric loss for the polymer PM_5. Only at the highest measuring temperature are the α- and the δ-relaxation middling visible. At lower temperatures the δ-relaxation appears only as a weak shoulder strongly overlapped by a conductivity contribution. Moreover for the lowest temperature the α-relaxation is very broad and flat. So it is very hard to estimate the relaxation rate. Figure 7b gives the derivative of ϵ' versus log f for the same temperatures. From a practical point of view to analyze the measured data before the derivative was taken the ϵ'-data have been smoothed over 5 points. For all temperatures the δ-relaxation is very well resolved. Even for the lowest temperature where in the loss curve the peak can be hardly detected Δ displays a well defined minimum for the δ-process. Also the relaxation rate for the α-process can be identified unambiguously although the minimum is fairly broad. It should be noted that in general the α-process becomes broader with decreasing temperature.

Typical examples of the temperature dependence of the relaxation rates of the α- and δ-relaxation ($\log f_{p\alpha}$ and $\log f_{p\delta}$) are plotted in Figures 8 and 9 for the polymers PM_4 and PM_6, respectively. The high temperature δ-relaxation process shows a behavior which is also found for liquid-crystalline polymethacrylates carrying phenylbenzoate as mesophase-forming unit (*16,23*). This means a step-like change of the temperature dependence of $\log f_{p\delta}$ at the transition from the isotropic to the liquid-crystalline state. For the samples PM_6 and PM_5 which have a smectic high temperature phase no further step-like change at the phase transition from the smectic high temperature phase to the nematic low temperature phase is observed. This implies that

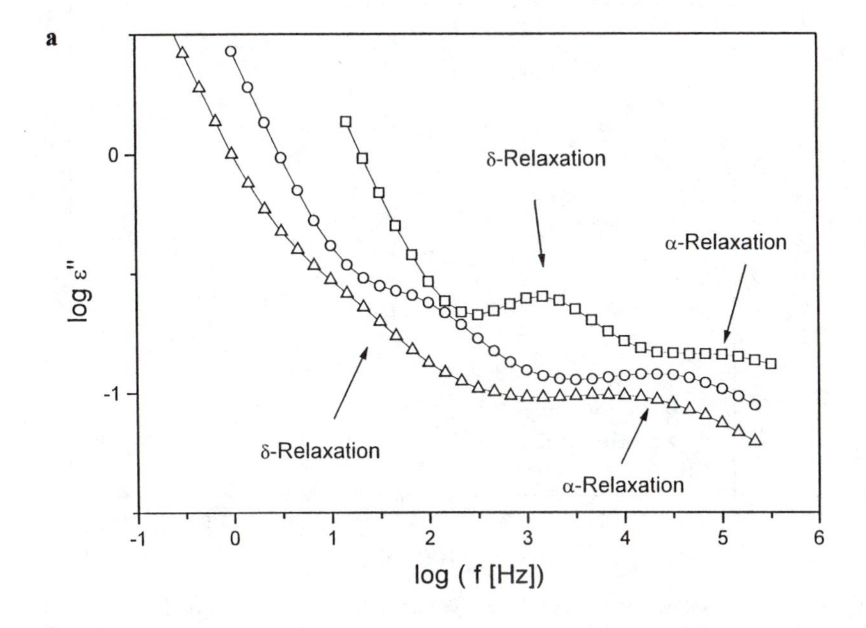

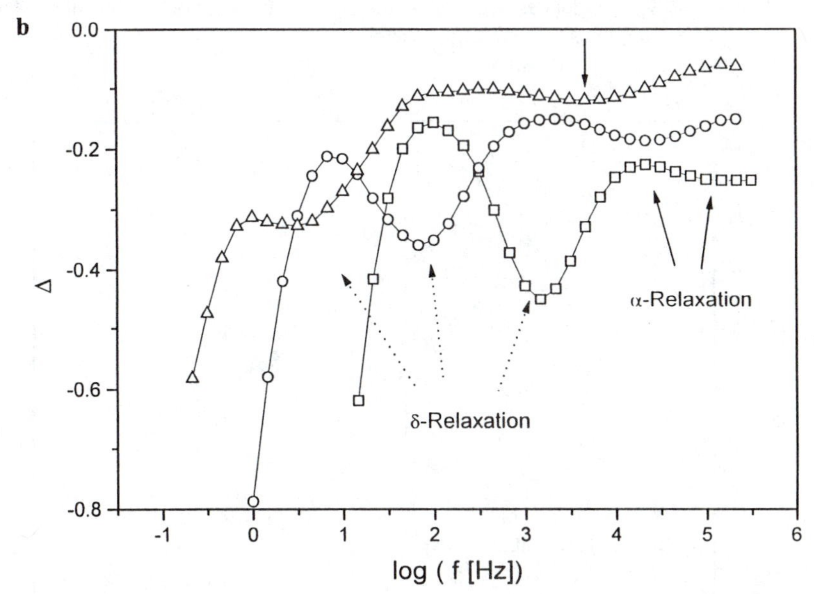

Figure 7. a: Dielectric loss vs. logarithm of frequency for the polymer PM_5 : □- 393.8 K, ○- 373.8 K, Δ-359.1 K. b: $\Delta = \partial \epsilon' / \partial \log f$ for the polymer PM_5 : □- 393.8 K, ○- 373.8 K, Δ-359.1 K.

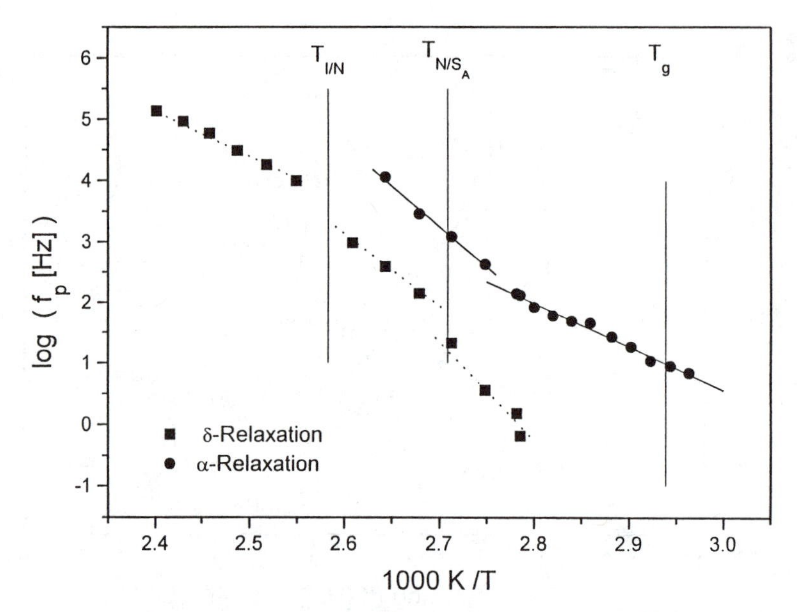

Figure 8. Relaxation rates of the α- (●) and of the δ-process (■) for the polymer PM_4. Solid lines are fits of the Arrhenius-equation to the data. The dotted lines are guides for the eyes.

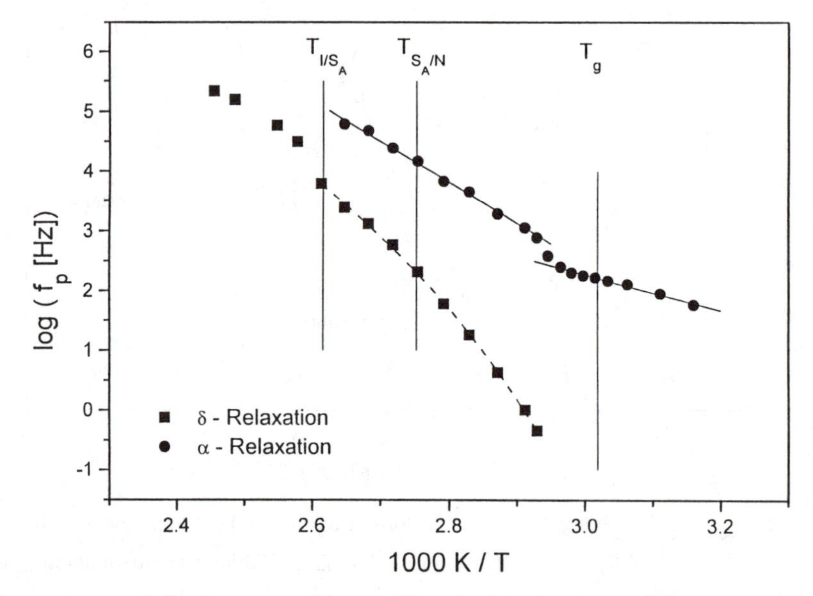

Figure 9. Relaxation rates of the α- (●) and of the δ-process (■) for the polymer PM_6. Solid lines are fits of the Arrhenius-equation to the data. The dashed line is the the fit of the VFT-equation to the data.

the molecular mobility is not changed so much at that phase transition. This is not the case for the sample PM_4 which has an opposite phase sequence. For this material a further step-like change of log $f_{p\delta}$ is observed at the transition from the nematic to the smectic phase, indicating a further change of mobility at that phase transition. Moreover, the dependence of log $f_{p\delta}$ is strongly curved in a plot over 1/T (see Figures 8 and 9) which is characteristic of glass-forming systems. In many cases such a behavior can be described by the Vogel/Fulcher/Tammann-equation (VFT) (*27*)

$$\log f_{p\delta} = \log f_{\delta\infty} - \frac{A}{T - T_0} , \quad (4)$$

where log $f_{\delta\infty}$ and A are constants and T_0 is the so called ideal glass transition temperature. For liquid -crystalline materials it was argued that equation (4) has to be modified to introduce the influence of molecular order (*8,16*). Nevertheless appropriate fits have been done for the samples PM_3, PM_5 and PM_6 for the low temperature relaxation data. (This means that only data at temperatures below the lowest phase transition temperature have been included.) To reduce the number of fit-parameters log $f_{\delta\infty}$ was fixed at 10 because this number was found as the mean value of the results obtained by free unrestricted fits. The estimated parameters are displayed in Table III and the quality of the fits can be judged from Figure 9.The difference between the glass transition temperature T_g and T_0 is around 60 K to 70 K for all three samples. Typically such a value have been found for the dynamic glass transition of simple amorphous polymers. For that reason this result suggests, that the δ-process in our systems seems to be closely related to the glass transition.

Compared to the temperature dependence of log $f_{p\delta}$ the temperature dependence of log $f_{p\alpha}$ is very unusual. For all spacer lengths (n=3-6) evaluated the temperature dependence of log $f_{p\alpha}$ has two different regimes (see Fig. 8 and 9). The transition between these two dependencies occurs just above the glass transition temperature at a relaxation rate log $f_{p,\,dc}$ of approximately 2.5 which seems to be a constant for all samples. Such a behavior has not been found for liquid-crystalline polymethacrylates with phenyl benzoate in the side group. For these systems the α-relaxation at low frequencies correlates nicely to the glass transition temperature (*16*). To be more quantitative for each region an Arrhenius equation is fitted to the data which yields two apparent activation energies $E_{\alpha H}$ and $E_{\alpha L}$ for the high and low temperature regimes, respectively. The results are given in Table IV. For the high temperature regions the estimated apparent activation energies are much too high for local motional processes. This means that the molecular motions responsible for this relaxation process should have some degree of cooperativity. The apparent activation energy for the low temperature regime is much lower where the ratio between both activation energies varies from 1.4 to 4.5 unsystematically (see Table IV). From the change in the apparent activation energy one can conclude that the character of the molecular motion involved in the α-relaxation changes. As to the molecular reason for such a behavior one can only speculate for now. One possibility is that just above T_g an additional phase transformation or modification of the liquid-crystalline structure takes place which is not visible in the X-ray experiments but changes the molecular mobility of the system.

Table III. Vogel/Fulcher/Tammann-Parameter for the δ-Relaxation

Spacer lengths	A [K]	T_0 [K]	T_g-T_0 [K]
n=3	822±15	290±10	58
n=4	-	-	-
n=5	828±15	271±5	68
n=6	743±15	273±5	58

Table IV. Apparent Activation Energies for the α-Relaxation

Spacer lengths	High temperature $E_{\alpha H}$ [kJ /mol]	Low temperature $E_{\alpha L}$ [kJ /mol]	$E_{\alpha H}$/ $E_{\alpha L}$
n=3	94.1	20.9	4.5
n=4	253.8	136.5	1.86
n=5	92.1	66.8	1.4
n=6	131.0	57.8	2.2

Another interpretation starts from the fact that the degree of cooperativity of molecular motions responsible for the α-process increases with decreasing temperature. Such an argument is often used to explain the Vogel/Fulcher/Tammann-behavior of the relaxation rate near the glass transition (see equation (4)) (*28*). It may be possible that at some point the degree of cooperativity of this process interferes with the liquid-crystalline structure. At that point the cooperativity cannot increase further and one can argue that the apparent activation energy should change. To decide which for an interpretation is correct additional investigations have to be carried out.

Conclusion

In conclusion we have done dielectric measurements on novel liquid-crystalline polymethacrylates with fluorine-containing azobenzene-chromophores (4'-trifluoromethoxy-azobenzene) as the mesogenic unit. The dielectric spectra show several relaxation processes, including the β-, α- and δ-relaxations which have been analyzed in detail. The temperature dependence of the relaxation rate of the β-process (rotational fluctuation of the mesogenic unit around the long axis) is Arrhenius-like and the activation parameters increase with the molecular order. The relaxation rate of the δ-process (rotational fluctuation of the mesogenic unit around the short axis) displays a curved behavior in a 1/T plot with step-like changes at phase transitions. A similar behavior was found recently for polymethacrylates with phenylbenzoate in the side chain. The temperature dependence of the relaxation rate of the α-process is very unusual compared to similar systems with phenylbenzoate as the mesogenic moiety. A change of its temperature dependence just above the glass transition temperature has been observed which has not been fully understood at present.

Further investigations will include related copolymer systems with fluorinated and non-fluorinated methoxy tail groups and liquid-crystalline side group polymers with longer, partially fluorinated systems.

Acknowledgements

The financial support of Th. Thiele by the German Sciences Foundation (DFG) is gratefully acknowledged. The work was also supported by the German Federal Minister of Education, Science and Technology and by the Berlin-State Senator for Science, Research and Cultural Affairs (project No.: 03C3005). A. Schönhals thanks S. Engelschalt for many suggestions and technical support. We further thank H.-E. Carius and D. Wolff for helpful discussions.

Literature Cited

(1) *Side Chain Liquid Crystal Polymers*, McArdele, C.B., Ed.; Blackie: Glasgow, 1989.

(2) Percec, V.; Tomazos, D.; Feiring, A. E. *Polymer* **1991**, *32*, 1897.

(3) Attard,G.S.; Dave, J.S.; Wallington, A. *Makromol. Chem.* **1991**, *192*, 1495.

(4) Kitazume, T.; Ohnogi,T.; Ito, K. *J. Am. Chem. Soc.* **1990**, *112*, 6608.

(5) LeBarny, P.; Ravaux, G.; Dubois, J.C.; Parnaix, J.P.; Njeumo, R.; LeGrand, C.; Levelut, A. M. *Proc. SPIE. Int. Soc. Opt. Eng (USA)* **1987**, *682*,56.

(6) LeGrand, C.; Bnunel, C.; Le Borgne, A.; Lacondre, N.; Spassky, N.; Vairan, J.-P. *Makromol. Chem.* **1990**, *191*, 2971.

(7) Schönhals, A.; Kremer, F.; Schlosser, E.; *Phys. Rev. Lett.* **1991**, 67, 999.

(8) Haws, C. M.; Clark, M.G.; Attard, G.S. in ref.1.

(9) Moscicki; J.K. in *Liquid Crystal Polymers- From Structures to Applications*; Collyer; A.A., Ed.; Elsevier, 1992.

(10) Simon, G.P. in Dielectric Spectroscopy of Polymers; Runt, J.; Fitzgerald, J., Eds.; American Chem. Society: Washington, D.C., in press.

(11) Kresse, H.; Talroze, V.R. *Makromol. Chem., Rapid Commun.* **1981**, *2*, 869 ; Kresse, H.; Kostromin, S.; Shibaev, V.P. *Makromol. Chem, Rapid Commun.* **1982**, *3*, 509.

(12) Zentel, R.; Strobel, G.; Ringsdorf, H. *Macromolecules* **1985**, 18, 960.

(13) Kremer, F.; Vallerien, S.U.; Zentel, R.; Kapitza, H. *Macromolecules* **1989**, 22, 4040.

(14) Pschorn, U.; Spiess, W.H; Hisgen, B.; Ringsdorf, H. *Makromol. Chem.***1986**, *187*, 2711

(15) Vallerien, S.U.; Kremer, F.; Boeffel, C. *Liquid Crystals* **1989**, *4*, 9.

(16) Schönhals, A.; Wolff, D.; Springer, J. *SPIE. Int. Soc. Opt. Eng (USA)* **1996**, *2779*,424.

(17) Attard, G.S.; Williams, G. *Polymer Communications* **1986**, *27*, 2.

(18) Wendorff, J.H.; Furmann, Th. *Dielectric Newsletter*; **1994**, July, 1.

(19) Ruhmann, R.; Thiele, T.; Prescher, D.; Wolff, D. *Makromol. Chem., Rapid Commun.* **1995**, 16, 161.

(20) Wolff, D.; Ruhmann, R.; Thiele, T.; Prescher, D.; Springer, J. *Liquid Crystals* **1996**, *20*, 553.

(21) Kremer, F.; Boese, D.; Meier, G.; Fischer E.W. *Prog. Colloid Polym. Sci.* **1989**, *80*, 129 ; Pugh, J.; Ryan, T. *IEE Conf. on Dielectric Material Measurements and Applications* **1979**, *177*, 404.

(22) Landau, L.; Lifschitz, E.M. *Statistical Physics*; Akademie Verlag: Berlin, 1979.

(23) Schönhals, A.; Gessner, U.; Rübner, J. *Makromol. Chem.* **1995**, *196,* 1671.

(24) Sillars, R.W. *J. Inst. Elect. Eng.* **1937**, *80*, 378.

(25) Havriliak, S.; Negami, S. *J. Polym. Sci.* **1966**, *C14*, 99.

(26) Böttcher, C.J.F.; Bordewijk, P. *Theory of Dielectric Polarization,* Elsevier: Amsterdam, Netherland, 1978, Vol. II

(27) Vogel, H. *Phys. Z.* **1921**, *22*, 645; Fulcher, G.S.; *J. Am. Chem. Soc.*, **1925**, *8*,339; Tammann, G.; Hesse W. *Z. Anorg. Allg. Chem.* **1926**, *156*, 245.

(28) Donth, E.J. *Relaxation and Thermodynamics in Polymers, Glass Transition*; Akademie Verlag: Berlin, 1992.

Chapter 20

The "Plastic Retina": Image Enhancement Using Polymer Grid Triode Arrays

Alan J. Heeger, David J. Heeger[1], John Langan, and Yang Yang

UNIAX Corporation, 6780 Cortona Drive, Santa Barbara, CA 93117

An array of polymer grid triodes (PGTs) connected through a common grid functions as a "plastic retina" which provides local contrast gain control for image enhancement. This device, made from layers of conducting polymers, functions as an active resistive network that performs center-surround filtering. The PGT array with common grid is a continuous analog of the discrete approach of Mead, with a variety of fabrication advantages and with a significant saving of "real estate" within the unit cell of each pixel.

When a person views a brightly lighted external scene through a window from inside a poorly lighted room, the individual has no difficulty seeing, simultaneously, the details of both the internal scene and the external scene. This is done by local contrast control; the visual system locally adjujsts the gain using lateral inhibition. An illustrative example is the office scene shown in Figure 1. Figure 1a is the original (14 bit) image, but displayed using only the dynamic range available on the printed page (approximately 8 bits). Gray regions greater than 255 were clipped (set to 255), simulating saturation in the region of highest brightness. In Figure 1b, the same image was rescaled (the intensity of each pixel was divided by 8) and displayed using the dynamic range available on the printed page (again, gray values greater than 255 were clipped). Figure 1b is thus analogous to the image shown on a video display with reduced gain; the features are visible only in the background (bright) regions. In both cases, a great deal of information contained in the original image has been lost; in Figure 1a, the viewer cannot see any detail in the bright regions of the image, in Figure 1b the viewer cannot see any detail in the darker regions of the image.

Local contrast control involves a combination of logarithmic compression and lateral inhibition, the latter provided by a horizontal resistive network (a "neural network") (*1*). After logarithmic compression, the output from a given pixel (V_i) is

[1]Current address: Department of Psychology, Stanford University, Stanford, CA 94305

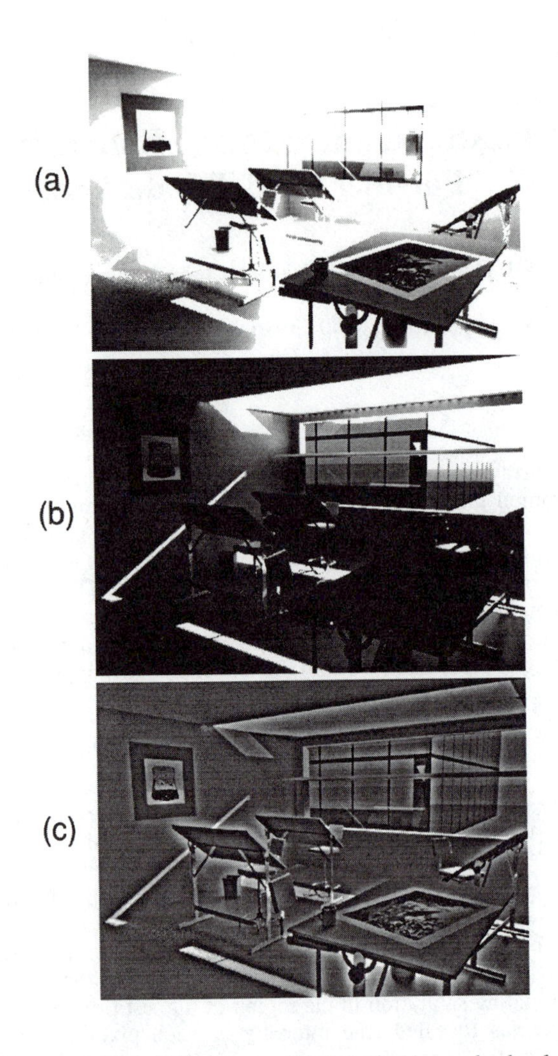

Figure 1:

a. Original (14 bit) image of an office scene displayed using only the dynamic range available on the printed page (approx. 8 bits). Gray values greater than 255 were clipped (set to 255). Features are visible only in the foreground (dark) regions.

b. The same image, rescaled (the intensity of each pixel was divided by 8), and again displayed using the dynamic range available on the printed page (again, gray values greater than 255 were clipped). Features are clearly visible only in the (bright) background.

c. The result of renormalizing the office scence image with local contrast gain control using Eq 3. Features are now visible throughout the image. (Reproduced with permission from ref. 2. Copyright 1995).

proportional to the log of the intensity (I_i) of that pixel:

$$V_i = V_o \log I_i \tag{1}$$

The lateral inhibition is implemented by subtracting from V_i the average of the surrounding values; thus, the renormalized image is defined by

$$v_i = V_o \log I_i - \langle V_o \log I_j \rangle_{ave} = V_o \log I_i - (V_o/N)\Sigma_j \log I \tag{2}$$

where the sum is over the neighboring pixels within the averaging range (the center surround or blurring range). Eq 2 is equivalent to:

$$v_i = V_o \log\left[\frac{I_i}{\left(\prod I_i\right)^{1/N}}\right] \tag{3}$$

The denominator is the geometric mean. Eq 3 is the mathematical expression which defines Mead's local contrast enhancement algorithm (1); the equation generalizes to allow for a weighted average by replacing 1/N by w_i where w_i are the weights.

The effect of processing the the original image of the office scene with the Mead algorithm is shown in Figure 1c. Figure 1c was obtained by renormalizing the 14 bit image with Eq 3 using a Gaussian weighted average over an 13x13 neighborhood. In Figure 1c, one can see details in the entire image even within the limited range available on the printed page. The simulations in Figure 1 demonstrate the power of local contrast enhancement.

Polymer Grid Triodes

The local contrast enhancement algorithm (Eqs. 2 and 3) can be implemented with a simple device made from layers of conducting polymers (2). The device consists of an array of polymer grid triodes (*3, 4*) connected through a common grid which serves as a resistive network.

The layered thin film PGT is shown schematically in Figure 2; the top layer **(5)** is the anode, the bottom layer **(1)** is the cathode. The third electrode **(3)**, analogous to the grid in a vacuum tube triode, is an open network of polyaniline (PANI) protonated to the highly conducting form with camphor sulfonic acid (CSA) (*3, 4*). Semiconducting polymer forms the two layers, **(2)** and **(4)**, between the anode and the polymer grid and between the polymer grid and the cathode, and it fills the void spaces within the porous PANI-CSA network **(3')**.

The tenuous network of conducting polyaniline that self-assembles in blends of PANI with insulating host polymers results from a compromise: the counter-ions want to be at the interface between the polar PANI (a salt) and the weakly polar (or nonpolar) host (*5,6,7,8*). On the other hand, the PANI and host tend to phase separate because there is no entropy of mixing for macromolecules. The result is a phase-separated structure with high surface area and with holes on every length scale (*9*).

Although the existence of the network morphology can be rationalized in these terms, a deeper understanding of the issues involved in the self-assembly of networks is lacking. The electron micrographs of the PANI networks indicate that, when the the surface-to-volume ratio of the PANI-CSA regions becomes too large, the connected network morphology cannot be maintained; i.e. there is a minimum diameter of order a few hundred Ångstroms for the connecting lengths. The origin of this minimum dimension is not understood.

The use of 'surfactant' counter-ions was introduced with the goal of making polyaniline processable in the conducting form (10). The self-assembly of the phase-separated network morphology was an unexpected - but very welcome - bonus. The generalization to a larger class of conducting polyblends with network morphology will depend on the ability to generate conducting polymers that can be processed in the conducting form. If this can be done, the possible materials for use as the 'grid' in PGTs and in arrays of PGTs can be expanded as well.

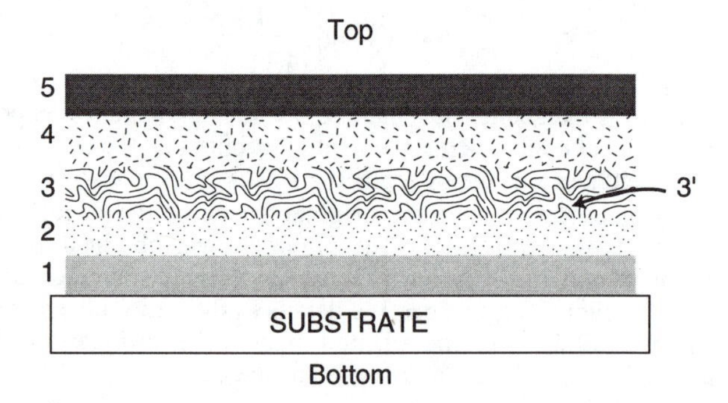

Figure 2: Structure of the polymer grid triode with the various layers. (**1**) and (**5**) are the cathode and anode (pixel) arrays, respectively. The other layers are continuous films common to all the PGTs within the array: (**2**) and(**4**) are semiconducting layers, poly(2-methoxy-5-(2'-ethyl-hexyloxy)-1,4-phenylene vinylene), MEH-PPV, and (**3**) is the common grid network filled with semiconductor (**3'**).

Polymer Grid Triode Arrays

An array of PGTs with a common grid (Figure 3) can perform local contrast enhancement like that simulated in Figure 1. Since the thin films which constitute the PGT array can be processed from solution, they can be layered directly on an array of photodetectors. Each node of the PGT array corresponds to one pixel of the image.

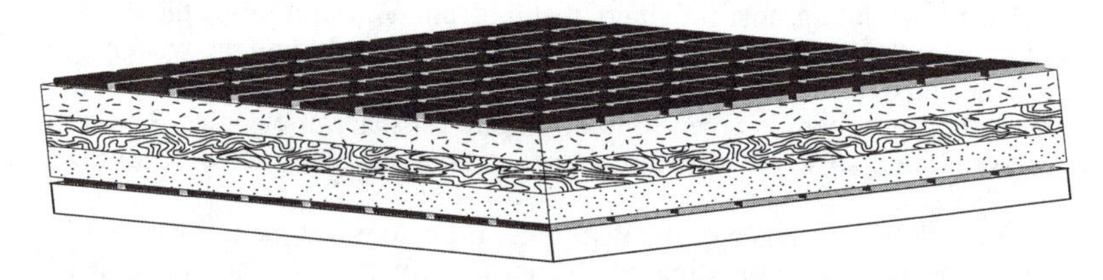

Figure 3: Schematic diagram of an array of PGTs with a common grid. Thickness (between anode and cathode) is approximately 0.3 μm. Each anode/cathode pad is, for example, 50 μm on a side.

The array of PGTs with common grid performs three important functions:

(i) The common grid functions as a resistive network that computes the averaging in Eq. 2.

(ii) The output current at one node of the PGT array is approximately the difference between the input anode-to-cathode voltage and the local grid voltage (see inset of Figure 6). Since the local grid voltage is the local average, the PGT array acts as a center-surround filter (*1, 2*) that computes the difference in Eq. 2.

(iii) The common grid PGT array provides the high input resistance needed for open circuit detector operation (*5*), which results in the logarithmic compression in Eq. 2.

Local Contrast Enhancement with PGT Arrays

Thus, local contrast enhancement, as described by Eqs. 2 and 3, can be directly implemented with the PGT array shown in Figure 3. The effective computation rates involved are impressive. The equivalent calculation implemented on a serial computer would require about 40 million mulitplications per second (256x256 pixels, 30 frames per second, 13x13 blurring using separable convolution).

Polymer grid triode arrays, with four triodes on a single substrate all with a common grid, have been fabricated. The structure of this linear array is shown schematically in Figure 4. The fabrication process for the triode array is similar to that of a single PGT (*3, 4*); the principal difference is that there are separate contact pads for each device in the array. For the arrays fabricated in this initial study, the sheet resistance of the common grid was approximately 20 KΩ/square.

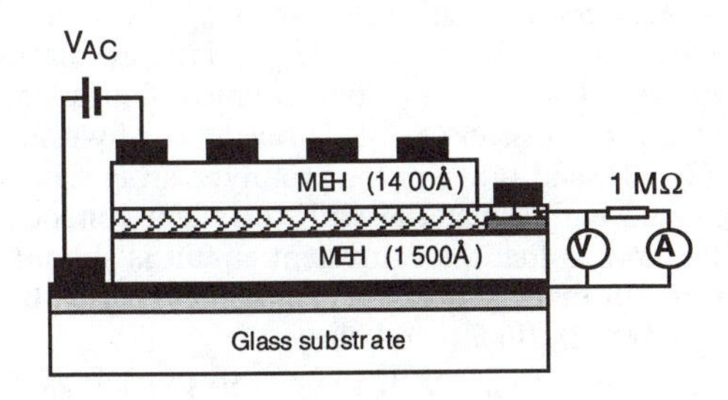

Figure 4. Four polymer grid triodes in an array with a common grid.

The voltage of the common grid (V_G) with respect to the anode was measured while a voltage (V_{AC}) was applied between the anode and the cathode on the left triode (see Figure 2c). The voltage of the common-grid was measured near the neighboring triode on the far right. As shown in Figure5 , V_G at this neighboring

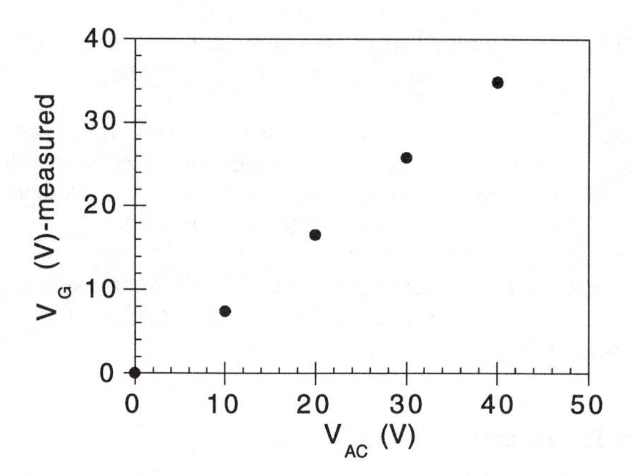

Figure 5: V_{AC} is voltage applied between the anode and cathode on the left of the array (see Figure 4). V_G is grid voltage measured near rightmost triode of the array. Grid voltage is proportional to input voltage applied at neighboring positions, i.e., the grid voltage provides a local average.

position responds in proportion toV_{AC}, showing that $V_G = \beta<V_{AC}>_{av}$; that is, the grid voltage responds to the local input and provides a local average. This result is reasonable and quite general; the common grid provides the center surround averaging that is required for lateral inhibition of response and local contrast control.

The final step required to demonstrate the validity of the polymer grid triode implementation of local contrast enhancement is to show that it computes the center-surround *difference* in Eq. 2. The equivalent circuit of the PGT demonstrated previously (*3, 4*) is that of two coupled diodes connected back-to-back, like that of a bipolar transistor (*5*). This is achieved by using *semiconducting* polymer in layers **(2)**, **(3')** and **(4)**. For the prototype array sketched in Figure 2c, layer **(2)** was fabricated with a material with sufficient conductivity to make an ohmic contact to the grid so that the equivalent circuit is simplified to a diode in series with a resistor. In the initial experiments, polyvinylcarbazole (PVK) was used for this resistor layer **(2)**. In forward bias,

$$I = I_o\exp[\gamma(V_{AC} - V_G)] + (V_{AC} - V_G)/R_i + V_G/R_s \tag{4}$$

where γ is a constant (*3, 4*), R_i is the internal series resistance of the diode (due to the bulk resistivity of the semiconducting material used in **(4)**) and R_s is the series resistance resulting from the PVK layer, **(2)**. Since the semiconducting layer is fabricated from a high resistivity, pure semiconducting polymer such as poly(phenylene vinylene), PPV, or one of its soluble deriviatives, $R_s << R_i$.

Thus, the output of the PGT is a function of (V_{AC} - V_G) only; that is,

$$I = F(V_{AC} - V_G). \quad (5)$$

Figure 6 demonstrates that this is in fact the case: I vs V_{AC} data are shown for different V_G. As expected, the I vs V_{AC} curves are sensitive to V_G; for example, at $V_{AC} \approx 5$ V, the current can be suppressed from 1 mA to zero by changing V_G. In the inset to Figure 4, we have replotted the curves from the two limiting data sets (V_G = -11 V and V_G = +13.2 V) as a function of (V_{AC} - V_G). Since the forward bias data collapse onto a single curve, Eq. 5 is indeed valid.

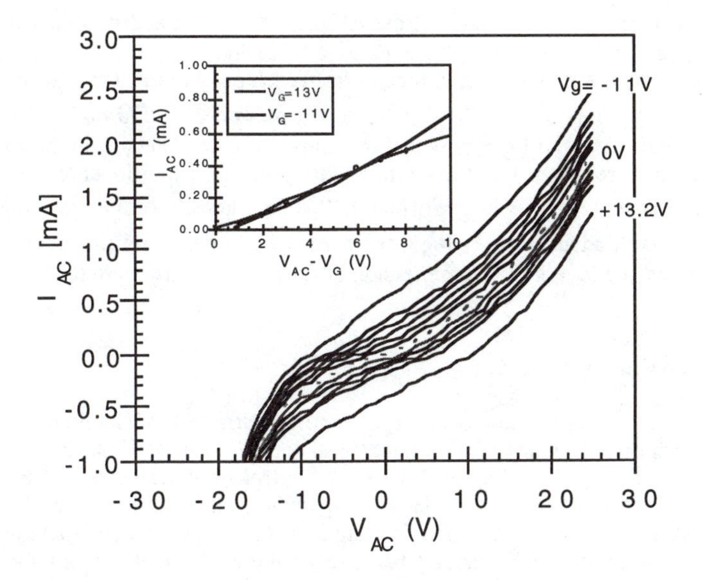

Figure 6: I vs V_{AC} for different grid voltages (V_G). The inset shows the two limiting data sets (V_G = -11 V and V_G = +13.2 V) as a function of (V_{AC} - V_G).

An analyic model of the device physics of the polymer grid triode (PGT) has recently been developed (*11*). The model demonstrates that the structure behaves as a three terminal device capable of current amplification, with the PANI network functioning as the control grid. An analysis of the generalized field-assisted carrier injection by tunneling, controlled by the grid voltage, was developed to model the charge injection and transport in the PGT. The results are in good agreement with the measured current-voltage characteristics (4). Furthermore, an effective diode model for the PGT was developed, and a simple, intuitive expression for charge transport in the presence of the grid network was obtained.

In the effective diode regime (*11*), the current through the PGT is a function of (V_{ac} - PV_{ag}); i.e. $I_{ac} \approx I(V_{ac} - PV_{ag})$ where is P is a geometric factor. This precisely the form required for use in image processing applications (see Eqn. 2 and

Eq. 5). The geometric factor, P, can be absorbed into the weighted averagimg range, w_i ; see the discussion following Eq. 3.

Consider then the array of PGTs, with common grid, sketched in Figure 2b, and assume, for simplicity, that the common grid is grounded at infinity. By utilizing the common grid,

$$V_G = \beta \langle V_{AC} \rangle_{ave} \tag{6}$$

where $\langle V_{AC} \rangle_{ave}$ denotes the average over neighboring pixels with a distance determined by the sheet resistance of the grid, the conductance to ground of the resistive layer and the geometric factor P. The characteristic length over which the average is taken is given by (1)

$$L = 1/[\rho\sigma]^{1/2} \tag{7}$$

where ρ is the sheet resistance of the PANI network grid and σ is the conductance per unit area to ground through the resistive PVK layer.

We conclude that the current output from each pixel of the array (I_i) is given by

$$I_i = F\,[V_{AC}^{(i)} - \beta \langle V_{AC} \rangle_{ave}] \approx [V_{AC}^{(i)} - \beta \langle V_{AC} \rangle_{ave}][\partial F/\partial V_{AC}] \tag{8}$$

Eq 8 is equivalent to Eq 3 provided the output of each detector on the focal plane array, which serves as input to an individual pixel, is proportional to the logarithm of the intensity; $V_A^{(i)} \propto \log I_L^{(i)}$ where $I_L^{(i)}$ is the intensity of the light incident on the ith pixel. Since $V_{out} \propto \log I_L$ for photovoltaic detectors under open circiuit conditions, the logarithmic compression of Eq 1 is straightforward (*5*).

Summary and Conclusion

The polymer grid triode array image processor differs fundamentally from those built with discrete silicon field-effect transistors (*1*). The PGT array makes use of the spreading resistance of the PANI control grid network to provide the interconnection of a given node to its neighbors; the conductivity of the PANI network enables center surround filtering as a result of lateral charge redistribution initiated by contrast differences. Charge redistribution through a continuous layer of material provides a natural means for averaging/blurring.

By controlling the concentration of polyaniline in the network, one can control the resistivity over many orders of magnitude (*6-10*). Similarly, by varying the thickness and the resistivity of layer **(2)**, by back biasing the grid-to-ground diode during operation, or by making **(2)** a bilayer which functions as a diode using conducting polymers (*7- 9*), one can vary the conductance of layer **(2)** over a wide range. The latter is particularly interesting since it allows dynamic control, *in situ*, of the spatial decay length. Thus, one can vary both ρ and σ in Eq 7 so as to be able to achieve values for L (either statically or dynamically) ranging from a few microns to 1 cm.

We have shown that this simple device, made from layers of conducting polymers, provides both logarithmic compression and lateral inhibition of response, as required for local contrast control. Nevertheless, the plastic retina is at an early stage of development. The utility of the PGT array for image enhancement will depend on a number of factors, including for example, sensitivity, noise, dynamic range, and matching from one pixel to the next, etc. that must be tested on an engineering prototype.

To build a full plastic retina, the polymer grid triode image enhancement array would be fabricated directly onto the output (back) side of a photodetector array (for example an infrared detector array) using each detector output pad as the anode or cathode of the PGT at that node. The semiconductor layers would be cast sequentially from solution and applied onto the detector array much like an anti-reflection coating. The final contrast enhanced output would be connected to a demultiplexer by "bump-bonding"; that is by cold-welding indium bumps arrayed reciprocally on the PGT array output and on the demultiplexer input.

Alternatively, the PGT array could be utilized to process the image after analog to digital conversion and integrated directly into a display (such as an LCD display). In this case, the PGT array would be fabricated directly on, as an integral part of, the display; for example, between the control circuits and the liquid crystal layer. The data would be logarithmically compressed digitally and in-put into the PGT array to process the image; the output from the pixels of the array of polymer grid triodes would serve as the input to the pixels of the display.

Acknowledgement: The office scene image in Figure 1 was generated using the Radiance computer graphics rendering program, developed by Greg Ward, funded by the Lighting Group at Lawrence Berkeley Laboratory, the U.S. Department of Energy, and the Laboratory d'Energie Solaire et de Physique du Batiment (LESO-PB) at the Ecole Polytechnique Federale de Lausanne (EPFL) in Switzerland.

References

1. Mead, C.; *Analog VLSI and Neural Systems*, (Addison Wesley, New York 1989).
2. Heeger, A. J.; Heeger, D. J.; Langan, J.; Yang, Y. Science 1995, *270,* 1642.
3. Yang,Y.; Heeger, A.J.; US patent application, Serial No. 08/227,979.
4. Yang,Y.; Heeger, A.J. Nature, 1994, *372*, 244.
5. Sze, S.M., *Physics of Semiconductor Devices*, Wiley, New York, 1981.
6. Reghu, M.; Yoon, C.O.; Yang, C.Y.; Moses, D.; Smith, P; Heeger, A.J.; Cao,Y.; *Phys. Rev.* B, 1994, *50*, 13931.
7. Yang, C.Y.; Cao, Y.; Smith, P.; Heeger, A.J.; Synth. Met. 1993, *53*, 293.
8. Heeger, A. J.; Trends in Polymer Science, 1995, *3,* 39.
9. Reghu, M.; Yoon, C.O.; Yang, C.Y.; Moses, D., Heeger, A.J.; Cao, Y. Macromolecules, 1993, *26*, 7245.
10. Cao, Y.; Smith, P.; Heeger, A.J., U.S. Patent 5 232 631
11. McElvain, J.; Heeger, A.J., J. Appl. Phys. (in press).

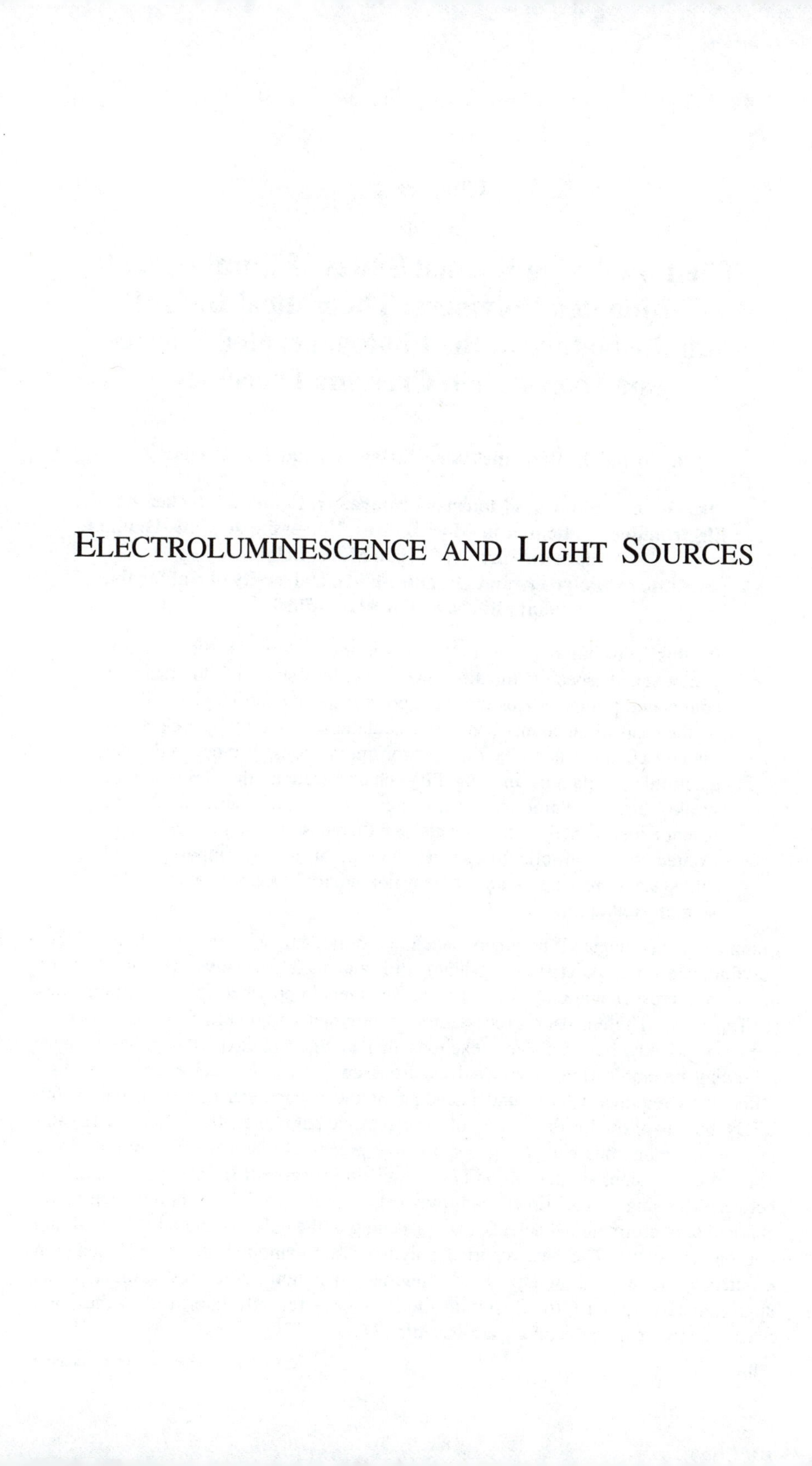

ELECTROLUMINESCENCE AND LIGHT SOURCES

Chapter 21

The Low-Lying Excited States of Luminescent Conjugated Polymers: Theoretical Insight into the Nature of the Photogenerated Species and Intersystem Crossing Processes

J. Cornil[1], D. Beljonne[1], Alan J. Heeger[2], and J. L. Brédas[1,2]

[1]**Service de Chimie des Matériaux Nouveaux, Centre de Recherche en Electronique et Photonique Moléculaires, Université de Mons-Hainaut, Place du Parc 20, B–7000 Mons, Belgium**
[2] **Institute for Polymers and Organic Solids, University of California, Santa Barbara, CA 93106–5090**

In this contribution, we first discuss the nature of the primary photoexcited species in the lowest excited state of luminescent conjugated polymers (for instance, polyparaphenylene vinylene, PPV) on the basis of quantum-chemistry calculations including both electron-phonon coupling and electron-electron interactions. We conclude that electronic excitations in long PPV chains lead to the formation of weakly bound polaron-excitons, in agreement with a large number of recent experimental measurements. We then describe singlet and triplet excited states in oligothiophenes and point to the importance of intersystem crossing as an efficient nonradiative decay process of the singlet excitations.

Luminescent conjugated polymers, such as poly(paraphenylene vinylene), PPV, polythiophene, PT, poly(paraphenylene), PPP and their derivatives (Figure 1) have attracted interest as materials for use as active layers in polymer light-emitting diodes (LEDs) (*1-5*). The emission process in such polymers originates from the radiative decay of weakly bound polaron-excitons that are generated in the polymer layer following recombination of injected electrons and holes. The achievement of high efficiencies requires a basic understanding of the phenomena occurring within the LEDs, both to allow for fine tuning of the electronic parameters of importance (6) and to limit the numerous competing nonradiative processes that arise for example from internal conversion, singlet fission (*7*), interchain processes (*8*), intersystem-crossing between the singlet and triplet manifolds (*9*), quenching of the polaron-excitons by charged or conformational defects, and quenching of the polaron-excitons by low-lying two-photon states. The best reported polymer LED exhibited an external quantum efficiency of approximately 4 % (internal quantum efficiency estimated as approximately 12 %) (*10*). Operating lifetimes have recently improved to the point where commercial applications are feasible (*11*).

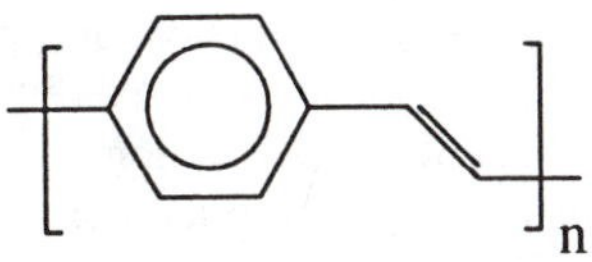

I 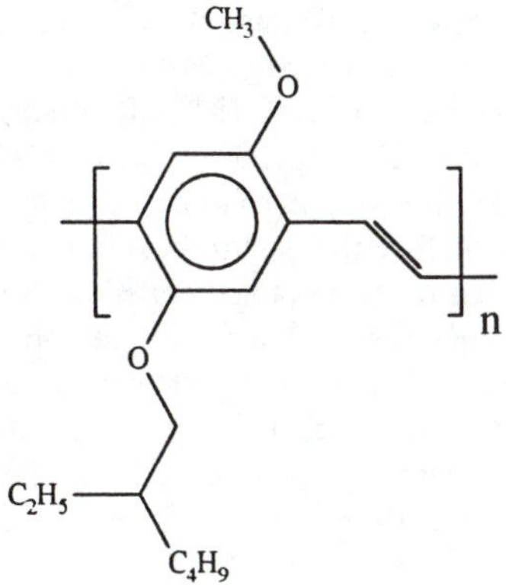 II

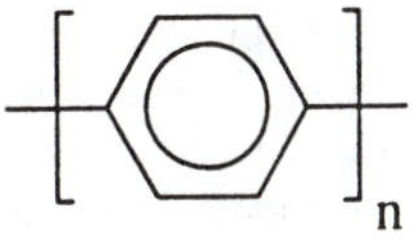

III 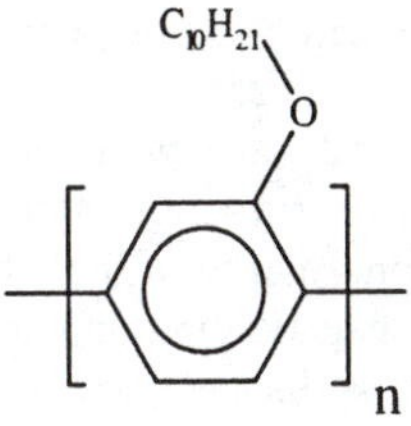 IV 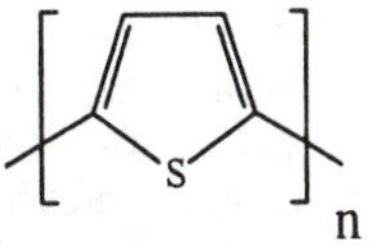 V

Figure 1. Molecular structure of: I) poly(paraphenylene vinylene), PPV; II) poly[2-methoxy-5-(2'-ethyl-hexoxy)-paraphenylene vinylene], MEH-PPV; III) poly(paraphenylene), PPP; IV) poly(2-decyloxy-paraphenylene), DO-PPP; V) polythiophene.

Quantum-chemistry calculations can prove to be useful in the field of organic light-emitting diodes. In particular, such calculations clarify the nature of the low-lying excited states of conjugated systems, and hence suggest new strategies yielding enhanced efficiencies. Two examples in this contribution illustrate that valuable information is obtained through analysis of theoretical models. First, we discuss the nature of the emitting species in the lowest excited state of luminescent conjugated polymers, including both electron-phonon and electron-electron interactions. Specifically, we address the following propositions that have been reported in the literature: (i) free charge carriers are generated in the excited state (*12*) and emission is an interband process; (ii) emission originates from a tightly bound electron-hole pair with a binding energy larger than 1 eV (*13*); (iii) emission is from the radiative decay of weakly bound polaron-exciton with a binding energy of a few tenths of an eV (*14-16*). Note that the polaron-exciton terminology implies that lattice relaxations are associated to the photogenerated electron-hole pair. Second, we refer to fluorescence measurements reported for oligothiophenes and substituted derivatives for which the main nonradiative decay channel originates in intersystem-crossing processes. We demontraste there that correlated calculations help to uncover the nature of the lowest-lying singlet and triplet excited states and to provide a meaningful interpretation of the evolution in fluorescence efficiency upon substitution or increase in chain length.

Discussion of the nature of the photogenerated species

Vibronic structure. Any Hamiltonian used to characterize the nature of the photogenerated species has to incorporate electron-phonon contributions since these correspond to a basic feature of π-conjugated compounds. A typical manifestation of lattice relaxation in the excited state is the appearance of vibronic progressions in the optical absorption spectra. Vibronic features could not be observed if the equilibrium positions in the ground and excited states were identical. This is sketched in Figure 2a (where we assume the existence of a single active vibrational mode coupled to the electronic excitation); the orthonormality of the vibrational wavefunctions would then exclusively allow transitions between vibrational levels of the two states with the same quantum number. In contrast, a displacement of the equilibrium geometry in the excited state (Figure 2b-2c) leads to the appearance of several vibronic features, usually between the zeroth vibrational level of the ground state and various levels of the excited state; when treated within the Franck-Condon approximation, the intensities are weighted by the overlap of the vibrational wavefunctions. We emphasize that the lowest energy transition (the 0-0 transition) is from the ground state to the relaxed excited state (*i.e.*, the excited state stabilized by lattice relaxation).

In a rigid band model, the excited-state relaxations (and thus the vibronic effects) are expected to decrease linearly with the number of atoms in the chain; hence, lattice relaxation would be insignificant for long conjugated molecules. The existence of a vibronic progression in the absorption spectra of conjugated macromolecules indicates, however, that self-localization phenomena (with associated lattice relaxation) occur in the excited states.

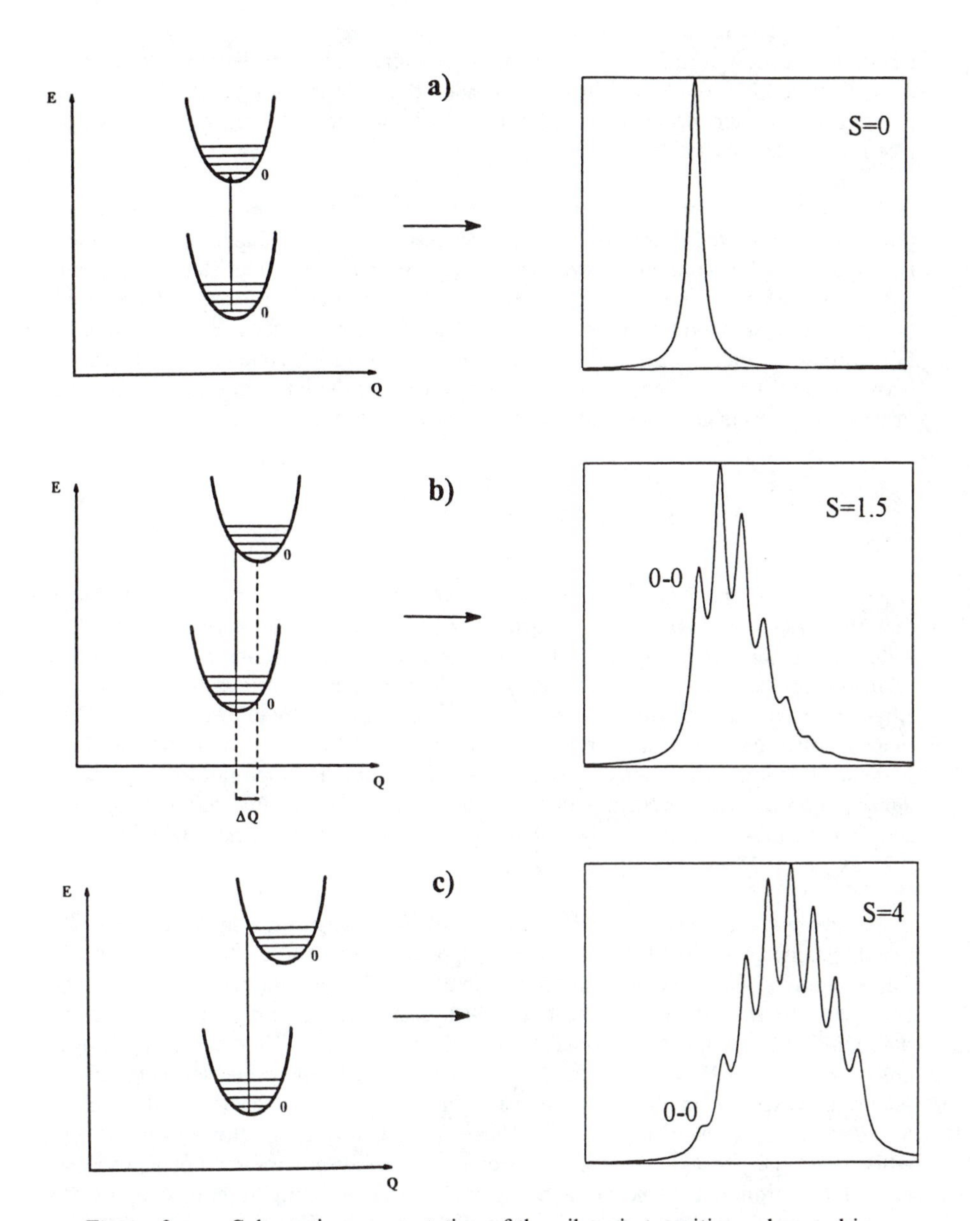

Figure 2. Schematic representation of the vibronic transitions observed in a situation where the equilibrium geometries of the ground and excited electronic states are: (a) identical (S=0); (b) weakly displaced (S=1.5); and (c) strongly displaced (S=4). S is the Huang-Rhys factor as defined in the text.

On the other hand, given the ease of delocalization and polarization of π electrons, electron correlation is another major ingredient to be incorporated in the theoretical model. We have thus to consider the contribution to the binding of photogenerated electron-hole pairs due to electron-lattice coupling and to electron-electron interactions (*17*).

We have recently set up a theoretical strategy to simulate the vibronic progressions observed in well-resolved optical absorption (*18*) and photoluminescence (*19*) spectra of PPV oligomers; we then utilize the optimized parameters to estimate the lattice relaxations occurring in the excited states. This theoretical approach is based on the INDO/SCI formalism (semiempirical Hartree-Fock Intermediate Neglect of Differential Overlap/ Single Configuration Interaction (*20*)), coupled to a displaced harmonic oscillators model to treat the vibronic contributions; in this context, the Franck-Condon integrals associated to vibrational mode x are given by:

$$<0|v^x>^2 = \frac{\exp^{-S_x} S_x^v}{v!} \qquad (1)$$

where S_x denotes the Huang-Rhys factor relative to the x-mode. The total value of this factor describes the extent to which geometry deformations take place in the excited state and actually corresponds to the average number of phonons involved in the relaxation process; summing over effective modes weighted by their associated Huang-Rhys factor therefore allows us to estimate the relaxation energy in this state (*18*). Our model further assumes that the relevant vibrational levels correspond to the so-called Я modes that induce geometry deformations going along the same line as the π-bond density modifications occurring in the excited state (*21*). In PPV oligomers, site-selective fluorescence spectra show that these are two effective modes with energies 0.16 eV and 0.21 eV in a 1:2 ratio (*22*).

Analysis of the *absorption spectra* of PPV oligomers indicates that the relaxation energy in the $1B_u$ excited state is approximately 0.30 eV for the 3, 4 and 5-ring oligomers; the relaxation energy is, therefore, relatively insensitive to oligomer chain length. In contrast, we have established on the basis of the experimental *photoluminescence spectra* reported for the same oligomers (*19*) that, in the case of emission, the total Huang-Rhys factor of the $1B_u$ excited state decreases as the chain length increases; the evolution is linear as a function of inverse chain length, tending to 0.2 eV at the limit of long chains. These contradictory results for absorption and emission suggest that the absorption process is affected by conformational disorder; the latter is strongly reduced in the emission spectra due to migration of the polaron-excitons towards the most ordered conjugated segments. This was demonstrated by Hagler et al for blends of MEH-PPV in polyethylene; the most ordered samples showed the smallest Stokes' shift (*23*). We emphasize that the relaxation energy estimates provided by the analysis of the photoluminescence spectra are in excellent agreement with direct geometry optimizations of the $1B_u$ excited state performed on the 2, 3, and 4-ring oligomers (*24*) within the AM1/CI formalism (semiempirical Hartree-Fock Austin Model 1 method (*25*) coupled to a configuration interaction scheme).

Polaron-exciton binding energy. Recent experimental measurements have been carried out on luminescent conjugated polymers to determine the polaron-exciton binding energy (defined as the difference between the creation energy of two noninteracting polarons of opposite charge and the formation energy of a neutral polaron-exciton (*17*)).

Internal photoemission experiments have been performed on polymers in a LED architecture; the measurements give access to the energy difference between the electron and hole injections and thus the energy gap for creation of two polarons of opposite charge; the data provide a value of 2.45 eV in the case of poly(2-methoxy-5-(2'-ethyl-hexyloxy) paraphenylene vinylene) - MEH-PPV (see Figure 1), taking into account image charge effects and extrapolating to zero photon energy (*26*). Since the 0-0 transition of the polymer peaks at 2.25 eV (*27*), the binding of the electron-hole pair is estimated to be 0.2 eV (±0.1 eV). Similar experiments using internal field emission report binding energies of 0.2 ± 0.2 eV in the case of MEH-PPV and poly-2-decyloxy-paraphenylene, DO-PPP (see Figure 1) (*27*).

The recent fabrication of light-emitting electrochemical cells (LECs) has enabled independent measurements of the binding energy (*28*); the emission process in such devices is expected to turn on at an applied voltage equal to the energy gap. The results collected for various luminescent polymers indicate that the turn-on voltage is essentially identical to the measured optical gap; the LEC data appear thus to be consistent with the semiconductor model in which the exciton binding energy is at most a few times k_BT at room temperature.

Ultraviolet Photoelectron Spectroscopy (UPS) measurements performed on a PPV sample indicate that the valence band edge is located at 1.55 ± 0.10 eV below the Fermi energy (*29*); assuming that the Fermi level is located in the middle of the gap, the energy gap corresponds to twice this value. Since UPS spectroscopy incorporates neither relaxation effects nor interactions between the emitted electron and the remaining hole, we then subtract from this value twice the polaron relaxation energy (2x0.15 eV from our AM1 calculations (*30*)) and obtain a value of 2.80±0.20 eV for the creation energy of two polarons of opposite signs. Since the 0-0 transition of the same PPV sample, *i.e.*, the formation energy of a neutral polaron-exciton, is measured at 2.45 eV, the binding energy of the polaron-excitons is estimated to be on the order of 0.35±0.20 eV. Since these estimates of the polaron lattice relaxation are lower limits (for example, they do not include ring-rotations), this estimate of the binding energy is an upper limit.

Based on these direct experimental measurements, we conclude that the binding energy of the singlet polaron-excitons in conjugated polymers might be relatively small

(≤0.20 eV) compared to earlier estimates (0.4 eV) (*14,15*). The data certainly demonstrate that binding energies as high as 1eV are not reasonable.

It is important to distinguish the contributions to the polaron-exciton binding energy arising from electron correlation effects and electron-lattice coupling. The results derived from both the analysis of the vibronic structures observed in photoluminescence spectra of long PPV oligomers and direct AM1/CI geometry optimizations (*24*), indicate that the relaxation energy in the lowest neutral excited state is approximately 0.20 eV. The amplitude of the polaron relaxation energy cannot be evaluated in a simple way by experimental means; as noted above, AM1 calculations provide an estimate on the order of 0.15 eV for each polaron. The comparison of the relaxation energy of two polarons to that of a neutral polaron-exciton leads to the conclusion that the lattice contribution to the binding energy is very weak and could actually even be negative. It is worth stressing that such a conclusion would not be expected in the framework of one-electron models where the relaxation energy of a neutral polaron-exciton is found to be equivalent to that of a doubly charged bipolaron and is thus much larger than in a single polaron (up to some 0.5 eV (*31*)). On the other hand, very recent highly correlated Density Matrix Renormalization Group (DMRG) calculations performed at the Extended Hubbard level by Shuai and co-workers (*32*) conclusively demonstrate that the electron-electron contribution to the polaron-exciton binding energy is in the range from 0.1 eV up to at most 0.3 eV. Thus, the sum of the two contributions (electron-electron plus lattice relaxation) is consistent with the relatively small singlet binding energy obtained from experiments.

We emphasize that the small value of the polaron-exciton binding energy results from a cancellation (indirect) of the electron-electron and electron-lattice contributions. This cancellation demonstrates the need for correlation effects to be taken into account when describing the excited state wavefunctions.

Note that the binding energy is defined with respect to the single particle energy gap which evolves linearly with the inverse number of rings (1/n). Thus, the 0-0 transition should also evolve (approximately) as 1/n even though the polarons and polaron-excitons wavefunctions extend over 3-4 rings (*24,30*).

Stokes' shift. We conclude this section by discussing the implications of the presence of a Stokes' shift. By definition, the Stokes' shift corresponds to the energy separation between the 0-0 peaks in absorption and emission. There is actually much confusion in the literature; indeed, some authors associate the Stokes' shift to the separation between the most intense peaks, even if they differ from the 0-0 bands; others have incorrectly claimed that the absence of a Stokes' shift indicates an absence of relaxation in the excited state, even when clear vibronic effects are observed.

The appearance of a Stokes' shift can actually be observed in situations where: (i) the emission and absorption processes involve different electronic states due to, for instance, intersystem crossing to a lower triplet state or to the existence of a lower energy A_g excited state;

(ii) conformational disorder induces the photogenerated electron-hole pairs to travel via a random walk process to lowest energy chain segments before light is emitted (*33*);
or (iii) emission takes place in the same segment after time has allowed for an optimal conformation to be adopted, thus implying that the absorption was not to the absolute minimum.
In the case of PPV chains, the latter two aspects can be operational; for instance, for twisted chains, point (iii) might be due to ring rotations occurring after excitation in order to reach the optimal excited-state coplanar geometry (*24*).

Excitation energies and intersystem crossing in oligothiophenes

We have calculated the vertical electronic excitation energies in oligothiophenes ranging in size from two to six rings, abbreviated as Thn below, by means of the Intermediate Neglect of Differential Overlap / MultiReference Double-Configuration Interaction (INDO/MRD-CI) approach (*34*). The theoretical chain-length dependence of the singlet-singlet, $S_0 \rightarrow S_1$, transition energies is compared in Figure 3 to the experimental evolution, extracted from optical absorption measurements in solution (*35,36*). The agreement between the experimental and theoretical results is excellent, which shows the suitability of the INDO/MRD-CI technique for the description of the electronic excitations in these oligomers. Both the spectroscopic data and the calculated values indicate a red-shift of the first one-photon allowed electronic transition with the size of the oligomers, due to the extension of the π-delocalized system. Extrapolation to an infinite chain length of the calculated transition energies, assuming a linear evolution of $S_0 \rightarrow S_1$ with respect to 1/n (*i.e.*, the inverse number of thiophene rings), leads to an optical gap close to 2 eV, which again is in good agreement with the spectroscopic data on polythiophene; recent optical measurements performed on regioregular polythiophenes indeed show an absorption with an onset at ~1.7-1.8 eV and a peak at ~2.5 eV (*37,38*).

In Figure 3, we also show the dependence of the energy difference between the S_0 ground state and the T_1 lowest triplet excited state on the number of thiophene rings, as calculated at the INDO/MRD-CI level on the basis of the S_0 geometry (*38*). As illustrated in Figure 3, the evolution with chain length of the $S_0 \rightarrow T_1$ energy difference is much slower than that of the $S_0 \rightarrow S_1$ excitation: the singlet-triplet energy difference is only lowered by ~0.2 eV when going from the dimer to the hexamer while the corresponding singlet-singlet absorption is characterized by a bathochromic shift that amounts to ~1.4 eV. The slower evolution with the number of thiophene rings of the triplet excitation energy (with respect to that of the singlet) reflects its more localized character, which originates in the exchange term. The localized character of the triplet agrees with Optically-Detected Magnetic Resonance (ODMR) experiments on polythiophene, indicating that the T_1 triplet state extends over not much more than a single thiophene ring (*40*). Moreover, similar ODMR measurements performed on oligomers ranging in size from two to five rings (*41*) indicate a very small chain-length dependence of the average spin-spin distance, r, in the triplet, as deduced within the dipolar approximation from the zero-field splitting, D (using the relation

$D/hc=2.6017\ r^{-3}$, where D/hc is in cm^{-1} and r in Å, we find r=3.27, 3.36, and 3.49 Å, in Th3, Th4, and Th5, respectively).

Optical absorption measurements in terthiophene solution (using a heavy-atom containing solvent, *i.e.*, 1,2-dibromomethane, to increase the spin-orbit coupling) have enabled the identification of the T_1 excited state at ~1.71 eV, which is in very good agreement with the 1.68 eV calculated value. Recently, Janssen *et al.* have observed that addition of C_{60} to oligothiophenes (from Th6 to Th11) in solution results in a quenching of the thiophene oligomers triplet state via energy transfer to C_{60} and produces triplet-state C_{60} (*42*). Therefore, the T_1 state of oligothiophenes ranging in size from 6 to 11 rings can be estimated to lie between the triplet energy of C_{60} (1.57 eV) (*43,44*) and that of Th3 (1.71 eV) (*45*). Finally, we note that Xu and Holdcroft have measured a phosphorescence peak at ~1.5 eV in polythiophene (*46*), a value very close to the one we can extrapolate for an infinite chain length (1.49 eV) on the basis of the calculated oligomer excitation energies.

Recent time-resolved fluorescence studies on unsubstituted (*47,48*) and substituted (*49*) thiophene oligomers in solution indicate a sharp increase of the fluorescence quantum yield ϕ_F when : (i) the number of thiophene rings is increased from two to seven; or (ii) electroactive end-groups are attached to the terthiophene molecule. In both cases, no significant alteration in the radiative decay rate constant, k_R, was observed. The evolution of ϕ_F with chain length and substitution was related to the decrease of the nonradiative decay rate, k_{NR}, dominated by the contribution arising from intersystem crossing (*48-51*). The importance of ISC as a nonradiative decay route of the singlet excitations has been also demonstrated by recent experimental investigations on polythiophene (*9*).

Radiationless transition theory expresses the intersystem rate constant in terms of: (i) a state density factor; (ii) the matrix element of the spin-orbit coupling Hamiltonian between the singlet and triplet wavefunctions; and (iii) an overlap factor, which accounts for the decrease in rate with an increasing energy gap between the two states involved in the crossing. Although a precise description of the intersystem crossing process would require the calculation of the spin-orbit coupling interaction as well as the overlap between the excited-state vibrational levels, we wish, at this stage, to present a qualitative picture for the ISC process in oligothiophenes. As first suggested by Ponterini and co-workers (*49*), we assume that the evolution of k_{NR} with size of the oligomers and substitution is mainly related to the variation in the overlap factor. As pointed out by Robinson and Frosch in the case of a series of aromatic hydrocarbons (*52*), this factor is primarily controlled by the singlet-triplet energy separation.

The S_1-T_1 energy difference that we calculate in unsubstituted oligothiophenes is much too large to give efficient singlet-triplet overlap, and hence the probability for S_1-T_1 crossing is very weak. However, ISC can also occur through other channels involving higher-lying triplet states. The INDO/MRD-CI calculations indicate that there is one triplet excited state (T_4), for which the wavefunction involves excitations from deep π-levels with large weights on the sulfur atoms, that lies at an energy close to that of S_1. In Figure 4, we plot the evolution of the $S_0 \rightarrow S_1$ and $S_0 \rightarrow T_4$ excitation energies with respect to the inverse number of thiophene rings. In bithiophene, the triplet T_4 lies below the singlet S_1, while for longer chains, due to the stronger stabilization of the S_1 excited state with chain length, the state ordering is reversed, S_1 appearing below T_4 (note that the calculated energy of the triplet T_4 excited state of Th3 is overestimated by comparison to the value interpolated by considering a linear relationship between the T_4 excitation energies and inverse chain length; as the crossing between the evolutions of the singlet S_1 and triplet T_4 excited states occurs for a chain length around the trimer, we associate this discrepancy to the fact that we did not include in the calculations the spin-orbit coupling which, in the crossing region, is expected to mix efficiently the singlet and triplet wavefunctions.

Rossi *et al.* have pointed out the existence of a thermally activated decay route of S_1 (*49*); they express the total nonradiative decay rate constant as:

$$k_{NR} = k_1 + k_2(T) = k_1 + A_2 \exp^{\frac{-\Delta E_{ISC}}{kT}} \tag{2}$$

where k_1 is a nonactivated decay rate (including processes such as nonactivated ISC, internal conversion, singlet fission) and k_2 is the intersystem crossing temperature-dependent decay rate. On the basis of the temperature dependence of k_{NR}, they estimate the pre-exponential factor, A_2, as equal to $1.9 \times 10^{10} s^{-1}$, and an activation energy of ~0.05 eV, in excellent agreement with the interpolated S_1-T_4 INDO/MRD-CI energy difference for Th3 (Table I). Assuming that k_1 remains constant when elongating the chain, we have calculated the k_{NR} rate constants on the basis of the experimental k_1 and A_2 values and of the calculated activation energies, *i.e.*, the S_1-T_4 energy difference estimated by fitting the INDO/MRD-CI excitation energies with a linear relationship, as done in Fig. 4 (*39*). In view of the different assumptions considered, the calculated k_{NR} decay rates can be considered to be in excellent agreement with the measured values (Table I). In bithiophene (Th2), the triplet T_4 excited state is below the singlet S_1 state and the S_1-T_4 ISC process is nonactivated and very efficient, leading to a very low fluorescence yield. When the chain elongates, as the activation energy becomes larger, the probability for inter-system crossing to occur decreases and ϕ_F is raised.

We have calculated the influence of grafting electroactive moieties on Th3; our results indicate an increase in the S_1-T_4 energy difference, which is more pronounced in the case of substitution with electron-withdrawing groups than with electron-donating moieties (*39*). As a consequence, we expect a stronger increase in fluorescence quantum yield for acceptor derivatives than donor derivatives. This trend is fully consistent with the experimental evolution of k_{NR} when going from Th3 to Th3-OCH3 and Th3-CHO; the large increase in the singlet-triplet energy difference in Th3-CHO gives rise to a much lower nonradiative decay rate and therefore a higher fluorescence efficiency (*48*).

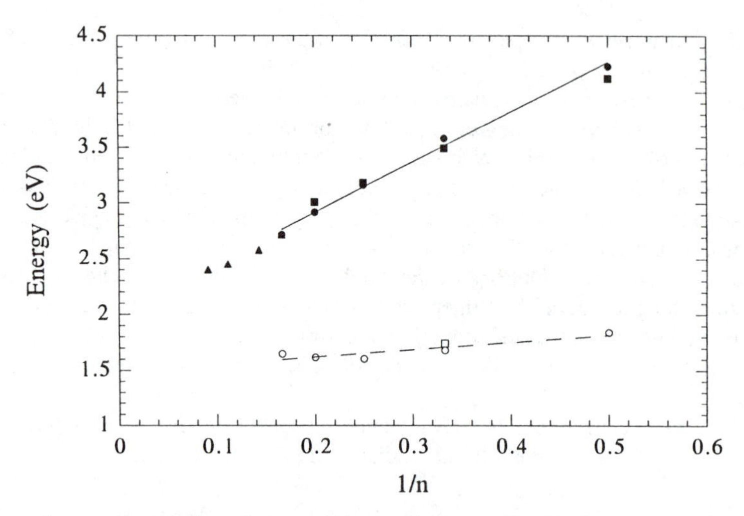

Figure 3. Comparison between the evolution with the inverse number of thiophene rings, 1/n: (i) the INDO/MRD-CI singlet-singlet $S_0 \rightarrow S_1$ (solid line, closed circles) and singlet-triplet $S_0 \rightarrow T_1$ (dashed line, open circles) energies; and (ii) the experimental $S_0 \rightarrow S_1$ (closed squares (*33*) closed triangles (*34*)) and $S_0 \rightarrow T_1$ (open squares (*37*)) energies obtained from measurements in solution.

Table I: Intersystem crossing rates in oligothiophenes

	$\Delta E(S_1\text{-}T_4)$	k_{NR}^{th} ($10^9 s^{-1}$)	k_{NR}^{exp}($10^9 s^{-1}$)	ϕ_F^{exp}
Th2	-0.10 (-0.10)	20.1	19.7[b]	0.018[b]
Th3	0.24 (0.05)	3.70	3.88[a] 3.7[c] 5.3[b]	0.07[a]0.05[c]0.07
Th4	0.16 (0.13)	1.23	1.54[a] 1.50[b]	0.20[a] 0.18[b]
Th5	0.18 (0.18)	1.14	0.82[a] 0.74[b]	0.28[a] 0.36[b]
Th6	0.22 (0.21)	1.13	0.70[a] 0.73[b]	0.42[a]

$\Delta E(S_1\text{-}T_4)$ is the S1-T4 energy difference (in eV) calculated, at the INDO/MRD-CI level, for the unsubstituted oligothiophenes; the values between parentheses have been extrapolated from a linear relationship between the excitation energies and the inverse number of rings. ϕ_F is the fluorescence yield; k_{NR}^{exp} denotes the nonradiative decay rate measured in solution and k_{NR}^{th} the value calculated on the basis of simple assumptions (see text).

a: Ref.47; b: Ref.48; c: Ref.49

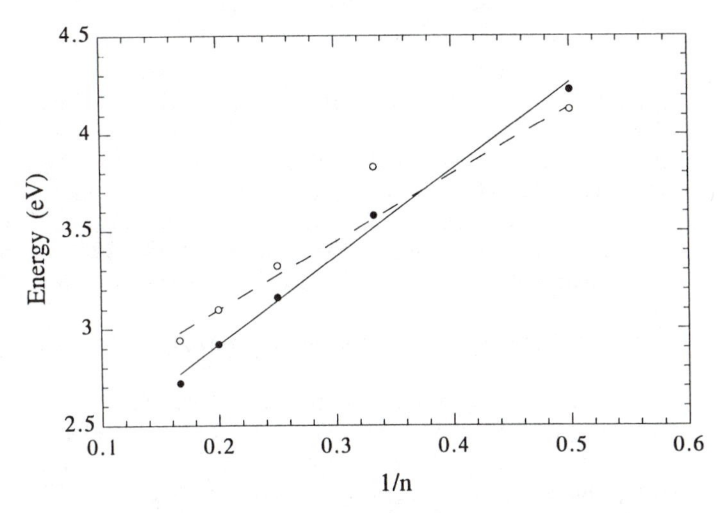

Figure 4. Evolution with the inverse number of thiophene rings, 1/n, of the INDO/MRD-CI $S_0 \rightarrow S_1$ (solid line, closed circles) and $S_0 \rightarrow T_4$ (dashed line, open circles) energies. Note that the transition energies calculated for Th3 were not included in the linear fits.

Synopsis

In this contribution, we have illustrated that quantum-chemical calculations can be efficiently exploited to help in the interpretation of experimental measurements on luminescent oligomers and polymers. We have first compared recent theoretical and experimental results dealing with the nature of the photogenerated species in the lowest excited state of luminescent conjugated polymers; we have concluded that these polymers sustain weakly bound polaron-excitons. We have also discussed the characteristics of the lowest-lying singlet and triplet excited states in oligothiophenes and have presented a model rationalizing fluorescence measurements; these show a huge increase in quantum efficiency when increasing the chain length and upon substitution with acceptor groups.

Acknowledgments

The work in Mons is conducted in the framework of the Belgian Federal Government "Pôle d"Attraction Interuniversitaire en Chimie Supramoléculaire" and is partly supported by the European Commission (ESPRIT Program LEDFOS-8013 and the Human Capital and Mobility Network SELMAT), the Belgian National Fund for Scientific Research (FNRS), and an IBM Academic Joint Study. JC is Aspirant and DB Chargé de Recherches of the FNRS. The work at UCSB was supported by the Office of Naval Research under Grant No. N00014-91-J-1235

Literature cited

[1] Burroughes, J.H.; Bradley, D.D.C.; Brown, A.R.; Marks, R.N.; Mackay, K.; Friend, R.H.; Burn, P.L.; Holmes, A.B. *Nature* **1990**, *347*, 539; Greenham, N.C.; Moratti, S.C.; Bradley, D.D.C.; Friend, R.H.; Holmes, A.B. *Nature* **1993**, *365*, 628.
[2] Braun, D.; Heeger, A.J., *Appl. Phys. Lett.* **1991**, *58*, 1982; Gustafsson, G.; Cao, Y.; Treacy, G.M.; Klavetter, F.; Colaneri, N.; Heeger, A.J. *Nature* **1992**, 357, 477.
[3] Grem, G.; Leditzky, G.; Ullrich, B.; Leising, G. *Adv. Mater.* **1992**, *4*, 36.
[4] Gill, R.E.; Malliaras, G.G.; Wildeman, J.; Hadziioannou, G. *Adv. Mater.* **1994**, *6*, 132.
[5] Ohmori, Y.; Uchida, M.; Muro, K.; Yoshino, K. *Solid State Commun.* **1991**, *80*, 605.
[6] Cornil, J.; Beljonne, D.; Brédas, J.L. *Synth. Met.*, **1996**, *78*, 209.
[7] Klein, G.; Voltz, R.; Schott, M. *Chem. Phys. Lett.* **1973**, *19*, 391.
[8] Jenekhe, S.A.; Osaheni, J.A. *Science* **1994**, *265*, 7653.
[9] Kraabel, B.; Moses, D.; Heeger, A.J. *J. Chem. Phys.* **1995**, *103*, 5102.
[10] Baigent, D.R.; Greenham, N.C.; Grüner, J.; Marks, R.N.; Friend, R.H.; Holmes, A.B.; Moratti, S.C.; Holmes, A.B. *Synth. Met.* **1994**, *67*, 3.
[11] Yu, G.; Zhang, C.; Cao, Y. *Appl. Phys. Lett.*, in press.
[12] Pakbaz, K.; Lee, C.H.; Heeger, A.J.; Hagler, T.W.; McBranch, D. *Synth. Met.* **1994**, *64*, 295; Lee, C.H.; Yu, G.; Heeger, A.J. *Phys. Rev. B* **1993**, *47*, 15543; Hagler, T.W.; Pakbaz, K.; Heeger, A.J. *Phys. Rev. B* **1994**, *49*, 10968.
[13] Leng, J.M.; Jeglinski, S.; Wei, X.; Benner, R.E.; Vardeny, Z.V.; Guo, F.; Mazumdar, S. *Phys. Rev. Lett.* **1994**, *72*, 156; Chandross, M.; Mazumdar, S.; Jeglinski, S; Wei, X.; Vardeny, Z.V.; Kwock, E.W.; Miller, T.M. *Phys. Rev. B* **1994**, *50*, 14702.
[14] Friend, R.H.; Bradley, D.D.C.; Townsend, P.D. *J. Phys. D: Appl. Phys.* **1987**, *20*, 1367.
[15] Gomes da Costa, P.; Conwell, E.M. *Phys. Rev. B* **1993**, *48*, 1993.
[16] Kersting, R.; Lemmer, U.; Deussen, M.; Bakker, H.J.; Mahrt, R.F.; Kurz, H.; Arkhipov, V.I.; Bässler, H.; Göbel, E.O. *Phys. Rev. Lett.* **1994**, *73*, 1440.
[17] Brédas, J.L.; Cornil, J.; Heeger, A.J. *Adv. Mater.*, **1996**, *8*, 447.
[18] Cornil, J.; Beljonne, D.; Shuai, Z.; Hagler, T.W.; Campbell, I.H.; Spangler, C.W.; Müllen, K.; Bradley, D.D.C.; Brédas, J.L. *Chem. Phys. Lett.*, **1996**, *247*, 425.
[19] Heller, C.M.; Campbell, I.H.; Laurich, B.K.; Smith, D.L.; Bradley, D.D.C.; Burn, P.L.; Ferraris, J.P.; Müllen, K. *Phys. Rev. B*, in press.
[20] Zerner, M.C.; Loew, G.H.; Kichner, R.F.; Mueller-Westerhoff, U. *J. Am. Chem. Soc.* **1980**, *102*, 589.
[21] Hernandez, V.; Castiglioni, C.; Del Zoppo, M.; Zerbi, G. *Phys. Rev. B* **1994**, *50*, 9815.
[22] Heun, S.; Mahrt, R.F.; Greiner, A.; Lemmer, U.; Bässler, H.; Halliday, D.A.; Bradley, D.D.C.; Burn, P.L.; Holmes, A.B. *J. Phys.: Condens. Matter* **1993**, *5*, 247.
[23] Hagler, T.W.; Pakbaz, K.; Heeger, A.J. *Phys. Rev. B* **1995**, *51*, 14199.
[24] Beljonne, D.; Shuai, Z.; Friend, R.H.; Brédas, J.L. *J. Chem. Phys.* **1995**, *102*, 2042.
[25] Dewar, M.J.S.; Zoebisch, E.G.; Healy, E.F.; Stewart, J.J.P. *J. Am. Chem. Soc.* **1985**, *107*, 3902.

[26] Campbell, I.H.; Hagler, T.W.; Smith, D.L.; Ferraris, J.P. *Phys. Rev. Lett.* **1996**, *76*, 1900.
[27] Yang, Y.; Pei, Q.; Heeger, A.J. *Synth. Met.* **1996**, *78*, 263.
[28] Pei, Q.; Yu, G.; Zhang, C.; Yang, Y.; Heeger, A.J. *Science* **1995**, *269*, 1086.
[29] Fahlman, M.; Lögdlund, M.; Stafström, S.; Salaneck, W.R.; Friend, R.H.; Burn, P.L.; Holmes, A.B.; Kaeriyama, K.; Sonoda, Y.; Lhost, O.; Meyers, F.; Brédas, J.L. *Macromolecules* **1995**, *28*, 1959.
[30] Cornil, J.; Beljonne, D.; Brédas, J.L. *J. Chem. Phys.* **1995**, *103*, 842.
[31] Choi, H.Y.; Rice, M.J. *Phys. Rev. B* **1991**, *44*, 10521.
[32] Shuai, Z.; Pati, S.K.; Su, W.P.; Brédas, J.L.; Ramasesha, S. *Phys. Rev. B*, in press.
[33] Kersting, R.; Lemmer, U.; Mahrt, R.F.; Leo, K.; Kurz, H.; Bässler, H.; Göbel, E.O. *Phys. Rev. Lett.* **1993**, *70*, 3820.
[34] Buenker, R.J.; Peyerimhoff, S.D. *Theoret. Chim. Acta* **1974**, *35*, 33.
[35] ten Hoeve, W.; Wynberg, H.; Havinga, E.E.; Meijer, E.W. *J. Am. Chem. Soc.* **1991**, *113*, 5887; Havinga, E.E.; Rotte, I.; Meijer, E.W.; ten Hoeve, W.; Wynberg, H. *Synth. Met.* **1991**, *41-43*, 473.
[36] Charra, F.; Fichou, D.; Nunzi, J.M.; Pfeffer, N. *Chem. Phys. Lett.* **1993**, *192*, 566.
[37] Chen, T.A.; Rieke, R.D. *Synth. Met.* **1993**, *60*, 175.
[38] McCullough, R.D.; Lowe, R.D.; Jayaraman, M.; Anderson, D.L. *J. Org. Chem.* **1993**, *58*, 904.
[39] Beljonne, D.; Cornil, J.; Friend, R.H.; Janssen, R.A.J.; Brédas, J.L. *J. Am. Chem. Soc.* **1996**, *118*, 6453.
[40] Swanson, L.S.; Shinar, J.; Yoshino, K. *Phys. Rev. Lett.* **1990**, *65*, 1140.
[41] Bennati, M.; Grupp, A.; Bauërle, P.; Mehring, M. *Mol. Cryst. Liq. Cryst.* **1994**, *256*, 751.
[42] Janssen, R.A.J.; Moses, D.; Sariciftci, N.S. *J. Chem. Phys.* **1994**, *101*, 9519; Janssen, R.A.J.; Smilowitz, L.; Sariciftci, N.S.; Moses, D. *J. Chem. Phys.* **1994**, *101*, 1787.
[43] Wei, X.; Vardeny, Z.V.; Moses, D.; Srdanov, V.I.; Wudl, F. *Synth. Met.* **1993**, *54*, 273.
[44] Zeng, Y.; Biczok, L.; Linschitz, H. *J. Phys. Chem.* **1992**, *96*, 5237.
[45] Scaiano, J.C.; Redmond, R.W.; Mehta, B.; Arnason, J.T. *Photochem. Photobiol.* **1990**, *52*, 655.
[46] Xu, B.; Holdcroft, S. *J. Am. Chem. Soc.* **1993**, *115*, 8447.
[47] Chorosvian, H.; Renstch, S.; Grebner, D.; Dahm, D.U.; Birckner, E. *Synth. Met.* **1993**, *60*, 23.
[48] Becker, R.S.; de Melo, J.S.; Maçanita, A.L.; Elisei, F. *Pure&Appl. Chem.* **1995**, *67*, 9.
[49] Rossi, R.; Ciofalo, M.; Carpita, A.; Ponterini, G. *J. Photochem. Photobiol A: Chem.* **1993**, *70*, 59.
[50] Periasamy, N.; Danieli, R.; Ruani, G.; Zamboni, R.; Taliani, C. *Phys. Rev. Lett.* **1992**, *68*, 919.
[51] Egelhaaf, H.J.; Oelkrug, D. *SPIE* **1994**, *2362*, 398.
[52] Robinson, G.; Frosch, R. *J. Chem. Phys.* **1963**, *38*, 1187.

Chapter 22

Synthesis and Properties of Novel High-Electron-Affinity Polymers for Electroluminescent Devices

Xiao-Chang Li[1], Andrew C. Grimsdale[1], Raoul Cervini[1], Andrew B. Holmes[1,2,4] Stephen C. Moratti[2], Tuck Mun Yong[1], Johannes Grüner[3], and Richard H. Friend[3]

[1]University Chemical Laboratory, Department of Chemistry, University of Cambridge, Lensfield Road, Cambridge CB2 1EW, United Kingdom
[2]Melville Laboratory for Polymer Synthesis, Department of Chemistry, University of Cambridge, Pembroke Street, Cambridge CB2 3RA, United Kingdom
[3]Cavendish Laboratory, Department of Physics, University of Cambridge, Madingley Road, Cambridge CB3 0HE, United Kingdom

New conjugated polymers with high electron affinities for use in polymer light-emitting diodes (LEDs) have been synthesized, and their optical and electrochemical properties studied. Incorporation of electron-withdrawing trifluoromethyl substituents was found to markedly enhance the electron affinity of poly(*p*-phenylene vinylene) (PPV). Polymers incorporating electron-transporting oxadiazole units either in the main-chain or as side-chains have been prepared and evaluated as emissive and charge-transporting materials in LEDs. Copolymers have been made containing both emissive and charge-transporting units as side-chains. The electrochemical oxidation and reduction potentials of all these materials have been measured and compared with those of other conjugated polymers.

Since the first report of polymer light-emitting devices (LEDs) using poly(*p*-phenylene vinylene) (PPV) as an emissive layer (*1*), much progress has been achieved in the understanding of the physics and materials chemistry of LEDs, and also in the search for new electroluminescent polymers. Colors ranging from infrared to violet can be achieved for polymer LEDs by using different emissive polymers. A variety of emissive polymers, such as substituted poly(phenylene vinylene) (*2,3*), poly(3-alkylthiophene) (*4,5*) and polyphenylene derivatives (*6-8*) have been described. These polymers are usually sandwiched between a transparent layer of indium-tin oxide (ITO) and a metal electrode. Singlet excitons are formed under double charge injection and decay radiatively to produce light emission at a wavelength in accordance with the band gap energy of the emissive polymers.

Electron Injection and Charge Transport in Polymer Light Emitting Devices

In addition to the considerable advances in the tuning of emission colors of polymeric LEDs, recent attention has focussed on improving the efficiency of devices. In order

[4]Corresponding author

to achieve higher device efficiency (photons emitted per injected electron), highly luminescent polymers must be chosen, and balanced charge (hole and electron) injection is considered to be crucial. Enhancement of electron injection and transport has been found particularly important for a balanced charge injection in polymer LEDs since most emissive polymers are more easily *p*-doped and have hole-transporting rather than electron transporting properties. One way to enhance electron injection and transport is to choose a lower work function metal such as calcium as a cathode (*9*). Another way is chemically to enhance the electron affinity of the emissive polymers. A good example is the introduction of electron-withdrawing groups, e.g. cyano groups along a poly(phenylene vinylene) chain (*10,11*). While efforts have been made to design new luminescent polymers with high electron affinity, great attention has also been directed to the design and synthesis of electron transporting polymers, particularly aromatic oxadiazole-containing compounds and polymers. Recent results have revealed that aromatic oxadiazole polymers can be used both as an electron transporting layer in multi-layered LEDs and as a light emitting layer.

In this article, we review the most recent research on such issues as the design and synthesis of novel electroluminescent polymers with high electron affinity, and of polymeric oxadiazoles with a strong tendency for *n*-doping, as well as those having high photoluminescence efficiency. Although most of this work was done in our laboratory, other important work is also included.

Trifluoromethyl-containing Oligomers and Polymers as Light Emitting Materials

To date there has been relatively little investigation into the effect of electron-withdrawing substituents on the efficiency of PPV derivatives. We decided to investigate the effect of an electron-withdrawing trifluoromethyl aryl substituent, as this group, in addition to being a strong electron acceptor, shows high chemical and electrochemical stability. Trifluoromethyl-substituted PPV has previously been obtained by Heck coupling of 2,5-dibromo-1-trifluoromethylbenzene with ethylene (*12*) which afforded low molecular weight (3500 Da) virtually insoluble material. We synthesized the trifluoromethyl-substituted PPV derivatives by means of a bromo-precursor route shown in Scheme 1 (*13*).

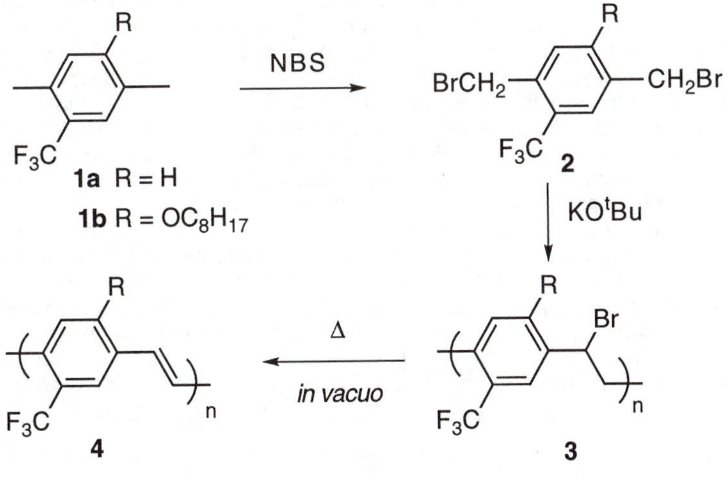

Scheme 1. Synthesis of CF_3 substituted poly(phenylene vinylene) (CF_3-PPV) **4** (adapted from *13*)

Bromination of trifluoromethyl *p*-xylenes **1** gave bisbromomethyl compounds **2** which on treatment with 1 equivalent of base gave the bromoprecursor polymers **3** which were soluble in chloroform, toluene and THF. Weight average molar masses of 84 and 252 kDa were obtained for **3a** and **3b** respectively. Thermogravimetric analysis (Figure 1) showed that the precursor polymers lost HBr commencing at *ca.* 150 °C and that this was complete by 300 °C. Conversion was accordingly performed at temperatures in this range to give the insoluble conjugated polymers **4** which showed orange fluorescence.

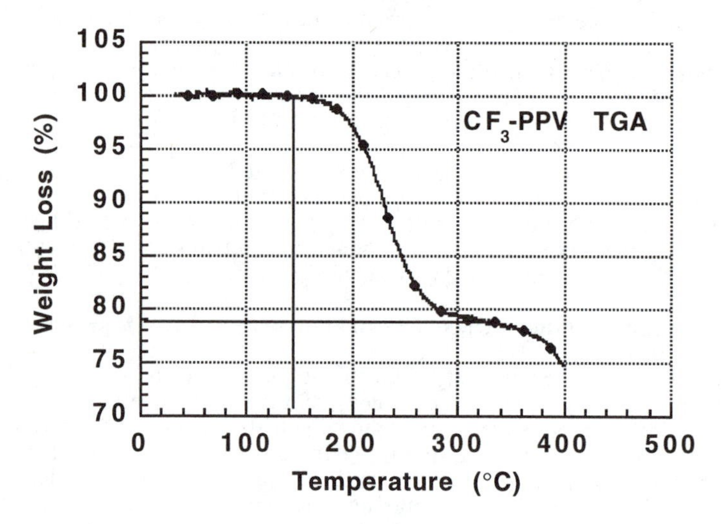

Figure 1. Thermogravimetric analysis curve of CF_3-PPV **3b**

The emission maxima were at 2.39 and 2.21 eV for **4a** and **4b** respectively with photoluminescence (PL) quantum efficiencies of 5-8%. Monolayer devices using these materials showed non-uniform emission, but two-layer devices with PPV as hole-transport layer and aluminum or calcium cathodes showed bright and uniform emission with internal efficiencies of 0.01%. The EL emission is slightly red-shifted with respect to the PL emission. No increase in efficiency was detected from substitution of calcium for aluminum, indicating that the polymers were efficient electron acceptors, and the low efficiencies reflect the modest fluorescence efficiencies of the polymers **4**.

Measurement of the reduction potentials of the polymers by cyclic voltammetry (Figure 2) using ITO coated electrodes showed that the reduction potentials of the polymers **4** were comparable with that of CN-PPV (*14*), indicating that the electron accepting properties of the trifluoromethyl group are comparable with those of the nitrile group. All cyclic voltammetry measurements were carried out using a thin film of polymer on a platinum disc electrode immersed in acetonitrile in the presence of Bu_4NClO_4 (0.1 M) as electrolyte. The reference electrode was Ag/Ag^+ (external reference ferrocene, Fc/Fc^+ 0.21 V), and the sweep rate was 20 mV/s.

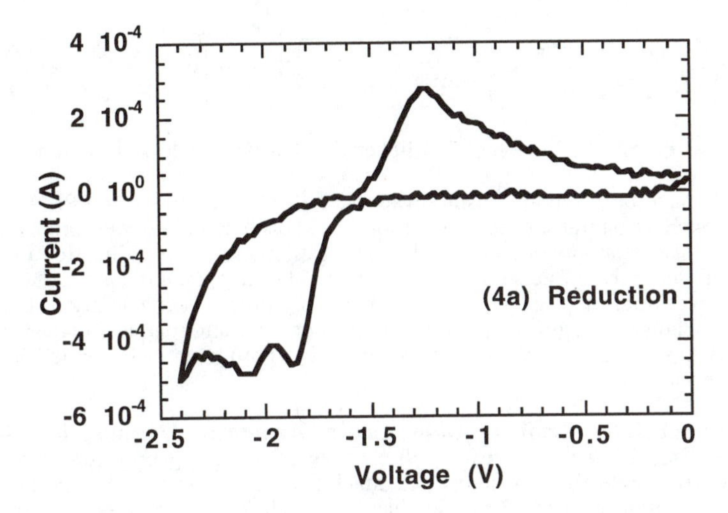

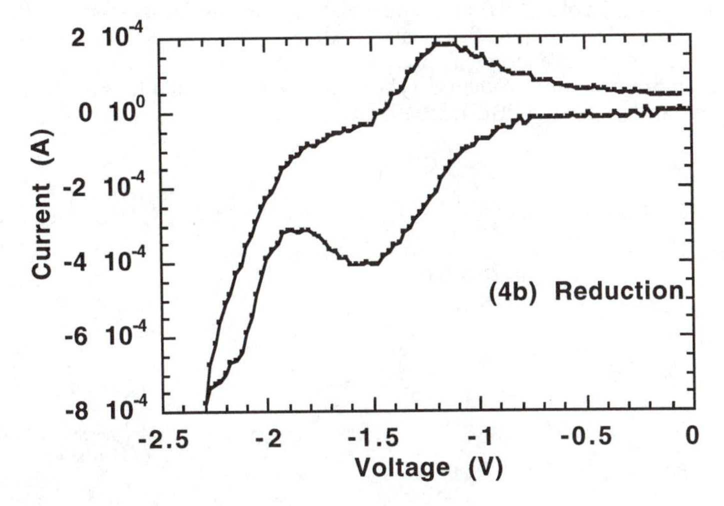

Figure 2. CV reduction plots for polymers **4a** and **4b**

Oxadiazole-based Polymers as Hole Blocking and Luminescent Materials

1,3,4-Oxadiazole is a stable electron deficient heteroaromatic compound. It is well known that 2,5-diphenyl-1,3,4-oxadiazole reacts with sodium to yield a radical anion which disproportionates to form a dianion (Scheme 2).

Aromatic oxadiazole compounds such as 2-(4-biphenyl)-5-(4-*tert*-butylphenyl)-1,3,4-oxadiazole (PBD) have been used as electron transporting materials in organic light emitting diodes (*15-17*). Multi-layered polymeric LEDs with improved

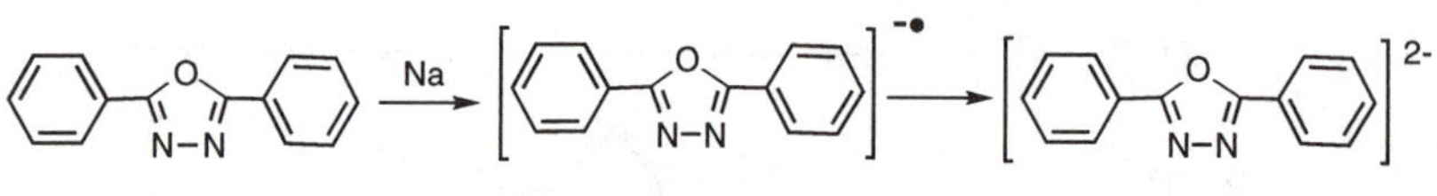

Scheme 2. Reduction of 2,5-diphenyl-1,3,4-oxadiazole with sodium

efficiency have been reported using evaporated PBD or a spin-coated PBD/PMMA blend as an electron transporting layer (*18,19*). In each case, however, problems that lead to device break-down (such as the aggregation and re-crystallization of PBD) may occur under the influence of an electrical field or temperature increase while the device is working. A possible way to tackle this problem is covalently to bond the aromatic oxadiazole segments to an amorphous or semi-amorphous polymer. These segments may be either attached to pendant side groups or incorporated into the polymer backbone.

Polymethacrylates Bearing Side Chain Aromatic Oxadiazole Charge Transporting Chromophores. A direct way of designing polymers containing aromatic oxadiazole chromophores is to attach PBD or its derivatives onto an easily processible polymer chain. This idea has been realized and reported by our group (*20*). Two typical aromatic oxadiazole compounds, namely 2-[4-tert-butylphenyl]-5-phenyl-1,3,4-oxadiazole (PPD) and 2-phenyl-5-biphenyl-1,3,4-oxadiazole (PBD) are covalently bonded to a polymethacrylate backbone. The polymethacrylate backbone was chosen for its excellent optical transparency in combination with its good stability and solution processibility. Scheme 3 shows the synthetic route for polymethacrylate polymers bearing PPD and PBD respectively.

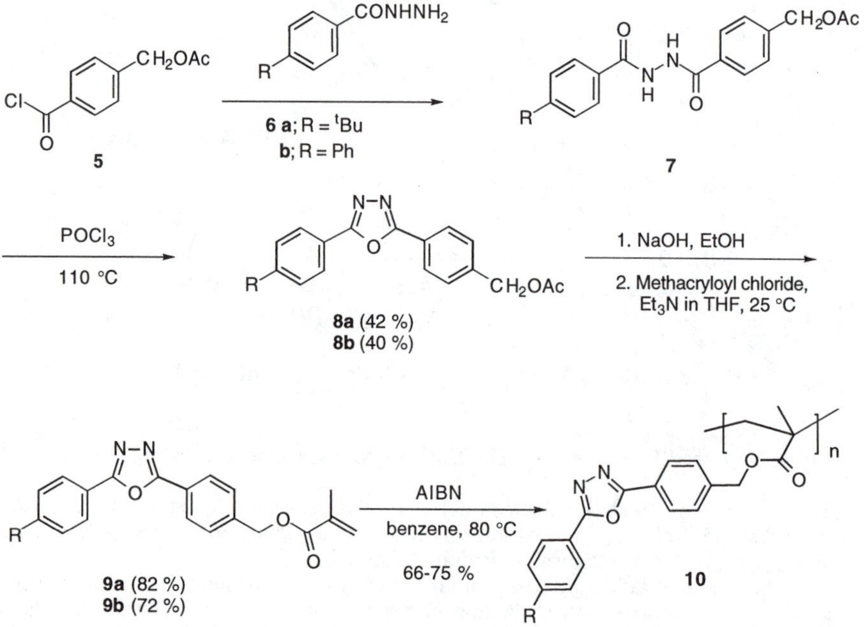

Scheme 3. Synthesis of polymethacrylate polymers **10** bearing PPD and PBD (adapted from *21*)

The synthesis of the two monomers is achieved through a multi-step route starting from terephthalaldehyde mono(diethylacetal) (*21*). The polymerization is accomplished by a radical method using AIBN as initiator, and the polymerization degree can be effectively controlled by choosing different monomer/initiator ratios. A higher ratio will lead to higher molar mass polymers. Both polymers have been fully characterized by NMR, FTIR and microanalysis, and all the analyses were in very good agreement with the proposed structures. Table I lists the properties of two polymers (PMA-PPD, **10a**; PMA-PBD, **10b**).

Table I. Properties of side chain oxadiazole polymers **10**

Polymers	M_n/M_w	Solubility [a]	T_g (°C)	UV-VIS λ_{max} (nm)	E_g (eV) [b]
PMA-PPD **10a**	52 /127 x 10^3	Toluene (s) $CHCl_3$ (s)	162	294	3.77
PMA-PBD **10b**	40 / 67 x 10^3	Toluene (n) $CHCl_3$ (s)	169	304	3.46

[a] s (soluble); n (insoluble) [b] Estimated from onset of absorption

The polymers have been used in different multi-layered devices using PPV as emissive layer. Typical devices were prepared on glass substrates precoated with patterned indium-tin oxide (ITO) electrodes (resistance < 20 Ω/square). In a two layer LED, the oxadiazole polymethacrylates (**10a** or **10b**) were spin-coated on top of poly(*p*-phenylene vinylene) (PPV), prepared on an ITO glass substrate by the sulfonium precursor route (*22*). Calcium was used as the top metal contact. A comparable device, but without the polymethacrylate, was fabricated as a reference. Both devices emitted green yellow light under forward bias potential (15 V).

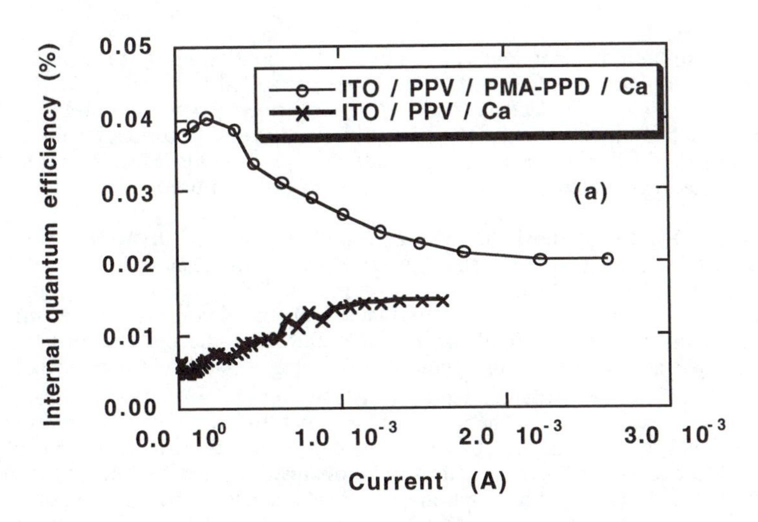

Figure 3a. Comparison of efficiency-current characteristics for the EL device ITO/PPV/polymer PMA-PPD **10a**/Ca with those for the single layer reference device ITO/PPV/Ca

As shown in Figure 3a, the internal efficiency (0.04%) of the two-layer PPV device at a current density of 0.5 mA cm^{-2} was increased by a factor of four by using **10a**, compared with a single layer PPV device. It was also observed that the turn-on voltage for the two-layer device was significantly lower than that of an analogous single layer PPV device as shown in Figure 3b.

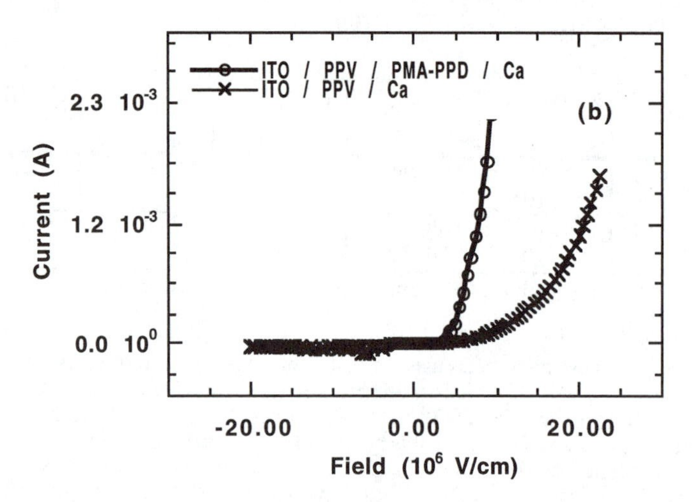

Figure 3b. Comparison of current-electric field characteristics for the EL device ITO/PPV/polymer PMA-PPD **10a**/Ca with those for the single layer reference device ITO/PPV/Ca (adapted from *21*)

A related polymethacrylate, covalently bonded with PPD and several polyethers containing PPD moieties have been reported by Strukelj *et al.* (*23,24*). LEDs containing these electron transporting (ET) polymers were found to be much more stable than those without. In terms of diode efficiency, no clear benefit resulting from the use of ETs has emerged; in contrast, the most important factor appeared to be the PPV conjugation length and not the type of ET polymer used.

Precursor Route Synthesis for Main Chain Oxadiazole Polymers. Main chain aromatic oxadiazole polymers tend to be highly crystalline and sparingly soluble due to their rigid rod-like chain. In order to suppress the crystallinity and enhance the solubility of poly(aromatic oxadiazole)s a 1,3-phenyl linkage has been introduced within the main chain so as to obtain the polyaromatic oxadiazole as shown in Scheme 4. The one-pot synthesis was accomplished using a published method (*25*). The NMR (^{1}H and ^{13}C), FT-IR and microanalysis data were found to be in good agreement with the assigned structure. The poly(oxadiazole) **14b-1** was found to be insoluble in conventional solvents such as chloroform though soluble in trifluoroacetic acid which enabled free standing polymeric films to be obtained by spin coating or casting. By contrast, **14a-1** was insoluble even in trifluoroacetic acid. Obviously a tertiary butyl substituent improves the processibility of the material. An alternative approach to the synthesis of **14** is the thermal conversion of the precursor polyhydrazide **13** which can be obtained under mild conditions by the reaction of the diacid **11** with the dihydrazide **12** (Scheme 4). The polyhydrazides **13a** and **13b** are soluble in dimethylsulfoxide and are slightly soluble in *N*-methylpyrrolidone (NMP). Polymer

13b has been found to be completely soluble in dimethylformamide (DMF) and NMP owing to the *tert*-butyl substituent. Intrinsic viscosities have been measured as 0.5 dL g^{-1} for **13a** (30 °C, DMSO) and 0.4 dL g^{-1} for **13b** (30 °C, DMF). Transparent free standing films of both polymers can be obtained by a casting technique.

120 °C, NMP/Py
$HO-P(OC_6H_5)_2$
77-92%

11a: X = H
11b: X = ^{t}Bu

12

13

275 °C
24 h
ca. 100%

100 °C
P_2O_5/CH_3SO_3H
96%

14
Via one-pot 14a-1 and 14b-1 | **Via precursor 13 to 14a-2 and 14b-2**

Scheme 4. Synthesis of processible main chain oxadiazole polymers **14** (adapted from *36*)

Aromatic acyl hydrazides can be converted into aromatic oxadiazoles simply by thermal conversion. From the study of the thermal gravimetric analysis (TGA) of polyhydrazides **13a** and **13b** (*26*), it is known that the cyclodehydration of the hydrazide to the oxadiazole commences at approximately 300 °C for both polymers. Interestingly the introduction of the *tert*-butyl group appears to increase the kinetics of thermal conversion, as the TGA profile of **13b** is observed to be much sharper than that of **13a**. Lower conversion temperatures over longer time intervals are recommended in order to prevent possible thermal degradation. The structural change of the polyhydrazide polymers during thermal conversion with time can be monitored using FT-IR spectroscopy. With thermal conversion, the strongest absorption peak at 1654 cm^{-1}, attributable to the carbonyl of the polyhydrazide, diminishes and two new peaks at 1084 cm^{-1} and 962 cm^{-1}, attributable to the oxadiazole ring, emerge and become more pronounced. Elemental analyses after conversion of **13a, 13b** into poly(oxadiazole)s **14** showed that conversion is complete within 36 h.

Figure 4 shows the UV-VIS spectra of the polymers **13a, 13b, 14a-2, 14b-1, 14b-2** and the photoluminescence spectrum of **14b-1**. The polyoxadiazole synthesized by the one-pot synthesis (**14b-1**) has a smaller HOMO-LUMO gap (3.76 eV) compared with **14b-2** (3.87 eV). This implies that the one-pot reaction leads to a more fully converted polyaromatic oxadiazole than thermal conversion of the polyhydrazide precursor.

The polyoxadiazoles **14** were found to fluoresce purple-blue under ultraviolet irradiation (quantum efficiency of film 11%). The photoluminescence emission spectrum of the polymer **14b-1** is shown in Figure 4.

The potential of poly(aromatic oxadiazole)s for electron injection has been studied by cyclic voltammetry. When swept anodically, the polymer **14b-1** was found to exhibit an irreversible oxidation peak at 1.67 V (with onset of 1.54 V); this value is

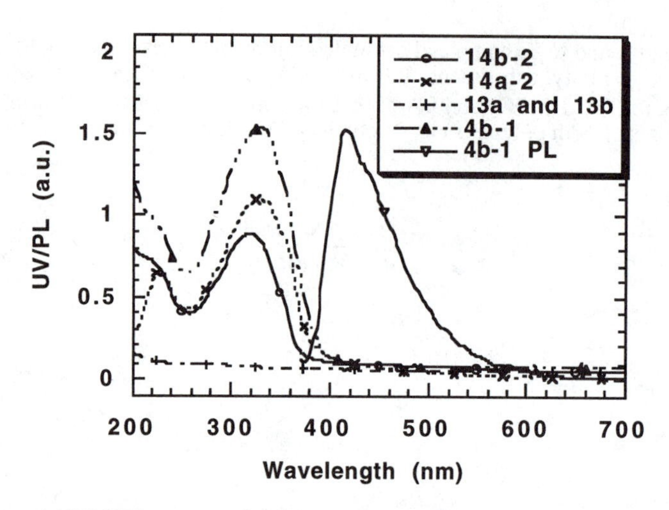

Figure 4. UV-VIS spectra of polyhydrazides and polyoxadiazoles **13** and **14**

high in comparison with other conjugated polymers. For example our measurements for poly[2-methoxy-5-(2'-ethylhexyloxy)-1,4-phenylene vinylene] (MEH-PPV) show that the first and second oxidations take place respectively at 0.5 V and 1.0 V. The high oxidation potential implies that polymer **14b-1** should possess good hole blocking properties within a solid state device.

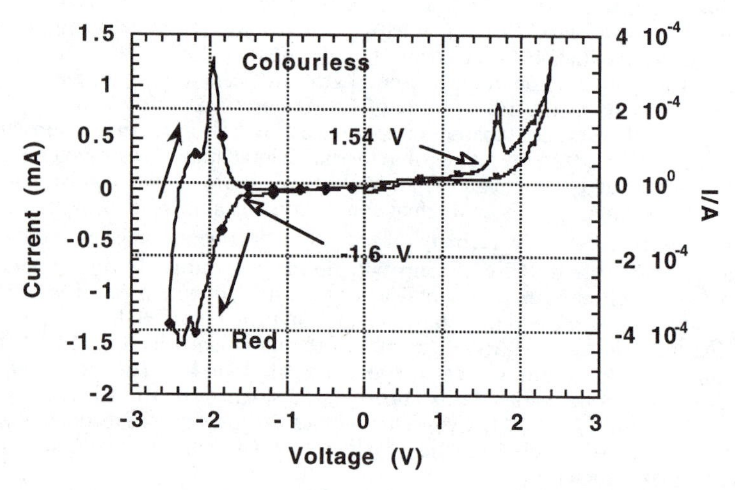

Figure 5. CV of polymer **14b**

When the polymer is swept cathodically (Figure 5) two reduction peaks at -2.16 V and -2.33 V are observed (with onset of reduction at -1.60 V). These correspond to the reduction of the neutral state to the anion (negative polaron) and dianion (negative bipolaron) respectively (Figure 6). Further, the two reduction peaks can be chemically reversed, as indicated by the two oxidation peaks at -1.96 V and -2.21 V. Therefore it is more difficult to reduce **14b-1** than a dialkoxypolycyanoterephthalylidene (CN-PPV) (-1.7 V) (*11*), but it is more easily reduced than MEH-PPV (-2.4 V). This allows the relative electron affinity of such polymers to be estimated.

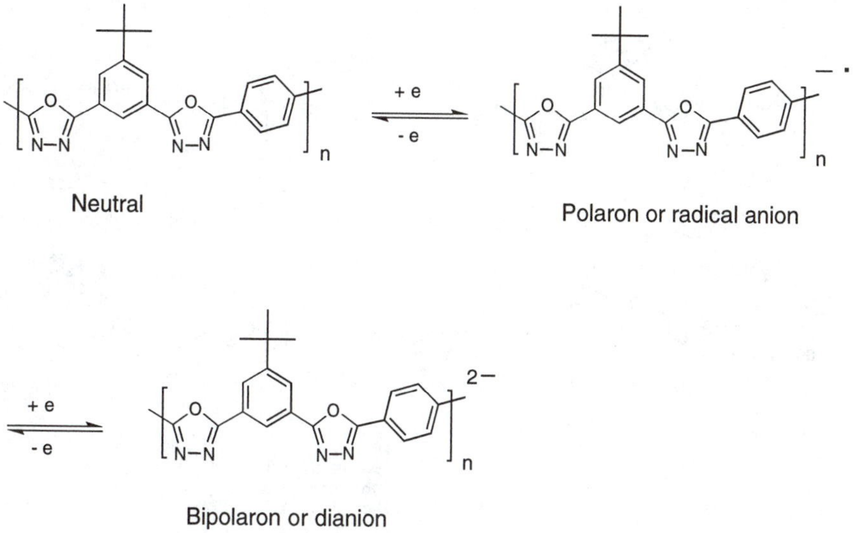

Neutral

Polaron or radical anion

Bipolaron or dianion

Figure 6. Proposed two step reduction of polymer **14b**

Processible Main Chain Aromatic Oxadiazole Polymers with Flexible Linkage. Fully conjugated main chain oxadiazole polymers have been widely studied as themally stable and high modular high strength materials (*27-31*). A pre-requisite for a polymer for application as an electron transporting or electroluminescent material, is solution processibility to form thin layers. In order to obtain processible and colorless main chain oxadiazole polymers, the introduction of a flexible linkage on the main chain has proven effective, as the flexible linkage not only interrupts the conjugation, but also induces torsion of the molecular chain (*32*). A typical example is poly(phenylene-1,3,4-oxadiazole-phenylene-hexafluoroisopropylidene) (PPOPH) (Figure 7) which is soluble in chlorinated solvents (*33,34*).

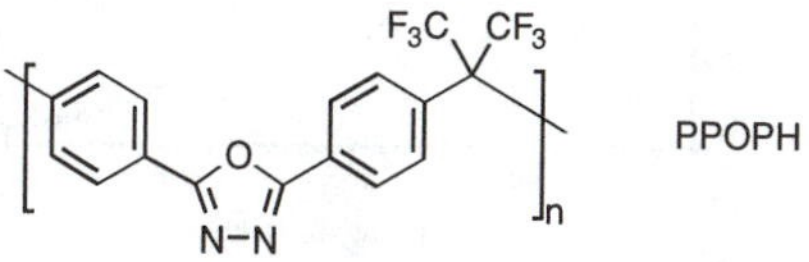

PPOPH

Figure 7. The structure of poly(phenylene-1,3,4-oxadiazole-phenylene-hexafluoroisopropylidene) (PPOPH)

Yang and Pei reported the use of PPOPH as the electron injection layer in MEH-PPV based LEDs (*35*). They found that the device performance and quantum efficiency were improved by a factor of 40 when using PPOPH as an electron injection layer between an emissive MEH-PPV layer and an aluminium cathode.

A wide range of polyoxadiazole polymers with hexafluoropropylidene linkages have been synthesized in Cambridge (*36*). As illustrated in Scheme 5, aromatic oxadiazole polymers **18** and **20** have been synthesized by a one pot reaction of the diacid **15** with the dicarboxylic hydrazide **16** and hydrazine sulphate **17** respectively at 100 °C in the presence of P_2O_5/CH_3SO_3H which serves both as solvent and cyclodehydration agent.

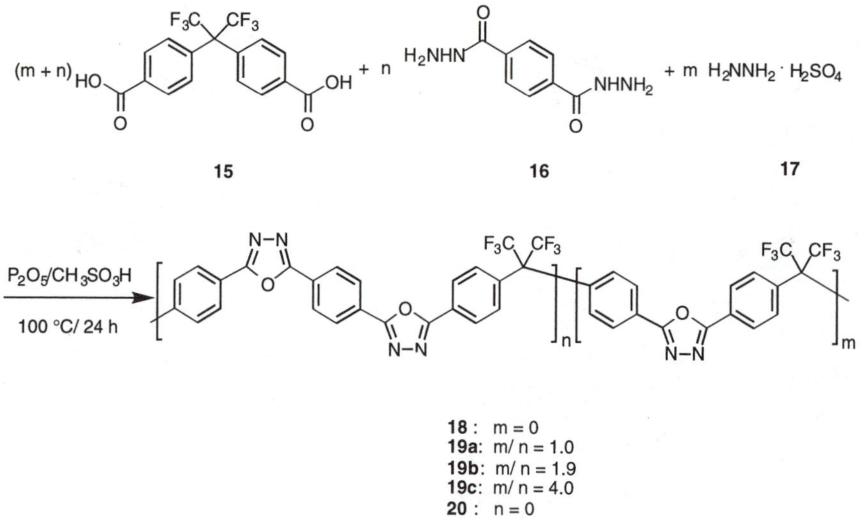

(m + n) **15** + n **16** + m $H_2NNH_2 \cdot H_2SO_4$ **17**

P_2O_5/CH_3SO_3H, 100 °C/ 24 h

18 : m = 0
19a: m/ n = 1.0
19b: m/ n = 1.9
19c: m/ n = 4.0
20 : n = 0

Scheme 5. Synthesis of aromatic oxadiazole polymers with hexafluoropropylidene linkage **18-20** (adapted from *36*)

The polymer **18** is a white powder with a band gap of 3.24 eV. It fluoresces blue under long wavelength UV light as shown in Figure 8. Photoluminescence

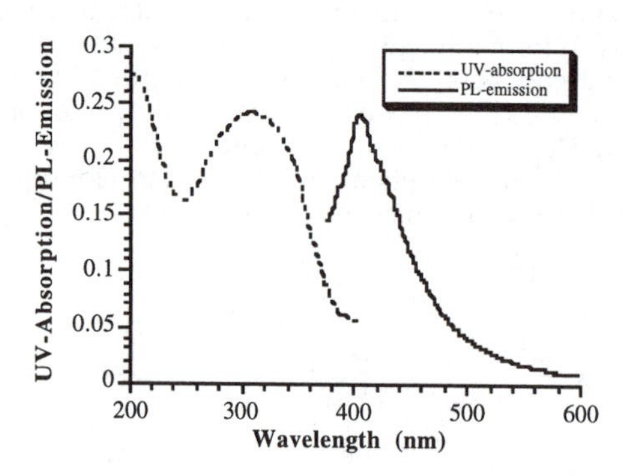

Figure 8. UV-VIS absorption and photoluminescence emission spectra of polymer **18**

measurements revealed that its solid state luminescence quantum efficiency is as high as 50%. The fiber-like polymer **20** has also been found to fluoresce under UV, but in the purple region owing to a larger band gap (3.73 eV). Since the shortest excitation wavelengths available for our photoluminescence measuring equipment are in this region, the photoluminescence efficiency could not be measured (Table II). The

Table II. Optical properties and solubilities of poly(aromatic oxadiazoles) **18-20**

Polymers	Solubility[a]	UV-VIS (film) λ_{max} (nm)	E_g (eV)[b]	Φ_{PL}
18	$CHCl_3$ (n) CF_3COOH (s)	306	3.24	0.50
19a	$CHCl_3$ (n) CF_3COOH (s)	302	3.43	0.37
19b	$CHCl_3$ (p) CF_3COOH (s)	294	3.44	0.30
19c	$CHCl_3$ (s)	290	3.47	0.28
20	$CHCl_3$ (s)	286	3.73	0[c]

[a] n (not soluble); s (soluble); p (partially soluble) [b] Estimated from onset of UV-VIS absorption [c] Not measured, as fluorescence outside instrument range

major advantage of **20** is its good solubility in chlorinated solvents, whereas polymer **18** is only soluble in strong acids such as trifluoroacetic acid, but is not soluble in common organic solvents. In order to adjust the properties, copolymers with different ratios of the blocks **18** and **20** have been synthesized. As shown in Table II, the properties of polymers **19** can be apparently tailored according to the ratio of **18** to **20**. With increasing proportions of the repeat unit **20**, the copolymers become more soluble in chloroform (**19b** is partially soluble and **19c** is completely soluble in chloroform). The band gap and UV absorption maxima of the polymers also change according to the ratio between the two repeat units (Figure 9). The two

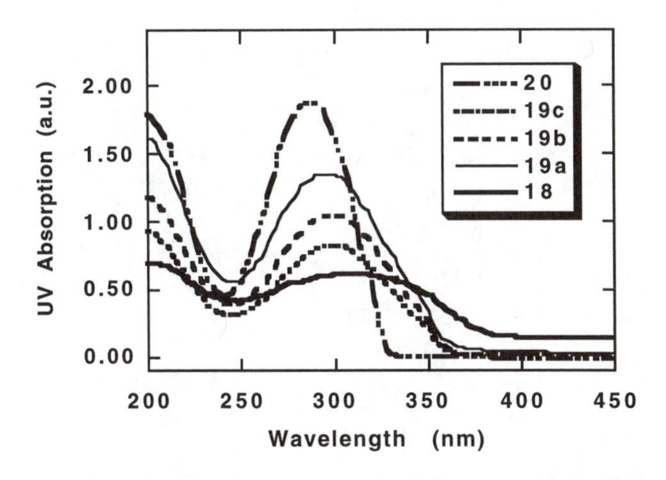

Figure 9. UV-VIS absorption spectra of the polymers **18-20**

homopolymers and their copolymers with different ratios are all colorless. Transparent thin films can easily be obtained by a solution casting technique. All the polymers were found to be very stable even in strong acid, such as concentrated sulfuric acid and not to melt below 220 °C. The molecular weights of **19c** and **20** have been measured by gel permeation chromatography (GPC) analysis using polystyrene as standard, indicating M_w/M_n = 177,000/41,000 for **19c** and 143,000/12,000 for **20**. All the polymers have been characterized by NMR, FT-IR and microanalysis.

In order to understand the charge injection and transporting/blocking abilities of this class of materials cyclic voltammetry has been carried out. This was accomplished in a typical three electrode cell with a polymer film on a working platinum electrode (against Ag/Ag^+, external reference ferrocene, Fc/Fc^+ 0.21 V) in acetonitrile with Bu_4NClO_4 as electrolyte. It was found that all the polymers are difficult to oxidize electrochemically even at high potential (1.6 V versus Ag/Ag^+). The results suggest that these polymers exhibit excellent hole blocking properties and possible resistance to oxidation. The reduction sweep showed that the polymers were reduced in a reversible electrochemical cycle. It was observed that upon reduction the polymers changed from colorless to red which was again reversed on oxidation. Using the reduction peak position, the polymer **18** was shown to be more easily *n*-doped (or reduced) than polymer **20**, with the copolymers having intermediate properties. It is interesting to note that there are two apparent cathodic (and corresponding anodic) peaks for polymers **18** and **19a-c** around -1.9 to -2.0 V. For polymer **20**, there is a weak but detectable cathodic (and corresponding anodic) peak at -2.29/-2.15 V. Figure 10 shows the cyclic voltammetry (CV) of polymer **18**. The two reduction processes have rarely been observed in conjugated polymers [e.g. CN-PPV(*14*)] probably because the first reduced state is stabilized by π-delocalization. In conjugated heteroaromatic alternating copolymers, a cyclic voltammogram with a main peak and a shoulder has been observed *(37,38)* which has been assigned to the *n*-doping of two different heterocyclic rings. As the conjugation in poly(phenylene-oxadiazole)s and in polyphenylenes is similar, we suggest another mechanism (Scheme 6). The two reduction processes may possibly correspond to the formation of a relatively localized radical anion (polaron) and dianion (bipolaron) respectively, as in the case of polymer **20**.

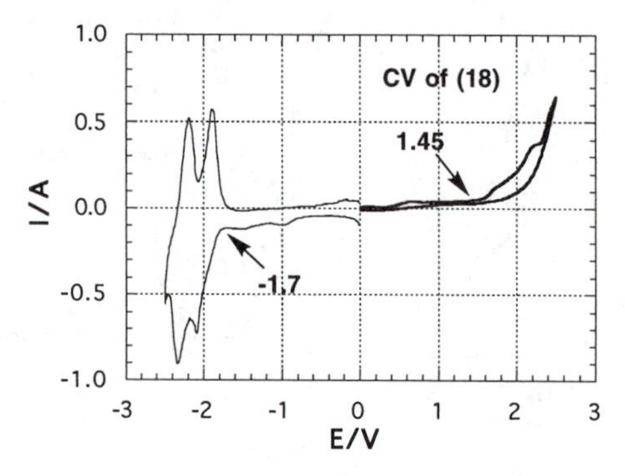

Figure 10. Cyclic voltammogram of polymer **18** (adapted from *45*)

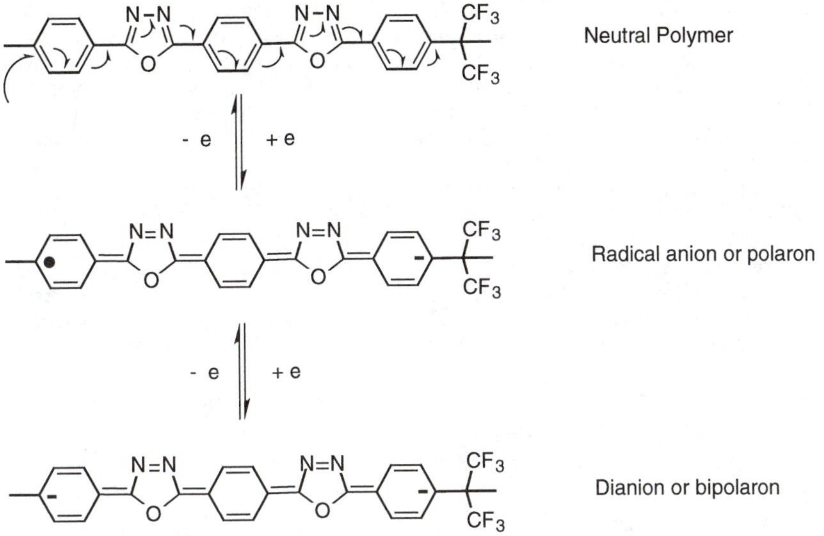

Neutral Polymer

- e +e

Radical anion or polaron

- e +e

Dianion or bipolaron

Scheme 6. Proposed reduction (*n*-doping) mechanism for polymer **18**

Different two layer light emitting diodes (LEDs) have been tested using the aromatic oxadiazole polymers as electron transporting layer, with PPV as emissive layer. It was found that the use of the aromatic oxadiazole polymers can realise efficient light emission using stable aluminum metal as cathode. No definite enhancement of electroluminescence efficiency has yet been observed. This can probably be explained by the fact that PPV and the oxadiazole polymers **18-20** have similar reduction potentials, and therefore by analogy similar LUMO levels.

Table III. Comparative optical and electrochemical properties of various polymers

Polymer	Optical bandgap onset (eV)	Oxidation onset (V)	Reduction onset (V)	Electrochem bandgap (eV)
PPV[a]	2.5	0.85	-1.7	2.55
CN-PPV[a,b]	2.1	1.05	-1.6	2.65
MEH-PPV[a]	2.1	0.34	-1.95	2.29
MEH-CN-PPV[a]	2.3	1.0	-1.4	2.4
Poly(oxadiazole) **18**	3.24	1.45	-1.7	3.15
Poly(oxadiazole) **14b**	3.19	1.54	-1.6	3.14
Poly(oxadiazole) **20**	3.73	1.60	-1.8	3.40

[a] Data measured on ITO-glass. Ag/Ag^+ as internal reference and ferrocene as external reference, (Fc/Fc^+ 0.21 V). [b] Electrolyte Me_4NClO_4

Table III summarizes the CV results measured on ITO glass for four major electroluminescent polymers. Several points emerge from the data in Table III : (i) CN-PPV (*10*) and MEH-CN-PPV (*39*) are better electron acceptors than MEH-PPV (*ca.* 0.35-0.55 V respectively); (ii) CN-PPV and MEH-CN-PPV are better electron acceptors than PPV (*ca.* 0.1-0.3 V) whereas PPV is a better hole acceptor (0.8-0.85 V); (iii) MEH-PPV (*39*) should be a much better hole acceptor than PPV (0.5 V); (iv) The oxadiazoles (measured on Pt) are worse electron acceptors than all except MEH-PPV and are extremely poor hole acceptors. They therefore probably enhance electron injection in MEH-PPV based LEDs whereas with PPV or CN-PPV they act as hole-blocking materials by providing a very large band offset between the HOMOs at the interface in multilayer devices.

Interestingly, polymer **18** was found to emit blue-purple light. Aromatic oxadiazole compounds are known to show blue electroluminescence (*40-42*). Main chain oxadiazole polymers can also be made fluorescent by proper modification of the chemical structure. Pei and Yang (*43,44*) have reported a new oxadiazole polymer **25** with both a flexible linkage and solubilizing alkoxy side-chains (Scheme 7). An LED of structure ITO/polyaniline/polymer **25**/Al has an external quantum efficiency close to 0.1 % and a turn-on voltage around 4.5 V.

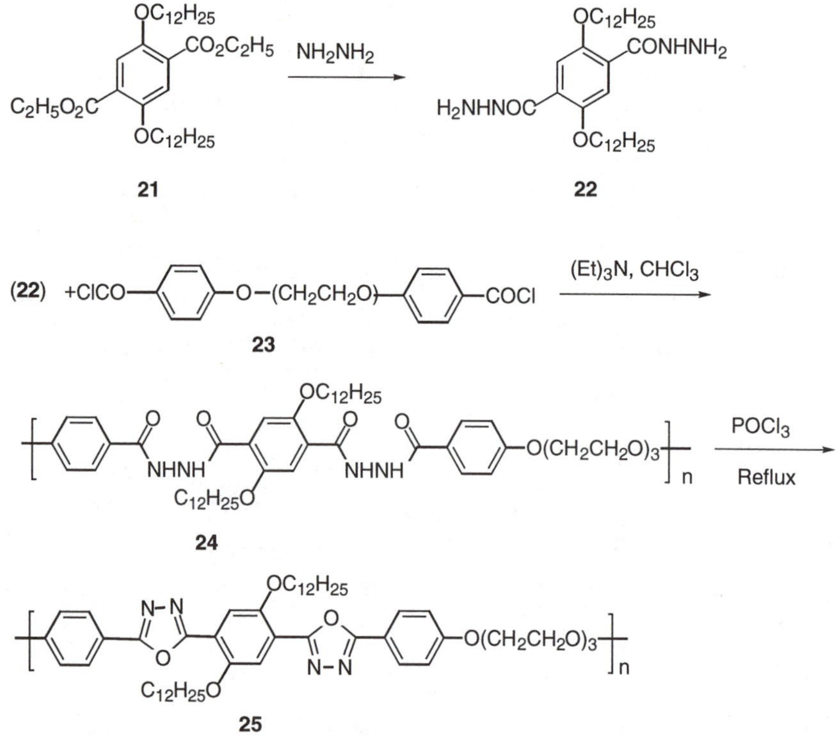

Scheme 7. Synthesis of oxadiazole polymer **25**

"Star-burst" Oxadiazole Polymers. When a 1,3,5-trifunctional monomer is introduced in the synthesis (Scheme 8), a highly branched or star-burst polymer, or even a crosslinked polyaromatic oxadiazole may be expected to form according to the ratio of the monomers (*45*). A 1:1:1 ratio of **26**: **27**: **28** has been chosen to form a star-burst polyaromatic oxadiazole polymer **29**.

The polymer obtained is a white powder with a band gap of 3.32 eV. Although it is not soluble in conventional solvents, it is easily soluble in trifluoroacetic acid, and a thin film can thus be formed by casting techniques, but it is quite brittle owing to its

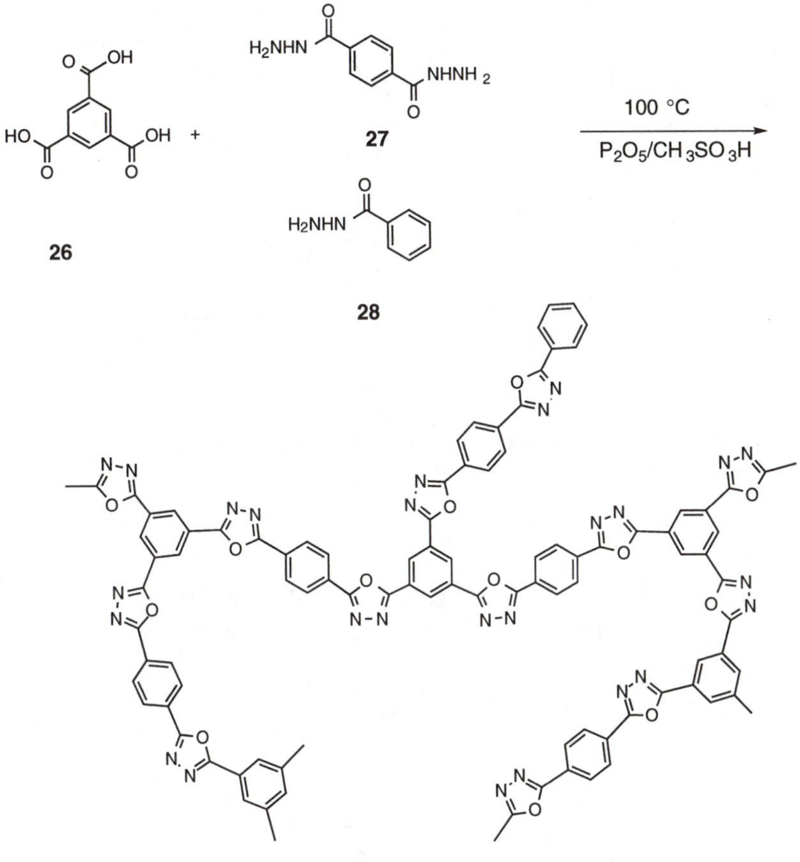

Scheme 8. Synthesis of a star-burst polyaromatic oxadiazole polymer **29** (adapted from *45*)

low molar mass and rigidity. The polymer was found to show blue fluorescence with UV light and has a maximum absorption peak at 310 nm as shown in Figure 11.

A two-layer EL device using PPV with polymer **29** and Ca as cathode was fabricated. The device characteristics are illustrated in Figure 12. A good working device which exhibited emission largely characteristic of PPV was obtained, and an internal efficiency of 0.1% was measured. Therefore the use of hyperbranched charge transporting materials may well offer a new strategy for enhancing device efficiency in the future.

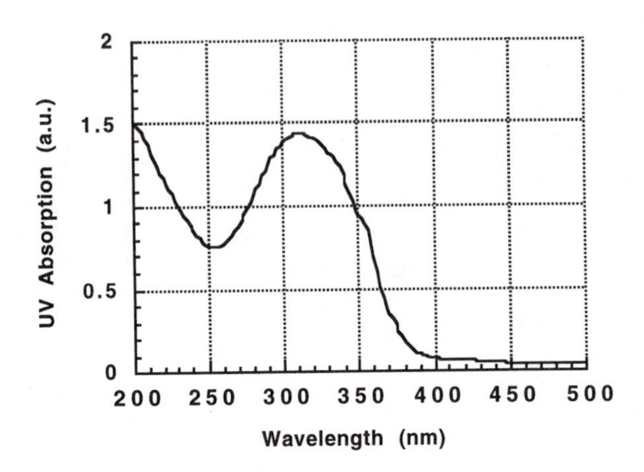

Figure 11. UV-VIS absorption spectrum of starburst polyaromatic oxadiazole **29** (adapted from *45*)

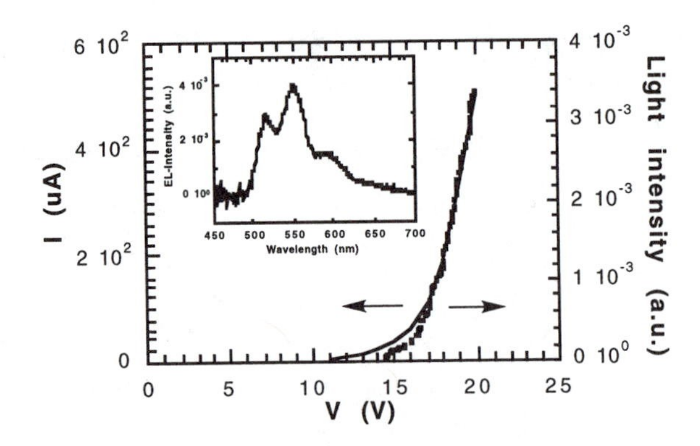

Figure 12. Current-voltage, luminance-voltage and EL emission spectrum of two-layer device ITO/PPV (100 nm)/polymer **29** (50 nm)/Ca (adapted from *45*)

Multi-functional Polymers for Polymer LEDs

One of the many advantages of a polymer over other materials is the possibility to combine multifunctionality into the chain. We have synthesized homopolymethacrylates carrying distyrylbenzene blue emitting side chains and charge transporting oxadiazole side chains. We have also shown the synthesis of copolymethacrylates bearing both blue light emitting blocks and charge transporting blocks (Scheme 9).

All the copolymerizations were carried out using radical conditions (AIBN as initiator) at elevated temperature (80 °C) in either toluene or benzene for 2-16 hours. The polymer yields were quite high (> 85%). The ratio of the two units in the copolymers was found to be very close to the feed ratio of the starting monomers (as confirmed by NMR spectra and microanalysis).

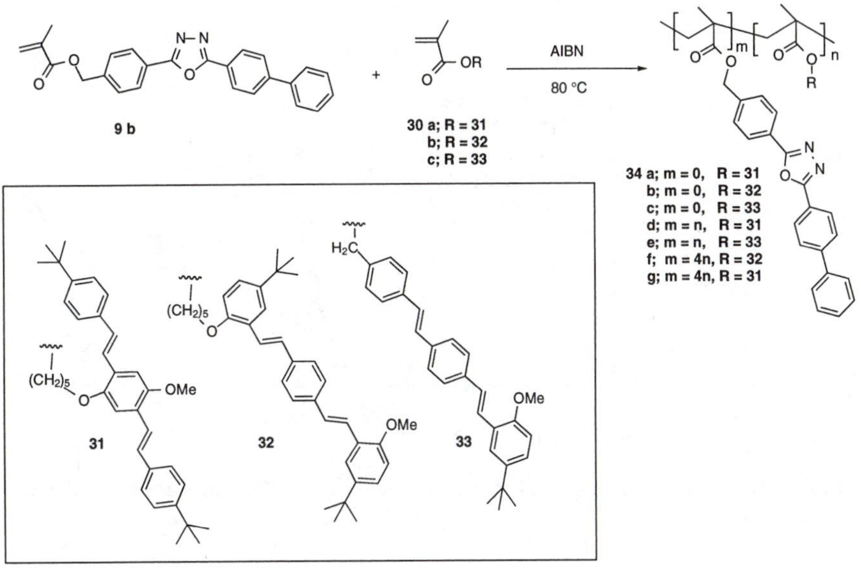

9 b + 30 a; R = 31, b; R = 32, c; R = 33 → AIBN, 80 °C → 34 a; m = 0, R = 31; b; m = 0, R = 32; c; m = 0, R = 33; d; m = n, R = 31; e; m = n, R = 33; f; m = 4n, R = 32; g; m = 4n, R = 31

31 32 33

Scheme 9. Synthesis of copolymethacrylates **34** bearing blue chromophores and oxadiazoles (adapted from *21*)

The copolymethacrylates produced LEDs having better stability and brighter blue light emission than those obtained from the corresponding homopolymers, which may be attributed to the better electron injection by the oxadiazole units. Figure 13 shows the optical properties of a typical copolymer **34e** (PMA-DSB-PBD) and its electroluminescence in a double layer device using PPV as hole-injection layer. The device has good shelf stability and an internal quantum efficiency (0.04%). It begins to emit blue light (475 nm) at a forward bias potential of 17 V (Figure 14).

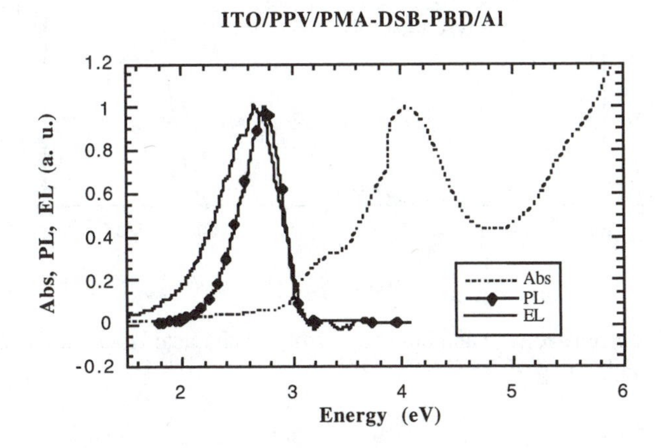

Figure 13. UV-VIS, PL and EL emission spectra of the triblock copolymer **34e**, the EL device configuration being ITO/PPV/polymer **34e**/Al

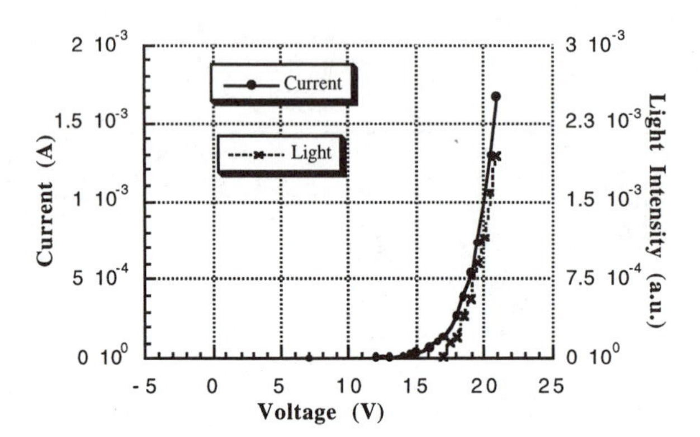

Figure 14. Current-voltage and luminance-voltage characteristics for the EL device shown in Figure 13

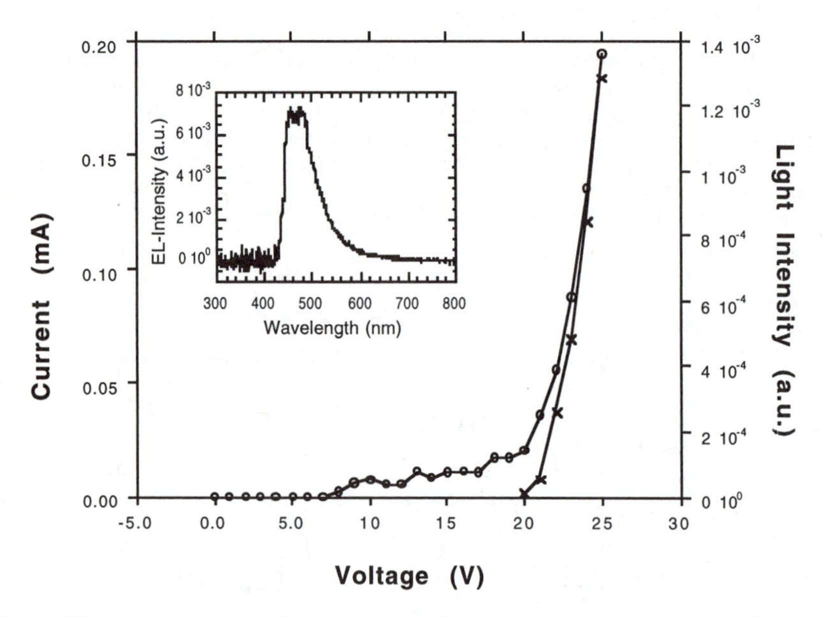

Figure 15. Current-voltage and luminance-voltage characteristics for the single layer device ITO/polymer **37**/Ca

In addition to preparing copolymers with blue chromophores and charge transporting segments, we have also prepared a ternary copolymer **36** as shown in Scheme 10. The ternary copolymer is highly fluorescent in the blue region (PL peak

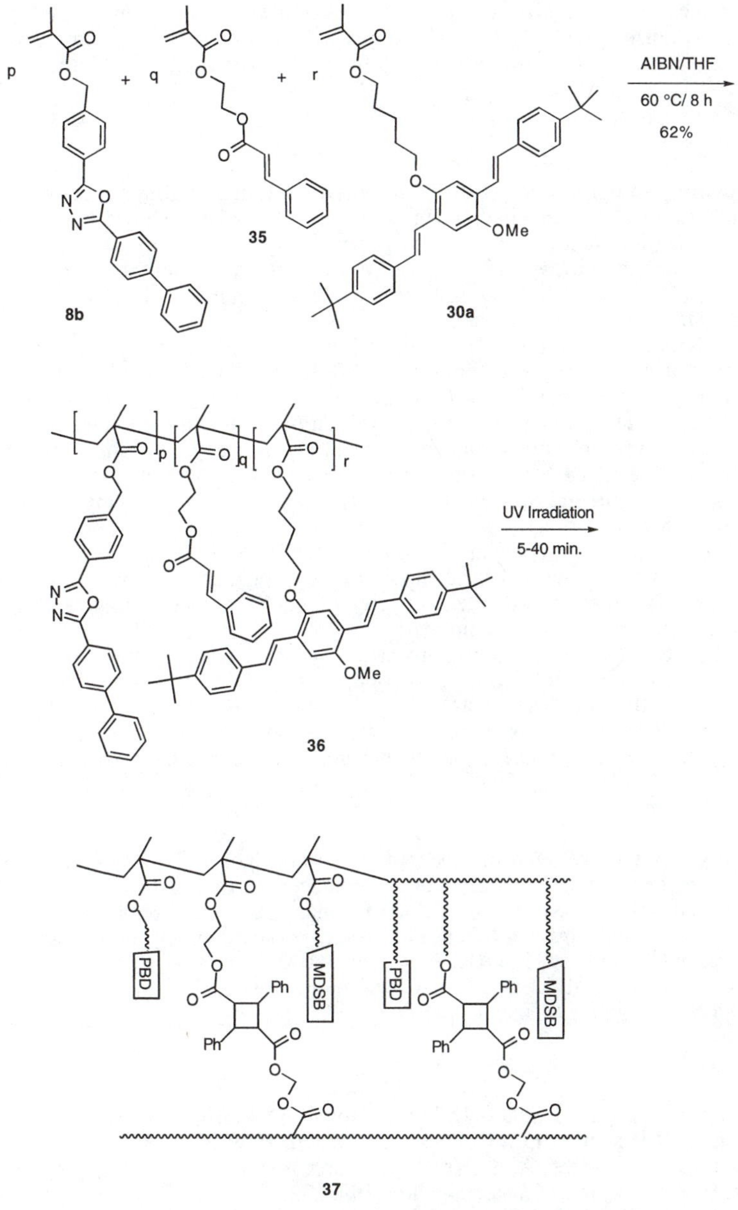

p **8b** + q **35** + r **30a** —AIBN/THF, 60 °C/ 8 h, 62%→ **36** —UV Irradiation, 5-40 min.→ **37**

Scheme 10. Synthesis of a ternary copolymethacrylate **36** bearing a blue-emitting chromophore, an electron injection oxadiazole and a photocrosslinkable unit

at 495 nm, solid state quantum eficiency 39%) and has a high molecular weight (11,200/53,500 M_n/M_w). It also exhibits excellent solubility in conventional solvents, such as toluene and chloroform. When the polymer is irradiated with UV-light photochemically-induced cross linking takes place and it becomes insoluble, while the photoluminescence is not markedly affected. All these properties make this material ideal for use in polymer LEDs. In a single layer device of ITO/copolymer **36**/Ca, emission of light was observed at 20 V with a peak wavelength of 490 nm, as shown in Figure 15.

Summary

The syntheses and properties of different polymers containing high electron affinity CF_3 and oxadiazole groups have been reviewed. It has been shown that when CF_3 groups are incorporated into conjugated polymers, such as poly(phenylene vinylene) derivatives, the polymers with CF_3 substituents show better electron accepting properties. This makes it possible to use stable aluminum in place of air sensitive calcium as metal cathode in polymer LEDs.

Polymers containing oxadiazoles either as pendant side groups or in the polymer main chain have been designed and synthesized. Functional polymethacrylates bearing oxadiazoles have very good solution processibility. They have been found to be useful as electron injection /hole blocking polymers in polymer LEDs. Lower turn-on voltage, improved quantum eficiency, and improved stability were observed. The ready functionalization of side chain polymers makes it possible to combine blue light emitting, electron injecting, and photocrosslinking moieties in a single copolymer.

Soluble main chain oxadiazole polymers can be realised using solubilizing side groups such as *tert*-butyl and long flexible alkoxy or alkyl groups. Alternatively, flexible spacers can be incorporated between aromatic oxadiazole units, and finally a soluble precursor route may be employed.

All the polymers containing aromatic oxadiazoles were found to be easily *n*-doped whereas they were difficult to *p*-dope as revealed by cyclic voltammetry, indicating that they possess electron injection and hole blocking properties which make them suitable for use as charge transporting layers in multilayer polymer LEDs. It was also shown that oxadiazole polymers can be used as emissive materials.

Acknowledgements

This work has largely been carried out in Cambridge, and we acknowledge the contributions of our colleagues Drs G.C.W. Spencer, A. Lux and F. Cacialli. We are also indebted to Dr. A. Kraft (Düsseldorf) for many valuable discussions. We thank the Engineering and Physical Sciences Research Council (UK), the Commission of the European Community (ESPRIT Project No. 8013 'LEDFOS', Brite Euram Project, BRE2-CT93-0592 'PolyLED'), the ICI Strategic Research Fund and Cambridge Display Technology for financial support.

Literature Cited

(1) Burroughes, J. H.; Bradley, D. D. C.; Brown, A. R.; Marks, R. N.; Mackay, K.; Friend, R.H.; Burn, P.L.; Holmes, A.B. *Nature* **1990**, *347*, 539.
(2) Burn, P. L.; Holmes, A. B.; Kraft, A.; Bradley, D. D. C.; Brown, A. R.; Friend, R. H. *J. Chem.Soc., Chem. Commun.* **1992**, 32.
(3) Zhang, C.; Braun, D.; Heeger, A.J. *J. Appl. Phys.* **1993,** *73*, 5177.
(4) Ohmori, Y.; Uchida, M.; Muro, K.; Yoshino, K. *Jpn. J. Appl. Phys., Part 2* **1991**, *20*, L1938.

(5) Malliaras, G. G.; Herrema, J. K.; Wildeman, J.; Wieringa, R. H.; Gill, R. E.; Lampoura, S. S.; Hadziioannou, G. *Adv. Mater.* **1993**, *5*, 721.
(6) Grem, G.; Leditzky, G.; Ullrich, B.; Leising, G. *Adv. Mater.* **1992**, *4*, 36.
(7) Huber, J.; Müllen, K.; Saalbeck, J.; Schenk, H.; Scherf, U.; Stehlin, T.; Stern, R. *Acta Polym.* *1994*, *45*, 244.
(8) Holmes, A. B.; Bradley, D. D. C.; Brown, A. R.; Burn, P. L.; Burroughes, J. H.; Friend, R. H.; Greenham, N. C.; Gymer, R. W.; Halliday, D. A.; Jackson, R. W.; Kraft, A.; Martens, J. H. F.; Pichler, K.; Samuel, I. D. W. *Synth. Met.* **1993**, *57*, 4031.
(9) Braun, D.; Heeger, A.J. *Appl. Phys. Lett.* **1991**, *58*, 1952.
(10) Greenham, N. C.; Moratti, S. C.; Bradley, D. D. C.; Friend, R. H.; Holmes, A. B. *Nature* **1993**, *365*, 628.
(11) Moratti, S.C.; Bradley, D.D.C.; Friend, R.H.; Greenham, N.C.; Holmes, A.B. *Mat.. Res. Soc. Symp. Proc.* **1994**, *328*, 371.
(12) Greiner, A.; Martelock, H.; Noll, A.; Siegfried, N.; Heitz, W. *Polymer* **1991**, *32*, 1857.
(13) Grimsdale, A.C.; Cacialli, F.; Grüner, J.; Li, X.-C.; Holmes, A.B.; Moratti, S.C. Friend, R.H. *Symth. Met.* **1996**, *76*, 165.
(14) Moratti, S. C.; Bradley, D. D. C.; Cervini, R.; Friend, R. H.; Greenham , N. C.; Holmes, A. B. *Proc. SPIE-Int. Soc. Opt. Eng.* **1994**, *2144*, 108.
(15) Adachi, C.; Tokito, S.; Tsutsui, T.; Saito, S. *Jpn. J. Appl. Phys.* **1988**, *27*, L269.
(16) Adachi, C.; Tokito, S.; Tsutsui, T.; Saito, S. *Jpn. J. Appl. Phys.* **1988**, *27,* L713.
(17) Hamada, Y.; Adachi, C.; Tsutsui, T.; Saito, S. *Jpn. J. Appl. Phys.* **1992**, *31*, 1812.
(18) Burn, P. L.; Holmes, A. B.; Kraft, A.; Brown, A. R.; Bradley, D.D.C.; Friend, R.H. *Mat. Res. Soc. Symp. Proc.* **1992**, *247*, 647.
(19) Brown, A. R.; Bradley, D. D. C.; Burroughes, J. H.; Friend, R. H.; Greenham, N. C. *Appl. Phys. Lett.* **1992**, *61*, 2793.
(20) Li, X.-C.; Giles, M.; Grüner, J.; Friend, R. H.; Holmes, A. B.; Moratti, S. C. *Proc. Am. Chem. Soc.: Polymeric Materials Science and Engineering* **1995**, *72*, 463.
(21) Li, X.-C.; Cacialli, F.; Giles, M.; Grüner, J.; Friend, R. H.; Holmes, A. B.; Moratti, S. C.; Yong, T. M. *Adv. Mater.* **1995**, *11*, 898.
(22) Burn, P. L.; Bradley, D. D. C.; Friend, R. H.; Halliday, D. A.; Holmes, A. B.; Jackson, R. W.; Kraft, A. *J. Chem. Soc., Perkin Trans. 1* **1992**, 3225.
(23) Strukelj, M.; Papadimitrakopoulos, F.; Miller, T. M.; Rothberg, L. J. *Science* **1995**, *267*, 1969.
(24) Strukelj, M.; Miller, T. M.; Papadimitrakopoulos, F.; Son, S. *J. Am. Chem. Soc.* **1995**, *117*, 11976.
(25) Dobinson, F.; Pelezo, C. A.; Black, W. B.; Lea, K. R.; Saunders, J. H. *J. Appl. Polym. Sci.* **1979**, *23*, 2189.
(26) Li, X.-C.; Spencer, G. C. W.; Holmes, A. B.; Moratti, S. C.; Cacialli, F.; Friend, R. H. *Synth. Met.* **1995**, *76*, 153.
(27) Frazer, A. H.; Sarasohn, I. M. *J. Polym. Sci., Part A-1* **1966**, *4*, 1649.
(28) Varma, I. K.; Greetha, C. K. *J. Appl. Polym. Sci.* **1978**, *220*, 411.
(29) Ueda, M.; Oda, M. *Polym. J.* **1989**, *21*, 193.
(30) Bach, H. C.; Dobinson, F.; Lea, K. R.; Saunders, J. H. *J. Appl. Polym. Sci.* **1979**, *23*, 2125.
(31) Kummerloewe, C.; Kammer, H. W.; Malincomico, M.; Martuscelli, E. *Polymer* **1991,** *32,* 2505.
(32) Fitch, J. W.; Cassidy, P. E.; Weikel, W. J.; Lewis, T. M.; Tril, T.; Burgess, L.; March, J. L.; Glowe, D.E.; Rolls, G. C. *Polymer* **1993**, *34*, 4796.
(33) Livshits, B. R.; Vinogradova, S. V.; Knunyants, I. L.; Berestneva, G. L.; Dymoshits, T. K. L. *Vysokomol. Soedim. Ser. A* **1973**, *15*, 961 [*Chem. Abstr.* **1973**, *79*, 105603h].

(34) Hensema, E. R.; Sena, M. E. R.; Mulder, M. H. V.; Smolders, C. A. *J. Polym. Sci., Polym. Chem.* **1994**, *32*, 527.
(35) Yang, Y.; Pei, Q. *J. Appl. Phys.* **1995**, *77* , 4807.
(36) Li, X.-C.; Holmes, A. B.; Kraft, A.; Moratti, S. C.; Spencer, G. C. W.; Cacialli, F.; Grüner, J.; Friend, R. H. *J. Chem. Soc., Chem. Commun.* **1995**, 2211.
(37) Janietz, S.; Schulz, B.; Törönen, M.; Sundholm, G. *Eur. Polym. J.* **1993**, *29*, 545.
(38) Kanbara, T.; Miyazaki, Y.; Yamamoto, T. *J. Polym. Sci., Polym. Chem.* **1995**, *33*, 999.
(39) Baigent, D. R.; Greenham, N. C.; Grüner, J.; Marks, R. N.; Friend, R. H.; Moratti, S. C.; Holmes, A. B. *Synth. Met.* **1994**, *67* , 3.
(40) Adachi, C.; Tsutsui, T.; Saito, S. *Appl. Phys. Lett.* **1990**, *56*, 799.
(41) Tsutsui, T.; Aminaka, E.; Fujita, Y.; Hamada, Y.; Saito, S. *Synth. Met.* **1993**, *57*, 4157.
(42) Berggren, M.; Gustafsson, G.; Inganäs, O.; Andersson, M. R.; Hjertberg, T.; Wennerström, O. *J. Appl. Phys.* **1994,** *76*, 7530.
(43) Pei , Q.; Yang, Y. *Adv. Mater.* **1995**, *7*, 559.
(44) Pei, Q., Yang, Y. *Chem. Mater.* **1995**, *7,* 1568.
(45) Li, X.-C.; Kraft, A.; Cervini, R.; Spencer, G.C.W.; Cacialli, F.; Friend, R.H.; Grüner, J.; Holmes, A.B.; De Mello, J.C.; Moratti, S.C. *Mater. Res. Soc. Symp. Proc.*, **1996**, *413*, 13.

Chapter 23

Polymer Light-Emitting Diodes Utilizing Arylene–Vinylene Copolymers as Light-Emitting Materials

Toshihiro Ohnishi, Shuji Doi, Yoshihiko Tsuchida, and Takanobu Noguchi

Tsukuba Research Laboratory, Sumitomo Chemical Company, Ltd., 6 Kitahara, Tsukuba, Ibaraki 300–32, Japan

Polymer light-emitting diode (P-LED) devices have successfully been fabricated using highly luminous poly(arylene vinylene) derivatives. Structural and energetic irregularity introduced into the conjugated polymers gave us a highly luminous polymer due to confinement of excitons. The irregularity can be generated by copolymerization of conjugated/non-conjugated segments, *m/p*-phenylene vinylenes and alkyl/alkoxy-substituted phenylene vinylenes. Among the copolymers, the *m/p*-phenylene vinylene copolymer gave the most highly efficient P-LED device because of the good balance of the exciton confinement and the charge transporting property . The optimized P-LED showed a maximum luminance of 55,000cd/m^2.

Organic and polymer light-emitting diodes, O-LED and P-LED, have attracted much attention as an accessible flat panel display and have shown good progress in the last few years. Since bright and low-voltage-driven O-LED devices were reported by Tang et al.(1), many light-emitting and charge-transporting materials and their devices have been reported(2). Some of the devices showed high luminance and long lifetime(3). Catching up with O-LED, a P-LED device was reported in 1990(4). Since many kinds of conjugated polymers previously studied as a conducting polymer have been studied as light-emitting materials. P-LED devices using PPV derivatives(5,6), poly(*p*-phenylene)(7) and poly(3-alkyl- thiophene)(8) have been reported since 1990.

We have already reported soluble light-emitting poly(2,5-dialkoxy-*p*-phenylene vinylene), RO-PPV, and its side chain length dependence of fluorescence intensity(9). Intensity of photo-luminescence, PL, of the RO-PPV film increased with the alkoxy chain length, which is attributable to the decrease of intermolecular interaction and exciton-confinement in the film.

High efficiency and long lifetime are essential factors for commercialization of O- and P- LEDs. There have been many attempts made to improve device performance so far(10). Device efficiency, η_d, is defined in the following equation(11).

$$\eta_d = \gamma \cdot \eta_{e\text{-}h} \cdot \phi_{fl} \quad [1]$$

where γ is the ratio of the number of minority carrier to that of major carrier, $\eta_{e\text{-}h}$ is the efficiency of electron-hole recombination in the device and ϕ_{fl} is the fluorescence efficiency of the light-emitting material.

The value of γ is related to the injection process and depends on electrode materials. In P-LED, a calcium electrode is used to increase the electron injection efficiency due to its low work function(12, 13). The value of $\eta_{e\text{-}h}$ is related to not only materials but also device structure. Multi-layer structure is commonly used to increase the value of $\eta_{e\text{-}h}$, generating the hole and/or electron accumulation near a light-emitting layer surface(1,14). The maximum value of $\eta_{e\text{-}h}$ has been suggested to be 25% because of the spin statistics of singlet exciton formation (11). To improve ϕ_{fl}, we have to use highly efficient materials or dye-doped materials(15).

After fundamental studies on soluble RO-PPV derivatives(9), we have focused on the synthesis of highly luminous polymers and fabrication of highly efficient devices to commercialize P-LED. Taking exciton confinement and charge transporting into account, we have copolymerized various kinds of arylene vinylene units to introduce structural and energetic irregularity into poly(arylene vinylene).

In this paper we will discuss the relation between light -emitting properties and the chemical structure of arylene vinylene copolymers.

Experimental Section

Poly(arylene vinylene) derivatives were prepared by Wittig reaction of diphosphonium compounds(I and V), and dialdehyde compounds(II, III,and IV) as shown in Figure1. When I and II, I and III , I and IV, and V and II are used, soluble alternating copolymers of (A), (B), (C), and (D) are obtained respectively(16,17,18). The diphosphonium compounds were prepared using triphenylphosphine and 2,5-dioctyloxy(dioctyl)-*p*-xylylene dichloride, and polymerized by addition of dialdehyde compounds and tert-BuOK as a catalyst in tert-BuOH at 90℃ for 7 hours to give a copolymer. Random copolymers having (A) and (B), (A) and (C), and (A) and (D) (abbreviated as A/B, A/C and A/D respectively) were also synthesized in a similar manner to the copolymers using corresponding three monomers selected among (I), (II) , (III), (IV), and (V)(19).

The copolymers were obtained by precipitation in methanol from reactant solution and purified by reprecipitation. Poly(2,5-dioctyloxy-p-phenylene vinylene), RO-PPV-8, used as a standard polymer was prepared by dehydrohalogenation of 2,5-dialkoxy-p-xylylene dibromide as reported previously(9). The number-average molecular weight, measured by GPC using polystyrene as a standard, was 5×10^3 - 1×10^4 for all copolymers and 8×10^4 for RO-PPV-8.

The P-LED device consists of a transparent electrode, a light-emitting polymer film, an electron-transporting or hole-blocking layer, and a negative electrode as shown in Figure 2. A buffer or additional layer(9) was used between the transparent electrode and the light- emitting polymer film in the case of a three-layer structure.

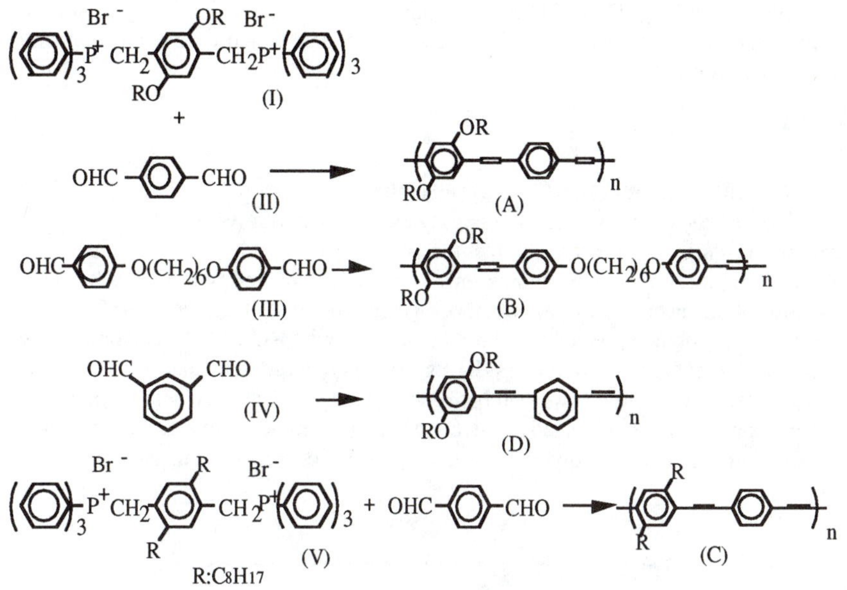

Br⁻ OR Br⁻
$(C_6H_5)_3P^+CH_2$–Ar–$CH_2P^+(C_6H_5)_3$
RO (I)
\+
OHC–⟨⟩–CHO (II) → (A) OR, RO, n
OHC–⟨⟩–$O(CH_2)_6O$–⟨⟩–CHO (III) → (B) OR, RO, $O(CH_2)_6O$, n
OHC CHO (IV) → (D) OR, RO, n
Br⁻ R Br⁻
$(C_6H_5)_3P^+CH_2$–Ar–$CH_2P^+(C_6H_5)_3$ + OHC–⟨⟩–CHO → (C) R, R, n
R (V)
$R: C_8H_{17}$

Figure 1. Synthetic route for alternating and random copolymers of arylene vinylenes

A polymer film, 50 to 100 nm in thickness, was spin-cast from the toluene solution onto an indium tin oxide, ITO, transparent electrode. A tris(8-quinolinol) aluminum, Alq_3, layer as an electron-transporting layer was then deposited onto the polymer film followed by co-evaporation of Li-Al or Mg-Ag alloy as a negative electrode under a vacuum of 10^{-6} Torr.

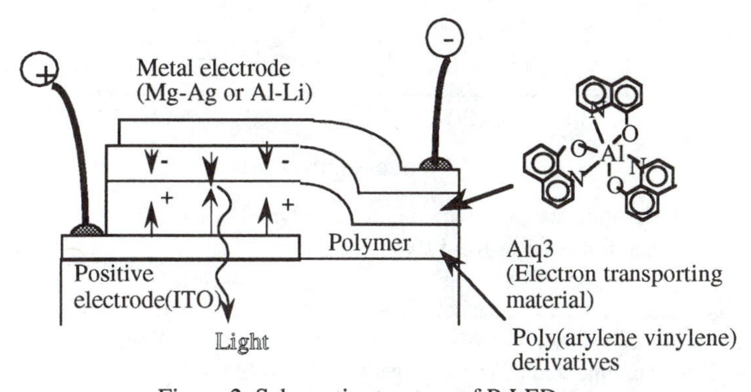

Figure 2. Schematic structure of P-LED

Absorption and luminescence spectra for thin films and devices using copolymers were measured with a Shimadzu UV-365 spectrophotometer and a Hitachi 850 fluorescence spectrophotometer, respectively. Luminance and current-voltage curves were obtained with a TOPCON BM-8 luminance meter, a Keithley 990 digital multimeter and a Takasago GP050-2 DC voltage source which were controlled with a personal computer.

Results and Discussion

Spectroscopic properties of Copolymers.

Alternating copolymers. All copolymers obtained in yellow or yellowish-orange powder are soluble in organic solvents such as toluene and chloroform and show strong fluorescence in both solution and film. Figure 3 shows the absorption spectra of the copolymers together with that of RO-PPV-8. The spectra of copolymers were blue-shifted compared with RO-PPV-8. The band gaps or absorption edges of copolymers (B) and (D) were larger than that of RO-PPV-8, while those of (A) and (C) were in between as shown in Table I. These results indicate that the non-conjugated segment and the *m*-phenylene structure shorten the conjugation length in the polymer due to low resonance effect and thus enlarge the band gap.

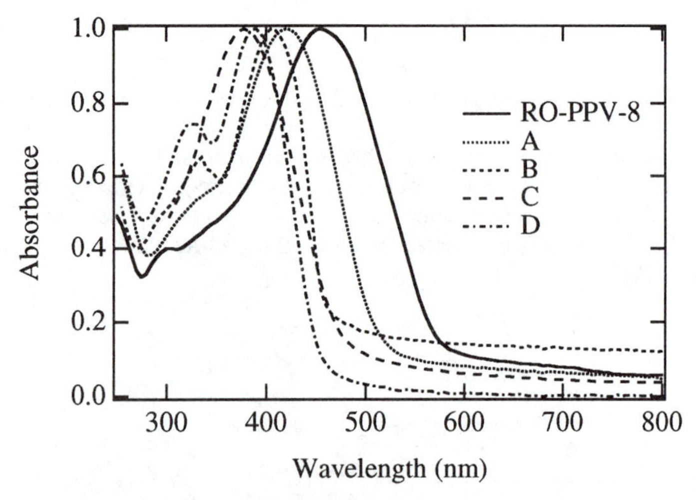

Figure 3. Absorption spectra of alternating copolymer of (A), (B), (C) and (D) together with that of RO-PPV-8 in film.

The fluorescence spectra of the copolymers were also blue-shifted similarly to the absorption spectra as shown in Figure 4. Spectral properties of all the alternating copolymers are summarized in Table I together with RO-PPV-8. The fluorescence intensity or the relative quantum efficiency of the fluorescence was estimated by dividing the signal intensity divided by the absorbance at excitation wavelength. The

(B) copolymer having the ether group in the polymer chain shows the strongest fluorescence in all the copolymers although it has the shortest conjugation length. The *m*-phenylene structure((D) copolymer) gives stronger fluorescence than *p*-phenylene structure((A) copolymer) because of less resonance effect. These results indicate that the fluorescence intensity depends on the degree of conjugation in the polymer chain. Since the (C) copolymer having the alkyl substituent shows stronger fluorescence than the alkoxy substituted copolymer (A), the alkoxy substituent suppresses the fluorescence. In fact methoxy-substituted poly(p-phenylene vinylene), PPV, shows weaker fluorescence than methyl-substituted PPV(20).

Table I. Optical properties of alternating copolymers in film

Polymer	repeating unit	Absorption peak(nm)	Band gap (eV)	Fluorescence peak(nm)	Fluorescence intensity (a.u.)
RO-PPV-8	RO-PV-8	455	2.12	582	3.6
Alternating	(A)	420	2.25	534	7.5
Ether type	(B)	405	2.66	504,528	50.5
R-PPV	(C)	375	2.53	496	9.8
m/p-phenylene	(D)	390	2.66	504	15.5

Band gap was estimated from wavelength of absorption edge
Fluorescence intensity=(area of fluorescence peak plotted against wave number) /(absorbance at excitation wavelength)

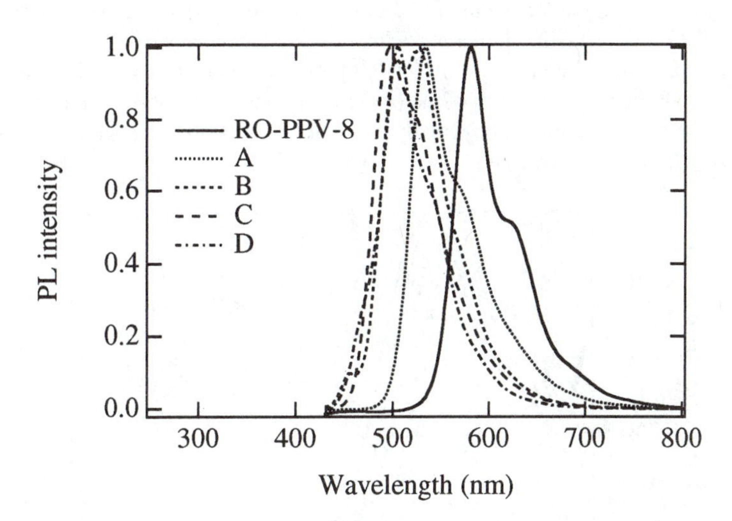

Figure 4. Fluorescent spectra of alternating copolymers of (A), (B), (C) and (D) together with that of RO-PPV-8

In these copolymers the substituent effect on the fluorescence intensity was observed similarly to that in RO-PPV reported previously(9). The effect is explainable on the basis of exciton migration(21).

Random copolymer. Figures 5 and 6 show the absorption and fluorescence spectra of the random copolymers consisting of (A) and (D) units. The absorption peak shifts proportionally to the ratio of (A) and (D), while the absorption edge shifts less than the peak. Similar shifts in absorption and fluorescence spectra were observed in the copolymers of A /B, and A/C.

The fluorescence spectra of random copolymers were, on the contrary, similar to that of (A) copolymer and were hardly shifted with the increasing component of (D) unit as shown in Figure 6. Such small spectral shifts in fluorescence were also observed in A/B and A/C copolymers. This result indicates that randomly copolymerizing the (A) unit and the others generates the low energy segment consisting of a sequence of (A) which has the lower bandgap than other segments.

Quantum efficiency of PL. Figure 7 shows the relation between the fluorescence intensity in film form and the ratio of (A) unit in the copolymers for A/B, A/C, and A/D. In the A/B and A/D copolymers, the fluorescence intensity increased with the decreasing ratio of (A) . Since wavelengths of fluorescence peak in the spectra are close to that of (A) polymer as shown in Figure 6, the fluorescence is emitted from the lower energy segments consisting of the (A) unit which are incorporated with segments having higher energy. This means that the exciton generated by light absorption migrates to the lowest energy state and is confined efficiently to generate radiation. The exciton confinement thus takes place more easily in copolymers having less content of (A) and leads to strong fluorescence. In the case of the A/C copolymers, the fluorescence intensity basically increased with the decreasing ratio of (A) unit except for the (C) polymer which shows smaller intensity than the copolymer having 10% of the (A) unit. This is attributable to smaller confinement of the exciton in (C) copolymer than random copolymers because there is no energetic irregularity in the (C) copolymer.

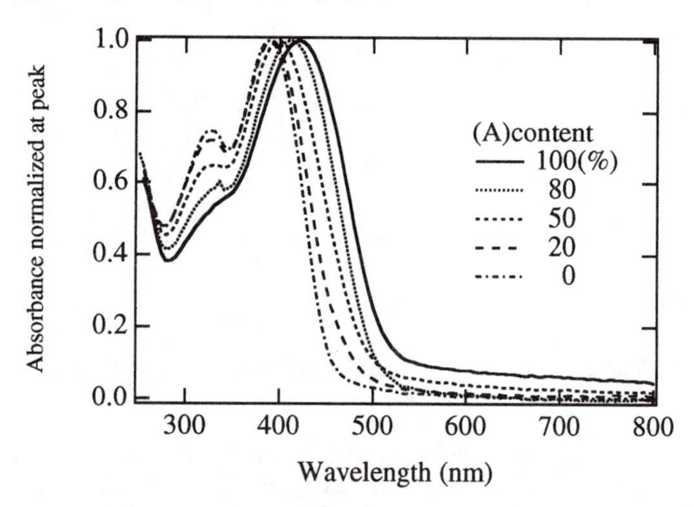

Figure 5. Absorption spectra of random copolymers consisting of (A) and (D) at various ratio.

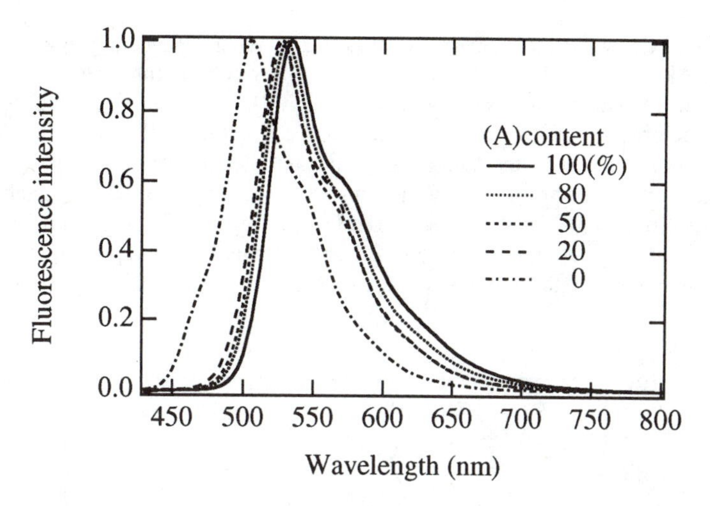

Figure 6. Fluorescent spectra of random copolymers consisting of (A) and (D) at various ratio.

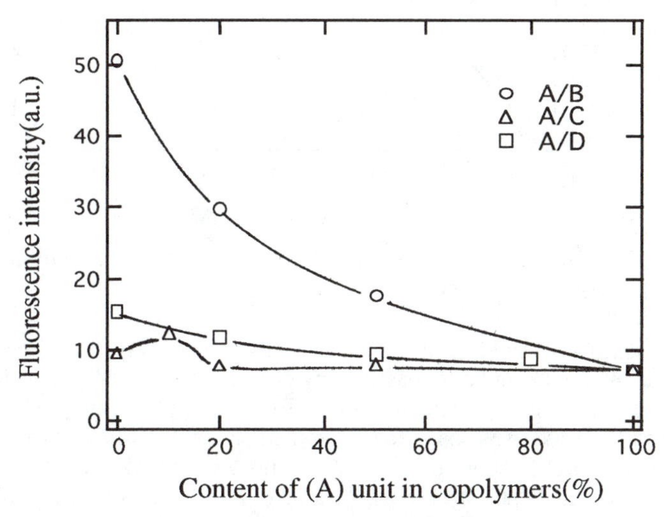

Figure 7. Relationship between content of (A) unit and fluorescence intensity for random copolymers, A/B, A/C and A/D.

Electroluminescence of P-LED-using copolymers.

Multi-layer device. High performance of the first O-LED(1) was achieved by employing a multi-layer structure of Alq_3 and a diamine compound. In the double-layer O-LED, the electron-accumulation takes place and hole and electron recombines to generate excitons more easily than a single-layer device. In the P-LED, Brown et al. reported that the double-layer structure of an oxadazole compound and

PPV was effectively used to increase luminance (22). We independently found out that using Alq_3 between RO-PPV layer and a negative electrode enhanced the luminance of a P-LED device as shown in Figure 8 (23). It is noteworthy that the light-emitting property of Alq_3 should be taken into account in the devices using Alq_3 since Alq_3 possesses not only light-emitting but also electron- transporting properties. An EL spectrum of the device fabricated is shown in Figure 9 together with PL spectrum of RO-PPV and Alq_3. The EL spectrum is in agreement not with PL spectrum of Alq_3 but with that of RO-PPV and no emission of Alq_3 is observed. Electroluminescence of the double-layer device is attributed to radiative decay of excited RO-PPV.

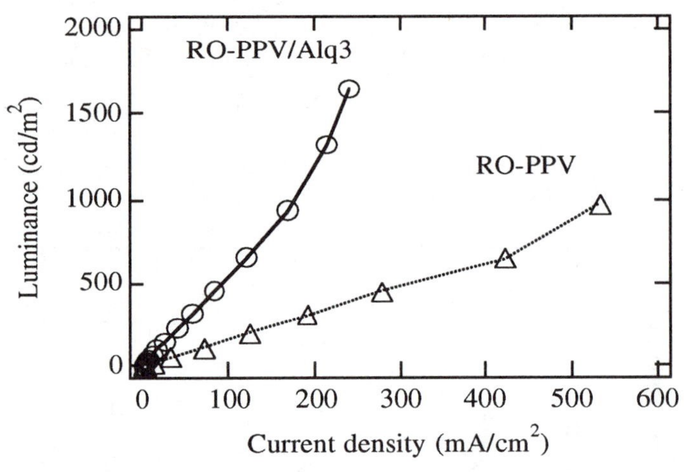

Figure 8. Current density-luminance curve of single- and double-layer device using RO-PPV.

The excited state of RO-PPV is formed generally in three ways: the direct recombination of holes and electrons, energy transfer from excited Alq_3 and re-absorption of the light emitted from Alq_3. The re-absorption is negligible because no emission from Alq_3 was observed as shown in Figure 9. To clarify which is excited, Alq_3 or RO-PPV, we measured the energy levels of Alq_3 and RO-PPV-8 electrochemically and optically(The energy level of HOMO(highest occupied molecular orbital) was calculated from the threshold oxidation potential in cyclic voltamogram measured against the potential of standard hydrogen electrode(4.5eV). The LUMO level was estimated from a threshold wavelength of an absorption spectrum.). As shown in Figure 10, there is an energy barrier for the hole-injection from RO-PPV to Alq_3, with no barrier for the electron-injection from Alq_3 to RO-PPV. This energy diagram strongly suggests that Alq_3 acted not only as an electron-transport material in the device but also as a blocking layer for the hole-injection from the RO-PPV layer and that the recombination of the holes and electrons mainly takes place in the RO-PPV layer.

In the copolymers, we also found out that Alq_3 also acted as a good electron-transporting and hole -blocking material and gave us highly luminous devices. Figure 11 shows the EL spectra of the devices using (A), (B), and A/D copolymers together with the PL spectrum of Alq_3. Both EL spectra differ from the PL spectrum of Alq_3 and are in good agreement with PL spectra of copolymers shown in Figures 4 and 6. In the devices using the copolymers and Alq_3, there is an energy barrier of about 0.3eV at the polymer-Alq_3 interface for the hole injection and no barrier for the electron injection as shown in Figure 10.

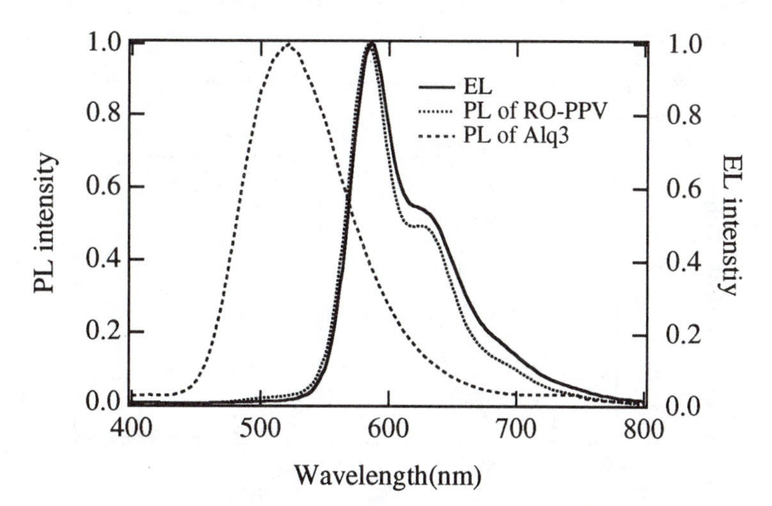

Figure 9. EL spectrum of a double-layer device using RO-PPV and Alq_3 together with PL spectra of RO-PPV and Alq_3

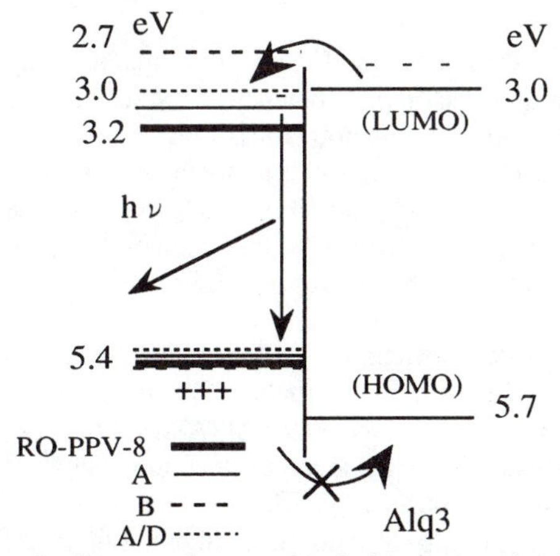

Figure 10. Schematic diagram of energy levels for the junction between the copolymers and Alq_3

In the case of copolymer (B), the double-layer device using Alq_3 showed an EL spectrum in the wavelength region where PL of the Alq_3 is seen. The EL spectrum of the device is in agreement with PL of the copolymer (B) rather than with that of Alq_3, which indicates that the electron-injection predominantly takes place in the device. The energy barrier height for the electron-injection was, however, about 0.3eV which is as high as that of hole-injection at the interface of Alq_3 and the copolymer. These results imply that the electron-injection takes place more easily than the hole-injection at the organic interface although the electrochemical estimation of the energy level is fairly rough.

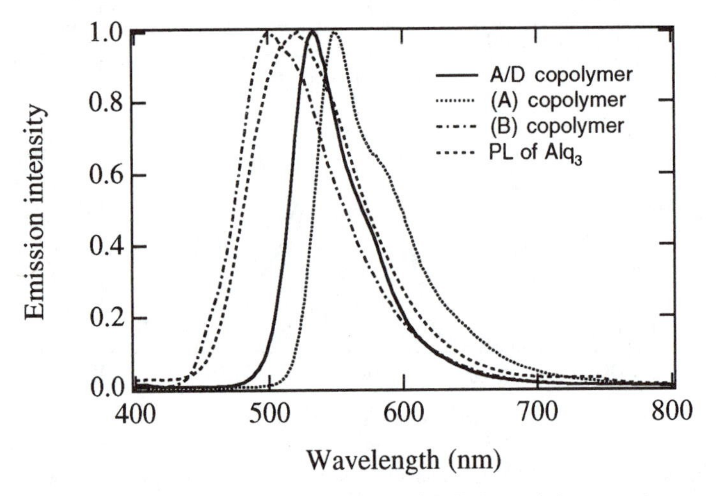

Figure 11. EL spectra of multi-layer P-LED using light-emitting copolymers and Alq_3 as an electron-transporting material together with PL spectrum of Alq_3

Device efficiency. The device efficiency depends not only on PL efficiency of the emitting polymer but also on the efficiency of hole-electron recombination which is related to the charge transporting properties of materials used in the device and to the device structure as described above. Employing a multi-layer device(23) and a Mg-Ag or Li-Al alloy electrode(9), we have already improved the device efficiency but have to further improve the efficiency for commercialization. We, thus, have studied the relationship between composition of the random copolymer and the device efficiency.

The EL efficiency of the devices using all the random copolymers showed a maximum around 50% of (A) unit as shown in Figure 12, while the fluorescence intensity increased with the decreasing ratio of (A) as shown in Figure 7. The increase of the device efficiency in the large ratio region of (A) unit is due to the increase of the fluorescence efficiency. The decrease in the small content region of (A) unit is attributable to the decreasing conductivity of the copolymer because the structural and energetic irregularity increased in the copolymer chain. These results suggest that the device efficiency depends not only on the fluorescence intensity but also on the charge carrier mobility of the copolymers.

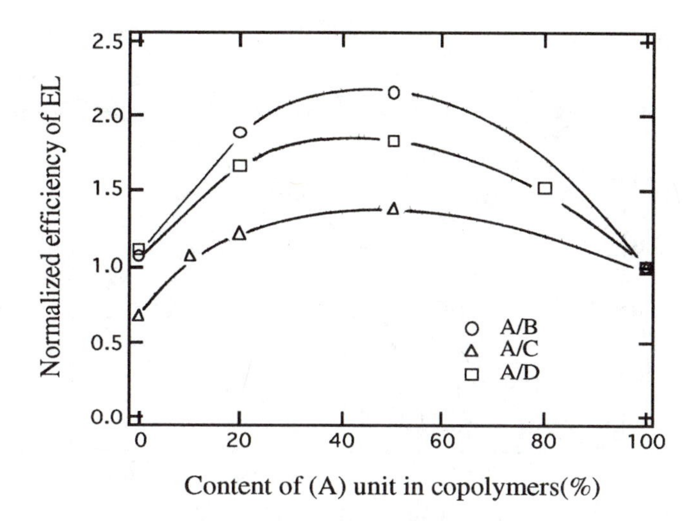

Figure 12. Relation between content of (A) unit and EL efficiency for random copolymers, A/B, A/C and A/D. Device structures are a bi-layer device for A/B and a multi-layer device for A/C and A/D.

Figure 13 shows the current density-luminance, I-L, curves of the P-LED devices using the random copolymers together with that for the (A) copolymer. The luminance of the devices increased almost linearly with the current-density for all devices. The maximum luminance reached 55,000 cd/m^2 for the A/D copolymer, 40,000cd/m^2 for A/C and (A) copolymers, and several thousands cd/m^2 for the A/B copolymer before break-down of the devices. In these devices except for that of A/B copolymer, Li-Al electrode was used because of its low work function.

Since the slope of the I-L curve represents the device efficiency, the A/D polymer-based device having the steepest slope showed the highest efficiency in the four devices. The maximum external device efficiency reached about 6 cd/A.

All of the devices can be driven at below 15V at high current density without break-down but they break-down at over 20V. In the case of the A/B copolymer, high current density was obtained although the luminance of the device is lower than those of the other devices. This is attributable to employing a two-layer structure and a Mg/Ag electrode.

Conclusion. The structural and energetic irregularity generated by copolymerization is effective in enhancing fluorescence intensity of poly(arylene vinylene) derivatives and gives highly luminous P-LED devices. The *m*-phenylene structure induces the irregularity effectively and gives the good balance of the exciton confinement and the charge transporting property for high performance devices. The multi-layer structure plays an important role in the P-LED using copolymers of arylene vinylenes to enhance the efficiency of hole-electron recombination. The optimized P-LED device using *m*- and *p*-phenylene vinylene copolymers showed high efficiency and maximum luminance as high as 55,000cd/m^2.

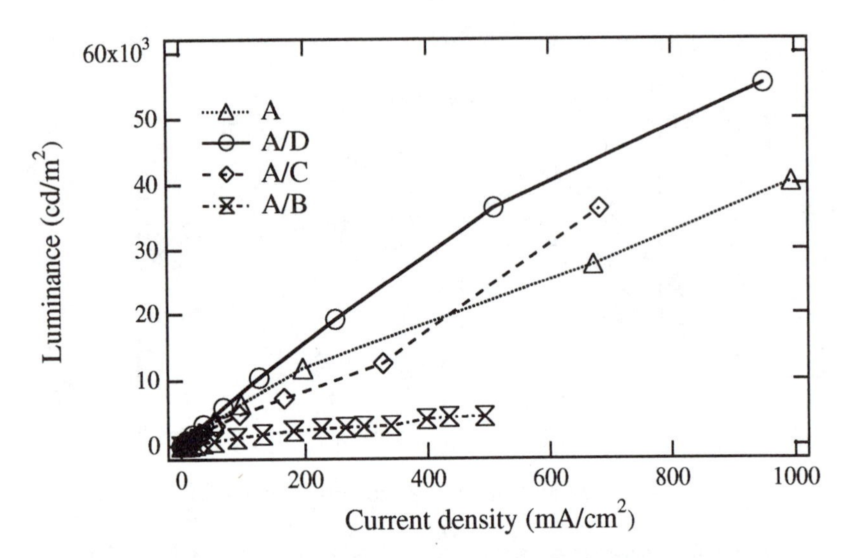

Figure 13. Luminance-current density curves of the P-LEDs using (A) alternating copolymer and A/B, A/C, A/D random copolymers. Li-Al electrode was used for (A) , A/C and A/D, while Mg-Ag was used for A/B.

Literature cited

1. Tang, C.W.; VanSlyke,S.A. *Appl. Phys. Lett.*, **1987**, *51*, 913.
2. Adachi, C.; Tokito, S.; Tsutsui, T.; Saito, S. *J pn J. Appl. Phys.* **1988**, *27*, L269.
3. Hamada, Y. ; Sano, T. ; Shibata,K. ; Kuroki, K. *Jpn. J. Appl. Phys.* **1998**, *34*, L824.
4. Burroughes, J.H.; Bradley, D.D.C.; Brown, A.R.; Marks, R.N.; Mackay,K.; Friend, R.H.; Burns,P.L.; Holmes, A.B. *Nature*, **1990**, *347*, 539.
5. Nakano, T. ; Doi, S. ; Noguchi, T. ; Ohnishi, T. ; Iyechika, Y. *European Patent* 0443861A3 ,1991.
6. Braun,D.; Heeger, A. *Appl. Phys.Lett.*, **1991**, *58*, 1982.
7. Grem, G.; Leditzky, G.; Ullrich, B.; Leising,G. *Adv. Mater.*, **1992**, *4* , 36.
8. Ohmori, Y. ; Uchida, M. ; Muro, K.; Yoshino, K. *Jpn J. Appl. Phys.*, **1991**, *30*, L1938.
9. Doi, S.; Kuwabara, M.; Noguchi, T.; Ohnishi, T. *Synth. Met.*, **1993**, *55-7*, 4174.
10. Saito, S.; Tsutsui, T.; Era, M.; Takada, N.; Aminaka, E.; Wakimoto, T. *Mol. Cryst. Liq. Cryst.*, **1994**, *253*, 125.
11. Tsutsui, T. ; Saito,S. *NATO ASI Ser., Ser. E*, **1993**, *246*, 123-134.
12. Braun, D.; Heeger, A.J. *J. Electronic Mater.*, **1991**, *20*, 945.
13. Burn, P.L.; Holmes, A.B.; Kraft, A.; Brown, A.R.; Bradley, D.D.C.; Friend, R.H.*Material Research Society Symposium Proceedings*, **1992**, *247*, 647.
14. Greenham, N.C.; Moratti, S.C.; Bradley, D.D.C.; Friend, R.H.; Holmes, A.B. *Nature*, **1993**, *365*, 628.

15.Tang, C.W.; VnSlyke, S.A.; Chen, C.H. *J. Appl. Phys.*, **1989**, *65*, 3610-3616.
16.Ohnishi, T.; Noguchi, T.; Doi, S. *Japanese Patent* 5-202355,1993.
17.Ohnishi, T.; Noguchi, T.; Doi, S. *Japanese Patent* 5-320635,1993
18.Doi, S.; Kuwabara, M.; Noguchi, T.; Ohnishi, T. ; Ishitobi, M. *European Patent* 0672741A1, 1995
19.Ohnishi, T.; Noguchi, T.; Doi, S. *EuropeanP atent* 0637621A1,1995
20.Authors, Unpublished data.
21.Rauscher,U.; Shüts, L.; Greiner,A.; Bässler, H. *J. Phys.:Condensed. Matter,* **1989**, *1*, 9751.
22.Brown,A.R. ; Bradley,D.D.C.; Burn,P.L.; Burroughes, J.H.; Friend, R.H.; Greenham, N.;Homles, A.B.; Kraft, A. *Appl. Phys. Lett.*, **1992**, *61*, 2793.
23.Ohnishi, T. ; Noguchi, T.; Doi,S. *Japanese Patent* 5-247460,1993.

Chapter 24

Novel Conjugated Polymers: Tuning Optical Properties by Synthesis and Processing

Ullrich Scherf and Klaus Müllen

Max-Planck-Institut für Polymerforschung, Ackermannweg 10, D–55128 Mainz, Germany

A major challenge in the design of novel conjugated polymers is to synthesize structures with tailor-made electronic properties, e.g. a well-defined and highly efficient emission behavior (photo- and electroluminescence). Three representative examples of poly(*para*-phenylene)-type (**PPP**-type) polymers demonstrate that the chromophoric properties of conjugated π-systems sensitively depend upon their molecular and supramolecular architecture. An important aspect on the molecular level is the interaction between the *para*-phenylene subunits. On the other hand, a specific supramolecular ordering (aggregation) of the individual molecules can drastically influence electronic properties. "Step-ladder" and ladder-type **PPP**'s as well as star-shaped oligophenylenes are characterized by promising light emitting properties. Poly(*para*-phenylene) based ribbons exhibit a highly efficient blue or yellow electroluminescence.

Conjugated polymers can now claim a considerable and uninterrupted degree of attraction over a period of several decades [1]. In the initial years, research into the synthesis of the first representatives of the new substance class of conjugated polymers (polyacetylene, poly(*para*-phenylene), poly(*para*-phenylenevinylene), polythiophene, polypyrrole, polyaniline) was important. The resulting polymers were characterized in most cases by their insolubility and infusibility, properties that considerably hindered their structural characterization and their processing. The majority of such compounds possessed no fully defined structure and the physical properties were influenced by structural defects. Moreover, it was often difficult to distinguish between neutral molecules and the "doped" species resulting from oxidation or reduction.

The last few years have now brought about a qualitative new development, as a consequence of considerable advances in the available synthetic methods. In the 70s and 80s purely synthetic aspects were in the forefront, whereas during the following years the effective physical function of conjugated polymers has progressively become

the main topic of research. However, in order to be able to draw a significant, definite correlation between the unique π-conjugated structure and a specific physical property (e.g., electrical- or photoconductivity, electroluminescence, photovoltaic effect), crucial new demands must be made of the materials being investigated including the following:
(1) First, the polymers should be as free of defects as possible. This requirement is to exclude the possibility that the physical function is influenced by structural defect. Reproducibility in the synthesis of the polymers regarding the properties obtained is included by definition.
(2) Second, the materials used must be processable, to be able to work them into the desired form (e.g., thin films or -layers, fibers). Processability can be achieved, for example, by rendering the polymers soluble. An established strategy for this involves the introduction of solubilizing groups (alkyl-, alkoxy- or aryl substituents) [2]. Another important principle is the processing of essentially insoluble polymers at the stage of still soluble precursors [3,4]. The precursors are brought into the necessary processing form and then converted in the solid state (most often thermally) into the corresponding conjugated polymers (precursor route). Other strategies involve the use of solubilizing counter-ions for doped species (polyaniline) [5] or processing via the intermediate formation of soluble charge-transfer complexes [6].

The requirements outlined represent a considerable challenge for polymer synthesis. The following article will show the appealing development, based on a central substance class of conjugated polymers, poly(*para*-phenylene)s.

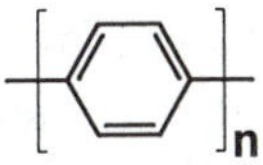

$[-C_6H_4-]_n$ **1** PPP

As a conjugated polyhydrocarbon, poly(*para*-phenylene)s represent a structure class that is currently being intensively investigated [7]. In turn, this is the result of important advances made in aromatic chemistry during the last few years. The availability of more effective, newer methods for aryl-aryl coupling represents an important driving force in the development of new synthetic strategies for poly(*para*-phenylene)s. In particular, the Pd(0)-catalyzed aryl-aryl coupling after Suzuki [8] (arylboronic acid and aryl halide or -tosylate) and the nickel(0)-catalyzed coupling according to Yamamoto [9] (aryl halide or -tosylate and aryl halide or -tosylate) have been employed most successfully.

Structurally Defined Poly(*para*-phenylene)s with "Step-Ladder" Structure

Unsubstituted poly(*para*-phenylene) **PPP** **1** as parent system of the substance class is an insoluble and infusible material, available by a variety of synthetic methods [10,11]. The lack of solubility and fusibility hinder both unequivocal characterization and the processing of **PPP**. In addition, **PPP** **1** appears as an ill-defined material possessing a considerable number of structural defects (branching, 1,2-coupling). Entries to the preparation of structurally defined, processable **PPP**s have evolved based on numerous synthetic principles.

One of these was aimed at the elaboration of a precursor route. Kaeriyama et al. [12] reported on the Yamamoto coupling of 1,4-dibromo-2-methoxycarbonylbenzene to poly(2-methoxycarbonyl-1,4-phenylene) (2) as a processable **PPP** precursor. The aromatic polyester precursor 2 is then saponified and thermally decarboxylated to 1 with CuO catalysts. However, the reaction conditions of the final step are quite drastic and cannot be carried out in the solid state (film).

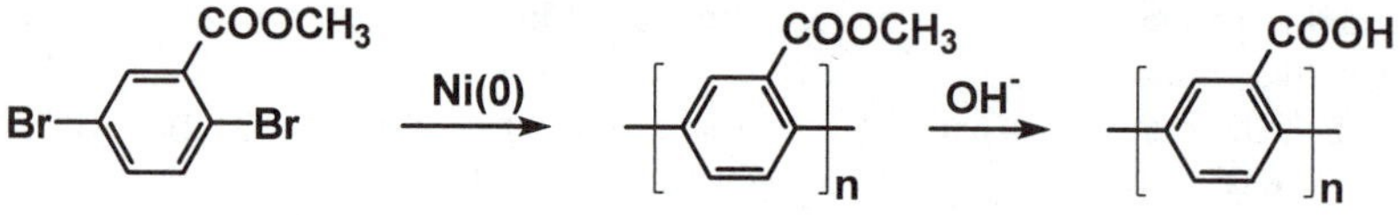

COOCH$_3$ Br Br Ni(0) COOCH$_3$ n OH$^-$ COOH n

2

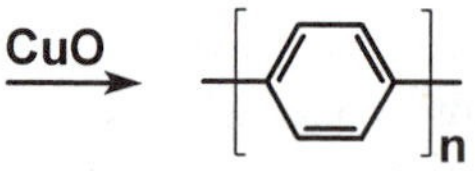

CuO n PPP

1

Grubbs [13] and MacDiarmid [14] and their respective groups described in 1992/1994 an improved precursor route starting from 3, the final step of which is the thermal, acid-catalyzed elimination of acetic acid to give **PPP** 1. They obtained free standing **PPP** films of defined structure, although these disadvantageously still contained large amounts of the acidic reagent employed (polyphosphoric acid). Nevertheless, in this work a reliable value for the longwave absorption maximum λ_{max} of **PPP** 1 could be obtained, about 336 nm. This value is of utmost importance in the interpretation of the optical properties of substituted **PPP**s.

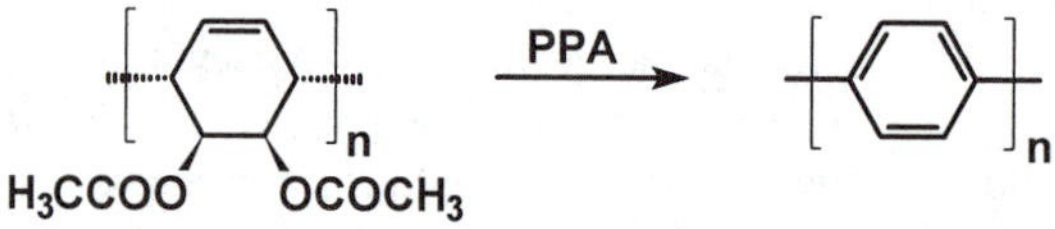

n H$_3$CCOO OCOCH$_3$ PPA n PPP

3 1

A second synthetic principle for structurally homogeneous **PPP**s involves the preparation of soluble **PPP** derivatives 4 by the introduction of solubilizing side groups. The pioneering work here was carried out at the end of the 80s by Schlüter, Wegner, et al. [15], who prepared for the first time soluble poly(2,5-dialkyl-1,4-phenylene)s 4 by way of a Suzuki cross-coupling. This method permitted the synthesis of polymers with an average of 100 1,4-phenylene units. In addition to alkyl substituents, soluble **PPP**s 4 are also known today that contain alkoxy groups as well as those with ionic side groups (carboxy, sulfonic acid functions). A disadvantage in regard to the electronic properties is caused by the changes in conformation effected by the substituents. The substituents in the 2- and 5-positions lead to a marked mutual twisting of the aromatic subunits, which manifests in a drastic reduction in the conjugative interaction. Thus, for poly(2,5-dialkyl-1,4-phenylene)s, neglegable absorption can be detected in the wavelength region above 300 nm.

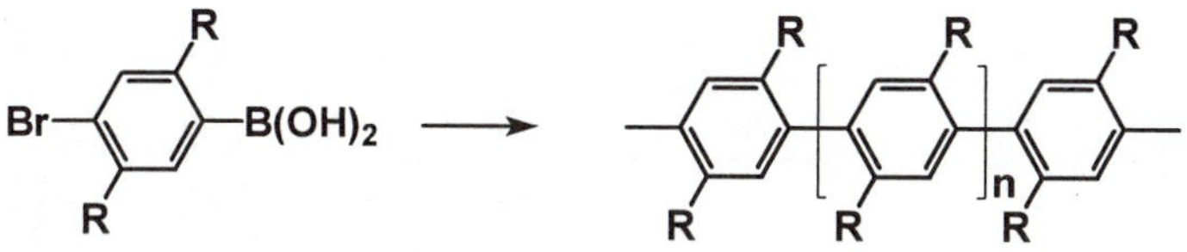

R

Br— —B(OH)$_2$ ⟶

R R R

n

R R R

4 R: - alkyl, -alkoxy

The results described thus far sketch the synthetic demands for being able to prepare processable, structurally defined **PPPs**, in which the π-conjugation remains fully intact or is even increased compared with that of the parent **PPP 1** system. The decisive step in the realization of this principle is the preparation of a **PPP** in which the aromatic subunits could be obtained in a planar or only slightly twisted conformation in spite of the introduction of substituents. One of the first examples was from Yoshino et al. [16], who prepared 9,9-substituted poly(fluorene)s 5, in which the solubilizing substituents are introduced in the form of a di-n-hexylmethylene bridge that spans the neighboring rings in pairs and forces a planar arrangement. The soluble and fusible poly(9,9'-dihexylfluorene-2,7-diyl)s (5) are obtained by oxidative coupling of 9,9'-dihexylfluorene with iron(III) chloride and are characterized by a value of M_n up to a maximum of 5000. The value of λ_{max} is about 388 nm, a result of the partial flattening of the **PPP** polymer to a "step-ladder" polymer so that the longwave absorption maximum is shifted batho-chromically by about 50 nm relative to that of the parent **PPP 1** structure.

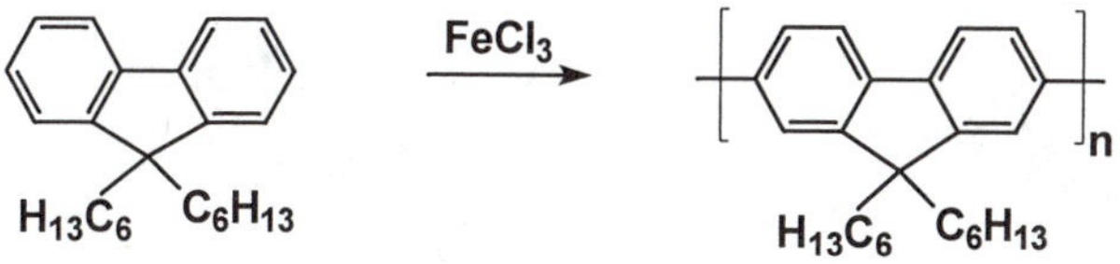

$FeCl_3$ ⟶

n 5

$H_{13}C_6$ C_6H_{13} $H_{13}C_6$ C_6H_{13}

An unsatisfactory aspect of this synthesis is the quite low degree of polymerization of a maximum 20 aromatic rings. Moreover, in addition to the predominant 2,7-coupling of the building blocks, other types of coupling can occur as structure defects.

Thus, it was logical to combine the principle for preparation of "step-ladder" structures with new, efficient and selective methods for aryl-aryl coupling. A first attempt at this was from Yamamoto et al. [17]. They coupled 2,7-dibromo-9,10-dihydrophenanthrene to give an ethano-bridged poly(*para*-phenylene) derivative (poly(9,10-dihydrophenanthrene-2,7-diyl)) (6) by way of low-valent nickel complexes, which were used either stoichiometrically as reagent ($Ni(COD)_2$) or were generated electrochemically in the reaction mixture. As a result of the lack of solubilizing substituents only the oligomer fraction with $M_n < 1000$ is soluble, the polymeric products precipitating out as an insoluble powder. The value of λ_{max} for the soluble fraction of **6** is about 360 nm.

Br Br Ni(0) n

6

Building on this, an advance was expected by combining the synthetic procedure of Yamamoto [17] with the introduction of solubilizing substituents. Accordingly, alkyl-substituted dihydrophenanthrens or tetrahydropyrenes offered themselves as starting monomers for the preparation of soluble "step-ladder" **PPP**s of this type [18].

2,7-Dibromo-4,9-dialkyl-4,5,9,10-tetrahydropyrenes **10** represent suitable starting monomers for the realization of this synthetic route. These difunctionalized tetrahydropyrene monomers were first prepared by Müllen et al. [18] as a precursor to 2,7-dibromo-4,9-dialkylpyrenes in a multi-step synthetic sequence. The synthesis included a Wittig olefination starting from the bis(triphenylphosphonium) salt of 2,2'-

$\overset{\oplus}{P}(Ph)_3 Br^{\ominus}$ $Br^{\ominus}(Ph)_3\overset{\oplus}{P}$ R-CHO R R hν (254 nm)

7 **8**

R R Br_2 / Pd / C Br Br R R

9 **10**

R: -alkyl

bis(bromomethyl)biphenyl (7) and aliphatic aldehydes, for example, nonyl aldehyde, to give 2,2'-bis(vinyl)biphenyl derivatives **8**. These then lead in a double photochemical cyclization to the corresponding 4,9-dialkyltetrahydropyrenes **9** as *cis/trans* isomer mixture. The key observation was that these tetrahydropyrenes **9** could be dibrominated very selectively and in high yield (> 90%) in the 2,7-positions with bromine in DMF and palladium on charcoal. This result was very unusual and extremely useful, as bromination usually takes place in the electronically favored 1-, 3-, 6- and 8-positions. Hitherto, 2,7-disubstituted tetrahydropyrenes were mostly prepared in tedious, multi-step synthetic sequences from meta-cyclophane precursors. The monomers **10** obtained are suitable for use as the dibromo component in a Yamamoto coupling.

Reaction of the dibromide **10** with $Ni(COD)_2$ in DMF/toluene gave a poly(4,9-dialkyl-4,5,9,10-tetrahydropyrene-2,7-diyl) **PTHP 11** [18] as a completely soluble, new type of **PPP** derivative with a "step-ladder" structure, in which each pair of neighboring aromatic rings is doubly bridged with ethano bridges. The solubilizing alkyl substituents appear to such a degree on the periphery of the molecule that they cannot cause twisting of the main chain. **PTHP 11** possesses a relatively high number average molecular weight, up to M_n = 20000, which corresponds to coupling of 46 **THP** units.

R R Br Br Ni(0) n R R

10 **11**

R: -alkyl

In our coupling experiments we use the monomer **10** as a *cis/trans* diastereomeric mixture. If the diastereoisomers are separated by fractional crystallization or chromatography in **10a/b** at the stage of the dibromo monomers, then stereoregular **PTHPs 11** are accessible (*cis*- or *trans*-polymer).

The polymers obtained after coupling of the *cis/trans* monomers **10a/b** should be investigated further, especially with regard to their (different?) packing behavior, which is of much interest from the point of view of the dependence of macroscopic electronic properties on the morphology of the solid state.

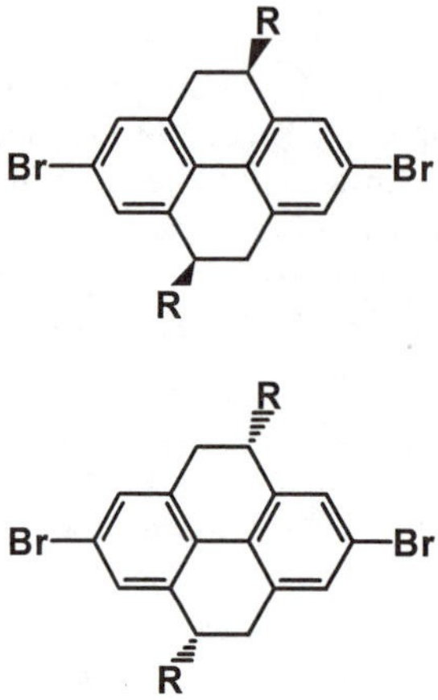

10a

***cis* - diastereomer**

(pair of enantiomers)

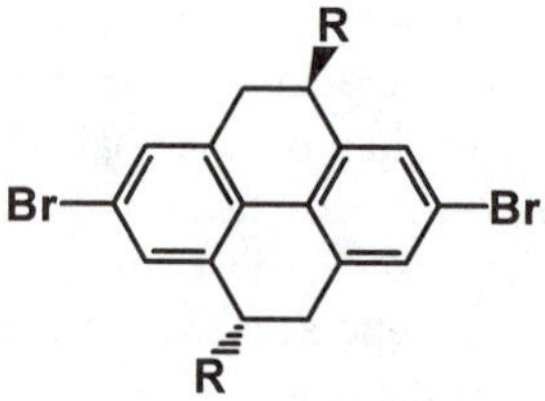

10b

***trans* - diastereomer**

PTHP 11 possesses a longest wavelength absorption λ_{max} of 385 nm, almost identical with the value for the "step-ladder" polyfluorenes of Yoshino et al. [16]. Thus, two independent confirmations exist for the correctness of the "step-ladder" concept: The introduction of solubilizing groups with simultaneous bridging of the subunits to guarantee the highest possible degree of conjugative interaction. With the aid of preparative gel permeation chromatography it was also possible to separate polydisperse mixtures of short-chain **PTHP** oligomers into their (monodisperse) individual components 12 [19].

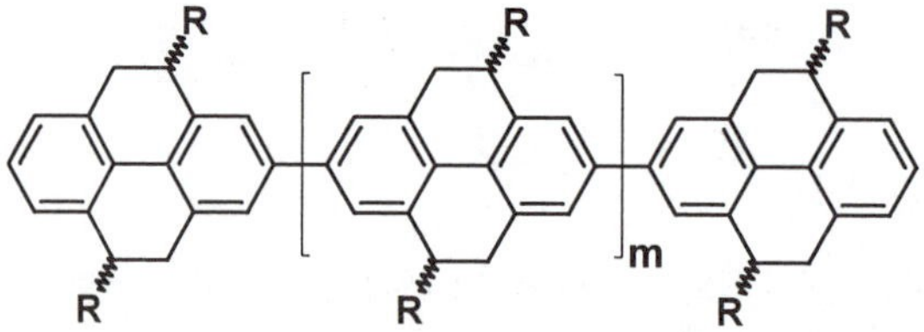

m: 0 - 8

12

R: -alkyl

With such a series of oligomers, the convergence of optical properties with increasing chain length can be followed [20] and the conjugation length in **PTHP 11** to be determined to about 10 monomer building blocks (i.e., 20 aromatic rings). In solution **PTHP** possesses an intense blue photoluminescence (PL) with a quite small Stokes shift between absorption and emission (λ_{max} absorption: 385 nm; λ_{max} emission:

425 nm). In the solid state the PL is shifted slightly bathochromically to λ_{max} ca. 457 nm, which is probably the result of aggregation in the solid state (Fig. 1).

The luminescence characteristic of **PTHP** suggests an investigation into its suitability as the active component in organic light-emitting diodes (LED) based on polymers. Such investigations showed the appearance of a quite intensive electroluminescence (EL) with a quantum yield of 0.1-0.15% (single layer construction ITO/**PTHP** **11**/Ca), with blue-green emitted light. Blue polymer LEDs represent an attractive target, as such building blocks derived from inorganic materials are not easily accessible. As a result of their band gap energy of 2.7-3.2 eV, **PPP** derivatives are particularly suitable as blue emitters.

In contrast to low molecular mass materials, polymeric emitters possess the advantage that they can be easily worked into transparent films with a low degree of scattering. In addition, they show a higher morphological stability compared with that of vapor-deposited low molecular weight compounds (low tendency toward recrystallization). The EL spectrum of **PTHP 11** is almost congruent with the PL spectrum, the applied voltage being relatively low at 13-15 V compared with that of other **PPP** derivatives described hitherto as active component (Fig. 2).

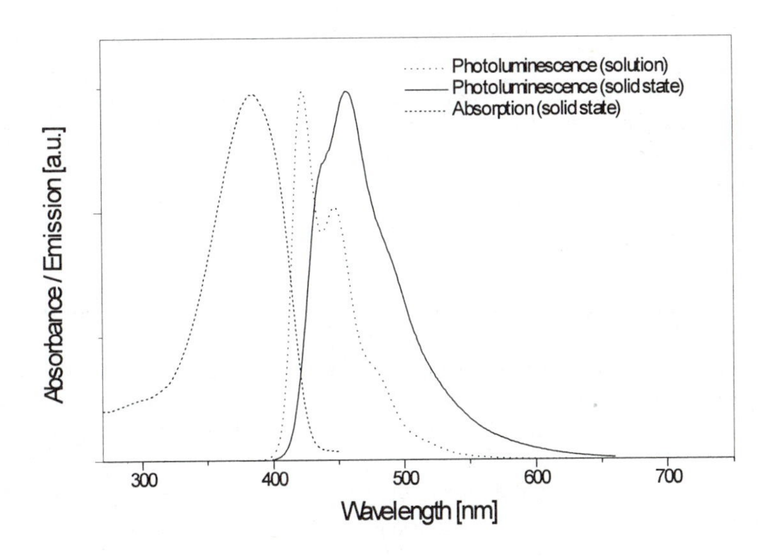

Fig. 1: Absorption (solid state, film: dashed line) and photoluminescence spectra (solution, in methylene chloride: dotted line; solid state, film: solid line) of **PTHP 11**

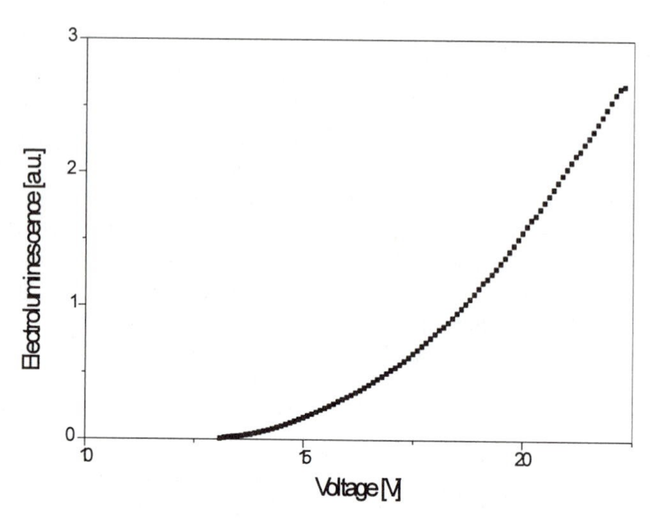

Fig. 2: Electroluminescence (Luminance)/Voltage - Characteristics of a **PTHP**-based blue LED (device configuration: ITO / **PTHP** **11** (ca. 150 nm) / Ca)

The light emitting diodes have shown themselves during operation to be surprisingly stable: initial experiments showed only a 10% drop in intensity (I_{const}) during 12 h (device configuration: ITO / **PTHP** **11** (ca. 150 nm) / Ca; voltage: 18 V; inert atmosphere). Further experiments are aimed at the extension of this new type of synthetic concept to other monomer structures (**PPP** derivatives from dihydrophenanthrene building blocks).

Ladder Polymers of the PPP Type

The logical continuation of the strategy outlined above was now the incorporation of the complete **PPP**-parent chromophore into the network of a completely planar ladder polymer. The complete flattening of the conjugated π-system by bridging of all the subunits should then lead to a maximum conjugative interaction. As with the **PTHP** systems, alkyl- or alkoxy side chains should lead to solubilization of the polymers. The impressive realization of this idea culminated in 1991 in the first synthesis of a soluble and conjugated ladder polymer of the **PPP** type [21]. This **PPP** polymer **LPPP** **15** was prepared according to a so-called classical route, in which an open-chain, single-stranded precursor polymer was closed to give a double-stranded ladder polymer. The synthetic potential of the so-called classical multi-step sequence was doubted for a long time; in the 1980s synchronous routes were strongly favored as preparative method for ladder polymers. In a multi-step route the main point is to be able to conduct the ring closure quantitatively and regioselectively. In the synthesis of **LPPP**, the precursor polymer **13** is initially prepared in an aryl-aryl coupling from an aromatic diboronic acid and an aromatic dibromoketone.

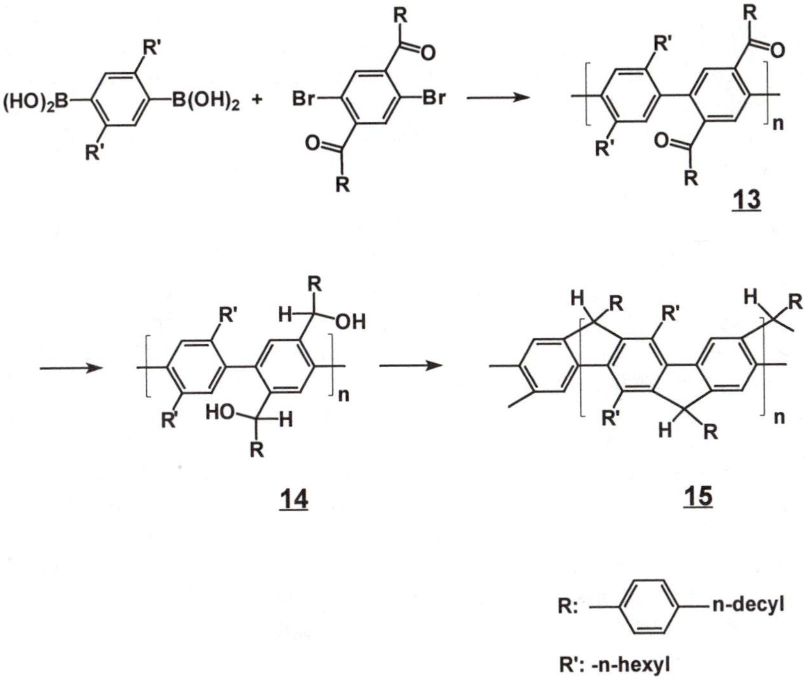

13

14 **15**

R: —C₆H₄—n-decyl

R': -n-hexyl

The cyclization to structurally defined, soluble **LPPP** then takes place in a two-step sequence, consisting of reduction of the keto group followed by ring closure of the secondary alcohol groups of **14** in a Friedel-Crafts-type alkylation.

The resulting ladder polymer **LPPP** **15** possesses a number average molecular weight of 25000, which corresponds to the incorporation of 65 phenylene units. No structure defects could be detected using NMR spectroscopy. **LPPP** **15** is characterized by unusual electronic and optical properties: the absorption maximum undergoes a marked bathochromic shift as a consequence of planarization of the chromophore, to a λ_{max} value of 440-450 nm. In addition, the longest wavelength π-π* absorption band possesses an unusually sharp absorption edge.

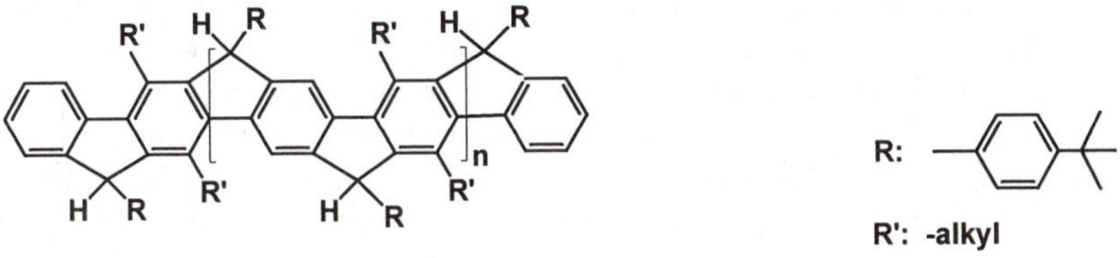

R: —C₆H₄—C(CH₃)₃

R': -alkyl

16

From investigations with a series of monodisperse, oligomeric model compounds **16** of the **LPPP** type (tetra, penta, and hexaphenyl derivatives) the effective conjugation length in **15** (convergence limit of the absorption) could be determined as ca. 11 phenyl rings [20]. Remarkably, the convergence of the optical absorption with increasing chain length occurs much more rapidly in the planar **LPPP** molecule than in the only partially bridged **PTHP** structures.

The photoluminescence (in solution) of **LPPP 15** is very intense and blue (λ_{max} emission: 450-460 nm). The Stokes shift between absorption and emission is extremely small (ca. 150 cm^{-1}), a consequence of the geometric fixation of the chromophore in the ladder structure. The PL quantum yields are high, in comparison with those of many other conjugated polymers: in solution values between 60 and 86% have been measured, in the solid state up to 30% [22]. In comparison, **PPP 1** synthesized by a precursor route shows a PL quantum yield of only 4%. Thus, it was obvious to investigate the suitability of this new type of material for application as active component in blue LEDs. Initially, this led to the surprising result that although efficient LEDs can be prepared with **LPPP**, the color of the emission in the solid state (film) is nevertheless yellow (PL and EL). In addition to primary emission of the **LPPP 15** chromophore in the blue region, the PL and EL spectra show an additional and unstructured broad emission band in the yellow region (ca. 600 nm; Fig. 3) [23,25]. The relative intensity ratio of these two bands is strongly dependent on the process used for preparation of the films (solvent, film thickness, preparation of the film). Thus, the blue emission band of the isolated chromophore disappears almost completely on annealing the film at about 150°C. The yellow emission band could then be characterized unequivocally in photophysical experiments as aggregate emission [24,25]. This result supports (1) PL investigations with a "site-selective" excitation [24], (2) time-resolved PL studies [26], and (3) photovoltaic experiments with **LPPP 15** [27]. The experiments provided proof that weak aggregate absorption, the "site-selective" excitation of which leads to emission of light from the emission band in the yellow region, can be detected even in the ground state above the absorption edge. Time-resolved PL spectroscopy established an unusually short lifetime of about 50 ps for the blue primary emission, which can only be explained by an energy relaxation of the initial excited state into the lower energy aggregate state. The lifetime of the latter is, at ca. 450 ps, roughly an order of magnitude greater. The photovoltaic experiments furnished a highly efficient sensitivity for **LPPP** films in the region of the aggregate absorption/emission bands at 600 nm.

EL experiments showed that the yellow-emitting LEDs prepared from **LPPP 15** exhibit quite remarkable characteristics (single layer construction ITO/**LPPP 15**/Ca; quantum efficiency: ca. 0.8%, applied voltage: 4-6 V, drop in intensity < 5% in 24 h operation at constant current [25]). These figures are in the range of the best values described hitherto for polymeric emitters in a single layer arrangement, for example, poly(*para*-phenylenevinylene) **PPV** and **PPV** derivatives. The comparison of **LPPP 15** and **PPV** is also interesting for another reason. In addition to the spontaneous emissions (PL or EL) conjugated polymers are also of interest as materials for stimulated emission (optically or electrically pumped). For effects of this type, the ratio of stimulated emission to photoinduced absorption as a result of the formation of charge-separated mid-gap states in particular is of interest for conjugated polymers. Initial experiments allow the conjecture of a significant superiority of **LPPP 15** [28], as the stimulated emission in **LPPP** is markedly more intense than that of **PPV** under comparable conditions.

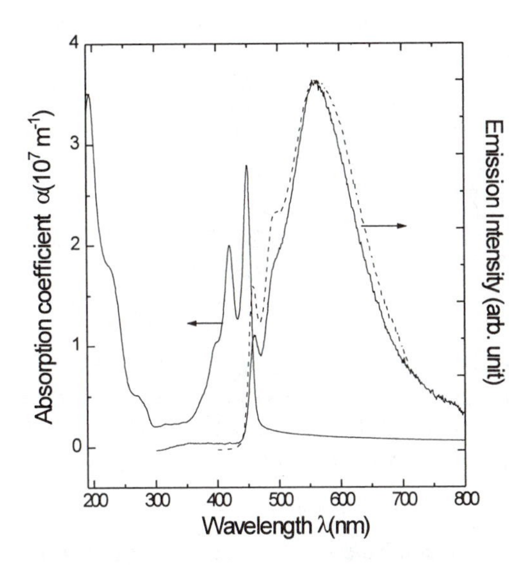

Fig. 3: Absorption and photoluminescence (dashed line) spectra of a thin film of **LPPP 15** and electroluminescence (solid line) spectrum of a ITO / **LPPP 15** (60 nm) / Al device [29].

From the perspective of the strategies aimed at fabrication of efficient *blue* LEDs, the results outlined above regarding yellow **LPPP** light emitting diodes are, nevertheless, unsatisfactory. In order to prepare blue LEDs from **LPPP** materials, it is necessary to efficiently mask out or suppress the dominant yellow aggregate emission. Precisely the preparation of blue LEDs would use the inherent advantage of **LPPP 15** (band gap energy ca. 2.75 eV). With other conjugated polymers with smaller band gap energy, such as **PPV**, blue LEDs cannot normally be prepared. In this case only the transition to very short chain, oligomeric chromophores brings the shift in the color of emission to blue, although the change to oligomers also brings disadvantages in handling, processing, and stability (recrystallization tendency of amorphous films, poor film forming properties, low mechanical stability).

Various means are available to mask out or suppress the aggregate emission. Masking out of the yellow aggregate emission can be achieved by using the "microcavity" effect. Here, one uses special layers that act as mirrors above and below the active zone; individual regions of the emission bands are amplified by resonance.

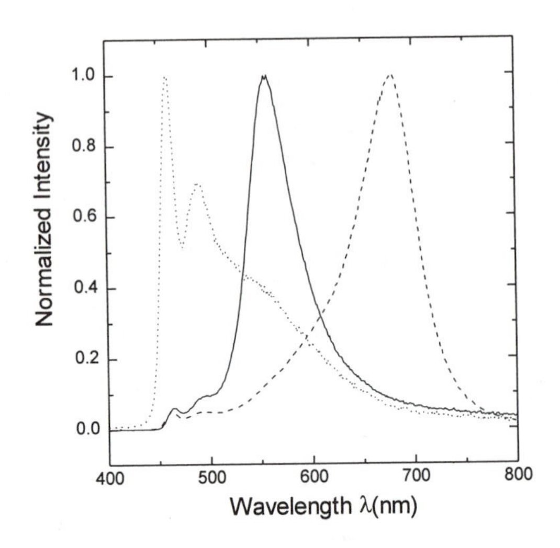

Fig. 4: Normalized electroluminescence spectra of Au/ **LPPP 15** /Al light emitting devices with different thickness of the **LPPP 15** layer: 60 nm (dotted line), 150 nm (dashed line) and 250 nm (solid line). The thickness of the aluminium and gold electrode was 55 nm and 22 nm, respectively [29].

In this case the broad emission band of **LPPP 15** offers the possibility of also being able to prepare red and yellow LEDs, as well as blue emitting LEDs, out of the same material, simply by variation in the film thickness (Fig. 4) [29].

Suppression of the aggregate emission is now possible by two quite different means. At first, the aggregate emission could be almost completely shut out by simply mixing the **LPPP 15** with a matrix polymer. LEDs with 1% **LPPP 15** in poly(vinyl carbazole) **PVK** as emitter material are characterized by a pure blue light emission with a quantum efficiency of ca. 0.15% in layer formation (ITO/1% **LPPP 15** in **PVK**/Ca). Polystyrene (PS) is unsuitable as matrix: in this case the emitter **LPPP** must be used with a much greater dilution (0.001%) in order to suppress the yellow emission band. At this level of dilution of the chromophore, the formation of efficient, stable emitting LEDs was no longer possible.

If suppression of the aggregate emission in this first example was based on a purely physical principle, chemical modification of the **LPPP 15** structure for the same aim is also possible. Thus, one option is the introduction of additional substituents into the **LPPP** skeleton. Thus, the introduction of an additional methyl group to the methylene bridge in **LPPP** (addition of $LiCH_3$ in place of reduction of the keto groups) leads to ladder polymers 17 whose solid state PL spectra show only very weak aggregate emissions; the solution- and solid state PL spectra are almost identical (Fig. 5; [30]).

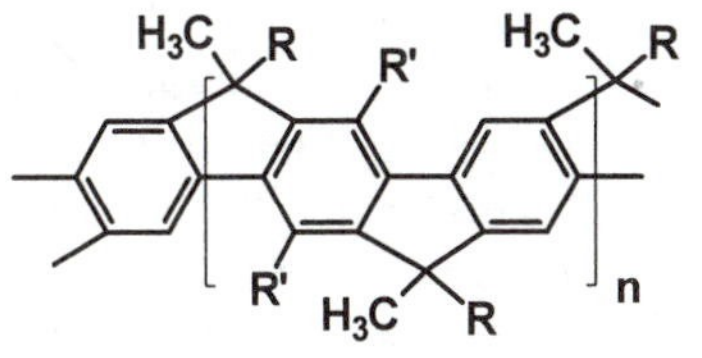

R: —C6H4— n-decyl

R': -n-hexyl

17

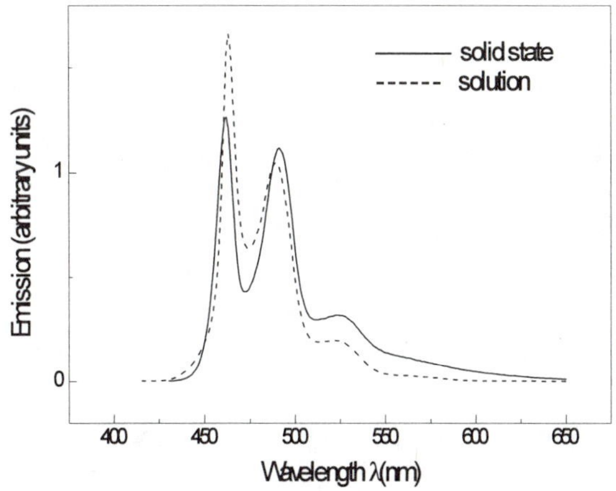

Fig. 5: Photoluminescence spectra of the metyl-substituted **LPPP** derivative **17** (solid line: solid state; dashed line: solution, methylene chloride)

A further possibility is the transition from completely planar **LPPPs** **15** resp. **17** to statistical "step-ladder" copolymers **18**, composed of planar **LPPP** emitter segments that are connected by spacer units. The intermediate groups ("spacers") effect a twisting of the subunits, which decisively hinder aggregation. The principle could be realized by adding a dibromo comonomer (2,5-dihexyl-1,4-dibromobenzene), which does not form methylene bridges during the cyclization, to the monomer mixture for the Suzuki aryl-aryl coupling.

LPPP copolymers of this type, **CoLPPP** **18** [31], show, above a "spacer" fraction of about 50%, the disappearance of the aggregate emission band in the solid state, and also after thermal annealing of the samples. In this context it is worth mentioning that the position of the blue emission in **CoLPPP** **18** hardly changes as a function of the copolymer composition or when compared with that of the parent system **LPPP** **15**, although in the absorption spectrum the absorption bands of the oligomeric ladder segments of differing length can be detected independently. Thus, they function as almost independent (isolated) chromophores, the spacer blocks acting as efficient barriers to conjugation. The almost unaffected position of the blue emission band can be explained in that in the solid state, effected by an efficient energy relaxation to the lowest energy state, the emission always occurs as the result of optical excitation of the longest ladder segment in the region of the band edge [31]. Thus, **CoLPPPs** **18** are highly efficient luminophores in the blue region of the spectrum (λ_{max} emission: ca. 450 nm) both in the photo- and in the electroluminescence. The PL quantum yields are, at 85%, remarkably high [22]. In a single-layer construction ITO/**CoLPPP** **18**/Ca unusually efficient pure blue LEDs can be constructed (external quantum yields of up to 1%) [32]. Regarding the applied voltage (ca. 15-20 V), the light intensity (ca. 1000 C/m^2), and the stability of the light emission (half-life times of about 10 h) the results are also particularly promising.

In particular the transition to multi-layer devices should bring additional possibilities for improvement in these figures, especially regarding the applied voltage and the long-term stability. Moreover, the first promising experiments have been conducted using soluble poly(2,5-dialkyl-1,4-phenylene-1,3,4-oxadiazole)s **19** as electron transport layer [33].

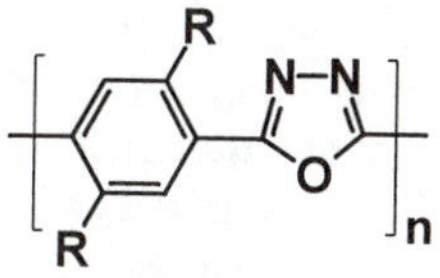

19 R: -alkyl

Furthermore, there is the possibility of integrating the electron-deficient diaryl-oxadiazole building block directly into the **CoLPPP** structure, as a component of the main chain or in side chains [34]. In this fashion, the preparation of a tailor-made emitter material was achieved with main-chain modified **CoLPPPs** **20**. These materials showed a quantum yield in the electroluminescence of up to 2.5% with the single-layer structure ITO/**20** (diaryloxadiazole-modified CoLPPP)/Ca [34].

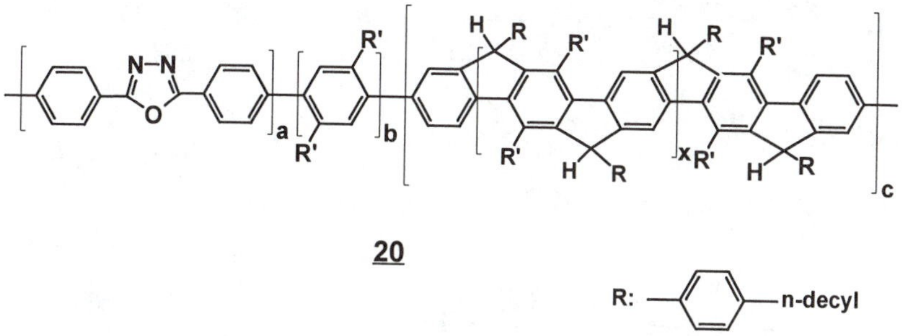

20

R: -C6H4-n-decyl

R': -n-hexyl

The results just described demonstrate impressively that the transition from homopolymer **LPPP** **15** to segmented ladder polymers **18** with a partially interrupted conjugative interaction leads to blue LEDs without impairment of the specific characteristics. On the contrary, at least in our example, a very desirable improvement in individual parameters has been established. These results suggested the investigation also of other copolymers formed from oligomeric emitter segments of the **PPP** type and suitable "spacer" units for their suitability as active components in polymeric LEDs.

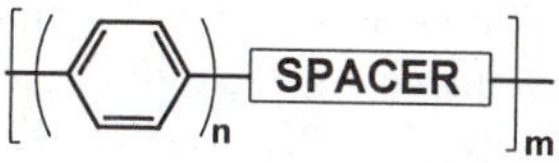

SPACER

PPP Copolymers with Tetra- and Hexaphenyl Segments

The question of using **PPP** as unsubstituted root system for use as emitter in polymeric LEDs can hardly be considered given the unsolved problems of processing the material into thin films. Nevertheless, the aim of preparing processable **PPP** copolymers composed of short-chain **oligophenyl** emitter units coupled via suitable aliphatic "spacer" segments, in order to permit processing from solution or in the melt, is enticing. Our synthetic strategy thereto contains an aryl-aryl coupling of appropriate monomers in the final step, so that copolymers are formed from oligophenyl- and aliphatic spacer sequences. Thus, from 4-bromo-4'-hydroxybiphenyl (21) and epichlorohydrin we synthesized a monomer with α,ω-placed 4-bromobiphenyl-4'-yl groups, which was etherified to 22 with long-chain aryl halides at the aliphatic hydroxy group to increase the solubility.

$ClH_2C\text{-}CH\text{-}CH_2$ (O)

Br–C6H4–C6H4–OH → (OH^-) R–CH2–CH(OH)–CH2–R

21

$C_{12}H_{25}Cl$

R–CH2–CH($OC_{12}H_{25}$)–CH2–R

22

-R = –O–C6H4–C6H4–Br

The dibromo monomer 22 was then converted (1) in a homo coupling after Yamamoto to copolymers with tetraphenyl units 23 and (2) in an aryl-aryl cross-coupling after Suzuki with 4,4'-biphenylbis(ethylene glycole boronate) to give the corresponding copolymers with hexaphenyl blocks 24. These oligomeric copolymers have a number average molecular weight M_n of ca. 3000 [35]. The insolubility of longer chain molecules is the factor limiting the molecular weight, as the coupling is carried out as precipitation polycondensation. The initially precipitated products, however, can be re-dissolved, albeit only in halogenated hydrocarbons such as tetrachloroethane. The UV/Vis spectroscopic characterization gave a longwave absorption maximum λ_{max} for the tetraphenyl copolymer 23 of 313 nm, for the hexaphenyl derivative 24 λ_{max} of 335 nm. As a result of the 4,4'-oxy substituents, the value of λ_{max} for the copolymer is shifted bathochromically compared with that of the tetra- and hexaphenyl derivatives (292 and 308 nm, respectively). The PL-emission maxima lie in the UV region (396 nm, tetraphenyl polymer 23) or in the blue region of the visible spectrum (429 nm, hexaphenyl derivative 24). The advantage is that no aggregation phenomena are ob-served here in the solid state, so that the PL spectra are almost identical in solution and in the solid state. A light emitting diode could be prepared from the tetraphenyl deri-vative 23. However, it was not possible to process the polymer into defect-free films because of the poor film-forming properties. Blending with poly(vinylcarbazole) (0.045 mol-% tetraphenyl copolymer 23) led to homogeneous films, which could be used as the active layer in LEDs. The light emission is blue-white; surprisingly, in contrast to the PL, a considerable fraction of the EL emission takes place in the spectral region of lower emission energy (Fig. 6; [35]). An explanation for this would be the formation of exiplexes between the tetraphenyl copolymer 23 and the matrix polymer poly(vinyl-carbazole).

$OC_{12}H_{25}$ n

23

Ni(0) Yamamoto homocoupling

Br Br $OC_{12}H_{25}$

22

Pd(0) Suzuki coupling

$OC_{12}H_{25}$ n

24

The described incorporation of the rigid oligophenyl chromo- and luminophores in the skeleton of a flexible copolymer represents *one* possibility for the synthesis of processable material with oligophenyl building blocks. *Another*, non-polymer strategy consists of the gathering of numerous oligophenyl chromophores to a star-shaped or strongly branched molecule. Such a structure should be characterized by a considerable increase in solubility as a result of its steric construction.

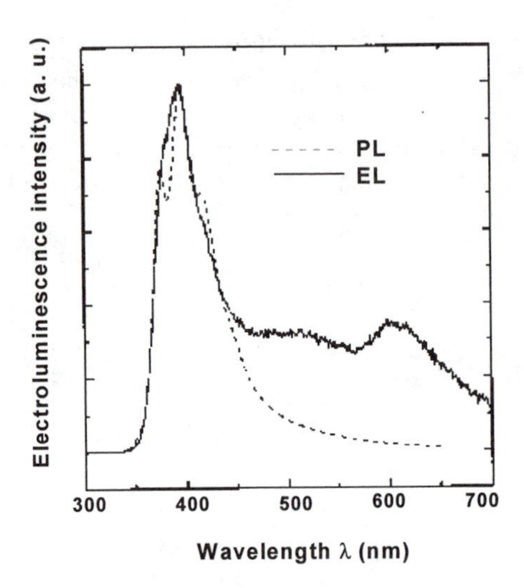

Fig. 6: Photo- (solid state; dashed line) and electroluminescence spectra (single layer-device ITO / PVK + 23 / Ca; solid line) of copolymer 23 containing quaterphenyl segments [35]

Hexa(oligophenyl)benzenes

Hexa(oligophenyl)benzenes (e.g. 25 or 26) represent a most promising substance group in the sense of the aim just described. Two efficient synthetic routes have been elaborated for the preparation of hexa(terphenyl)- and hexa(quaterphenyl)benzene. The first involves a palladium-catalyzed trimerization of diarylacetylenes [36] as the key step, as demonstrated by the synthesis of a hexakis-alkylated hexa(terphenyl)benzene derivative 25 from the corresponding bis(terphenyl)acetylene (27). The peripheral *tert*-alkyl substituents serve to solubilize the molecule.

The second synthetic strategy consists of the coupling of hexa(4-iodophenyl)-benzene (28) with an oligophenyl boronic acid to a hexa(oligophenyl)benzene by extending the aromatic chain [36]. This route is demonstrated by the reaction of hexa(4-iodophenyl)benzene (28) with an alkylated terphenyl boronic acid under formation of the hexa(tetraphenyl)benzene derivative 26. Once again, the aliphatic substituents serve to guarantee sufficient solubility.

The new type of star molecules constructed in this fashion are now also attractive as blue luminophores as active component in LEDs, in particular also because of their ability to be worked into transparent, amorphous films. The quaterphenyl derivative 26 shows, for example, a long wavelength absorption maximum at 314 nm, and an intensive blue photoluminescence with an emission maximum at 401 nm (Fig. 7).

Similar star-shaped molecules with oligophenyl arms derived from 9,9-spirobifluorene as the central unit have been synthesized by Salbeck et al. and suggested as potential emitter materials for blue LEDs [37].

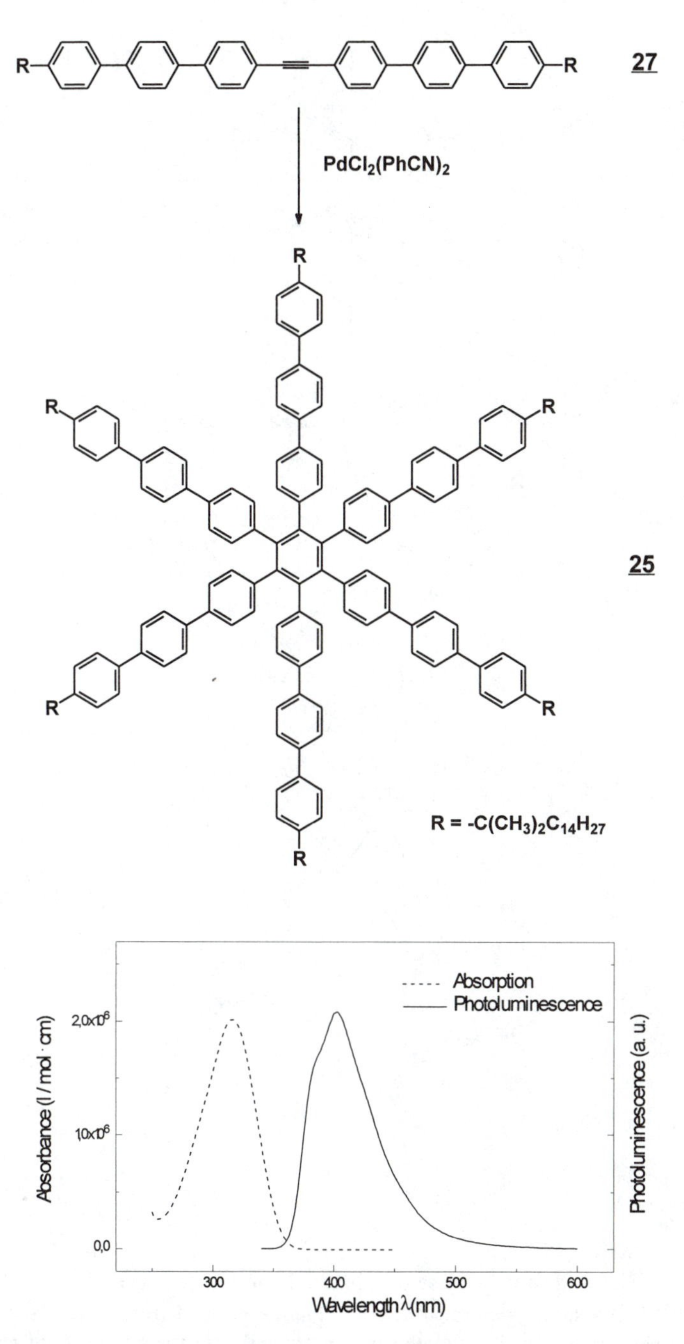

Fig. 7: Absorption (solution, in methylene chloride: dashed line) and photoluminescence spectra (solution, in methylene chloride: solid line) of the hexa(quaterphenyl)benzene derivative **26**

28

Pd(0)
Suzuki coupling

$R = -C(CH_3)_2C_{14}H_{27}$

26

Conclusion

The results discussed in the course of this article on the synthesis of new, structurally defined **PPP** derivatives demonstrated the qualitative leap from a purely structurally motivated search for new conjugated macromolecules to the synthesis of structurally defined, processable **PPP** materials with properties tailor-made for a particular application. Precisely the combination of the solubilization required for processability *and* the complete retention of the conjugated character led to a new generation of **PPP** derivatives in which a comprehensive correlation of structure and properties is possible. The new quality is in particular the result of the availability of new, efficient methods of aryl-aryl coupling, which permit chemo- and regioselective synthesis of **PPP** derivatives. A series of these new **PPP** derivatives has proven its value in the very topical search for polymeric emitters for organic light emitting diodes and thereby obtained a concrete position. The performance figures attained with these new **PPP** luminophores in some cases represent the best values to date, and thereby encourage a far-reaching extension of the synthetic and photophysical work in a genuinely interdisciplinary approach.

Acknowledgments

Special thanks to Heinz Bässler, Vera Cimrová, Richard Friend, Wilhelm Graupner, Julian Grimme, Johannes Grüner, Joachim Huber, Bert Keegstra, Martin Kreyenschmidt, Günther Leising, Ulli Lemmer, Rainer Mahrt, Dieter Neher, Michael Ravenscroft, Hermann Schenk, Jutta Schnee, Frank Uckert und Uwe Wiesler for collaboration and various support during the synthesis, physical measurements and the preparation of this manuscript.

Literature Cited

[1] *"Handbook of Conducting Polymers"*, Vol. 1 and 2; Skotheim, T. A.; Ed.; M. Dekker, New York **1986**.

[2] Heitz, W.; *Chem.-Ztg.* **1986,** 110, 385; Ballauf, M.; *Angew. Chem., Int. Ed. Engl.* **1989,** 28, 253.

[3] Ballard, D. G. H.; Courtis, A.; Shirley, I. M.; Taylor, S. C.; *J. Chem. Soc., Chem. Comm.* **1983,** 954.

[4] Wessling, R. A.; *J. Polym. Sci., Polym. Symp.* **1985,** 72, 55.

[5] Cao, Y.; Smith, P.; Heeger, A. J.; *Synth. Met.* **1993,** 55-57, 3514.

[6] Jenekhe, S. A.; Johnson, P. O.; Agrawal, A. K.; *Macromolecules* **1989,** 22, 3216.

[7] Tour, J. M.; *Adv. Mater.* **1994,** 6, 190; Schlüter, A.-D., Wegner, G.; *Acta Polymer.* **1993,** 44, 59.

[8] Miyaura, M.; Yanagi, T.; Suzuki, A.; *Synth. Commun.* **1981,** 11, 513.

[9] Kanbara, T.; Saito, N.; Yamamoto, T.; Kubota, K.; *Macromolecules* **1991,** 24, 5883; Yamamoto, T.; Morita, A.; Miyazaki, Y.; Maruyama, T.; Wakayama, H.; Zhou, Z.; Nakumura, Y.; Kanbara, T.; Sasaki, S.; Kubota, K.; *Macromolecules* **1992,** 25, 1214.

[10] Kovacic, P.; Jones, M. B.; *Chem. Rev.* **1987,** 87, 357.

[11] Marvel, C. S.; Hartzell, G. E.; *J. Am. Chem. Soc.* **1959,** 81, 448.

[12] Chaturvedi, V.; Tanaka, S.; Kaeriyama, K.; *Macromolecules* **1993,** 26, 2607.

[13] Gin, D. L.; Conticello, V. P.; Grubbs, R. H.; *J. Am. Chem. Soc.* **1992,** 114, 3167.

[14] Gin, D. L.; Avlyanov, J. K.; MacDiarmid, A. G.; *Synth. Met.* **1994,** 66, 169.

[15] Rehahn, M.; Schlüter, A.-D.; Wegner, G.; Feast, W. J.; *Polymer* **1989,** 30, 1054; Rehahn, M.; Schlüter, A.-D.; Wegner, G.; Feast, W. J.; *Polymer* **1989,** 30, 1060.
[16] Fukuda, M.; Sawada, K.; Yoshino, K.; *J. Polym. Sci. A*; **1993,** 31, 2465.
[17] Saito, N.; Kanbara, T.; Sato; T.; Yamamoto, T.; *Polym. Bull.* **1993,** 30, 285.
[18] Kreyenschmidt, M.; Uckert, F.; Müllen, K.; *Macromolecules* **1995,** 28, 4577.
[19] Kreyenschmidt, M.; Müllen, K.; in preparation.
[20] Grimme, J.; Kreyenschmidt, M.; Uckert, F.; Müllen, K.; Scherf, U.; *Adv. Mater.* **1995,** 7, 292.
[21] Scherf, U.; Müllen, K.; *Makromol. Chem. Rapid Commun.* **1991,** 12, 489.
[22] Stampfl, J.; Graupner, W.; Leising, G.; Scherf, U.; *J. Lumin.* **1995**, 63, 117.
[23] Huber, J.; Müllen, K.; Salbeck, J.; Schenk, H.; Scherf, U.; Stehlin, T.; Stern, R.; *Acta Polymer.* **1994,** 45, 244.
[24] Mahrt, R. F.; Siegner, U.; Lemmer, U.; Hopmeier, M.; Scherf, U.; Heun, S.; Göbel, E. O.; Müllen, K.; Bässler, H.; *Chem. Phys. Lett.* **1995,** 240, 373.
[25] Grüner, J.; Wittmann, H. F.; Hamer, P. J.; Friend, R. H.; Huber, J.; Scherf, U.; Müllen, K.; Moratti, S. C.; Holmes, A. B.; *Synth. Met.* **1994,** 67, 181.
[26] Graupner, W.; Leising, G.; Lanzani, G.; Nisoli, M.; de Silvestri, S.; Scherf, U.; *Chem. Phys. Lett.,* **1996**, 246, 95.
[27] Köhler, A.; Grüner, J.; Friend, R. H.; Müllen, K.; Scherf, U.; *Chem. Phys. Lett.,* **1995**, 243, 456.
[28] Mahrt, R. F.; Haring Bolivar, P.; Pauck, T.; Wegmann, G.; Lemmer, U.; Siegner, U.; Hopmeier, M.; Hennig, R.; Scherf, U.; Müllen, K.; Kurz, H.; Bässler, H.; Göbel, E. O.; *Phys. Rev. B;* in print.
[29] Cimrová, V.; Neher, D.; Scherf, U.; *Appl. Phys. Lett.,* in print.
[30] Graupner, W.; Eder, S.; Tasch, S.; Leising, G.; Lanzani, G.; Nisoli, M.; de Silvestri, S.; Scherf, U.; *J. Fluorescence,* in print.
[31] Grem, G.; Paar, C.; Stampfl, J.; Leising, G.; Huber, J.; Scherf, U.; *Chem. Mater.* **1995,** 7, 2.
[32] Grüner, J. F.; Hamer, P.; Friend, R. H.; Huber, J.; Scherf, U.; Holmes, A. B., *Adv. Mater.* **1994,** 6, 748.
[33] Wiesler, U.; Huber, J.; Scherf, U.; in preparation.
[34] Grüner, J. F.; Friend, R. H.; Huber, J.; Scherf, U.; *Chem. Phys. Lett.,* **1996**, 251, 204.
[35] Cimrová, V.; Neher, D.; Keegstra, M. A.; Scherf, U.; *Macromol. Chem. Phys.,* in print.
[36] Keegstra, M. A.; De Feyter, S.; De Schryver, F. C.; Müllen, K.; *Angew. Chem.* **1996**, 108, 830.
[37] Saalbeck, J.; *Angew. Chem.*, submitted.

Chapter 25

Aromatic-Amine-Containing Polymers for Organic Electroluminescent Devices

J. Kido, G. Harada, M. Komada, H. Shionoya, and K. Nagai

Department of Materials Science and Engineering, Graduate School of Engineering, Yamagata University, Yonezawa, Yamagata 992, Japan

We investigated the suitability of aromatic amine-containing polymers as active layers in organic electroluminescent devices. Polymers used in this study include poly(*N*-vinylcarbazole) (PVK), poly(*N*-substituted methacrylamide)s, poly(methacrylate) and poly(arylene ether)s. The device structures are a double-layer-type and a single-layer type. The double-layer-type devices consist of a polymer hole transport layer and an electron-transporting emitter layer. High luminance of over 14,000 cd/m^2 was observed for some double-layer devices using the polymer hole transport layer. Single-layer devices with dye-dispersed PVK emitted white light with a luminance of over 4000 cd/m^2. This demonstrates that dye-dispersed polymer systems are quite useful to obtain white light.

Organic electroluminescent (EL) devices are the subject of study by many researchers because of their potential application as light-emitting devices which operate at low drive voltages. These devices are injection-type, in which carriers, such as electrons (radical anions) and holes (radical cations), are injected into the organic emitter layer where they recombine. It is, therefore, necessary for the component organic materials to possess carrier-transporting properties as well as fluorescence properties.

Although electroluminescence in organic molecules have been known since the 1960's (*1*), a major breakthrough was made in the 80's by Tang and VanSlyke who developed a device exhibiting practical brightness (*2*). Their device consisted of a double layer structure with an organic hole transport layer and a luminescent metal complex layer. The hole transport layer plays an important role in transporting holes and blocking electrons, thus preventing electrons from moving into the electrode without recombining with holes. Aromatic diamine derivatives, such as *N,N'*-diphenyl-*N,N'*-bis(3-methylphenyl)-1,1'-biphenyl-4,4'-diamine (TPD), have been widely used as a hole transport layer because of their high hole mobilities (10^{-3} cm^2/Vs) and good film-forming properties (*3–5*). In these devices, degradation is partly caused by the crystallization of the organic molecules (*6,7*), which destroys the contact between organic layers. Therefore, the use of a less crystalline polymer is desirable to minimize the degradation of the device. In addition, the use of polymers will simplify the device fabrication processes because solution casting methods can be used to fabricate organic thin layers.

In this paper, some of the aromatic amine-containing polymers synthesized in our group are introduced and the devices using these polymers are summarized. Examples of the devices using these polymers as a hole transport layer as well as an emitter layer are discussed.

Experimental Section

The structures of the polymers used are shown in Figure 1. Side-chain-type polymers, except poly(*N*-vinylcarbazole) (PVK), were synthesized by radical polymerizations of the corresponding monomers, and main-chain-type polymers were synthesized by the nucleophilic substitution reactions of the corresponding difluoro compounds and diphenols. All the polymers were purified at least three times by reprecipitation into methanol. The details of the syntheses will be published elsewhere.

The structure of double-layer-type devices is an anode/polymer/emitter layer/cathode (Figure 2a). Single-layer type was an anode/polymer/cathode (Figure 2b). The emitter material in the double-layer device is tris(8-quinolinolato)aluminum(III) complex, Alq, which has an electron drift mobility of 10^{-5} cm^2/Vs, and has been used as an emitter layer in organic EL devices (*2, 8–10*). Alq was synthesized from aluminum trichloride and 8-quinolinol using piperidine as a base, and was purified by recrystallization from *N,N*-dimethylformamide/ethanol. The anode was ITO (indium–tin–oxide) that is coated on a glass substrate, having a sheet resistance of 15 Ω/□, and the cathode was magnesium-silver (10:1). The polymer layer was formed by spin coating or dip coating. Alq was vacuum deposited at 2×10^{-5} Torr onto the polymer layer with a thickness of 700 Å. Finally a 2000-Å-thick magnesium and silver (10:1) layer was deposited on the Alq layer surface as the top electrode at 5×10^{-6} Torr. The deposition rates for Alq was 3 Å/s, and for magnesium:silver 10-11 Å/s. During the evaporation process, the thickness and evaporation rate were monitored with a thickness monitor (ULVAC CRTM 5000) having a quartz oscillator. The actual thicknesses of the layers were measured with a Sloan Dektak 3ST surface profiler. The emitting area was 0.5×0.5 cm^2.

In single-layer devices, the emitter layer was fabricated by the dip coating of the polymer layer. Dichloroethane solutions containing PVK, having a molecular weight of 150,000 purchased from Kanto chemical Ltd., and several dopant dyes were prepared and dip coated onto an ITO coated glass substrate. The thickness of the PVK layer was ca. 1000 Å. Then, a magnesium and silver cathode was codeposited.

Ionization potentials (Ip), or Highest Occupied Molecular Orbital (HOMO), of the polymers and the vacuum deposited films of organic dyes were determined from the wavelength dependence of photoemission of electrons using Riken Keiki AC-1, and the energy gap values, Eg, were determined from the lower energy threshold of the electronic absorption spectra of the thin films of the materials. Then, pseudo electron

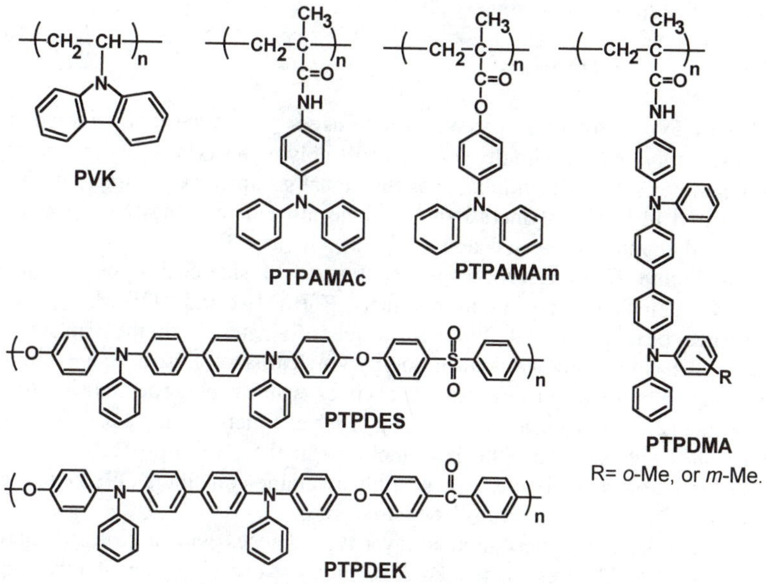

PVK

PTPAMAc

PTPAMAm

PTPDES

PTPDEK

PTPDMA

R= *o*-Me, or *m*-Me.

Figure 1. Structures of polymers used.

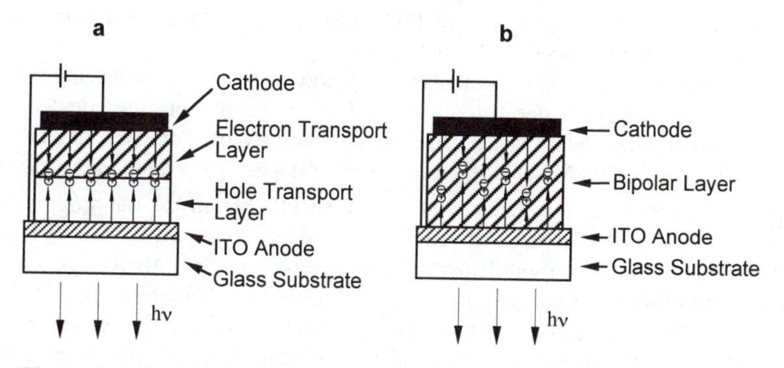

Figure 2. Structures of (a) double-layer and (b) single-layer EL devices.

affinities (Ea), or Lowest Unoccupied Molecular Orbital (LUMO), were determined from the Ip values (HOMO) and Eg values.

Luminance was measured with a luminance meter Topcon BM 8 at room temperature. EL spectra were taken with an optical multichannel analyzer (Hamamatsu Photonics PMA 10). Electronic absorption spectra were taken with a Shimadzu 2200A ultraviolet-visible spectrophotometer.

Double-Layer-Type Devices

Side-Chain-Type Polymers. We synthesized two types of side-chain-type polymers, polymethacrylamide (*11*) and polymethacrylate, both containing triphenylamine or tetraphenyldiamine as functional groups. As illustrated in Figure 3, hole transport in these polymers is due to hole movement along the triphenylamine moieties and hopping between chains.

In Figure 4, fluorescence spectra of some of side-chain-type polymers are shown. Triphenylamine-containing polymers, PTPAMAc and PTPAMAm, possess a fluorescence peak at around 390 nm, which originates from the triphenylamine moieties. It has been known that, in poly(*N*-vinylcarbazole), carbazole excimers are formed, and emission originates from the excimer, which causes spectral shift to longer wavelength (*12*). It has also been known that such excimer-forming sites serve as hole-trapping sites, which lowers the hole mobility in the PVK film (*13*). Since such excimer formation is negligible in these triphenylamine-containing polymers, it can be assumed that there is no hole traps by excimer sites.

Using these polymers, double-layer-type devices with a structure of ITO / polymer / Alq / Mg:Ag were fabricated. The thickness of the polymer and the Alq layer are 200 Å and 700 Å, respectively. When operated in a continuous dc mode for a forward bias with ITO at positive polarity, the devices exhibited bright green emission originating from the Alq layer. A photograph of the device is shown in Figure 5A. This suggests injection of holes from ITO to the polymer layer, and hole transport through the polymer layer to the emitter layer.

Luminance increased with increasing injection current, and the driving voltage is dependent on the polymer used. Luminance levels of the device using polymethacrylamide, PTPAMAm, are higher than those of the device using polymethacrylate, PTPAMAc, at the same drive voltages; for example, the maximum luminance of 4,800 cd/m^2 is reached at 15 V for PTPAMAm(*11*), but 4,600 cd/m^2 is reached at 18 V for PTPAMAc. The higher luminance observed for the PTPAMAm system is due to higher current densities compared with the PTPAMAc system. Such difference in current density may be due to higher hole drift mobility in PTPAMAm than in PTPAMAc, because the Ip value of the PTPAMAm is the same as that of PTPAMAc, 5.6 eV. It has been reported for molecularly doped polymers that hole mobility depends on the polymer matrix and that the glass transition temperature (Tg) as well as the dielectric constant of the polymer are important factors in the hopping transport (*14*). It can be concluded that polymers with amide bonds are more suitable than polymers with ester bonds.

The barrier height for hole injection from ITO to polymer can be lowered by using polymers with lower ionization potentials; thus, driving voltage can be lowered. For example, tetraphenyldiamine-containing polymers, PTPDMA, have Ip of ca. 5.4 eV which is smaller than that of PTPAAm, 5.6 eV. It is, therefore, expected that

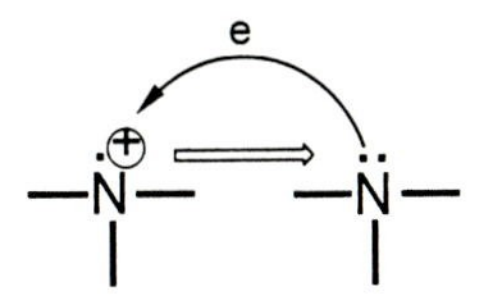

Figure 3. Hole transport along triphenyl amine moieties.

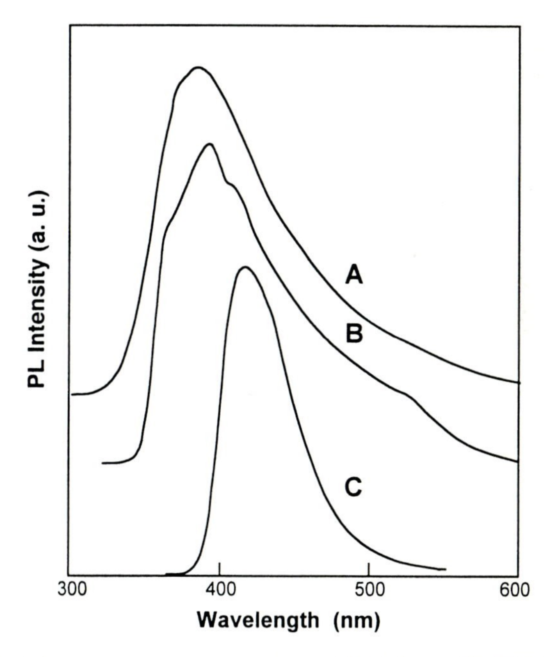

Figure 4. Photoluminescence spectra of (A) PTPAMAm, (B) PTPAMAc and (C) *o*-Me PTPDMA films.

A

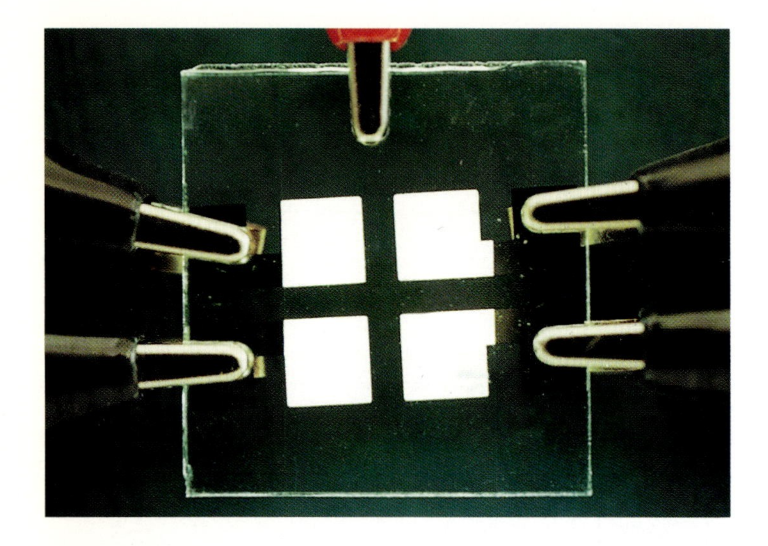

B

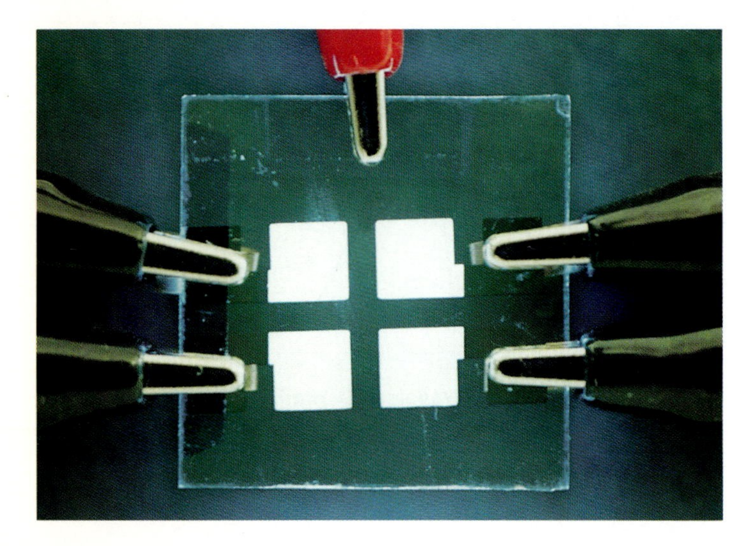

Figure 5. Photographs of (A) a green-light-emitting device and (B) a blue-light-emitting device.

devices using PTPDMAs will have lower driving voltages than those using PTPAAm. Figure 6 shows the luminance–current density-voltage curves for a device with a *o*-Me PTPDMA/Alq. The maximum luminance of ca. 20,000 cd/m^2 is reached at 14 V (*15*), and the external quantum efficiency of 1.1% photons/electron is observed at 10 V, which is similar to the value observed for the EL device using a small molecular weight aromatic amine and Alq(*2*). It was also found that the position of the methyl substituent, *o*- or *p*-, is not critical in these polymers although performance of the devices using vacuum-deposited tetraphenyldiamine derivatives depends highly on the position of the substituent (*16*).

Main-Chain-Type Polymers. In double-layer-type devices having a TPD-doped polymer and Alq, the device lifetime highly depends on the Tg of the host polymer. It has been reported that polymers with a higher Tg provide long device lifetimes (*17*). Among the polymer used as the host matrix, poly(ether sulfone) with a Tg of ca. 220 °C exhibits the longest device lifetime, which is 30 fold longer relative to that of a device having a vacuum deposited TPD. Therefore, the use of a polymer having a high Tg would give stable devices.

Since poly(arylene ether)s are known to be one of the thermally stable engineering plastics (*18*), we designed polymers having arylene ether structures and hole-transporting tetraphenyldiamine units, and synthesized poly(arylene ether sulfone) (PTPDES) (*19*) and poly(arylene ether ketone) (PTPDEK) (*19, 20*), shown in Figure 1. As expected, PTPDES and PTPDEK exhibit high Tgs, 190 °C and 173 °C, respectively. Figure 7 displays fluorescence spectra of the polymers. PTPDES exhibits an emission peak at 420 nm, originating from the diamine moiety. On the other hand, PTPDEK shows a broad emission peak at 500 nm, which indicates that the diamine moieties form exciplex with benzophenone moieties. Such exciplex formation is well known for molecular TPD, and various electron accepting molecules, such as 1,3,4-oxiadiazole derivatives, form exciplex with TPD. It was also found that Ip of PTPDES is lower, 5.5 eV, than that of PTPDEK, 5.6 eV. The higher Ips of these polymers compared with that of molecular TPD is attributed to the electron-withdrawing groups, such as sulfone and ketone groups, of the polymers.

Figures 8 and 9 show luminance–voltage curves and current density-voltage curves for polymer (200Å) /Alq (700Å) devices. The maximum luminance of 14,000 cd/m^2 is reached at 14 V for PTPDES/Alq, while that of 9,400 cd/m^2 is reached for PTPDEK/Alq. Compared with the device with PTPDES, the driving voltage is higher for the device with PTPDEK, which may be due to the slightly higher Ip of PTPDEK (5.6 eV) than that of PTPDES (5.5 eV). In this case, the barrier height for hole injection from ITO to polymer layer is higher for PTPDEK. It is also likely that the exciplex formation sites act as hole-trapping sites and consequently hole mobility in PTPDEK is decreased, resulting in the higher driving voltages.

Single-Layer-Type Devices

Single-layer devices have the simplest device structure, and the fabrication process is simpler than that of the multilayer type. In these devices, both electrons and holes should be injected to the emitter layer. To this end, we have used dye-dispersed polymer systems in which polymers are molecularly dispersed with carrier-transporting

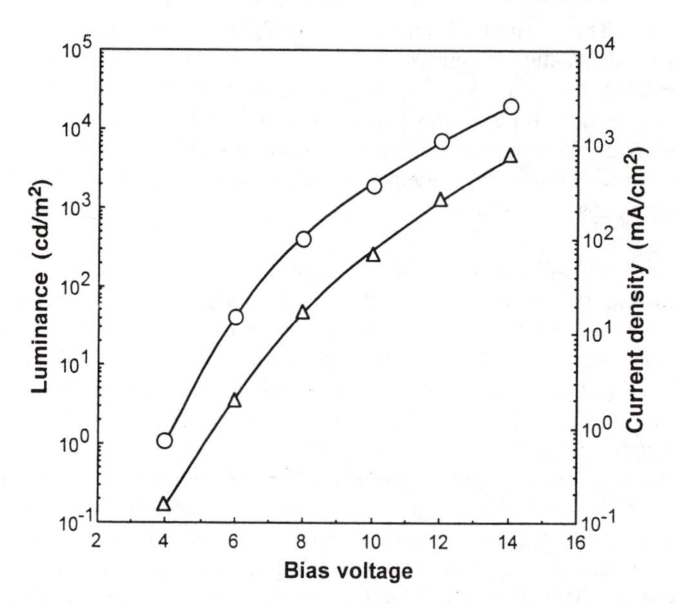

Figure 6. Luminance–voltage, current density-voltage characteristics of *o*-Me PTPDMA (200Å) / Alq (700Å) devices.

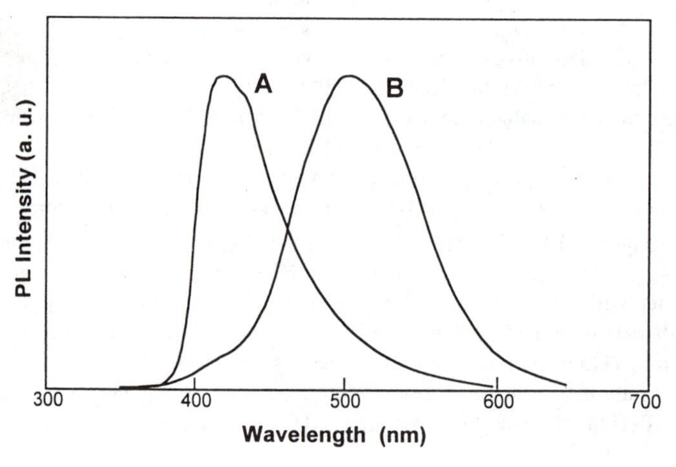

Figure 7. Photoluminescence spectra of (A) PTPDES and (B) PTPDEK films.

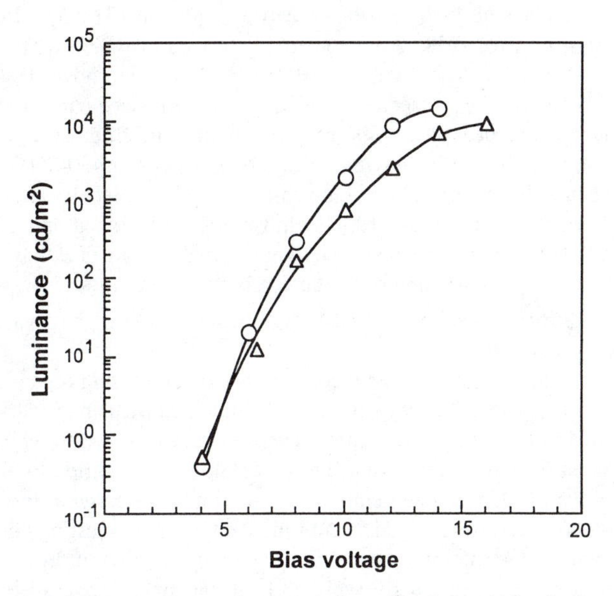

Figure 8. Luminance–voltage characteristics of polymer/Alq devices. (circles) PTPDES and (triangles) PTPDEK.

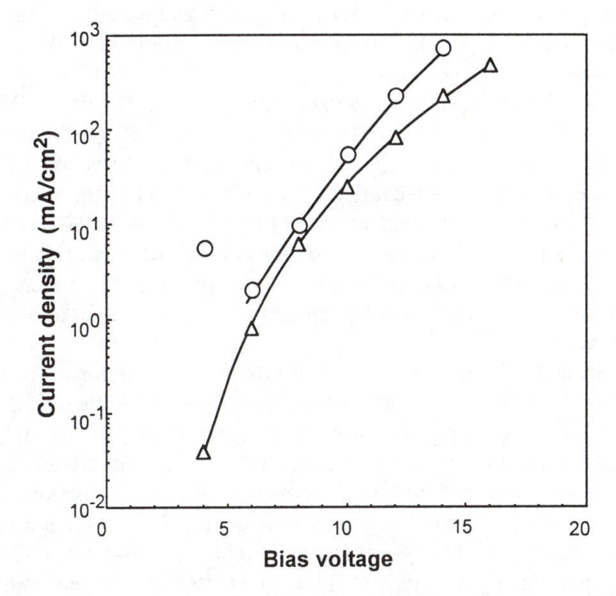

Figure 9. Luminance–current density characteristics of polymer/Alq devices. (circles) PTPDES and (triangles) PTPDEK.

low molecular weight additives to balance carrier injection (*21–23*). In one of the devices, hole-transporting PVK was molecularly doped with 30 wt% of electron-transporting 1,3,4-oxadiazole derivative (PBD) (*23*). PBD is known to be electron-transporting (*24*) and has been used as a dopant in single-layer devices (*25*). Because PBD has a fluorescence peak at ca. 390 nm, a blue emitting fluorescent dye, 1,1,4,4-tetraphenyl-1,3-butadiene (TPB), having a fluorescence peak at 430-450 nm was used as an emitting center in order to obtain blue emission. The molecular structures of the materials used are shown in Figure 10. With 3 mol% TPB doped and 30 wt% PBD doped to PVK, the device emitted pure blue light originally from TPB, having a peak at around 440-450 nm. A photograph of the device is shown in Figure 5B. A luminance as high as 450 cd/m^2 was achieved at 18 V (*23*), which is the highest value observed from single-layer blue light-emitting devices.

Since the emitted light can be tuned with ease by dispersing organic dyes of the desired emitting color (*26, 27*), white light can be obtained by using several dyes (*27*). The color of the emitted light of this single-layer device was tuned to white by adding several fluorescent dyes with a different fluorescent color. In addition to blue-emitting TPB, fluorescent dyes such as green-emitting Coumarin 6, yellow-emitting DCM 1 and orange-emitting Nile Red were added as emitting centers. By adjusting concentrations of these dyes, white electroluminescence was obtained as shown in Figure11A. Four peaks from the dopants were clearly seen: TPB at 440 nm, Coumarin 6 at 490 nm, DCM 1 at 520 nm, and Nile Red at 580 nm (*23*).

Luminance-voltage curve of the device is shown in Figure 12. Maximum luminance of 4100 cd/m^2 was reached at 20 V, which is the highest value ever reported for white-light-emitting organic EL devices. Such high luminance was realized because of the low concentration of the dopant dyes, which minimize the concentration quenching of the dopant fluorescence.

When the dopant concentrations are higher, PBD (30 wt%), TPB (3 mol%), coumarin 6 (0.08 mol%), DCM 1 (0.04 mol%), Nile Red (0.03 mol%), peaks originating from TPB, coumarin 6, DCM 1 become weaker relative to Nile Red, Figure 11B. In this case, the color of the emitted light is yellow. When the concentrations are PBD (30wt%), TPB (4 mol%), coumarin 6 (0.4 mol%), DCM 1 (0.2 mol%), Nile Red (0.15 mol%), the emitted light is orange-red mostly from Nile Red, Figure 11C. These results indicate that energy transfer among dopant dyes is quite efficient and that the concentrations of each dopants should be appropriately low in order to obtain emission from all the dopants.

There are at least two possible mechanisms available for dopant excitation. One is energy transfer from the carrier transporting molecules (carbazole or PBD, or both) to the dopants via the Förster type resonance energy transfer (*28*). In this case, the excited energy is transferred from the host to the dopants that have appropriately lower excited energy levels relative to that of the host. The other is the carrier trapping mechanism, in which dopants serve as carrier trap and provide recombination sites (*29*). The excitation mechanism of this type becomes highly possible when the following conditions are met; that is, Ip of dopant is lower than that of host, and electron affinity (Ea) of dopant is higher than that of host. In the case of dye-dispersed PVK, the orders of Ip and Ea values are: Ip, TPB (6.0 eV) > PBD = BBOT (5.9 eV) > PVK (5.8 eV) > DCM 1 (5.6 eV) > Coumarin 6 (5.5 eV) > Nile Red (5.4 eV), and Ea, Nile Red = DCM 1 (3.5 eV) > Coumarin 6 = TPB (3.2 eV) > BBOT (3.0 eV) > PBD (2.4 eV) >

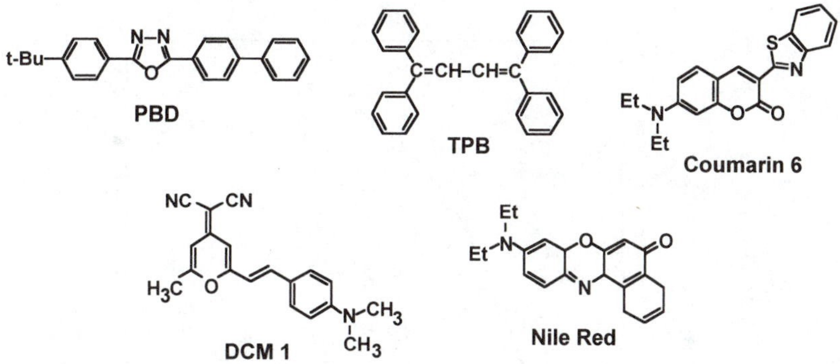

Figure 10. Structures of dopant dyes used.

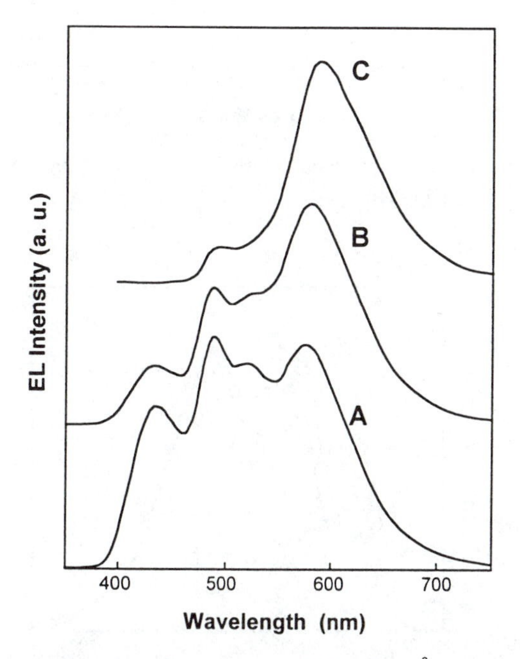

Figure 11. EL spectra for ITO/dye-dispersed PVK (1000Å)/Mg:Ag devices. PVK is molecularly dispersed with (A) 30 wt% PBD, 3 mol% TPB, 0.04 mol% coumarin 6, 0.02 mol% DCM 1, 0.015 mol% Nile Red, (B) 30 wt% PBD, 3 mol% TPB, 0.08 mol% coumarin 6, 0.04 mol% DCM 1, 0.03 mol% Nile Red, (C) 30 wt% PBD, 4 mol% TPB, 0.4 mol% coumarin 6, 0.2 mol% DCM 1, 0.15 mol% Nile Red. J = 20 mA/cm^2.

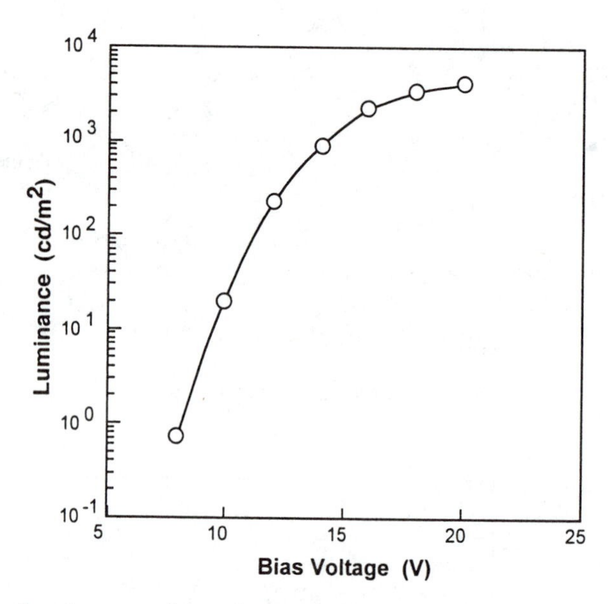

Figure 12. Luminance–voltage characteristics of white-light-emitting ITO/dye-dispersed PVK/Mg:Ag device.

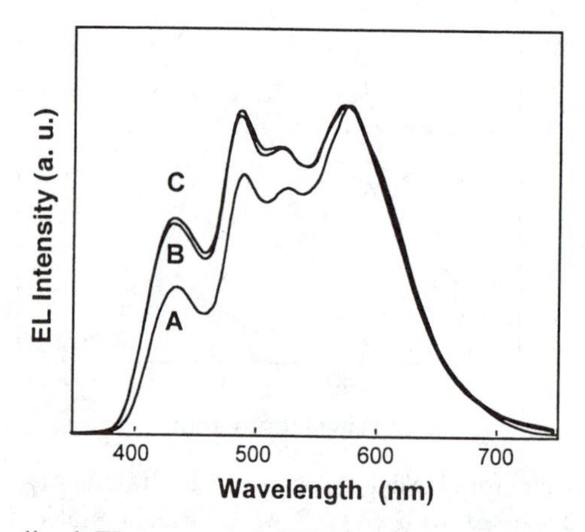

Figure 13. Normalized EL spectra of ITO/dye-dispersed PVK (1000Å)/Mg:Ag devices. PVK is molecularly dispersed with 30 wt% PBD, 3 mol% TPB, 0.04 mol% coumarin 6, 0.02 mol% DCM 1, 0.015 mol% Nile Red. (A) J = 10 mA/cm^2, (B) J = 20 mA/cm^2, (C) J = 200 mA/cm^2.

PVK (2.3 eV). Comparing Ip and Ea of the carrier transporters and the dopants, we can assume that TPB is mainly excited by the energy transfer because of its higher Ip than those of PVK and PBD. Holes are not likely trapped at TPB sites. In contrast, Nile Red can also be excited by the carrier trapping because it serves as a deep hole trap as well as an electron trap due to its low Ip and high Ea relative to those of PVK and PBD. Similarly, both energy transfer and carrier trapping can be operative for Coumarin 6 and DCM 1. In addition to these mechanisms, energy transfer among dopants should not be neglected.

Spectral variations in the white-light-emitting device under different current density are observed as shown in Figure 13. The emission peak of Nile Red is relatively higher at the current density below 20 mA/cm^2, and there is no significant difference in the spectrum at the current densities over 20 mA/cm^2. This result indicates that carriers are trapped at Nile Red, and the excitation of Nile Red by the carrier trapping mechanism is pronounced at low current density levels. At higher current density levels, the energy transfer mechanism dominates the excitation of the dopants because of the saturation of the carrier trapping sites.

Conclusions

In conclusion, we have demonstrated the application of aromatic amine-containing polymers in the fabrication of EL devices. Tetraphenyldiamine-containing polymers, PTPDMA and PTPDES, are quite useful as a hole transport layer, providing high luminance in bilayer devices. Hole-transporting PVK serves as an emitter layer in single-layer devices when it is molecularly dispersed with electron-transporting molecules. By using several emitting centers in these devices, bright white light can be obtained.

Literature Cited

1. Pope, M.; Kallmann, H. P.; Magnante, P. *J. Chem. Phys.* **1963**, *38*, 2042.
2. Tang, C. W.; VanSlyke, S. A. *Appl. Phys. Lett.* **1987**, *51*, 913.
3. Adachi, C.; Tokito, S.; Tsutsui, T.; Saito, S. *Jpn. J. Appl. Phys.* **1988**, *27*, L269.
4. Kido, J.; Nagai, K.; Ohashi, Y. *Chem. Lett.* **1990**, 657.
5. Kido, J.; Kimura, M.; Nagai, K. *Science* **1995**, *267*, 1332.
6. Han, E.; Do, L.; Yamamoto, N.; Niidome, Y.; Fujihira, M. *Chem. Lett.* **1994**, 969.
7. Han, E.; Do, L.; Fujihira, M. *Chem. Lett.***1995**, 57.
8. Hosokawa, C.; Tokailin, H.; Higashi, H.; Kusumoto, T. *Appl. Phys. Lett.* **1992**, *60*, 1220.
9. Kido, J.; Nagai, K.; Okamoto, Y.; Skotheim, T. *Appl. Phys. Lett.* **1991**, *59*, 2760.
10. Kido, J.; Ohtaki, C.; Hongawa, K.; Okuyama, K.; Nagai, K. *Jpn. J. Appl. Phys.* **1993**, *32*, L917.
11. Kido, J.; Harada, G.; Nagai, K. *Kobunshi Ronbunshu* **1995**, *52*, 216.
12. Itaya, A.; Okamoto, K.; Kusabayashi, S. *Bull. Chem. Soc. Jpn.* **1977**, *50*, 22.

13. Yokoyama, M.;Akiyama, K.; Yamamori, N.; Mikawa, H.; Kusabayashi, S. *Polymer J.* **1985**, *17*, 545.
14. Borsenberger, P. M. *J. Appl. Phys.*, **1990**, *68*, 5188.
15. Kido, J.; Komada, M.; Harada, G.; Nagai, K. *Polym. Adv. Technol.* 1995, *6*, 703.
16. Adachi, C.; Nagai, K.; Tamoto, N. *Appl. Phys. Lett.* **1995**, *66*, 2679.
17. Uemura, T.; Kimura, H.; Okuda, N.; Ueba, Y. *The 56th Autumn Meeting of The Japan Society of Applied Physics, Extended Abstracts*, No.3, 1029 (1995).
18. Hedrick, J. L.; Labadie, J. W. *Macromolecules* **1988**, *21*, 1883.
19. Kido, J.; Harada, G.; Nagai, K. *Polym. Adv. Technol.* **1996**, *7*, 31.
20. Kido, J.; Harada, G.; Nagai, K. *Polym. Prep. Jpn.* **1995**, *44*, 1848.
21. Kido, J.; Kohda, M.; Okuyama, K.; Nagai, K. *Appl. Phys. Lett.* **1992**, *61*, 761.
22. Kido, J.; Kohda, M.; Hongawa, K.; Okuyama, K.; Nagai, K. *Mol. Cryst. Liq. Cryst.* **1993**, *227*, 277.
23. Kido, J.; Shionoya, H.; Nagai, K. *Appl. Phys. Lett.* **1995**, *67*, 2281.
24. Adachi, C.; Tsutsui, T.; Saito, S. *Appl. Phys. Lett.* **1989**, *55*, 1489.
25. Mori, Y.; Endo, H.; Hayashi, Y.*Oyo Buturi* **1992**, *61*, 1044.
26. Kido, J.; Hongawa, K.; Nagai, K.; Okuyama, K. *Macromol. Symp.* **1994**, *84*, 81.
27. Kido, J.; Hongawa, K.; Okuyama, K.; Nagai, K. *Appl. Phys. Lett.* **1994**, *64*, 815.
28. Tang, C. W.; VanSlyke, S. A.; Chen, C. H. *J. Appl. Phys.* **1989**, *65*, 3610.
29. Utsugi, K.; Takano, S. *J. Electrochem. Soc.* **1992**, *139*, 3610.

Chapter 26

Application of Thin Films of Polyaniline and Polypyrrole in Novel Light-Emitting Devices and Liquid-Crystal Devices

Alan G. MacDiarmid[1] and Arthur J. Epstein[2]

[1]Department of Chemistry, University of Pennsylvania, Philadelphia, PA 19104–6323
[2]Department of Physics, Ohio State University, Columbus, OH 43210–1106

Light-emitting electroluminescent devices are described in which the conjugated light emitting polymer is separated from one or both of the device electrodes by a film of non-conducting polyaniline. Novel electrochemically-driven electroluminscent devices are also described. The effect on the properties of polypyrrole or polyaniline (deposited from aqueous polymerizing solutions of the monomer) caused by the hydrophilicity/hydrophobicity of the substrate surface is utilized by a "microcontact printing" technique to form patterned liquid crystal display devices.

The ability to cast high quality thin films of conducting polymers from their solutions in organic solvents (*1*,*2*) or to deposit them on selected substrates from aqueous solution (*3*-*5*) has permitted their use both in their lowly conducting (*6*) and also in their highly conducting forms (*7*) in novel devices. The non-doped semiconducting form of polyaniline (emeraldine base ; EB) can, for example, be conveniently spin-cast from its solution in N-methylpyrrolidinone (NMP) to produce high quality thin films (*4*,*5*). Thin cohesive films of polyaniline (*4*,*5*) and polypyrrole (*3*,*4*) in their doped, highly conducting form can readily be deposited on selected polymer or glass substrates by a "1-dip" *in situ* process from dilute aqueous solutions of the monomer where it is undergoing oxidative polymerization.

In this report, we describe the preparation and properties of certain types of the above films and their use in novel electroluminscent and liquid crystal display devices and in "microcontact printing".

Symmetrically Configured AC Light-Emitting (SCALE) Devices

Light-emitting "5-layer" devices having the configuration M/EB/P/EB/ITO when M=Al, Cu or Au, EB=polyaniline (emeraldine base), P=poly(2,5-dihexadecanoxy phenylene vinylene pyridyl vinylene), PPV.PPyV and ITO=indium tin oxide glass show electroluminescent properties in both forward and reverse bias modes (*6*,*8*-*9*) (Figures 1-3). Furthermore, as shown in Figure 2, the devices can operate with an AC applied potential; two light pulses are observed in each cycle. At appropriately

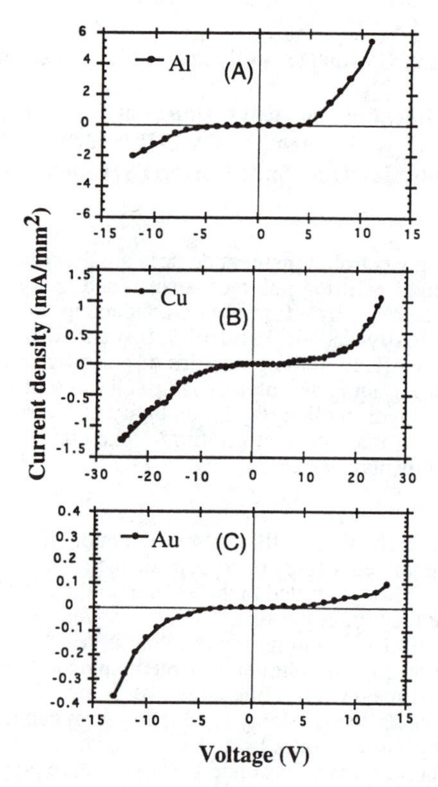

Figure 1. I-V characteristics of SCALE devices using (A) Al, (B) Cu and (C) Au as the electrode. Reproduced with permission from reference 15. Copyright 1996 Materials Research Society.

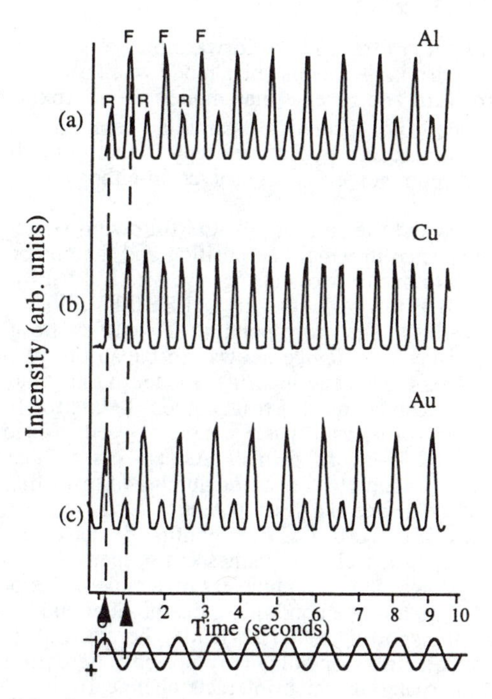

Figure 2. Electroluminescence intensity as a function of time for a metal/EB/PPV.PPyV/EB/ITO device driven by a 1 Hz sinusoidal voltage: metal (a) Al from ± 8 V, (b) Cu from ± 27 V and (c) Au from ± 8 V.
Reproduced with permission from reference 15. Copyright 1996 Materials Research Society.

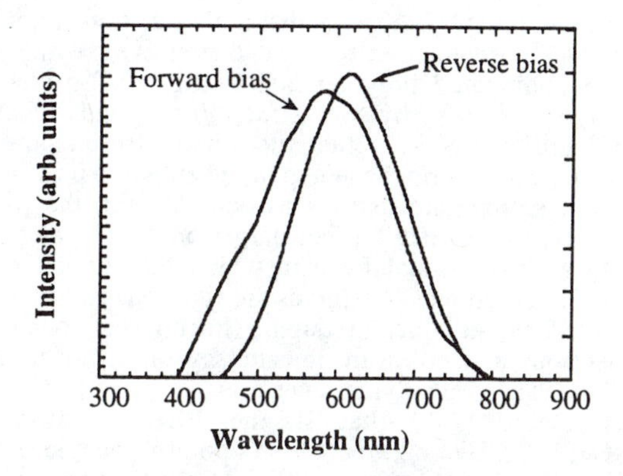

Figure 3. Electroluminescent spectra of a Au/EB/PPV.PPyV/EB/ITO device in forward and reverse bias modes at ~ ± 8 V. For clarity, the intensity in the forward mode (λ_{max} = 585 nm) has been normalized to approximately the same intensity as in the reverse bias mode (λ_{max} = 616 nm).
Reproduced with permission from reference 15. Copyright 1996 Materials Research Society.

selected potentials, light emission in the forward bias mode is more intense when M=Al but is more intense in the reverse bias mode when M=Au. When M=Cu the intensities in the forward and reverse bias modes are approximately the same at ±~27 V. In the absence of the insulating emeraldine base, in the case of aluminum and copper, electroluminescence is observed only in the forward bias mode; in the case of gold no electroluminescence is observed in either forward or reverse bias modes.

In order to understand the role of EB, the following devices involving only aluminum were constructed in which the position and the number of layers of EB were varied from zero to one to two, viz., "3-layers":Al/PPV.PPyV/ITO; "4-layers-1": Al/PPV.PPyV/EB/ITO; "4-layers-2": Al/EB/PPV.PPyV/ITO; "5-layers": Al/EB/PPV.PPyV/EB/ITO. The corresponding I/V curves are given in Figure 4. Only the SCALE ("5-layers") device shows the capability of operating in both forward and reverse bias modes and in an AC mode. In the "5-layers" device only, both holes and electrons can be injected from both ITO glass and from Al electrodes. A similar phenomenon is observed when copper is used instead of aluminum in analogous "3-layers" and "5-layers" devices. As can be seen, these devices exhibit most unusual electrical properties, viz., as the number of insulating layers *increases*, the total resistance of the device at a given potential *decreases*.

Unless the electrical properties of light emiting devices are first understood it seems most unlikely that their electroluminescent properties, which are dependent on the electrical properties, can be completely understood. The possibility must be considered that under appropriate conditions, EB might act both as a good electron and as a good hole transporting material. It may be concluded that reduction in injection barriers for electrons or holes may possibly be optimized by judicious matching of electrode material which interacts chemically in a favorable manner with the polymer with which it is in contact and that the nature of the polymer/polymer interface may also play a critical role.

Electrochemically-Driven Light-Emitting Cells

Electrochemically-driven light-emitting cells have been reported very recently although there is not yet complete agreement as to the exact mechanism or processes by which they operate (*10-14*). It appears that a p/n junction is initially formed (by electrochemical p- and n- doping of a conjugated emissive polymer) in the center of a thin film of the conjugated polymer containing a solid electrolyte such as $LiCF_3SO_3$ dissolved in polyethylene oxide, PEO, sandwiched between two electrodes. It is postulated (*10-12*) that the small turn-on potentials observed (approximately equal to the energy band gap of the emissive polymer) are due to the fact that the dopant anions and cations associated with the p- and n- doping processes respectively, compensate for the charges on the polymer chains. This may be compared with a conventional LED in which the semiconducting emissive polymer layer is oxidized on one side (holes are injected) and reduced on the other (electrons are injected) but in which no doping (injection of counter dopant ions) is involved. The electrons and holes are injected by tunneling through the energy barriers formed at the electrode/polymer interfaces.

It has been stated (*12*) that if the PEO is eliminated from a Al/PPV;$LiCF_3SO_3$;PEO/ITO cell where PPV=poly(*p*-phenylene vinylene) that it behaves similarly to the conventional Al/PPV/ITO LED device. However, we find that in an Al/MEH-PPV;TBATS/ITO cell (*15*) where TBATS=tetrabutylammonium p-toluenesulfonate (Figure 5) that it behaves very differently from an Al/MEH-PPV/ITO device (Figure 6). In particular, the presence of TBATS: (i) involves light-emission in the reverse bias mode, (ii) results in a much greater light intensity

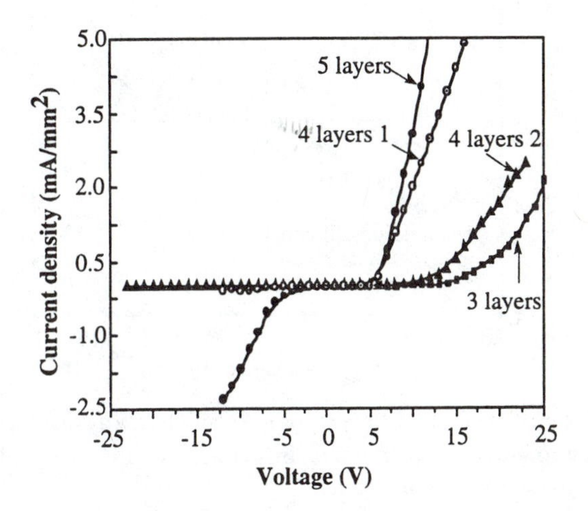

Figure 4. I-V characteristics of 3-layered, 4-layered and 5-layered devices. Reproduced with permission from *SPIE Proc.* vol 2528. Copyright 1995 Photo-optical Instrumentation Engineers.

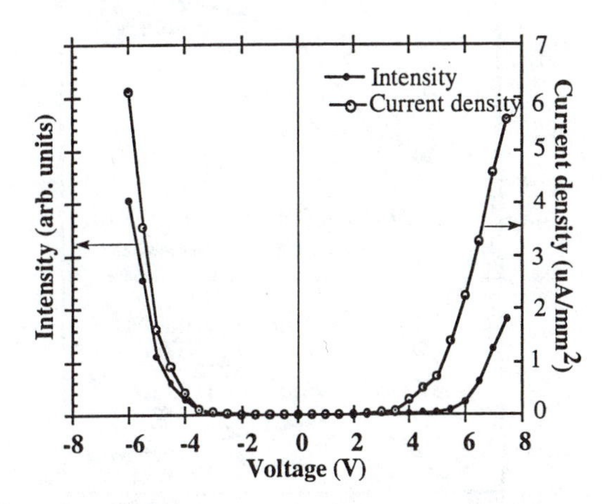

Figure 5. Current density and intensity of light emission vs. voltage in a Al/MEH-PPV+TBATS/ITO device.
Reproduced with permission from reference 15. Copyright 1996 Materials Research Society.

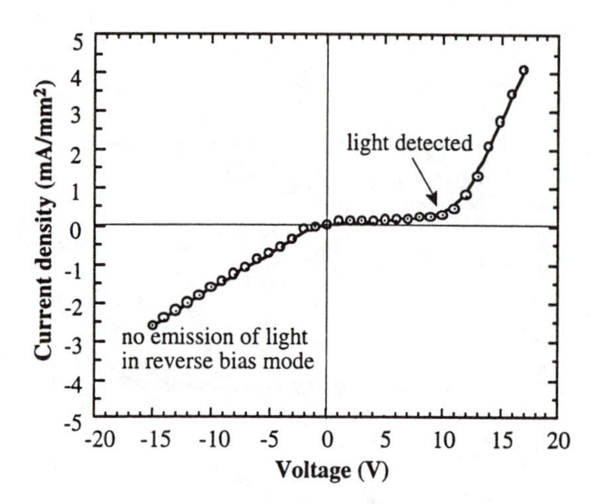

Figure 6. I-V characteristics of a Al/MEH-PPV/ITO device. Reproduced with permission from reference 15. Copyright 1996 Materials Research Society.

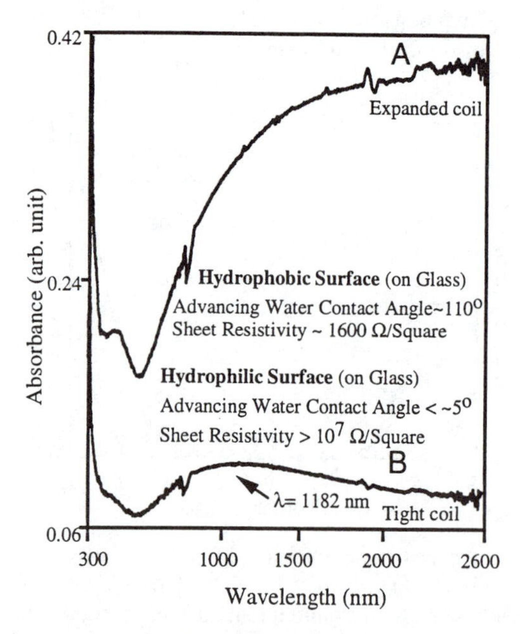

Figure 7. Vis/uv spectra of polypyrrole anthraquinone-2-sulfonate deposited (dipping time: 15 minutes) on (A) a hydrophobic glass surface (film thickness ~ 400 Å) and (B) a hydrophilic glass surface. (Spectrum A was recorded vs. a hydrophobic glass slide in the spectrometer reference beam and spectrum B was recorded vs. a hydrophilic glass slide in the spectrometer reference beam).

clearly visible in the presence of a direct overhead fluorescent light, (iii) results in a lower turn on voltage and (iv) that in some, but not all devices, the current is in the microampere range rather than in the milliampere range normally associated with a conventional LED. We by no means pretend to understand at the present time the relative importance of the many possible variable parameters involved, but we believe that such devices are of very great scientific interest and of potential technological importance.

"1-Dip" *in situ* Deposition of Polypyrrole and Polyaniline on Hydrophobic and Hydrophilic Glass Surfaces.

High quality thin films of doped polypyrrole and doped polyaniline can be conveniently deposited during a few minutes at room temperature on glass and plastic substrates from dilute aqueous solutions of the respective monomer as it undergoes oxidative polymerization (*3-5,15*). We find that the deposition rate and the properties of the films are greatly dependent on the nature of the substrate surface, e.g., whether deposited on hydrophilic or hydrophobic surfaces.

Glass microscope slides may be readily rendered hydrophilic (advancing water contact angle $< \sim 5^0$) by treatment with a hot concentrated sulfuric acid/30% hydrogen peroxide mixture or hydrophobic (advancing water contact angle $\sim 110^0$) by a standard treatment involving exposure to ~ 0.4 wt% hexane solution of $C_{18}H_{37}SiCl_3$.

Figures 7 and 8 show the result of deposition studies involving polypyrrole (*15*) and polyaniline respectively in which treated glass microscope slides were dipped in the same solution of polymerizing monomer for the same length of time. In both cases the rate of deposition of polymer on the hydrophobic surface is very much greater than on the hydrophilic surface. We believe that this may be related to the "like dissolves like" principle, i.e., some of the covalent monomer is preferentially adsorbed from the aqueous solution on to the covalent $C_{18}H_{37}$- film coating the surface more so than on to the polar hydrophilic glass surface, thus favoring more rapid polymerization on the hydrophobic surface. For both polymers, adhesion is stronger to the hydrophilic surface, the films passing the "Scotch Tape" test.

Not surprisingly, the surface resistance of the thinner films on the hydrophilic surfaces is very much greater than that of the thicker films deposited on the hydrophobic surfaces. It is also possible that the conductivity of the films deposited on the hydrophilic and hydrophobic surfaces may differ from each other. This possibility is presently being investigated.

As can be seen from Figure 7, the spectra of the polypyrrole films deposited on hydrophilic and hydrophobic surfaces differ greatly, the peak at 1182 nm for example in the former spectrum being absent in the latter spectrum which instead shows a well defined free carrier tail extending to 2600 nm. By analogy with studies on polyaniline (*16*), we believe that the polymer deposited on the hydrophilic surface might have a tight coil molecular conformation while that deposited on the hydrophobic surface might have an expanded coil molecular conformation.

It can be seen from Figure 8 that the spectra of the polyaniline films differ, although not as greatly as for the polypyrrole films, according to whether it is deposited on hydrophilic or hydrophobic surfaces. It should be noted that different dopant anions were used for the polypyrrole and polyaniline films. On-going studies suggest that the two polymers may behave even more similarly when they both have the same dopant anion.

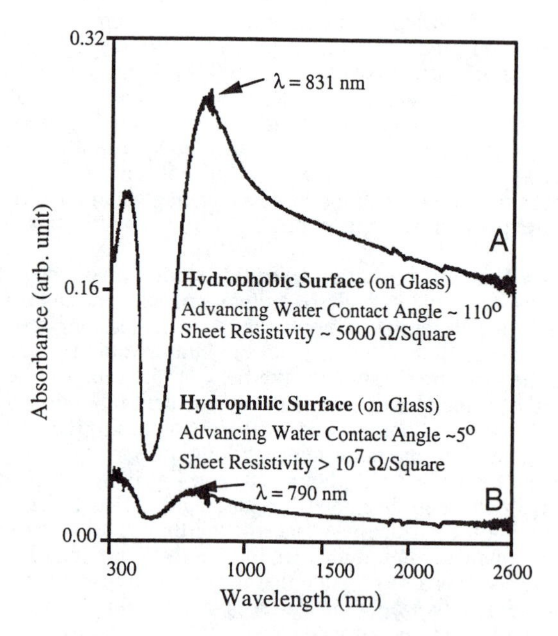

Figure 8. Vis/uv spectra of polyaniline.HCl deposited (dipping time: 5.5 minutes) on (A) a hydrophobic glass surface (film thickness ~ 1200 Å) and (B) a hydrophilic glass surface. (Both spectra were recorded vs. "as-received" glass microscope slides in the spectrometer reference beam).

Flexible Liquid Crystal "Light Valves"

Novel, flexible, completely organic, polymer dispersed liquid crystal (PDLC) "light valves" were fabricated using two flat pieces of commercial overhead transparency substrates (Nashua *xf-20*) coated with polypyrrole between which a film of commercial PDLC material (Norland Products Co. NOA 65 optical adhesive and BDH Ltd. E7 liquid crystal fluid together with EM. Ind. 15 micron polystyrene spacers) was sandwiched. The optical adhesive was polymerized by exposure to UV light. Thin conducting polypyrrole films of varying controllable thickness were deposited on the overhead transparency.

For use in flat screen liquid crystal displays it is necessary to optimize the thickness of the polypyrrole deposit so as to simultaneously obtain the maximum transmittance and minimum resistance necessary for satisfactory devices. For example, a 10 minute dip of Nashua *xf-20* overhead transparency produces a polypyrrole film having a thickness of ~250 angstroms, a surface resistivity of 7,200 ohms/square and a transmittance centered near the middle (600 nm) of the visible region (~400 nm to ~800 nm) of 89% using an uncoated substrate in the reference beam of the spectrometer. Figure 9 illustrates preliminary results obtained to date with a completely flexible, all organic light valve using polypyrrole as the conducting medium for both electrodes (*17*). A PDLC device using conducting ITO glass for both electrodes was used as a standard for comparison. The results are satisfactory for certain applications such as light-weight, non-breakable windows of variable transmittance.

Application of "Microcontact Printing" for the Production of Patterned Polypyrrole and Polyaniline Films

We are combining the selective deposition of polypyrrole and polyaniline on hydrophilic/hydrophobic surfaces as described in the preceeding section with the recent microcontact printing technique (*18-20*) to produce patterned conducting polymer films which we have demonstrated can be used in PDLC display-type devices.

A key objective of the collaborative research with G. M. Whitesides and Y. Xia (Harvard University) is to determine the maximum resolution of polyaniline and polypyrrole patterns attainable using only simple, commonly available equipment, i.e., a desk-top computer and a standard spectrometer plotter, (resolution, ~ 25 μm). The steps comprise: (i) designing any desired pattern on the computer, (ii) reducing the pattern on the computer to any desired size, e.g. > ~ 1 cm x 1cm; (iii) transferring the design to a floppy disk, (iv) inserting the disk into the computer driving the plotter of the spectrometer, (v) replacing the pen in the plotter by an object with a sharp point, e.g. a sewing needle, (vi) covering , e.g. by spinning or other means a heated microscope slide or silicon wafer with a thin layer (~ 20-40 μm) of low melting (working temperature ~ 52^0C) wax such as Amaco Flexwax 120, (vii) scratching the pattern on the thin layer of wax using the needle in the pen holder of the plotter, and placing it (attached to its substrate) in a petri dish, (viii) pouring a well stirred mixture of Dow Corning Sylgard 184 silicone elastomer (10 parts by weight) and Sylgard 184 curing agent (1 part by weight) on top of the wax engraving to a depth of 5-10 mm, (ix) allowing the mixture to polymerize to the silicone elastomer during ~ 3 days at room temperature, (x) removing the silicone stamp from the wax-engraved pattern and discarding (it has the wax "scrapings" produced in the engraving of the pattern adhering to it, (xi) repeating step (viii) and allowing polymerization , (xii) removing the silicone elastomeric stamp which now has the 3-D design imprinted on its lower face, (xiii) wiping the patterned face of the silicone stamp with a piece of cotton containing the "ink" as a ~ 0.4 wt% solution of $C_{18}H_{37}SiCl_3$ in n-hexane, (xiv) evaporating the n-hexane in a stream of nitrogen for

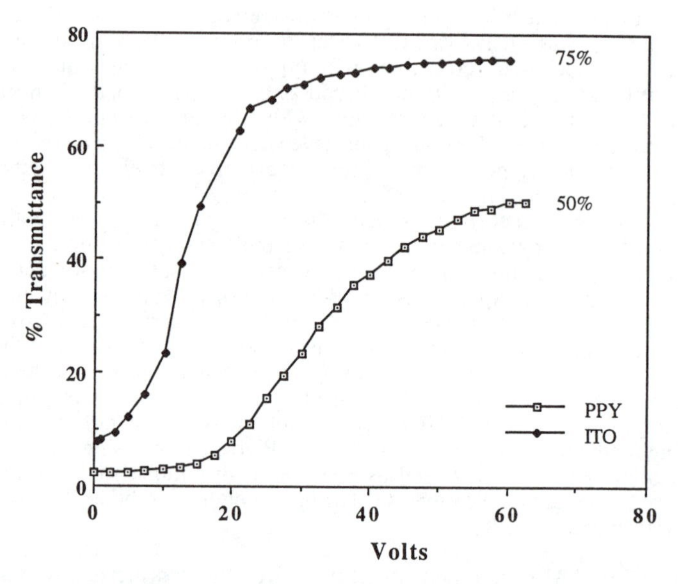

Figure 9. Relationship between transparency (% transmittance in air at 600 nm) and applied voltage of polymer dispersed liquid crystal display devices constructed using two ITO glass electrodes and two polypyrrole film (on plastic) electrodes as the conducting transparent substrates.

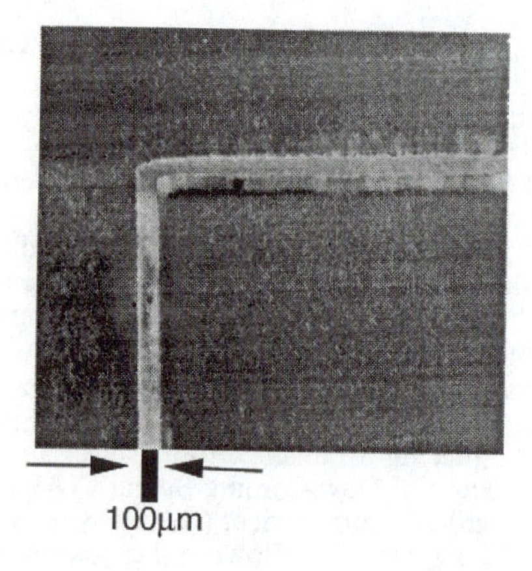

Figure 10. SEM of polyaniline.HCl selectively deposited by a "1-dip" process on a ~100 µm wide line of hydrophobic $C_{18}H_{37}SiCl_3$ substrate.

~ 1min., (xv) pressing the stamp firmly for ~ 10 seconds on the substrate surface so as to imprint a pattern of a thin hydrophobic $C_{18}H_{37}$- film on the substrate, (xvi) immediately cleaning the stamp by rinsing it with cotton soaked in n-hexane, (xvii) placing the substrate having the imprinted $C_{18}H_{37}$- layer in to the appropriate dipping solution.

An example of an SEM of polyaniline.HCl selectively *in situ*-deposited (5.5 min. dipping time) on an ~ 100 μm wide hydrophobic $C_{18}H_{37}$- line imprinted on a clean , hydrophilic microscope slide is given in Figure 10.

Examples of the selectivity of polypyrrole 1-dip, *in situ*, deposition are given in the SEMs in Figure 11. Dark lines are polypyrrole selectively deposited on hydrophobic $C_{18}H_{37}$- surfaces imprinted on a clean, hydrophilic microscope slide as substrate. The upper three figures (Figures 11a) originated from a desk-top computer-drawn pattern used for producing the wax engraved master; the deposition time in the polymerizing pyrrole solution was 12.0 min. Preliminary studies directed towards optimizing both the selectivity of deposition and resolution using the wax engraving technique are very encouraging. The lower two figures (Figures 11b) demonstrate the resolution attainable using a commercial relief master from which the silicone stamp was made. Our objective is to attain these types of results using the (non-lithographic) wax engraving technique. Since the silicone stamps can be used repeatedly without loss of resolution (*18-19*), this technique holds potential promise for the inexpensive production of semi-micro circuitry and liquid crystal displays on rigid or flexible substrates in certain types of "throw away" devices such as, e.g., sensors.

As an example, an effective polymer dispersed liquid crystal interdigitated array display has been fabricated by combining the concepts and techniques given in the "Flexible Liquid Crystal Light Valves" section with those given in this section. A thin interdigitated polypyrrole display pattern was deposited on a glass microscope slide as described above from the computer designed pattern given in Figure 12. Its thickness was chosen so that its optical transmittance and surface resistance were appropriate for liquid crystal display purposes. The PDLC mixture was sandwiched between the microscope slide (the polypyrrole pattern acting as one electrode) and ITO coated glass which acted as the other electrode. When placed on an overhead projector, the whole device produced a dark image on the screen. Application of an AC potential to the ITO/glass electrodes and one half of the pattern, produced a clear bright image of that half of the pattern, strong light passing through the semi-transparent polypyrrole line electrodes. Application of the potential to both polypyrrole electrodes resulted in both halves of the interdigitated polypyrrole array appearing as a white pattern on a black background on the projector screen. This dramatically demonstrates the large difference in surface resistance between polypyrrole deposited on hydrophobic vs. hydrophilic surfaces as described in Figure 7.

Conclusions

It is concluded that thin films of conducting polymers such as polyaniline and polypyrrole, whether in their highly or lowly conducting forms, whether cast from their solutions in organic solvents or deposited from solutions of the polymerizing monomer are of both fundamental scientific interest and of possible technological use for certain applications. However, much still remains to be understood concerning the role of conjugated polymers in various light-emitting devices and the effect on their properties induced by the nature of the substrate surface on which they are deposited.

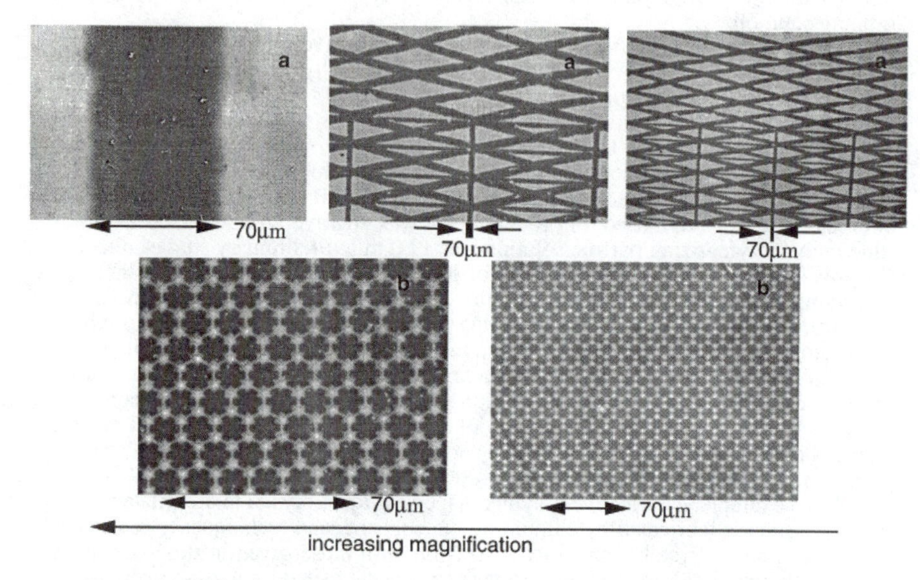

Figure 11. Selective deposition of polypyrrole on patterned hydrophobic surface.

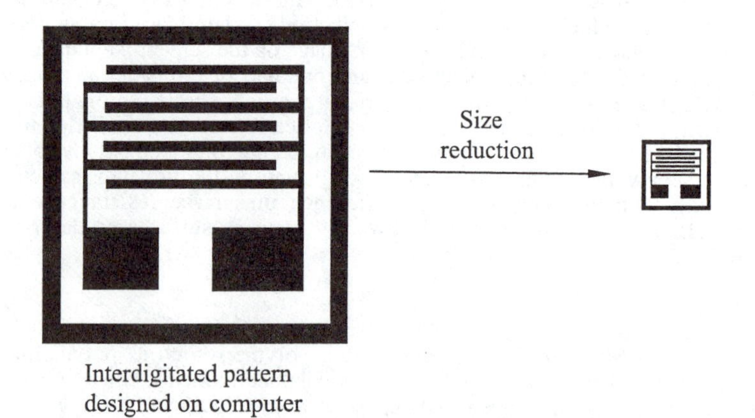

Figure 12. Computer constructed and computer reduced (17mm x 17mm) interdigitated array.

Acknowledgments

The authors gratefully acknowledge the University of Pennsylvania Materials Research Laboratory, NSF (Grant No. DMR-91-20668) for financial support of studies on light emitting devices and the Office of Naval Research (K. J. Wynne, Program Officer) for the remainder of the research described. The LED studies were performed by H. L. Wang and F. Huang; those on the electrochemically-driven light-emitting cells were conducted by F. Huang. Polypyrrole, polyaniline and microcontact printing studies were performed by Z. Huang and P.-C. Wang in collaboration with G. M. Whitesides and Y. Xia (Harvard University). The authors wish to thank Dr. B. R. Hsieh (Xerox Corp.) for the sample of MEH-PPV and Mr. F. Huang for his untiring efforts in producing this manuscript.

References

1. Manohar, S. K.; MacDiarmid, A. G.; Cromack, K. R.; Ginder, J. M. and Epstein, A. J. *Synth. Met.*, **1989**, *29*, E349.
2. Angelopoulos, M.; Ray, A.; MacDiarmid, A. G. and Epstein, A. J. *Synth. Met.*, **1987**, *21*, 21.
3. Kuhn, H. H.; Child, A. D.; and Kimbrell, W. *Synth. Met.*, *71*, **1995**, 2139 and references therein.
4. Manohar, S. K.; MacDiarmid, A. G. *Bull. Am. Phys. Soc.*, **1989**, *34, 582.*
5. Manohar, S. K.; MacDiarmid, A. G. and Epstein, A. J. *Synth. Met.*, **1991**, *41*, 711.
6. Wang, H. L.; MacDiarmid, A.G.; Wang, Y. Z.; Gebler, D. D. and Epstein, A. J. *Synth. Met.*, *78*, **1996**, 33.
7. Gustafsson, G.; Cao, Y.; Treacy, G. M.; Klavetter, F.; Colaneri, N.; and Heeger, A. J. *Nature*, *357*, **1992**, 477.
8. Wang, Y. Z.; Gebler, D.D.; Lin, L.B.; Blatchford, J. W.; Jessen, S.W.; Wang, H. L. and Epstein, A. J. *Appl. Phys. Lett.*, *68*, **1996**, 894.
9. Wang, H. L.; Park, J. W.; Fu, D. K.; Marsella, M. J.; Swager, T. M.; MacDiarmid, A.G.; Wang, Y. Z.; Gebler, D. D. and Epstein, A. J. *Polym. Prepr.*, *36*, **1995**, 45.
10. Pei, Q.; Yu, G.; Zhang, C.; Yang, Y. and Heeger, A. J. *Science*, *269*, **1995**, 1086.
11. Pei, Q.; Yu, G.; Zhang, C.; Yang, Y. and Heeger, A. J. *Science*, *270*, **1995**, 719.
12. Pei, Q.; Yang, Y.; Yu, G.; Zhang, C.; and Heeger, A. J. *J. Am. Chem. Soc.*, *118*, **1996**, 3922.
13. Bard, A. J. *Science*, *270*, **1995**, 719.
14. Richter, M. M.; Fan, F.R.F.; Klavetter, F.; Heeger, A. J.; and Bard, A. J. *Chem. Phys. Lett.*, *226*, **1994**, 115.
15. MacDiarmid, A. G.; Wang, H. L.; Huang, F.; Avlyanov, J. K.; Wang, P. -C.; Swager, T. M.; Huang, Z.; Epstein, A. J.; Wang, Y. Z.; Gebler, D. D.; Ranganathan, S.; Calvert, J. M.; Crawford, R. J.; Vargo, T. G.; Wynne, K. J.; Whitesides, G. M.; Xia, Y.and Hsieh, B. R. in *Materials Research Society, Symposium Proceedings* , Eds. Jen, A.K.-Y.; Lee, C.Y-C.; Dalton, L.R.; Rubner, M.F.; Wnek, G.E.; Chiang, L.Y.; *43*, **1996**, 3.
16. Avlyanov, J. K.; Min, Y.; MacDiarmid, A. G. and Epstein, A. J. *Synth. Met.*, *72*, **1995**, 65 and references therein.
17. Shashidhar, R.; Calvert, J. M.; Crawford, R. J.; Wynne, K. J.; Vargo, T. G.; MacDiarmid, A. G.; Avlyanov, J. K.; Navy Case 77,014, filed 3/1/95; US Patent Application 08/401, 912, filed March 9, 1995; pending.
18. Kim, E.; Xia, Y. and Whitesides, G. M. *Nature*, *376*, **1995,** 581.
19. Kumar, A.; Biebuyck, H. A. and Whitesides, G. M. *Langmuir*, *10* , **1994**, 1498.
20. Y. Xia, M. Mrksich, E. Kim and G. M. Whitesides, *J. Am. Chem. Soc.*, *117*, **1995**, 9576.

Chapter 27

Near-Edge, X-ray-Absorption, Fine-Structure Study of Poly(2,3-diphenylphenylene vinylene) Following the Deposition of Metal

E. Ettedgui[1,5], H. Razafitrimo[1], B. R. Hsieh[2], W. A. Feld[3], M. W. Ruckman[4], and Y. Gao[1,6]

[1]Department of Physics and Astronomy, University of Rochester, Rochester, NY 14627
[2]Webster Center for Research and Technology, Xerox Corporation, 800 Phillips Road, Webster, NY 14580
[3]Department of Chemistry, Wright State University, Dayton, OH 45435
[4]Department of Physics, Brookhaven National Laboratory, Upton, NY 11973

The interaction of Al, Ca and Au with poly(2,3-diphenylphenylene vinylene) during the metal/polymer interface formation is studied by means of near edge X-ray absorption fine structure (NEXAFS) spectroscopy. We found that Al does not create new unoccupied states within the band gap during interface formation. However Ca/ and Au/polymer interface formation provides clear evidence for the presence of new intra-gap states as well as a strong resonance at higher photon energies. The new unoccupied states evidenced by NEXAFS strongly suggest the existence of a bipolaron lattice at the Ca/polymer interface that may enhance charge injection through the interface.

The development and characterization of polymer-based light emitting diodes (LEDs) continues to attract much attention because of their exciting potential in the production of flat panel displays and other novel light emitting devices. The relatively short lifetimes of such devices impede their commercial development, however. Recent efforts to mitigate LED degradation occurring during device operation have resulted in reported lifetimes of order of 1000 hours (*1*). Additional improvements in the performance of polymer-based LEDs have focused on increasing the efficiency of electroluminescence (EL) by creating multilayer devices (*2,3*). The materials which compose these layers can be chosen based on their charge transport capabilities as well as by considering the properties of the interfaces resulting from their juxtaposition (*3*). Furthermore, modifications of the interface geometry to facilitate charge injection by increasing the contact area between the substrate and the polymer have yielded electroluminescent devices of efficiencies greater than 2% (*2*).

[5]Current address: B320/Polymer Building, National Institute of Standards and Technology, Gaithersburg, MD 20899

[6]Corresponding author

A better understanding of the mechanism responsible for EL will improve the performance and lifetime of polymer-based LEDs. Current models that describe the mechanism underlying this process draw analogies between photoluminescence and electroluminescence. EL is postulated to result from the recombination of positive and negative charge carriers across the band gap of conjugated polymers (*4,5*). Charge injection into the polymer precedes light emission and it has been widely reported that metals of low work function, such as Ca, inject electrons more efficiently into the polymer than do those of higher work function. This has been verified from studies of EL efficiency as well as transport measurements (*6-8*). More recent studies which monitored the effect of varying the cathode and anode materials independently strongly suggest that charge injection into the polymer proceeds via tunneling across interfacial barriers. The heights of the individual barriers equal the differences in work functions between the polymer and the electrode materials (*8,9*). Full depletion of the polymer as a result of its low charge carrier density constitutes a central tenet of the tunneling model. Such a depletion prevents charge transfer normally associated with Schottky barrier formation in semiconducting materials and leads to the conclusion that the electronic bands in the polymer are rigid. The model successfully predicts the dependence of such properties of polymer-based LEDs as turn-on voltage and device efficiency with slight discrepancies at low voltages.

Analysis of the interface between the metal electrode and the polymer layer would be useful in determining whether in fact charge transfer occurs during interface formation. In addition, it may prove important in understanding the discrepancy between the predicted and reported device efficiencies at low voltages. The current-voltage (I-V) characterization of polymer-based LEDs precludes such an analysis, since transport measurements require the deposition of a relatively thick metal overlayer, which may obscure some of the interfacial characteristics of the LEDs. Photoelectron spectroscopy obviates this difficulty by allowing a surface sensitive investigation of the occupied electronic states in a sample during the initial stages of the interface formation. X-ray photoelectron spectroscopy (XPS) studies have shown that, during metal deposition on their surfaces, conjugated polymers exhibit changes in the binding energies of their core level electrons akin to band bending in three-dimensional semiconductors. The changes observed are reminiscent of Schottky barrier formation, and as in crystalline semiconductors, depend on details of surface preparation (*10-16*). Unlike traditional semiconductors, however, polymers exhibit gradual band bending which depends on the thickness of the metal overlayer, and the process may continue beyond the deposition of 30 Å metal (*10-12*). Detailed studies of the energy levels near the valence band edge have shown the appearance of new electronic states resulting from charge transfer as a result of the deposition of Ca and Na (*17-19*). Theoretical work corroborates this interpretation by predicting the transfer of electrons from Na and Ca into the polymer leading to new valence states (*18-20*). These results, because they contradict the tunneling model detailed above, call for detailed studies of the interface formation between metals and conjugated polymers.

Studies of charge transport in polymer-based LEDs suggest that holes have a higher mobility than electrons and serve as the primary charge carriers (*21-23*). In order to gain a complete picture of the interface formation between metals and polymers, it would therefore be useful to consider in addition to occupied states the formation of unoccupied energy levels during metal deposition. Near edge X-ray absorption fine structure (NEXAFS) is a powerful tool to investigate the unoccupied states, complementing the information obtained from XPS (*24*). In our study of the interface formation between metals and conjugated polymers, we have probed with NEXAFS the changes in unoccupied states induced by the deposition of Al, Ca and Au on poly(2,3-

diphenylphenylene vinylene) (DP-PPV). Our results indicate a thickness dependent response which varies with the metal employed. Although we only report the evolution of DP-PPV during metal deposition, we have also studied the case of metal deposition on poly(*p*-phenylene vinylene) (PPV) and did not find any fundamental differences between the two polymers. Al does not significantly alter the population of unoccupied states. By contrast, Ca and Au depositionbroaden the π_1* resonance at the C K-edge. In addition, Ca and Au bring new features in the π_2* and C-H* region as well as in the σ* portion of the spectra. We attribute the broadening of the π_1* level during the interface formation of Ca and Au with DP-PPV to the formation of new intra-gap states. The new feature noted at higher photon energies appear to reflect a redistribution of electron densities along the backbone of the polymer. In the case of Ca/DP-PPV, the new unoccupied states may partially explain the improved charge injection reported for this low work function metal.

Experimental

This NEXAFS study was carried out at the National Synchrotron Light Source at Brookhaven National Laboratory using the U7a beamline. NEXAFS spectra are collected by exposing the sample to a monochromatic photon beam of varying energy and monitoring the dipole transitions from a fixed core-level energy to unoccupied states that result. Following the excitation of the core-level electron, the core hole decays usually via an Auger process. As a result, the photon beam indirectly induces a current that is recorded as a function of excitation energy. By varying the energy of the photon beam and monitoring the photocurrent and thus photon absorption, it is possible to estimate the population of unoccupied states (*24*). This technique is also useful in the detection of specific bonds in the molecules, since these tend to carry specific signatures. A grating monochromator enabled the selection of photons at desired energies. The photocurrent was then recorded as a function of photon energy. The photon energies used in this study ranged from 270 to 340 eV with a step size of 0.2 eV and probed transitions of C 1s core level electrons to unoccupied states (C_K edge). In addition to the sample photocurrents, we collected the photocurrent of a copper mesh in the path of the light beam to be able to normalize the photocurrent for the response of the system.

DP-PPV is composed of alternating vinylene and phenylene groups along its backbone, similarly to PPV, with the addition of phenyl side-groups on the phenylene ring. This polymer was selected because of its bright electroluminescence and chemical stability. In addition, the presence of phenyl side groups on DP-PPV increases the solubility of the precursor polymer and so facilitates its handling. DP-PPV samples 250-500 Å thick were deposited on gold coated Si wafers by spin coating, as described previously (*29*). Following their introduction into the ultrahigh vacuum (UHV) chamber, the samples were heated *in vacuo* to 200°C for two hours to remove surface contaminants from the surface prior to spectroscopy. Metals were evaporated from a resistively heated W boat (Ca and Au) or basket (Al) at an average rate of 1Å/min. We estimated the thickness of the metal overlayer with a quartz crystal thickness monitor, bearing in mind that, since the polymer surface is not atomically flat, this serves only as a nominal value. During the deposition, the polymer films remained at room temperature and NEXAFS spectra were recorded at pre-selected coverages. The time for the evaporation and data acquisition was about 1 hour for each coverage.

Results

In Figure 1, we show the C_K edge NEXAFS spectrum of DP-PPV. The character of the fine structure spectrum in the vicinity of the absorption edge provides important information about the electronic structure of the conduction band in the sample under study. Because of the relative paucity of NEXAFS studies of conjugated polymers, the assignment of the resonances found by NEXAFS is based, to a first approximation, on analyzing the features of complex molecules by considering the contributions of their various sub-groups, in this case benzene and ethylene (*24*). Compared to polyethylene, polystyrene, and polyphenylene data, our DP-PPV data reflect contributions primarily from the phenylene portion of the polymer (*24,29,30*). We ascribe the sharp resonance near 287 eV to the transition of the C1s core-level electron to π_1* directly above the band gap, following ~290 eV are the C-H* and π_2* structures and at 293-315 eV are the σ_1* and σ_2* regions. The π_1* level corresponds to the conduction band minimum.

Al deposition on DP-PPV does not alter the NEXAFS spectrum dramatically. We present in Figure 2 the evolution of the DP-PPV during the course of Al deposition. The spectra presented are normalized with respect to the mesh current and are offset for each of the different metal coverages. The structure of the clean sample matches that of Figure 1, taken for the same polymer. The NEXAFS signal intensity decreased after the deposition of 30 Å Al so the signal obtained was multiplied by 2 in order to facilitate comparison. Following the deposition of Al, the spectra reveal a slight decrease in intensity in their leading edges, at photon energies below π_1* resonance. Also, within the C-H* and π_2* region, we find a loss in signal intensity after the deposition of 30 Å of Al. Although noticeable, these changes are relatively minor, since the overall shape of the NEXAFS spectrum following the deposition of 30 Å Al remains mostly unchanged.

The evolution in NEXAFS spectra of DP-PPV following the deposition of Ca differ markedly from those observed for Al. Similarly to Figure 2, we present in Figure 3 the evolution of the DP-PPV spectrum as Ca is deposited. The clean sample exhibits the same spectrum as that shown in Figures 1 and 2. Following the deposition of Ca, we find dramatic changes in the NEXAFS resonances. Photocurrent within the leading edge of the π_1* resonance increases as Ca is deposited onto the polymer. The increase is noticeable from 2 to 4Å of Ca coverage. This indicates the formation of new intra-gap states. Beyond the π_1* resonance, at a photon energy of ~291 eV which is in the region of the π_2* and C-H* states, we note the presence of a strong new feature.

We have also studied the evolution of the metal/DP-PPV interface during the deposition of Au (in Figure 4). We find that the NEXAFS spectrum evolution of DP-PPV is quite similar to the one obtained from Ca deposition. The observed changes very closely mirror those reported in the case of Ca. We see an increase in the density of unoccupied states and we again find a strong resonance at ~291 eV.

Discussion

The overwhelming majority of published studies on the evolution of the electronic structure of conjugated polymers during metal deposition and the consequent interface formation report changes in occupied electronic states. Therefore, we view our results on the effects of Al deposition on DP-PPV in the context of these previous investigations. Our own XPS studies of the Al/PPV interface indicate that although the polymer does not appear to react with Al a rigid shift associated with band bending results from metal deposition (*10-11*). Molecular models of the interfacial reaction of Al with PPV reveal that the metal forms covalent bonds which disrupt the backbone of PPV (*17*). Conjugation of

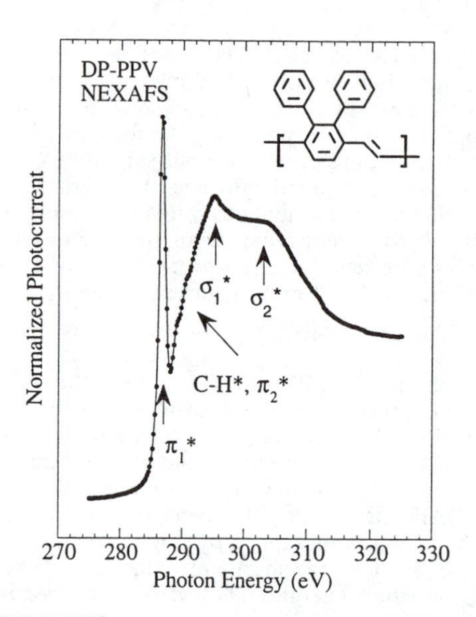

Figure 1. The NEXAFS spectrum obtained for a clean sample of ex-situ converted DP-PPV. The various regions of the spectrum are identified in terms of specific resonances, which are estimated based on the spectra of benzene and ethylene. DP-PPV structure is shown in the inset.

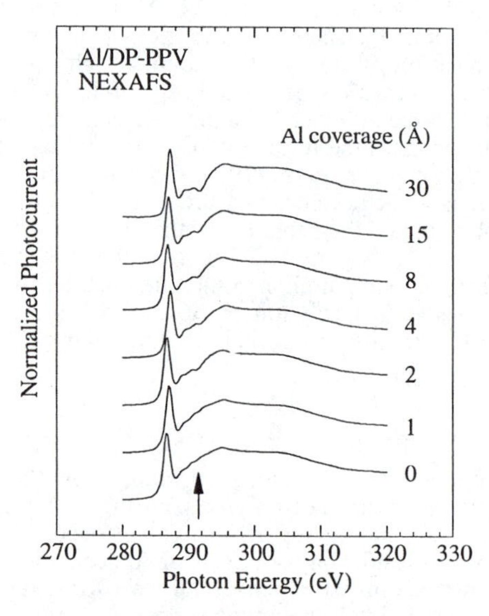

Figure 2. The evolution of NEXAFS spectrum during Al deposition on DP-PPV. Apart from a decrease in the π_2* and C-H* regions of the spectrum of unoccupied states as indicated by the arrow, Al deposition does not affect the spectrum. (Reproduced with permission from reference 35.)

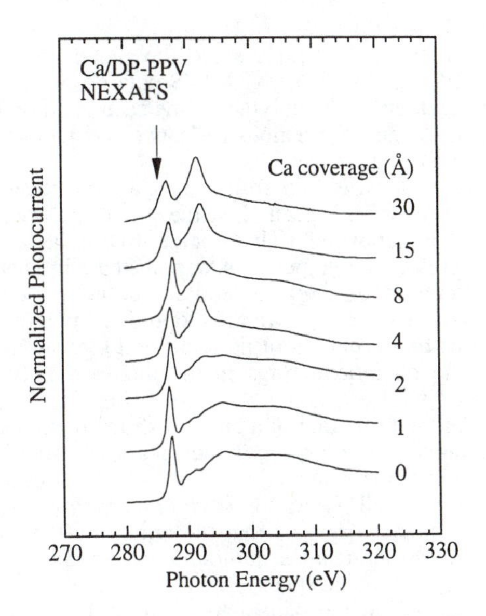

Figure 3. The evolution of NEXAFS spectrum during Ca deposition on DP-PPV. Changes after 4 Å of Ca deposition reveal the growing of the new intra-gap feature at the leading edge of the π_1* resonance as indicated by the arrow. Strong resonance peak emerges also at higher energy from the same Ca coverage. (Reproduced with permission from reference 35.)

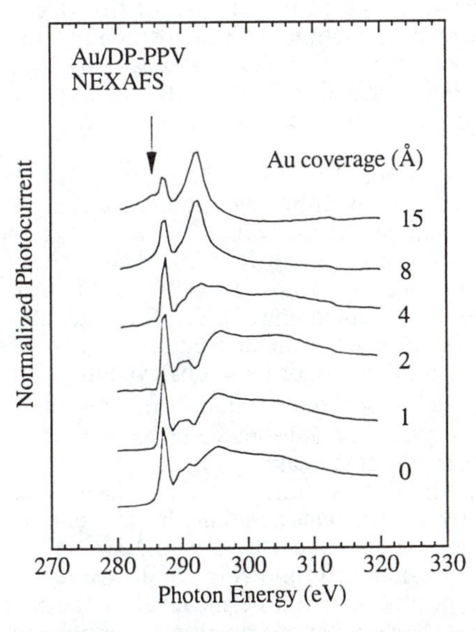

Figure 4. The evolution of NEXAFS spectrum during Au deposition on DP-PPV. Changes after 4 Å of Au deposition mirror those observed in Ca/DP-PPV interface formation.

the system is maintained, however, since the metal atoms create a bridge between conjugated regions (*31*). NEXAFS data do not point directly to the disruption of the backbone in the polymer in the case of Al on DP-PPV although the changes in the C-H* and π_2* regions indicate some interaction between metal and sample.

The evolution of NEXAFS following Ca deposition differs from that observed for Al. The observed spectra lose the structure associated with benzene and ethylene as Ca is deposited. This does not come as a surprise since the deposition of low work function metals on conjugated oligomers has been shown to result in deformation of their molecular geometry (*20*). Furthermore, vibrational studies of PPV have shown that doping this polymer results in charge transfer that induces the formation of polarons and bipolarons. As a result, the conformation of the phenylene rings in the undoped polymer takes on the character associated with a quinoid structure (*32*). The isomerization of the polymer resulting in the formation of a quinoid structure induces a redistribution of the electronic density along the polymer chain and modifies the available states. This evolution of the conjugation of the polymer following Ca deposition may explain the observed NEXAFS spectrum. The examination of the NEXAFS spectra of model molecules reveals that the redistribution of charge densities from a benzene-like structure to a quinoid structure alters and enhances the population of higher unoccupied states (*24*).

The changes within the band gap of the polymer, revealing the new unoccupied states at the leading edge of the conduction band, upon deposition of Ca likely derive from charge transfer from the metal into the polymer. Support for this comes from photoelectron spectroscopy studies on the interaction between Ca and conjugated polymers. We found a rigid shift in the binding energies of the core level binding energies with respect to the Fermi level following Ca deposition (*10,11*). These changes resemble band bending seen in three dimensional semiconductors and suggest charge transfer between Ca and the polymer at the interface (*16*). Studies of the valence band structure of conjugated polymers as a function of Ca deposition reported by Dannetun *et al.* have revealed the presence of new states at the leading edge of the valence band (*19*). This has been ascribed to the transfer of electrons from Ca into the polymer. By comparison, Al deposition results in relatively minor changes to conduction band structure.

The results obtained by NEXAFS during the Ca/DP-PPV interface formation agree well with published photoelectron spectroscopy studies. Measurements of changes in the valence band suggested the formation of bipolarons as a result of charge transfer (*18*). The formation of bipolarons and polarons implies the creation of new states inside the band gap (*30*). Some of these states may be unoccupied and NEXAFS reveals their presence after the deposition of Ca. The changes in the unoccupied states of DP-PPV following the deposition of Ca are reminiscent of those observed in poly (3-methylthiophene), where ClO_4^- doping caused the formation of new unoccupied states. The changes were attributed to the appearance of bipolarons which become narrow bands when the doping level increases (*31*).

A statistical mechanics model of the interface formation between metals and conjugated polymers provides additional insight into the charge transfer that occurs between metal atoms and DP-PPV (*32,33*). This model also helps to understand the results obtained following the deposition of Au. In this model, one considers the positions of the Fermi level of the metal, the valence and conduction bands of the polymer and polaron and bipolaron states relative to the vacuum level. It is found that if the Fermi level of the metal atom is either above or below the energy of the polaron or bipolaron states, then charge transfer between metal atoms and the polymer will tend to result in order to minimize the total energy of the system comprising the polymer and metal. In addition, the bipolaron states are found to be more stable than the polaron state, favoring their creation. As shown in Figure 5, Al's Fermi level falls between the two bipolaron levels, so that little charge transfer is expected at the interface. Ca, on the other hand, because of its low work function has a Fermi level which lies above the bipolaron state. Energy minimization of the system requires the transfer of electrons from Ca to the polymer. In the case of Au, the work function of the metal is so high that its Fermi level lies below the bipolaron state. Electrons are withdrawn from the polymer, again leading to the formation of a bipolaron lattice. The bipolarons resulting from the deposition of Ca are negative, while they are positive for Au. The isomerization of the polymer is similar, however, so that in both cases the backbone of DP-PPV is expected to take on a quinoid character. Therefore, the NEXAFS spectra of DP-PPV following the deposition of both metals are expected to be similar, as we found experimentally.

The underlying basis of this model is largely confirmed in the case of low work function metals, where molecular modeling of the interaction between Na, Ca and Al and PPV indicates that Na and Ca should both give rise to new valence states resulting from charge transfer, while Al should react with the polymer to form covalent bonds. The formation of covalent bonds is not addressed in the statistical mechanics model. This reveals one of its shortcomings. Nevertheless, it provides a good intuitive basis to understand the dynamics of the interface formation between metals and conjugated polymers. Complementing the models of the valence band structure, a more recent calculation of the evolution of the electronic structure of a 4-ring oligomer of poly (*p*-phenylene vinylene) reveals that the interaction of Ca with the conjugated system results in the formation of a new occupied states within the original forbidden band gap together with new unoccupied states at the leading edge of the conduction band and at higher energy (Choong V.-E., Park Y., Gao Y., and Hsieh B. R., unpublished).

Charge transfer at the interface, regardless of the details of the resulting states, is expected to lead to the formation of a built-in electric field. Such an electric field behaves very closely to the one that forms following the deposition of metals on semiconductors and results in a process akin to Schottky barrier formation. Using poly(2-methoxy-5-(2'-ethyl-hexoxy)-1,4-phenylene vinylene) (MEH-PPV), a PPV derivative, internal photoemission spectroscopy has confirmed the presence of bipolarons at the Ca/MEH-PPV and at the Au/MEH-PPV interfaces, while they do not occur at the Al/MEH-PPV interface (Campbell I. H., personal communication, 1995). It therefore appears likely that the changes in the NEXAFS spectra following Ca and Au deposition reflect bipolarons formation at the interface. Taking the formation of this built-in field into account, we have shown that the I-V behavior reported for conjugated polymer LEDs can be correctly modeled throughout the range where data have been reported (*34*). In our model, when the externally applied electric field exceeds the built-in electric field by a sufficient margin, the behavior proposed in the absence of a built-in electric field is recovered.

Conclusion

NEXAFS provides the first clear experimental evidence of the difference in the formation of the Al/DP-PPV, Ca/DP-PPV and Au/DP-PPV interfaces. We report that Al does not induce the formation of new unoccupied states in DP-PPV, whereas Ca and Au cause the creation of new intra-gap states as well as a higher lying resonance in DP-PPV. Valence band photoelectron spectroscopy of model

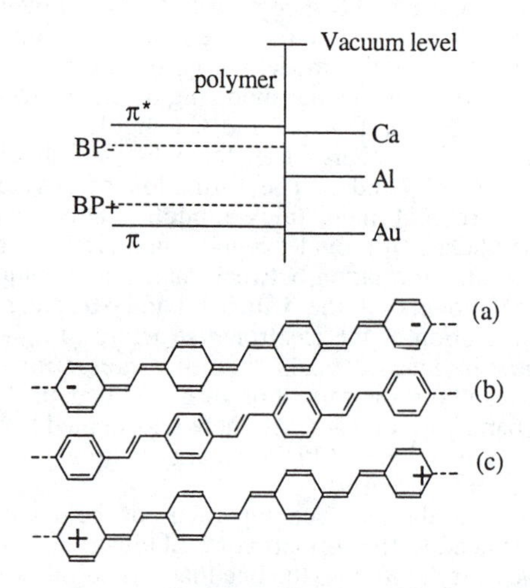

Figure 5. Energy level shematics at the metal/polymer interface. The formation of bipolaron lattice at the interface depends on the relative positions of the metal Fermi level and the bipolarons. At the interface, Ca will tend to form negative bipolarons (a), Al will not form bipolarons (b) and Au will tend to form positive bipolarons (c). Although bipolarons extend over several monomeric units, they span only 4 units in this diagram (*33*). We illustrated the bipolarons with PPV chains.

molecules confirmed that Ca induces new states within the band gap of the pristine material, whereas Al does not. Based on a published model of the metal/polymer interface, we suggest that the evolution of the NEXAFS spectra reported in this study reflect charge transfer between Ca and DP-PPV and Au and DP-PPV, whereas charge transfer does not occur significantly between Al and DP-PPV. The formation of intra-gap states complements the study of the evolution of occupied states previously reported and agrees with the notion of band bending at the interface. The appearance of new unoccupied states points to the appearance of bipolarons, which are considered to be the primary charge carriers in conjugated polymers. The presence of a polaron or bipolaron lattice at the interface may enhance charge transfer at the interface by the creation of unoccupied states at energies near the Fermi energy of the metal. The difference in the evolution of the interface formation with polymer by Al, and Ca may explain why Ca injects electrons more readily into the conjugated polymers during LED operation.

Acknowledgments

This research was supported in part by the National Science Foundation under grant DMR-9303019. Brookhaven National Laboratory is supported by DOE. One of us (HR) acknowledges the Graduate Fellowship from the African-American Institute.

Literature Cited

1. Cacialli, F.; Friend, R.H.; Moratti, S.C.; Holmes, A.B. *Synth. Met.* **1994**, *67*, 157.
2. Yang, Y.; Westerweele, E.; Zhang, C.; Smith, P.; Heeger, A.J. *J. Appl. Phys.* **1995**, *77*, 694.
3. Baigen, D.R.; Greenham, N.C.; Grüner, J.; Marks, R.N.; Friend, R.H.; Moratti, S.C.; Holmes, A.B. *Synth. Met.* **1994**, *67*, 3.
4. Burroughes, J.H.; Bradley, D.D.C.; Brown, A.R.; Marks, R.N.; Mackay, K.; Friend, R.H.; Burn, P.L.; Holmes, A.B. *Nature* **1990**, *347*, 539.
5. Colaneri, N.F.; Bradley, D.D.C.; Friend, R.H.; Burn, P.L.; Holmes, A.B.; Spangler C.W. *Phys. Rev. B* **1990**, *42*, 11671.
6. Braun, D.; Heeger, A.J. *Appl. Phys. Lett.* **1991**, *58*, 1982.
7. Braun, D.; Heeger, A.J.; Kroemer H. *J. Electron. Mater.*. **1991**, *20*, 945.
8. Parker, I.D.; *J. Appl. Phys.* **1994**, *75*, 1656.
9. Heeger, A.J.; Parker, I.D.; Yang Y. *Synth. Met.* **1994**, *67*, 23.
10. Ettedgui, E.; Razafitrimo, H.; Park, K.T.; Gao, Y.; Hsieh B.R. *J. Appl. Phys.* **1994**, *75*, 7526.
11. Ettedgui, E.; Razafitrimo, H.; Park, K.T.; Gao, Y.; Hsieh B.R. *Surface and Interface Analysis* **1995**, *23*, 89.
12. Hsieh, B.R.; Ettedgui, E.; Park, K.T.; Gao, Y. *Mol. Cryst. Liq. Cryst.* **1994**, *256*, 381.
13. Gao, Y.; Park, K.T.; Hsieh, B.R. *J. Chem. Phys.* **1992**, *97*, 6991.
14. Gao, Y.; Park, K.T.; Hsieh, B.R. *J. Appl. Phys.* **1993**, *73*, 7894.
15. Razafitrimo, H.; Park, K.T.; Ettedgui, E.; Gao, Y.; Hsieh, B.R. *Polym. International* **1995**, *36*, 147.
16. Sze, S.M. In *Physics of Semiconductor Devices*; 2nd ed.; Wiley: New York, 1981.

17. Dannetun, P.; Löglund, M.; Salaneck, W.R.; Fredriksson, C.; Stafström, S.; Holmes, A.B.; Brown, A.; Graham, S.; Friend, R.H.; Lhost, O. *Mol. Cryst. Liq. Cryst.* **1993**, *228*, 43.
18. Dannetun, P.; Löglund, M.; Fahlman, M.; Fauquet, C.; Beljonne, D.; Brédas, J.L.; Bässler, H.; Salaneck, W.R. *Synth. Met*. **1994**, *67*, 81.
19. Dannetun, P.; Fahlman, M.; Fauquet, C.; Kaerijama, K.; Sonoda, Y.; Lazzaroni, R.; Brédas, J.L.; Salaneck, W.R. *Synth. Met.* **1994**, *67*, 133.
20. Löglund, M.; Dannetun, P.; Fredriksson, C.; Salaneck, W.R.; Brédas, J.L. *Synth. Met.* **1994**, *67*, 141.
21. Hsieh, B.R.; Antoniadis, H.; Abkowitz, M.A.; Stolka, M. *Polym. Prepr.* **1992**, *33*, 414.
22. Antoniadis, H.; Hsieh, B.R.; Abkowitz, M.A.; Jenekhe, S.A.; Stolka, M. *Mol. Cryst. Liq. Cryst.* **1994**, *256*, 381.
23. Antoniadis, H.; Abkowitz, M.A.; Hsieh, B.R. *Appl. Phys. Lett.* **1994**, *65*, 2030.
24. Stöhr, J. In *NEXAFS Spectroscopy*; Springer Series in Surface Science; Springer-Verlag: New York, 1992, Vol. 25.
25. Ohta, T.; Seki, K.; Yokoyama, T.; Morisada, I.; Edamatsu, K. *Phys. Scr.* **1990**, *41*, 150.
26. Yokoyama, T.; Seki, K.; Morisada, I.; Edamatsu, K.; Ohta, T. *Phys. Scr.* **1990**, *41*, 189.
27. Dannetun, P.; Löglund, M.; Fahlman, M.; Boman, M.; Stafström, S.; Salaneck, W. R.; Lazzaroni, R.; Frederiksson, C.; Brédas, J.L.; Graham, S.; Friend, R.H.; Holmes, A.B.; Zamboni, R.; Taliani, C. *Synth. Met*. **1993**, *55-57*, 212.
28. Tian, B.; Zerbi, G.; Müllen, K. *J. Chem. Phys.* **1991**, *95*, 3198.
29. Hsieh, B.R.; Antoniadis, H.; Bland, D.C.; Feld, W.A. *Adv. Mater.* **1995**, *7*, 36.
30. Bradley, D.D.C.; Brown, A.R.; Burn, P.L.; Friend, R.H.; Holmes, A.B.; Kraft, A. in *Electronic Properties of Polymers*; Springer Series in Solid-State Sciences; Springer-Verlag: New York, 1992, Vol. 107, 304.
31. Tourillon, G.; Fontaine, A.; Jugnet, Y.; Tran, M. D.; Braun, W.; Feldhaus, J.; Holub-Krappe, E. *Phys. Rev. B* **1987**, *36*, 3483.
32. Brazovskii, S. A.; Kirova, N. N. *Synth. Met.* **1993**, *55-57*, 4385.
33. Davids, P. S.; Saxena, A.; Smith, D. L. *J. Appl. Phys.* **1995**, *78*, 4244.
34. Ettedgui, E.; Razafitrimo, H.; Gao, Y.; Hsieh, B.R. *Appl. Phys. Lett.* **1995**, *67*, 2705.
35. Ettedgui, E.; Razafitrimo, H.; Gao, Y.; Hsieh, B.R.; Feld W. A.; Ruckman, M. W. *Phys. Rev. Lett.* **1996**, *76*, 299.

Enhanced Properties and Device Performance Through Self-Assembly and Nanostructure Control

Chapter 28

Specificity and Limits of Organic-Based Electronic Devices

Francis Garnier

Laboratoire des Matériaux Moléculaires, Centre National de la Recherche Scientifique, 2 rue Henri-Dunant, 94320 Thiais, France

Various organic conjugated materials have been up to now proposed as active semiconducting layers in thin film transistors, based on conjugated polymers or on shorter conjugated oligomers and small π–electron rich molecules. Mode of operation of these devices shows that a high carrier mobility together with a low conductivity are required for their figure of merit. Experimental results from the literature show that as-grown conjugated polymers and other amorphous materials exhibit a low carrier mobility, of the order of 10^{-3} to 10^{-5} $cm^2V^{-1}s^{-1}$. All attempts to increase the mobility through slight doping of the organic semiconductor have failed due to a from the variable range hopping mechanism which describes charge transport in these materials. On the other hand, conjugated oligomers are well defined materials offering various physical and chemical ways for a control of the structural organization of thin films made out of them. It is thus shown that carrier mobility is directly related to the long range structural order in these films, i.e. to the decrease of grain boundaries, leading to values close to 10^{-1} $cm^2V^{-1}s^{-1}$ which is comparable to that of amorphous hydrogenated silicon. Additionally, conductivity in thin films of conjugated oligomers is mainly determined by the purity of the materials, allowing values lower than 10^{-7} Scm^{-1}. This independent control of mobility and conductivity allows the realization of oligomer-based thin film transistors showing characteristics close to those of classical a-Si:H based ones.

Organic semiconductors have been studied in the literature since the early 1950s *(1)*, but they raised a steady increasing interest since conjugated materials emerged in the literature in the late 1970s. As-synthesized, these materials possess a semiconducting state, with a conductivity in the range of 10^{-9} to 10^{-6} Scm^{-1}, which is generally p-type, although ion implantation, such as Li^+ in polyacetylene, also allows realization of n-type semiconduction. Although the first conjugated polymer synthesized, polyacetylene, showed low stability in air and very poor processability, new classes of environmentally stable conjugated polymers were soon proposed, such as polythiophenes, polyaniline, and poly(phenylenevinylene) *(2,3)*. Their processability was largely improved by the use of soluble polymer precursors, or by the grafting of solubilizing alkyl or alkoxy groups. But, due to the very low control of the polymerization reaction, these conjugated polymers mainly exist in an amorphous state, showing large distribution of conjugation lengths, together with significant concentration of chemical impurities and structural defects. Later, in the mid 1980s, conjugated oligomers were proposed as a new class of better defined materials, which raised a steady increasing interest both as model compounds for their parent polymers, and also for their promising electrical and optoelectrical properties in their own right *(4)*.

These semiconductors have been studied through the characterization of various devices, such as thin film transistors, TFTs, and light emitting diodes, LEDs, with a long term goal toward the development of a new area of organic-based electronics *(5-16)*. Among these electronic devices, TFTs operate in the most simple way, involving only charge injection and transport in a thin semiconductor layer. These devices are thus well adapted for analyzing and discussing the basic charge transport properties of these organic semiconductors.

MODE OF OPERATION OF TFTs

Various types of substrates have been proposed for the construction of TFTs, with the first ones involving a highly doped silicon wafer acting as the gate electrode, on which an SiO_2 layer formed the insulator, as shown in Figure 1 *(6-9)*. Other materials have been later used as substrate, such as glass, on which a gate electrode (Ag, Al) was vacuum evaporated, followed by the deposition of a thin insulating layer. The source, S, and drain, D, metal electrodes were then patterned on the insulator, using either conventional lithography techniques, or vacuum evaporation through a mask. These electrodes are generally made out of gold, in order to build an ohmic contact with the organic p-type semiconductor, whereas silver appears better fitted in the case of an n-

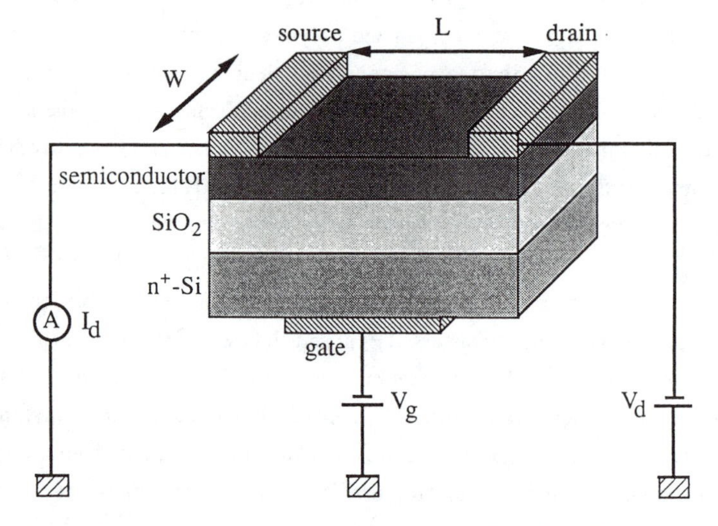

Figure 1. Schematic view of an organic-based Thin Film Transistor.

type semiconductor. The organic semiconductor was deposited in a last step, leading to a bottom gated planar device structure. In the case of most conjugated polymers, due to their low processability, thin films were obtained through the electropolymerization of a monomer, e.g., thiophene, or through the deposition of a soluble polymer precursor. Soluble conjugated polymers, obtained by the grafting of long alkyl chains, e.g., poly(3-alkyl)thiophenes, could be easily spin-casted *(8,17)*. Another interesting route for the construction of TFTs has been based on the use of the Langmuir-Blodgett technique *(11)* In the case of conjugated oligomers, phthalocyanines and other small molecules, the most convenient and rational route toward highly pure and defect free semiconducting films involved vacuum deposition from the semiconductor powder in a tungsten boat, heated to the semiconductor melting point *(10,12)*. The potentially interesting feature of an organic-based device originates from the fact that all its elements, including substrate (polycarbonate, polyimide,...), insulator (polymethylmethacrylate, PMMA, or polyimide), semiconductor, and even electrodes (conducting ink) can be realized out of organic materials, which require only low processing temperatures, thus opening interesting perspectives for flexible organic devices *(13,18,19)*.

These insulated-gate thin film transistors operate as unipolar devices, in which the majority positive charge carriers, in the case of a p-type semiconductor, are attracted to the semiconductor-insulator interface, through negative polarization of the gate electrode, forming a conducting channel through which current flows between source and drain electrodes. The geometrical parameters defining a TFT are the channel width, W, the channel length, L, and the capacitance per unit area of the insulator, C_i. Two main regimes are observed when plotting the drain current as function of drain voltage, V_d, at constant gate voltage, V_g. A linear regime, observed for low drain voltage, followed by a saturation regime, when the drain voltage exceeds the gate voltage. The variation of drain current, I_d, with applied gate voltage V_g and drain voltage V_d , is given by equation (1) *(20)*:

$$I_d = (W/L)\mu C_i \, [(V_g - V_T)^2 - (1/2)\, V_d^2] \qquad (1)$$

where μ is the field-effect carrier mobility, and V_T, the threshold voltage. For higher values of V_d, the saturation regime is described by equation (2)

$$I_{d,sat.} = (W/2L)\, \mu \, C_i \, (V_g - V_T)^2 \qquad (2)$$

Whereas an accumulation regime, equations (1, 2), has been clearly identified under negative gate bias for p-type organic semiconductors, the occurence of a depletion regime under positive gate bias has been seldom characterized. The thickness of this depletion layer, W_{sc}, varies following equation (3):

$$W_{sc} = (\varepsilon/C_i)[(1 + 2C_i^2(V_g - V_{fb})/(qN\varepsilon_s)^{1/2} - 1] \quad (3)$$

where ε_s is the dielectric constant of the semiconductor, V_{fb} the flat band potential and N the dopant concentration. Interestingly, the thickness of the depletion layer, W_{sc}, increases up to the total thickness of the semiconductor layer, d, which is reached at a "pinch-off" voltage V_p given by equation (4):

$$V_p = (qNd^2/2\varepsilon_0\varepsilon_s)(1 + 2C_s/C_i) \quad (4)$$

where C_s is the dielectric capacitance of the semiconducting layer. Equation (4) is particularly interesting, as it shows that the dopant concentration, N, can be obtained by the determination of the depletion pinch-off voltage V_p.

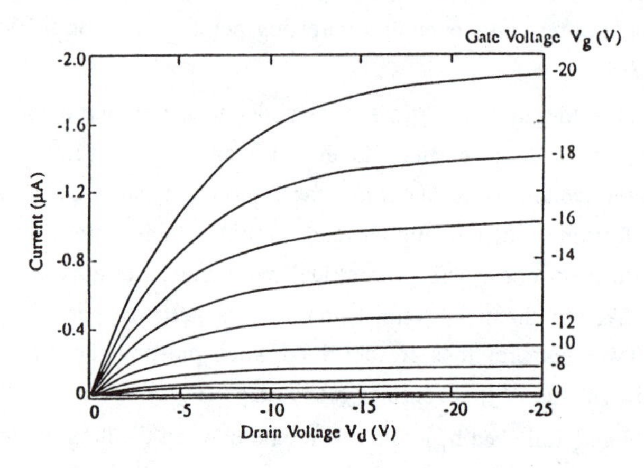

Figure 2. Experimental output $I_d = f(V_d)$ obtained with organic TFT (on glass substrate) with polymethylmethacrylate insulating layer (C_i = 10 nF), sexithiophene semiconducting layer (25 nm thick), and gold source and drain electrodes (W = 5 mm, L = 50 µm).

Experimental output curves, $I_d = f(V_d)$, obtained with sexithiophene-based TFTs in the accumulation regime, as shown in Figure 2, confirm the relevance of organic-based TFTs *(15)*. Besides the obtained drain current, a critical characteristic of a TFT device concerns the dynamic range, or Ion/Ioff ratio, expressed by equation (5), which must be as high as possible, and exceed some 10^7 for practical applications:

$$\mathrm{Ion/Ioff} = 1 + C_i V_d (\mu / 2 \sigma d) \quad (5)$$

These equations clearly define the relevance of using a semiconductor in TFT devices. Equations (1,2) show that a high carrier mobility value μ is indeed needed for reaching a high current (ca. μA) output in the "on" position of the device, as well as for obtaining short switching times. However, equation (5) also shows that a high field-effect mobility is not sufficient, and that a simultaneous low value of the conductivity σ is also required for reaching a high dynamic ratio Ion/Ioff. In fact, when considering amorphous hydrogenated silicon, it must be remembered that its mobility reaches 1 $cm^2V^{-1}s^{-1}$, and that its conductivity is very low, around 10^{-8} Scm^{-1}, which ensures a high dynamic I_{on}/I_{off} ratio, larger than 10^7, for a-Si:H based devices.

CONJUGATED MATERIALS USED IN ORGANIC-BASED TFTs.

Conjugated materials can be grouped into two main classes: (1) macromolecular and amorphous, i.e. conjugated polymers; and (2) molecular, i.e. conjugated oligomers and other π-electron rich molecules. The most studied conjugated polymers include the following:

Polyacetylene

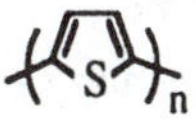

Polythiophene

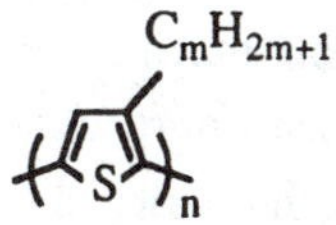

C_mH_{2m+1}

Poly(3-alkylthiophene)

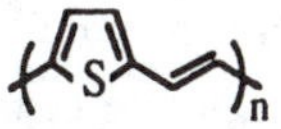

Poly(thienylene-vinylene)

Conjugated oligomers, e.g., pentacene and oligothiophenes, have been more recently used as active layers in TFTs, either unsubstituted or alkyl substituted, as side or end groups *(9,10,12,13,21)*. Unsubstituted oligothiophenes, from terthiophene 3T to octithiophene 8T, and α,ω-dialkyl substituted oligothiophenes, from dialkylterthiophene to dialkyloctithiophene, have been deposited by vacuum evaporation. Other classes of π-electron rich molecules have been studied during the

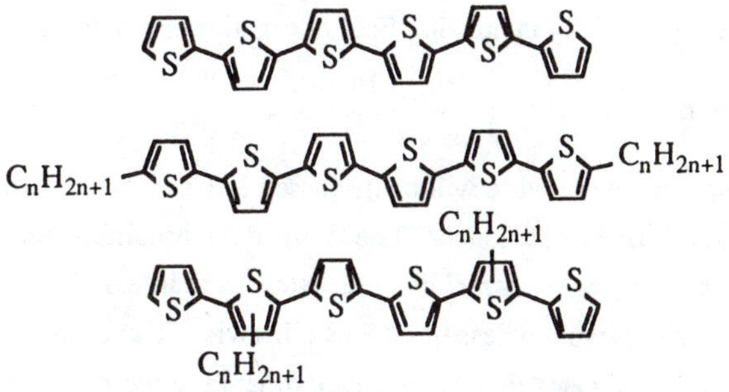

C_nH_{2n+1} C_nH_{2n+1}

C_nH_{2n+1}

C_nH_{2n+1}

last ten years, such as scandium, lutetium and thulium diphthalocyanines, nickel and zinc phthalocyanines *(23)*, fullerene *(24)*, and tetracyanoquinodimethane, TCNQ *(14)*, which were deposited by vacuum evaporation.

These organic semiconductors present charateristics distinct from inorganic covalent semiconductors. Inorganic semiconductors, e.g., silicon, have a three dimensional architecture, in which atoms are held together by strong covalent bonds, with energies on the order of 76 kcal $mole^{-1}$ such as in the case of Si--Si bonds. Semiconductivity appears then as a collective property, which develops with the constitution of the 3D architecture of the material. Owing to the strong interatomic bonds, the width of the conduction and valence bands is large, from which one may expect significant carrier mobility values. These materials are highly sensitive to chemical impurities and to surface states, owing to the presence of dangling bonds at their surface. On the other hand, organic molecular semiconductors are made out of molecules, which are only held together by weak van der Waal forces of about 10 kcal $mole^{-1}$. The electronic properties of the solid are already present in the individual molecules, as shown for instance by the similarity of the absorption spectra of individual molecules and of their assembly in the solid state. These features indicate that charge transport in molecular solids operates mainly through individual states. Furthermore, the width of the valence and conduction bands is small, which suggests that carrier mobility should be much lower in these solids. Finally, these organic molecular materials are less sensitive to chemical impurities, undergoing much fewer substitutions, and the essential absence of dangling bonds prevents them from being highly sensitive to surface states. These considerations show that organic molecular materials cannot be expected to present as high carrier mobilities as those of their inorganic counterparts, which reach 10^3 $cm^2V^{-1}s^{-1}$ for monocrystalline silicon. *(20)* On the other hand, it has been shown that carrier mobility in monocrystalline condensed aromatic hydrocarbons reaches values in the range of 1 to 10 $cm^2V^{-1}s^{-1}$, at room temperature *(25)*. Thus, owing to the ease of realizing highly structured films of organic molecular compounds, one can reasonably hope that organic semiconducting films will be able to reach mobility values in the range of 10^{-1} to 1 $cm^2V^{-1}s^{-1}$, close to those shown by a-Si:H, opening a potentially interesting field of organic-based devices.

DEVICE CHARACTERISTICS.

Electrical characteristics obtained in the literature will be discussed in terms of the two classes of conjugated materials defined previously. The main data obtained with *conjugated polymers and other amorphous materials* are listed in Table I. These semiconductors generally behave as p-type, unless quoted as otherwise. The range of values, given for some compounds, indicates that various experimental attempts have been carried out for improving their electrical properties. The lowest values of conductivity and mobility refers to as-deposited semiconducting films, which indicate that as-prepared semiconductors possess very low mobilities, in the range of 10^{-8} $cm^2V^{-1}s^{-1}$ to 10^{-4} $cm^2V^{-1}s^{-1}$. This is believed to be a result of poor efficiency of charge hopping in highly disordered materials, originating from self-localization and defects. In order to improve the field-effect mobility of these organic semiconductors, their doping level has been intentionally increased, either electrochemically in the case of electropolymerized polythiophene *(6)*, or chemically as in the case of poly(3-alkylthiophene) *(8)*, poly(DOT)3 *(17)*, C_{60} *(24)*, and TCNQ *(14)*. Some other attempts have been based on the "annealing" of the semiconducting film, often realized under oxygen atmosphere *(23)*.

Table I. Electrical characteristics of conjugated polymers and amorphous materials

Material	Deposition Technique	Conductivity (S/cm)	Mobility ($cm^2V^{-1}s^{-1}$)	Ref.
Polyacetylene	Polymer.	10^{-5}	10^{-5}	5
Polyacetylene	Precursor	10^{-5}	10^{-4}	7
Polythiophene	Electropoly.	10^{-6} to 10^{-5}	10^{-6} to 10^{-5}	6
Polyalkylthiophene	Spin coating	10^{-8} to 10^{-5}	10^{-8} to 10^{-5}	8
Poly(DOT)3	Spin coating	10^{-8} to 10^{-5}	10^{-6} to 10^{-3}	17
Polythienylene-vinylene	Precursor	10^{-6} to 10^{-4}	10^{-5} to 10^{-3}	22
Lu, Tm diphthalo-cyanine	Vac. evap.	10^{-4} to 10^{-3}	10^{-4} to 10^{-2}	23
Fullerene (n-type)	Vac. evap.	10^{-8} to 10^{-5}	10^{-5} to 10^{-2}	24
TCNQ (n-type)	Vac. evap.	10^{-10} to 10^{-6}	10^{-10} to 10^{-4}	14

The results listed in Table I confirm that a significant increase of field-effect mobility is indeed obtained by intentional doping by almost two orders of magnitude. This behavior also establishes the existence of a relationship between intentional doping and carrier mobility *(6,8,17,23)*. However, the data on Table I also indicate that simultaneously a very large enhancement is observed in the conductivity, which is increased by a factor of 10 to 10^2. This behavior can be easily understood on the basis of the mechanism of charge transport in these materials. Conductivity, which is given by the relation $\sigma = N_f q \mu$ where N_f is the density of free carriers, is well known to increase with the doping level, according to a $\sigma \sim N^\gamma$ relationship. It follows that a direct relationship between mobility and conductivity, $\mu \sim \sigma^{\gamma/\gamma-1}$, can be expected, which has indeed been experimentally observed (17). On the other hand, the hopping mechanism associated with amorphous organic semiconductors infers a strong dependence of charge transport on the hopping distance R, $\sigma = 2q^2R^2\nu_{ph}N(E_F)\exp(-2\alpha R)(-W/kT)$, where R is the hopping distance, $N(E_F)$ the density of states at the Fermi level, α the electronic wavefunction overlap, and W the energy difference between initial and final electronic states *(26)*. Increase of doping level leads to the increase of $N(E_F)$, and to the decrease of the hopping distance R, which results in the increase of conductivity and also of carrier mobilty.

Increase of the doping level, N, leads to a significant increase of mobility, up to $\mu = 10^{-3}$ $cm^2V^{-1}s^{-1}$, but also leads to a simultaneous increase of the conductivity, up to values of the order of 10^{-3} Scm^{-1}. Large ohmic currents are therefore expected, which increase I_{off} currents and dramatically reduce the dynamic ratio of the corresponding devices. Hence, simultaneous increase of mobility and conductivity invalidates this approach to the improvement of charge transport efficiency in conjugated polymers and disordered materials.

On the other hand, early work on *conjugated oligomers* revealed a significant increase in field-effect mobility, whereas conductivity remained low, suggesting another transport mechanism. These first results stimulated a large interest in these materials since conjugated oligomers offer the unique advantage of high purity, well defined structure, and monodisperse conjugation length, opening the way to the rationalization of structural effects on electrical properties of conjugated materials *(13,15)*. The first results on sexithiophene, 6T, and on pentacene showed a large increase of mobility, up to 2×10^{-3} $cm^2V^{-1}s^{-1}$, accompanied with a relatively low conductivity, 10^{-6} S cm^{-1} *(10)*. Structural characterization by X-ray diffraction (XRD) evidenced that thin films of 6T were highly ordered, made of stacks of planar 6T molecules, which suggested a transport mechanism analogous to that occuring in charge transfer complexes, e.g.,TTF-TCNQ. These complexes are also made of regular stacks of donors and acceptors molecules, with short intermolecular distances, allowing important overlap of the π orbitals of neighbouring molecules. The charge propagates preferentially along the stacking axis of donor (or acceptor) molecules, through the overlapping π orbitals with structural organization playing a major role. Following this model, charge transport efficiency in conjugated oligothiophenes would require the molecules to remain fully planar and parallel to each other, in the closest possible packing, with the longest possible range order, to avoid grain boundaries known to be very efficient traps for charges.

Various routes have been followed for an *à priori* control of the structural organization of molecules in the semiconducting film, considered as a molecular assembly. Physical approaches involve either the modification of the experimental conditions used for film deposition or a post-deposition film treatment. Another route uses a chemical approach which involves engineering semiconductor molecules to induce self-assembly properties. Finally, the ultimate step would be the growth of a single crystal of the organic semiconductor with properties considered as achievable limits for charge transport.

The structural organization of organic materials deposited as thin films on a substrate can be controlled by the *temperature of the substrate* as well as by the *rate of evaporation. (15)*. A detailed study has been carried out on sexithiophene, 6T, which was deposited as 2 to 3 μm thick films on Si substrates and analyzed by XRD in θ-2θ scanning symmetrical reflexion mode by polarized UV-visible spectroscopy, and also by scanning electron microscopy. Experimental results evidenced the polymorphism of 6T. When the substrate is held at 77 K, molecular motions are frozen and the first deposited 6T molecules nucleate and crystallize preferentially with their long axis lying parallel to the substrate plane. At room temperature and low deposition rate, 6T molecules adopt another crystallite orientation with their c axis perpendicular to the substrate plane. At room temperature and high deposition rate, some 6T molecules remain frozen on the substrate, and the 6T film displays the two preceeding populations, with crystallites having their c long axis either parallel or perpendicular to the substrate plane. This corresponds to the kinetically favored or thermodynamically favored orientation of the crystallites, respectively. When deposited at 190 °C, films of 6T exclusively show several orders of meridional 00l reflexions, indicating that a great majority of crystallites are grown with their c axis perpendicular to the substrate plane. When further heating the substrate to 260 °C, the XRD spectrum only reveals low angle 00l reflections, indicating that the 6T films are entirely crystallized, with the crystallite (a,b) face parallel, and c axis perpendicular to the substrate plane, respectively. Polarized UV-visible spectroscopy carried out on films deposited at various substrate temperatures confirmed these preferential organization of 6T molecules *(15)*. Finally, the morphology of these films has been analyzed by the use of scanning electron microscopy *(15)*. When deposited at room temperature, isotropically distributed 50-nm diameter crystallites are observed. When the substrate is held at 190 °C, the crystallites show an elongated shape in a close packing arrangement with larger dimensions reaching 30x200 nm^2. On the last sample realized at 260 °C, the 6T layer shows discontinuous surface, possibly arising from a cellular growth. Long interconnecting lamellae about 50 nm wide are observed, giving rise to a network over the film surface. These results confirm that the control of substrate temperature provides a way to monitor the grain size and shape, together with the homogeneity of structural organization, in order to achieve highly organized molecular layers with low concentration of grain boudaries.

More recently, another method has been described for improving the structural organization involving the micromelting of a film of conjugated material. As-sublimed polycrystalline films of 6T with isotropic grain size of about 50 nm have been submitted to short pulse heating, resulting in a significant increase of the crystallite size (up to microns) *(16)*. In conclusion, these structural characterizations confirm that the control of susbtrate temperature and evaporation rate for film deposition, together with the annealing of the film, allow control of the structural organization of oligothiophene molecules in their solid state.

Chemical engineering of these oligothiophenes represents an interesting alternative route for controlling the molecular organization in the film at a mesoscopic level *(13)*. Oligothiophene molecules have been substituted at their α,ω end positions with alkyl chains which are known to present intermolecular self-recognition properties *(13,21)*. Detailed spectroscopic studies under polarized light, together with XRD spectra, have confirmed that, even when deposited on substrate at room temperature, films of dihexylsexithiophene, DH6T, are highly structured. Numerous high-order 00l reflections are observed in the XRD spectrum, up to the 34th order, in agreement with results obtained on dimethylquaterthiophene, indicating that the crystallites have their long c axis perpendicular to the substrate plane. Furthermore, structural organization at the mesoscopic level, obtained from X-ray pole figures, confirmed the existence of almost one single population of molecules standing up on the substrate plane with their (a,b) face as contact plane. The molecular layer structure realized on a substrate at room temperature must be associated with the stacking properties brought by the terminal alkyl groups which are already known for inducing long range ordering and mesophases. These films can be described by a *liquid crystal like superstructure* imposed by the terminal alkyl groups for the whole molecular assembly as represented in Figure 3.

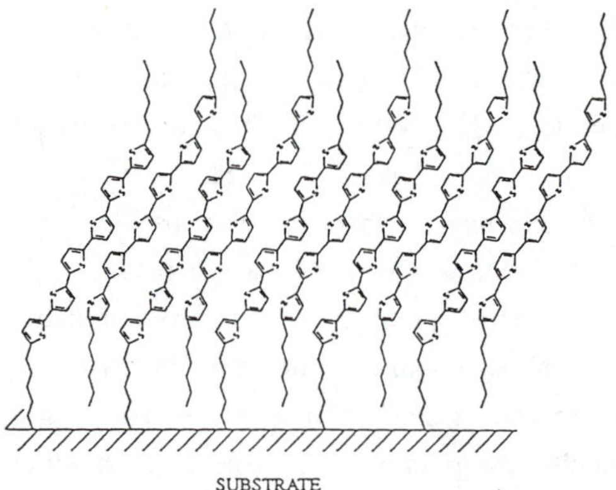

Figure 3. Schematic representation of an α,ω–dihexylsexithiophene monolayer on substrate.

Alkyl-alkyl recognition based on lipophilic-hydrophobic interactions is a strong driving force for a close packing of the conjugated sexithiophene backbones, and more importantly, for long range molecular ordering. Electrical properties of 6T films with various structural organization have been characterized and the results are shown in Table II. The conductivity parallel to the susbtrate surface, $\sigma_{//}$, was measured using a four probe technique and the perpendicular conductivity , $\sigma_{\perp}$, was measured from a sandwich type structure with two gold contacts. Parallel conductivity decreases with increasing temperature and this can be ascribed to the desorption of impurities on the heated substrate. On the other hand, the very large parallel conductivity shown by a 6T film deposited on a substrate maintained at 77 K can be attributed to the pollution of the film during vacuum deposition on the cold substrate.

Table II. Electrical characteristics of sexithiophene (6T) and α,ω–dihexylsexithiophene (DH6) films deposited on substrates at various temperatures *(13, 16)*

Substrate, Temperature	Morphology	Molecular Orientation[a,b]	Conductivity (Scm^{-1}) $\sigma_{\perp}$	$\sigma_{//}$	Mobility ($cm^2V^{-1}s^{-1}$) μ
6T, RT	isotropic grains 50-nm diam.	Para & Perp	$2x10^{-7}$	$1.5x10^{-6}$	$2x10^{-3}$
6T, 190 °C	long grains, $30x200\ nm^2$	Perp	--	$9x10^{-7}$	$9x10^{-3}$
6T, 260 °C	connected long grains, $50x400\ nm^2$	Perp	$4x10^{-9}$	$1.2x10^{-7}$	$2.5x10^{-2}$
DH6T, RT	liquid crystal	Perp	$5x10^{-7}$	$6x10^{-5}$	$8x10^{-2}$

a para = parallel to substrate, b perp = perpendicular to substrate

Likewise, the perpendicular conductivity decreases with increasing substrate temperature, indicating that the conditions used for film deposition allow a further purification of the molecular material which should significantly improve its electrical properties. Importantly, the anisotropy of conductivity increases with substrate temperature in agreement with the proposed model of charge transport along the stacking axis of the 6T molecules, i.e., parallel to the substrate plane. The increase of anistropy of conductivity $\sigma_{//} / \sigma_{\perp}$ is in accord with the increase in molecular ordering perpendicular to the substrate surface as previously shown from structural characterizations.

The field-effect mobility, μ , measured on TFTs realized on glass substrate was calculated in the linear regime of I_d-V_d curves, from equation (1). The lowest value was obtained for a film deposited at room temperature, which has been shown to be the less organized one, possessing two orientations of crystallites. When 6T films are deposited on a heated substrate, the field-effect mobility increases rapidly, up to one order of magnitude, reaching 2.5×10^{-2} $cm^2V^{-1}s^{-1}$. This mobilityt value, recently confirmed on TFT devices with micron-sized channels *(16)* clearly confirms the predominant role played by structural organization of the semiconducting film and also the control which can be exerted by the experimental conditions used for the film deposition.

Compared to unsubstituted oligothiophenes, α,ω–dialkyloligothiophenes, such as α,ω–dihexylsexithiophene DH6T, exhibit an increased parallel conductivity, $\sigma_{//}$, together with a larger anisotropy ratio, which reflects the longer range order and fewer defects existing in layers of these conjugated materials. More significantly, there is a large increase in the field-effect mobility, reaching $\mu = 8 \times 10^{-2}$ $cm^2V^{-1}s^{-1}$. This value which is close to that of a-Si:H gives reasonable hope for real applications of these materials as active layers in thin film transistors. Perhaps the most important conclusion concerns the potential use of the chemical approach for controlling the structural organization of oligothiophene films as molecular assemblies. Molecular engineering of oligothiophenes, realized by end-substitution with alkyl groups, is an elegant and powerful way for inducing self-assembly properties in oligothiophene molecules, leading to liquid crystal-like structured layers.

Single crystals represent the ultimate molecular organization and should set the limit of achievable electrical properties. Our group recently succeeded in growing single crystals of unsubstituted sexithiophene whose size and shape are compatible with the fabrication of field-effect transistors. A complete crystallographic study has been carried out on 6T single crystals *(27)* which showed that the unit cell is monoclinic and contains four molecules closely packed in a herringbone structure as shown in Figure 4.

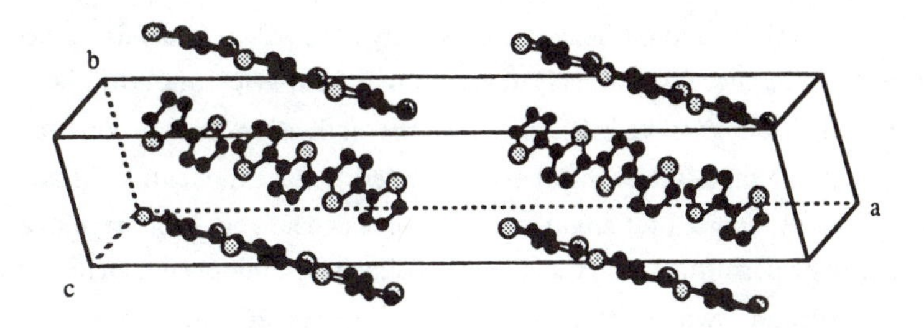

Figure 4. Crystal structure of sexithiophene

6T single crystals appear as small plates with dimensions of about 5x5 mm^2 area and 5 to 10 μm thickness. The long axis of the monoclinic unit cell is perpendicular to the crystal plane which means that the stacking axis of the 6T molecules runs parallel to the large area of the crystal. The most significant feature of the crystal structure of 6T is the complete planarity of the molecules, which is even more planar than observed in terthiophene single crystals. Furthermore, 6T molecules lie strictly parallel one to each other, ensuring a very large overlap of their π molecular orbitals. Electrical characterization of 6T single crystals has shown that the conductivity is very low, with an upper limit of 10^{-9} Scm^{-1}, which implies a very low doping level. Field-effect transistors have been fabricated from 6T single crystals *(27)*. Due to experimental difficulties in realizing a field-effect transistor on such crystals, the measured mobility, $\mu = 7x10^{-2}\ cm^2V^{-1}s^{-1}$, must be considered a largely underestimated value of the field-effect mobility in a single crystal of 6T.

Interestingly, the amplification characteristics of a 6T single crystal-based transistor can be used for the calculation of the dopant concentration by using equation (4). The observed pinch-off voltage of + 30 V corresponds to the completely depleted semiconductor when depletion width W_{sc} reaches the thickness of the semiconducting crystal. The dopant concentration calculated from equation (4), $N = 3x10^{14}\ cm^{-3}$, corresponds to 0.2 ppm, which can be comparated to the dopant concentration measured in non-intentionally doped conjugated materials, at a concentration of about 10^{17}-10^{18} cm^{-3}. The very low concentration is indicative of the high purity of the crystallized material.

CHARGE TRANSPORT IN CONJUGATED OLIGOMERS

Even if pratical applications of organic-based devices in electronics is potentially attractive, one of the most important goal of the electrical characterization of conjugated polymers and oligomers is the analysis and understanding of the charge transport mechanism in these materials. Studies have been performed on temperature effects on conductivity and on field-effect mobility *(11,21,28)*. Analysis carried out on thin films of fullerene, C_{60}, already revealed that field-effect mobility is thermally activated at a fixed gate voltage, and that the activation energy decreases with increasing gate voltage, which has been attributed to an exponential distribution of traps in the gap as reported for amorphous hydrogenated silicon. More recently, similar analysis has been performed on 6T and on α,ω–dihexylsexithiophene, DH6T, from 100 K to 300 K, revealing two temperature dependent regimes *(28)*.. At temperatures above 150 K, the conductivity is thermally activated, with an activation energy of 0.22 eV for DH6T and 0.26 eV for 6T. At lower temperatures, a change of the slope is observed in the Arrhenius plot of conductivity. The field-effect mobility also shows a strong gate-bias dependent activation energy and tends to saturate at both high gate bias and high temperature. These data were analyzed within the framework of a multiple trapping and release model and shown to fit a double exponential distribution. This behavior can be associated with the presence of deep and tail states near the transport level, comparable to the case of a-Si:H. Importantly, the microscopic mobility, μ_o, of both 6T and DH6T are found to be comparable. The lower effective mobility observed in the case of 6T is attributed to its higher density of deep traps. Following the results obtained in the structural analysis of conjugated oligomers, these traps can be associated with grain boudaries which have been shown in higher concentration for 6T films. When temperature decreases, the probability of thermal release from localized traps diminishes. A transition from trapping to thermally activated hopping has been proposed as the dominant transport mechanism which is in agreement with the significant change of slope of the Arrhenius plot. Furthermore, a back transition from hopping to trapping can be observed at low temperature when increasing the gate voltage. Under increasing gate bias, traps become filled with injected charges, and, all deep traps being filled, charge transport switches back to a multiple trapping and release mechanism. This is in agreement with the fast increase of saturation current with gate voltage. Eventually, all traps can become filled which means that any additional charge will then move freely with the microscopic mobility.

The proposed multiple trapping and release mechanism appears to bring a satisfying picture of the charge transport process in conjugated oligomers, indicating the critical role played by traps (grain boundaries, chemical impurities). This model accounts for the large increase in field-effect mobility observed when 6T is deposited on heated substrate and also when using liquid crystal-like structured DH6T. Thus, in the case of conjugated oligomers, the proposed trapping mechanism better elucidates how field-effect mobility of thin films can be increased to almost its upper limit corresponding to a single crystal. It also predicts that large I_{on}/I_{off} ratio for the drain current can be achieved through the decrease of dopant concentration.

CONCLUSION

Thin film transistors have been fabricated from various organic semiconductors including conjugated polymers and oligomers. Two categories of behavior can be differentiated. First, in most of the conjugated polymers and in a large number of "amorphous" molecular materials, conduction is governed by a hopping mechanism. Structural disorder and grain boundaries together with large density of chemical impurities impose a very low efficiency for charge transport, with carrier mobilities, μ, of about 10^{-5} $cm^2V^{-1}s^{-1}$. In agreement with the hopping mechanism, field-effect mobility depends on the doping level N which can be experimentally increased. However, conductivity also increases simultaneously upon doping and TFTs made with these materials present an inherently very poor I_{on}/I_{off} dynamic ratio. The second category corresponds to molecular materials such as short conjugated oligomers. In these materials, charge transport obeys a multiple thermal trapping mechanism and is hence only dependent on the density of traps, whereas conductivity depends on the doping level. Highly ordered and very pure materials allow achievement of both a high mobility, of the order of 10^{-1} $cm^2V^{-1}s^{-1}$, and also a low conductivity, of the order of 10^{-7} Scm^{-1}, which meet the requirements for efficient TFT devices. Highly crystalline films can be realized by adjusting the experimental conditions for film deposition. Long range molecular ordering in semiconducting films can also be easily achieved by an elegant chemical route involving the substitution at both ends of the conjugated molecule, of alkyl groups, which bring self-assembly properties to these molecules.

Acknowldgements

G. Horowitz, A. Yassar , R. Hajlaoui, and D. Fichou are acknowleged for their contribution to this work.

Literature Cited

1. Pope, M. ; Swenberg, C.E. *Electronic Processes in Organic Crystals*; Clarendon Press: Oxford, **1982.**
2. Tourillon, G.; Garnier, F. *J. Electroanal. Chem.* **1982,** *135,* 173.
3. MacDiarmid, A.G.; Chiang, J.C.; Mu, S.L.; Somasini, N.L.D.; Wu, W. *Mol. Cryst. Liq. Cryst.* **1985,** *121,* 187.
4. Fichou, D. ; Horowitz, G.; Nishikitani, Y.; Garnier, F. *Chemtronics* **1988,** *3,* 176.
5. Ebisawa, F. ; Kurosawa, T.; Nara, S. *J. Appl. Phys.* **1983,** *54,* 3255.
6. Tsumura, A.; Koezuka, H.; Ando, Y. *Synth. Met.* **1988,** *25,* 11.
7. Burroughes, J.H.; Jones, C.A.; Friend, R.H. *Nature* **1988,** *335,* 137.

8. Assadi, A.; Svensson, C.; Wilander, M.; Inganäs, O. *Appl. Phys. Lett.* **1988,** *53,* 195.

9. Horowitz, G.; Fichou, D.; Peng, X.Z.; Xu, Z.; Garnier, F. *Solid State Comm.* **1989,** *72,* 381.

10. Horowitz, G.; Peng, X.Z.; Fichou, D.; Garnier, F. *Synth. Met.* **1992,** *51,* 419.

11 Paloheimo, J.; Kuivaleinen, P.; Stubb, H.; Vuorimaa, E.; Yli-Lahti, P. *Appl. Phys. Lett.* **1990,** *56,* 1157.

12. Akamichi, H.; Waragai, K.; Hotta, S.; Kano, H.; Sakati, H. *Appl. Phys. Lett.*, **1991,** *58,* 1500.

13. Garnier, F.; Yassar, Y.; Hajlaoui, R.; Horowit, G.; Deloffre, F.; Servet, B.; Ries, S.; Alnot, P. *J. Amer. Chem. Soc.* **1993,** *115,* 8716.

14. Brown, A.R.; Deleeuw, D.M.; Lous, E.J.; Havinga, E.E. *Synth. Met.* **1994,** *66,* 257.

15. Servet, B.; Horowitz, G.; Ries, S.; Lagorse, O.; Alnot, P.; Yassar, A.; Deloffre, F.; Srivastava, P.; Hajlaoui, R.; Lang, P.; Garnier, F. *Chem. Mater.* ***1994,*** *6,* 1809.

16. Dodabalapur, A.; Torsi, L.; Katz, H.E. *Science* **1995,** *68,* 270.

17. Brown, A.R.; Deleeuw, D.M.; Havinga, E.E.; Pomp, A. *Synth. Met.* **1994,** *68,* 65.

18. Garnier, F.; Horowitz, G.; Peng, X.Z.; Fichou, D. *Adv. Mater.* **1990,** *2,* 592.

19. Garnier, F.; Hajlaoui, R.; Yassar, A.; Srivastava, P. *Science* **1994,** *265,* 1684.

20. Sze, S.M. *Physics of Semiconductor Devices*; John Wiley: New York, **1981**.

21. Waragai, K.; Akimichi, H.; Hotta, S.; Kano, H.; Sakaki, H. *Synth. Met.* ***1993,*** *55-57,* 4053.

22. Fuchigami, H.; Tsumura, A.; Koezuka, H. *Appl. Phys. Lett.* **1993,** *63,* 1372.

23. Guillaud, G.; Al Sadoun, M.; Maitrot, M.; Simon, J.; Bouvet, M. *Chem. Phys. Lett.* **1990,** *167,* 503.

24. Hoshimono, K.; Fujimori, S. ; Fujita, S. *Jpn. J. Appl. Phys., Part 2* **1993,** *32,* L1070.

25. Karl, N. *Mol. Cryst. Liq. Cryst.* **1989,** *171,* 157.

26. Mott, N.F.; Davis, E.A. *Electronic Processes in Non-Crystalline Materials*, 2nd Ed., Oxford: Clarendon Press, **1979**.

27. Horowitz, G.; Bachet, B.; Yassar, A.; Lang, P.; Deloffre; F., Fave, J.L; Garnier, F. *Chem. Mater.* **1995,** *7,* 1337.

28. Horowitz, G.; Hajlaoui, R. ; Delannoy, P. *J. Phys. III France* **1995,** *5,* 355.

Chapter 29

Self-Assembled Heterostructures of Electroactive Polymers: New Opportunities for Thin-Film Devices

M. Ferreira, O. Onitsuka, W. B. Stockton, and M. F. Rubner[1]

Department of Materials Science and Engineering, Massachusetts Institute of Technology, Cambridge, MA 02139

A process involving the alternate spontaneous adsorption of oppositely charged polymers onto substrate surfaces has been utilized to fabricate a number of new thin film multilayer heterostructures with electrical and optical properties that can be tuned at the molecular level. Examples of what can be accomplished with this new approach include the fabrication of thin film light emitting devices based on multilayer heterostructures of poly(p-phenylene vinylene) (PPV) and various polyanions and the fabrication of ultra-thin, electrically conductive multilayers of polyaniline. In the former case, light emitting devices with high brightness (>100 cd/m^2 in the range of 8-10 volts) and tunable emission wavelengths have been created through the use of multi-bilayer "slab" systems that are used to control the charge injection and transport characteristics of the device. It was also demonstrated that the presence of very thin (about 20 Å thick) insulating layers at the Al/polymer interface improves device efficiency by a factor of 2 - 4. In the latter case, the layer-by-layer processing of polyaniline was accomplished by using hydrogen bonding interactions. This represents the first time that such secondary forces have been used to fabricate multilayer structures of polyaniline.

The ability to control molecular and supramolecular structure at the nanoscale level is now recognized as one of the most promising means of exploiting the incredibly diverse electrical and optical properties of conjugated polymers and related electroactive organic materials. This is particularly true when one considers the use of these materials in thin film heterostructure devices where molecular-level control over layer thickness, interfaces (both internal and electrode interfaces) and thin film architecture is essential to realize the full potential of these systems. Recently, a very simple and yet powerful technique for manipulating charged polymers into multilayer thin films has been described (*1,2*). This approach, which involves the alternate deposition of oppositely charged polymers from dilute aqueous solutions, makes it possible to process conjugated polymers in a layer-by-layer manner with nanometer level control over the thickness of the individual layers. To date, our group has demonstrated that this process can be used to manipulate a wide variety of materials into multilayer thin films including, conjugated polyions (*3,4*), electrically conducting polymers (*5,6*), light

[1]Corresponding author

emitting polymers (*7-10*), derivatized fullerenes (*7,8*), precursor polymers (*11*), and molecular dyes (*12*). This paper reviews some of our more recent developments in this area.

Experimental Section

Poly(p-phenylene vinylene) (PPV) thin films were fabricated via the molecular self-assembly adsorption technique using an approximately $2x10^{-5}$ M solution of the tetrahydrothiophenium precursor polymer at a pH of around 4.5. The synthesis of this material is described elsewhere (*13*). Since the precursor polymer is positively charged, multilayer fabrication was carried out by using a suitable negatively charged material. The polyanions used in this work were prepared as follows. The poly(methacrylic acid) (PMA) (Polyscience, 15,000 g/mole) was diluted from its original solution (30% polymer in water) to a concentration of $1x10^{-2}$ M. The pH was adjusted to 3.5 by adding HCl. The poly(styrene-4-sulfonate) (SPS) (Aldrich, 70,000 g/mole) was dissolved in Millipore water to a concentration of $1x10^{-2}$ M ($1x10^{-3}$ M for SPS/PAH spacer studies) and the pH was adjusted to 4.0 with HCl. The ionic strength for both polyanion solutions was adjusted to a level of 0.1M by adding NaCl. For the top layer study, the PMA solution was diluted to $5x10^{-3}$ M and no ionic strength adjustment was performed. The polyallylamine hydrochloride (PAH) (Aldrich, 50,000 g/mole) solution was prepared by dissolving the polymer in Millipore water to a concentration of $5x10^{-3}$ M and the pH was adjusted to 3.5 by adding HCl. All solutions, except the PPV precursor, were filtered through fine paper filter (1-5 μm pore size) prior use.

Commercially available microscope glass slides were used as substrates for optical characterization of the PPV thin films. For the light emitting devices, glass substrates patterned with 2 mm wide and about 2000 Å thick lines of indium tin oxide (ITO) were used as transparent electrodes. Evaporated aluminum was used as the top electrode material in all devices.

The layer-by-layer adsorption was carried out by using an automatic programmable slide stainer (HMS programmable slide stainer from Zeiss Inc.). This unit can dip substrates into as many as 21 different solutions, providing a means to fabricate simple as well as complex multilayer structures. The PPV films were built-up by dipping the substrates first in a bin containing the PPV precursor solution for 10 minutes. Subsequently, the substrates were moved to a water bin for 1 minute, to a different water bin for another minute and finally to a flow through water bath for 5 minutes. The polyanion layers were deposited by following the same procedure. The desired number of bilayers was achieved by repeating this sequence as many times as necessary.

The PPV precursor polymer was converted to the fully conjugated form by thermal treatment. The films were dried under dynamic vacuum overnight and then heated up to 210°C. This temperature was maintained for 10 hours and the samples were left to cool down for a period of about 10 more hours. The whole procedure was carried out under vacuum. Details concerning the layer-by-layer processing of polyaniline via hydrogen bond interactions can be found in a previous publication (*14*).

Results and Discussion

Self-Assembly Process. The ability to readily manipulate a wide variety of electroactive polymers at the nanoscale level makes it possible to create thin film heterostructures with optical and electrical properties that can be fine-tuned at the molecular level. As indicated earlier, we have demonstrated that this process can be utilized with many different conjugated, nonconjugated and electrically conductive polymers. Any variety of complex multilayer heterostructures can be fabricated with these materials as long as the deposition process is always alternated between a polycation and a polyanion. Thus, one can fabricate complex multilayer heterostructures in which the sequence and types of layers deposited are controlled via a simple computer controlled dipping apparatus. The typical dimensions of these adsorbed layers are in the range of 5-50 Å, with the level of molecular interpenetration occurring between adjacent layers being dependent, among other things, upon the thickness of the individual layers and the relative stiffness of the polymer chains. Also note that it is possible to controllably manipulate polymers with a diverse collection of electrical and optical properties including high electrical conductivity (polypyrrole: 500 S/cm (*6*)) and efficient electroluminescence (poly(p-phenylene vinylene).

As indicated above, the layer-by-layer self-assembly of adsorbed polymer layers is accomplish via the use of electrostatic attractions. We have found, however, that with suitable polymer pairs, it is also possible to utilize other secondary forces to fabricate these novel heterostructures such as hydrogen bonding interactions. This is illustrated in Figure 1 which shows the layer-by-layer self-assembly of polyaniline with various hydrogen bonding polymers (*14*). The linear relationship observed between the absorbance of the polyaniline exciton band at 630 nm and the number of bilayers deposited clearly shows that the deposition process is reproducible from layer to layer. We have found that with polyaniline, the amount of conjugated polymer adsorbed per layer (as indicated by the slopes of the curves in Figure 1) is actually greater when hydrogen bonding interactions are used to drive the self-assembly process as opposed to electrostatic interactions. Note, for example, that the slope of the Pani/PVP bilayer combination is significantly higher than that obtained from the Pani/SPS bilayer combination. This reflects the strong tendency of the amine and imine sites of the polyaniline backbone to engage in hydrogen bond formation (*15*). The layer-by-layer self-assembly of polyaniline via hydrogen bonding interactions can be carried out with water soluble polyethers (PEO), polyalcohols (PVA) and cyclic polyamides (PVP). With suitable variations in molecular weight and solution conditions (pH, concentration etc.) the thicknesses of the bilayers deposited via this process can be adjusted from 30-125 Å. The resultant multilayer thin films can be doped with strong acids to achieve conductivities as high as 5 S/cm.

The ability to create ultrathin films of electrically conductive polymers such as polyaniline and polypyrrole makes it possible to utilize these materials as semi-transparent anodes in flexible light emitting devices based on organic electroluminescent materials. In collaboration with workers at the Naval Research Laboratory (NRL), for example, we have recently demonstrated that flexible light emitting devices can be fabricated from sheets of poly(ethylene terephthalate) microphoto-lithographically patterned with electrically conductive, semi-transparent electrodes of polypyrrole (Shashidhar, R.; Rubner, M. F., results to be published). The molecular level control exerted over electrode thickness and the nature of the electrode interface was critical to obtaining working devices.

Self-Assembled Light Emitting Devices based on PPV. In order to make efficient light emitting devices based on conjugated polymers, it is necessary to control carrier injection at the electrodes, the transport of holes and electrons within the device and the nature and spatial location of the carrier recombination zone. We have been utilizing the self-assembly technique to control these critical parameters at the molecular level. Such an approach not only leads to more efficient devices but also provides fundamental information about the mechanisms of charge injection and transport operating in these devices. Using self-assembled films fabricated with a cationic PPV precursor and different polyanions, for example, we have demonstrated that it is possible to dramatically manipulate device characteristics, including the color of light emitted and device efficiency, by simply varying the type of polyanion and sequence of PPV/polyanion bilayers used to self-assemble the PPV films (*9,10*).

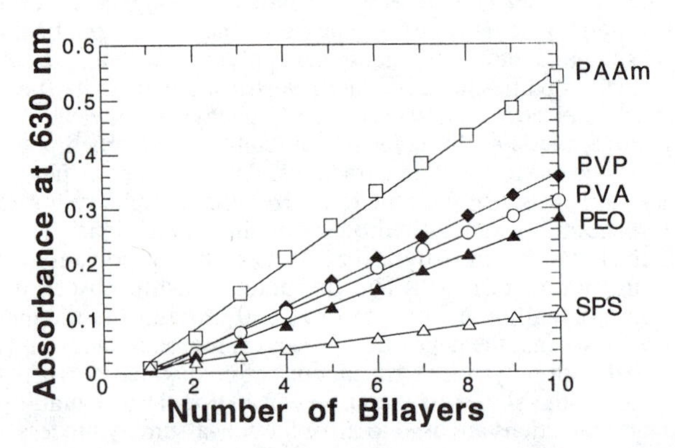

Figure 1. Absorbance at 630 nm versus number of deposited bilayers for polyaniline multilayer films assembled with six different polymers: polyacrylamide (PAAm), polyvinyl pyrrolidone (PVP), polyvinyl alcohol (PVA), polyethylene oxide (PEO) and polystyrene sulfonic acid (SPS).
(Reproduced with permission from reference 14. Copyright 1994 Materials Research Society.)

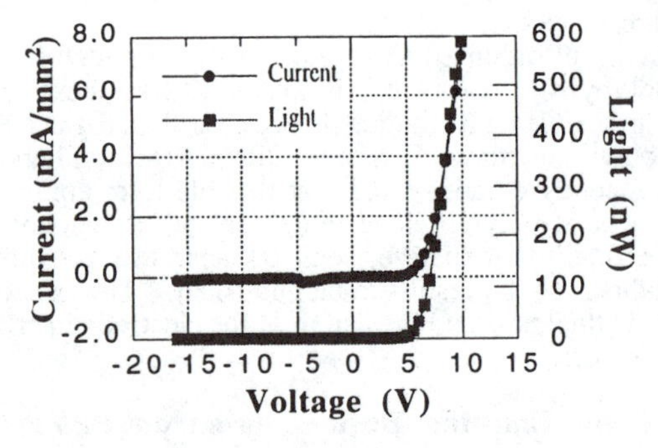

Figure 2. Current-Voltage and Light-Voltage plot for an ITO/$(PPV/SPS)_5$/$(PPV/PMA)_{15}$/Al heterostructure device.
(Reproduced with permission from reference 15. Copyright 1996 Materials Research Society.)

Figure 2 shows the device characteristics of a PPV-based multilayer heterostructure fabricated with 5 bilayers of PPV/SPS (about 40 Å thick) and 15 bilayers of PPV/PMA (about 240 Å thick). The final device has the following architecture: $ITO/(PPV/SPS)_5/(PPV/PMA)_{15}/Al$. A device of this type emits green light and exhibits luminance levels as high as 100 cd/m^2 (comparable to the brightness of a computer screen) and rectifying ratios as large as 10^6. This represents much better than typically observed performance for a PPV device utilizing a relatively low work function metal (aluminum) as the electron injecting electrode. The improved performance realized with this system is a direct consequence of the multilayer heterostructure created via the self-assembly process and the different PPV molecular environments established via the use functionally different polyanions.

In short, the PPV/SPS bilayers placed at the anode interface act to promote efficient hole injection into the light emitting PPV/PMA bilayers due to the fact that the sulfonic acid groups of the SPS molecules partially "acid dope" their surrounding PPV chains. The net result is a device that is significantly more efficient than a simple device fabricated from a single PPV bilayer combination (ie., either PPV/PMA or PPV/SPS) or a spin cast PPV film. Amazingly, improved device performance is observed even after only one PPV/SPS bilayer (about 8Å) is placed between the ITO electrode and the PPV/PMA bilayers. This demonstrates the advantages of being able to manipulate polymer organization at the molecular level.

The performance of this device can be further improved by modifying the nature of the aluminum/emitter interface. Since a molecular self-assembly approach is being used to fabricate these PPV based LEDs, it is possible to systematically control, at the molecular level, the type of bilayer system placed at the cathode interface and its thickness. For example, at the aluminum interface, it is possible to end the film with either a top layer of PPV or a PMA top layer. Previous studies (*9,10*) have shown that although the addition of one single layer means a difference of only 5 to 10 Å to the total film thickness, films having PMA as the outermost layer always yield devices 2 to 4 times more efficient than those of similar thickness and architecture having PPV as the layer in contact with the aluminum electrode. This is a very interesting observation considering that PMA behaves as an insulating material and is therefore not expected to improve carrier injection into the device. In order to understand the origin of this interesting effect, we have recently fabricated $ITO/(PPV/SPS)_{10}/(PPV/PMA)_{20}/Al$ heterostructure devices with extra insulating layers of PMA/PAH at the top of the heterostructure (*9*). Each PMA/PAH bilayer, in this case, contributes about 16 Å of insulating material to the total multilayer thickness. The relative efficiency of these devices as measured by a comparison of light versus current plots is shown in Figure 3.

The results obtained from this set of devices is quite interesting. We observe that the relative efficiency (as determined by the slopes of the L-I plots) initially increases when extra insulating layers are added to the top of the film (by about a factor of 3 when compared to a device with a PPV top layer) and then decreases dramatically once the total thickness of the insulating layers exceeds about 25-30 Å. Thus, the addition of very thin insulating layers significantly improves device efficiency until a critical thickness is reached after which electron injection from the Al electrode becomes hampered by the presence of what appears to be a very good insulating barrier. Studies are currently underway to determine the origin of this effect. Intriguing possibilities include a controlled decrease in the amount of exciton/cathode quenching taking place in the device and/or improved electron injection efficiency due to the presence of an effective insulating tunneling barrier at the cathode interface. Very recently, a similar effect has been observed in MEH-PPV films overcoated with thin insulating layers of poly(methyl methacrylate) (*16*). In this work, the thin insulating layers were deposited in a layer-by-layer manner by using the Langmuir-Blodgett technique. The authors of this work have attributed the observed device enhancement to a lowering of the effective barrier height for electron injection coupled with an increase in the effective barrier to hole injection.

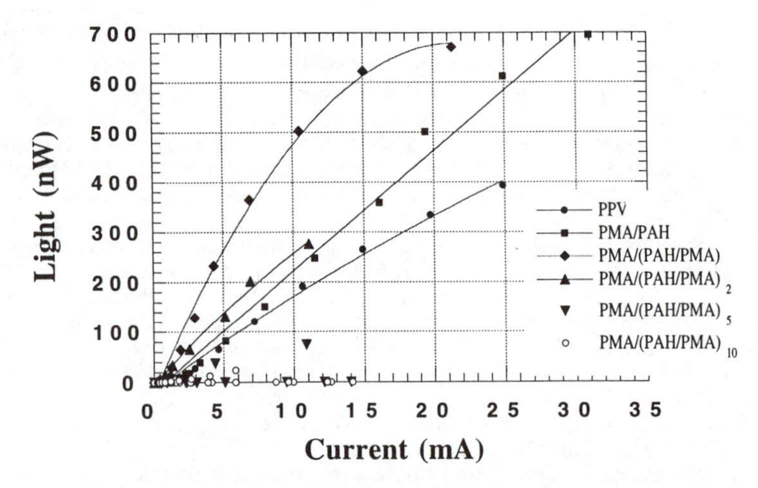

Figure 3. Light-Current plots for ITO/$(PPV/SPS)_{10}$/$(PPV/PMA)_{20}$/Al devices containing a different number of insulating PMA/PAH top layers. Legend shows outermost layers.
(Reproduced with permission from reference 15. Copyright 1996 Materials Research Society.)

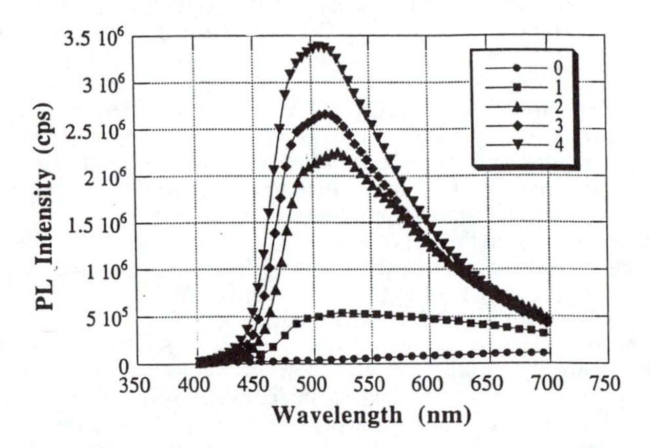

Figure 4. Photoluminescence spectra (excitation at 380 nm) of PPV/PTAA heterostructures with different numbers of SPS/PAH spacer bilayers placed between the PPV and PTAA (legend indicates the number of SPS/PAH spacer bilayers).

Implicit in the above discussion about molecular level control over device performance, was the assumption that the molecular architecture (sequence and dimensions of layers) created by the self-assembly process is preserved during the thermal conversion process used to create the light emitting form of PPV. Since the self-assembly process is carried out with a PPV precursor, it is necessary to thermally treat the final multilayer thin film to make the conjugated light emitting form of PPV. For the above films, this process was carried out at 210°C for about 11 hours. An important question to address is therefore what effect this thermal treatment has on the layered architecture established by the deposition process. At high temperatures, it is possible that some level of interlayer diffusion can occur thereby scrambling up the well defined architecture created by the layer-by-layer deposition process. To address this important issue, we have utilized energy transfer schemes to probe the integrity of multilayer structures containing PPV layers.

We have previously shown that when PPV is self-assembled with specific electronically active polyanions such as poly(thiophene acetic acid) (PTAA) or sulfonated fullerenes (S-C_{60})(*7*), the photoluminescence of the PPV is essentially completely quenched by the polyanion. The mechanism of this quenching is believed to be due to a photoinduced electron transfer process taking place between the excited PPV and the adjacent electroactive polyanion molecules. The quenching process, in this case, is not associated with a Forster type energy transfer since in both cases, the required spectral overlap of a donor emission band with an acceptor absorption band is not fulfilled. In addition, photo-induced electron transfer processes have previously been confirmed in PPV/C_{60} systems and can be exploited to fabricate thin film photovoltaic devices (*17*). In order to mediate this electron transfer process, we have constructed multilayer heterostructures in which the PPV donor and the polyanion electron acceptor are separated from each other with electronically inert spacer layers of known thickness. In addition to allowing studies of the electron transfer process, such structures provide important insights into the thermal stability of the multilayer structure. The "spacers" used in this study were bilayers of SPS/PAH with an experimentally determined bilayer thickness of 30 +/-5 Å.

Figure 4 shows the photoluminescence spectra of ...**PPV**/$(SPS/PAH)_x$/**PTAA**/$(SPS/PAH)_x$... heterostructures with different numbers of SPS/PAH bilayers between the PPV and PTAA layers. When x=0, that is no spacer layers exist between the PPV and the PTAA, it can be seen that the characteristic emission band of PPV is completely quenched by the adjacent PTAA molecules. However, with only one 30 Å spacer bilayer, the PPV luminescence begins to reemerge and is reestablished to its normal SPS/PPV multilayer emission intensity (in the range of 2-4 x 10^6 cps) after only 2 bilayers have been inserted. Thus, the heterostructures with 2, 3 and 4 bilayer spacers display photoluminescence intensities comparable to what is found in multilayers of PPV/SPS. The relatively small increase that occurs from 2 to 4 spacer bilayers is within the range of fluctuations normally observed for this system and most likely represents slight variations in PPV conversion efficiency in these different heterostructures. Similar results were obtained when sulfonated C_{60} was used as the electron acceptor, although with this system, because of the stronger electron acceptor character of fullerenes and possible higher levels of interpenetration, it was necessary to insert 5 spacer bilayers to reestablish the expected luminescence intensity of the PPV.

Given the expected level of interpenetration of adjacent layers of polycations and polyanions (about 5-10 Å), and the expected interaction radius of this electron transfer process (about 20-30 Å), these results suggests that the original multilayer architecture established by the layer-by-layer deposition process remains intact after thermal treatment (an overall thickness decreases does occur as a result of the thermal elimination process needed to create PPV). More recent studies using molecules that can participate in Forster energy transfer effects with PPV have also demonstrated that the multilayer organization is not disturbed to any large extent by the thermal treatment needed to create the light emitting form of PPV (Baur, J.; Rubner, M. F., results to be published). In short, these various energy and electron transfer schemes all suggest that large scale molecular interdiffusion is not taking place at the temperatures typically used to thermally treat these films.

Conclusions

The use of layer-by-layer deposition techniques to process electroactive polymers into thin films looks to be an extremely promising way to control optical and electrical properties at the molecular level. With this kind of control, it has already been demonstrated that it is possible to improve the efficiency of light emitting devices based on conjugated polymers and to create ultrathin, highly conductive thin film electrodes based on conducting polymers. It is anticipated that the complex multilayer heterostructures that are now possible via the use of this technique will also provide new opportunities for creating and exploring molecularly-based optical and electrical effects in thin film devices.

Acknowledgments

This work was partially supported by the National Science Foundation and by the MRSEC program of the National Science Foundation under award number DMR-9400334. The support of M. F. by CNPq of Brazil and O. O. by TDK Japan is also acknowledged.

Literature Cited

(1) Decher, G.; Hong, J.D.; Schmitt, J. *Thin Solid Films* **1992**, *210/211*, 831.
(2) Lvov, Y.; Decher, G.; Möhwald, H. *Langmuir* **1993**, 9, 520.
(3) Ferreira, M.; Cheung, J.H.; Rubner, M.F. *Thin Solid Films* **1994**, *244*, 806.
(4) Ferreira, M.; Rubner, M. F. *Macromolecules* **1995**, *28*, 7107.
(5) Cheung, J.H.; Fou, A.C.; Rubner, M.F. *Thin Solid Films* **1994**, *224*, 985.
(6) Fou, A. C.; Rubner, M. F. *Macromolecules* **1995**, *28*, 7115.
(7) Ferreira, M.; Rubner, M.F.; Hsieh, B.R. *Mat. Res. Soc. Symp. Proc.* **1994**, *328*, 119.
(8) Fou, A. C.; Onitsuka, O.; Ferreira, M.; Rubner, M. F.; Hsieh, B. *Mat. Res. Soc. Symp. Proc.* **1995**, *369*, 575.
(9) Ferreira, M.; Onitsuka, O.; Fou, A. C.; Hsieh, B.; Rubner, M. F. *Mat. Res. Soc. Symp. Proc.* **1996**, *413*, 49.
(10) Fou, A. C.; Onitsuka, O.; Ferreira, M.; Rubner, M. F.; Hsieh, B. *J. Appl. Phys.* **1996**, *79*, 7501.
(11) Baur, J. W.; Besson, P.; O'Connor, S. A.; Rubner, M. F. *Mat. Res. Soc. Symp. Proc.* **1996**, *413*, 583.
(12) Yoo, D.; Lee, J.; Rubner, M. F. *Mat. Res. Soc. Symp. Proc.* **1996**, *413*, 395.
(13) Hsieh, B.R. *Polym. Bull.* **1991,** *25*, 177.
(14) Stockton, W. B.; Rubner, M. F. *Mat. Res. Soc. Symp. Proc.* **1994**, *369*, 587.
(15) Angelopoulus, M.; Liao, Y.; Furman, B.; Graham, T. *Mat. Res. Soc. Symp. Proc.* **1996**, *413*, 637.
(16) Kim, Y-E.; Park, H.; Kim, J-J. *Appl. Phys. Letters* **1996**, *69*, 29.
(17) Sariciftci, N. S; Smilowitz, L.; Heeger, A. J.; Wudl, F. *Science* **1992**, *258*, 1474.

Chapter 30

Molecular Multilayer Films: The Quest for Order, Orientation, and Optical Properties

Gero Decher

Centre de Recherches sur les Macromolécules (UPR 022), Institut Charles Sadron, Université Louis Pasteur and Centre National de la Recherche Scientifique, 6 rue Boussingault, F–67083 Strasbourg Cedex, France

Organic thin film coatings of surfaces show potential for various applications including integrated optics. However, the unique characteristics of organic chromophores are exploited best if the films possess suitable architecture (layering), in-plane order and a fixed orientation of molecules with respect to the substrate. Good control over defects is required in order to avoid light scattering and to allow for optical waveguiding. This report describes three techniques which are used for the fabrication of molecular multilayer films and compares their properties: Langmuir-Blodgett films, transferred freely-suspended films, and layer-by-layer adsorbed polyelectrolyte films.

Ultrathin films composed of polymers or small organic molecules have attracted considerable attention in the last decades (*1-5*). The possibility of tailoring their architecture and thus their optical properties has prompted research towards potential applications in the areas of integrated optics, frequency doubling, light emitting devices or even sensors based partially on optical effects. After everything started out with Langmuir-Blodgett films in the 1930s (*6,7*), different approaches have been taken to fabricate molecular multilayer structures as films on solid supports (*8-17*). However, for potential applications several prerequisites have to be met and this is still a tremendous challenge for any such system. The major demands are clearly the optical quality of the film and an architecture with appropriate optical properties, but last not least also thermal and mechanical stability and the possibility for automated manufacturing. Just the first part of these tasks, expressed in other words, could be the quest for order, orientation and optical properties.

Techniques for the fabrication of molecular films. The existing techniques for multilayer fabrication address these requirements to a different extent. Langmuir-Blodgett films allow for excellent control of film architecture but typically lack stability. Self-assembled multilayers are extremely stable since they generally represent covalently crosslinked networks, but it seems to be difficult to obtain μ-thick films of high quality. Freely-suspended liquid crystalline films can be obtained as optical monodomains, but are extremely fragile. They can be transferred onto solid supports or, if prepared from polymeric liquid crystals, covalently cross-linked and thus tremendously stabilized, but the maximum areas of such films are restricted to a few cm^2 or even mm^2.

A relatively new approach is the consecutive adsorption of oppositely charged polymers or colloids onto solid supports, which yields films that possess internal structure on the nm scale and these films can also be homogenous over large areas. The control over the layer architecture is as straightforward as in the case of LB-films, but there seems to be limits with respect to control of molecular order and orientation. On the other hand, the process can easily be automated since it simply consists of multiple adsorption from solution. The resulting films have in some cases been observed to be stable for seven days at temperatures around 200 °C and at least for more than one year at ambient conditions.

Another advantage of the latter type of films is that the materials that can be incorporated into such films are not restricted to synthetic molecules, biopolymers such as proteins (*18-26*), DNA (*19,27* and Sukhorukov, G. B.; Möhwald, H.; Decher, G.; Lvov, Y. M. *Thin Solid Films* 1996, in the press) or inorganic colloids (*20,28-33* and Schmitt, J.; Decher, G.; Dressik, W. J.; Brandow, S. L.; Geer, R. E.; Shashidhar, R.; Calvert, J. M. *Adv. Mater.*, accepted for publication) have been employed for the fabrication of multilayer nanoheterostructures as well. It is even possible to carry out chemical reactions (thermal eliminations) in such films, thus allowing for the manipulation of chemical structures by using precursor materials for the film fabrication. The last possibility has prompted the preparation of polyelectrolyte multilayers incorporating precursors to electroluminescent polymers (*34-38*) which are typically intractable substances (*39*). For such applications the technique offers the promising possibility to construct film architectures composed of an electron injecting layer, an electroluminescent layer and a hole injecting layer with a total thickness of a few hundred Ångstroms or less and the first steps in this direction have already been taken.

The following paragraphs describe selected examples of Langmuir-Blodgett films, freely-suspended liquid crystal films, transferred freely-suspended films and layer-by-layer adsorbed films with respect to some of the desired properties described above.

Langmuir-Blodgett Films: Thermal Relaxation of Non-Centrosymmetric Structure, a Typical Case for Non-Polymeric Amphiphiles?

Whereas Langmuir-Blodgett (LB) films, which have dominated the area for over 50 years, allow for excellent initial control over the film architecture and partial control over molecular orientations, they typically lack thermal stability and undergo different structural relaxations with time (*3,40-42*). This is not unexpected since they are prepared as monomolecular layers on the surface of water and then transferred onto solid supports (Figure 1). Thus the environment of the molecules changes dramatically during deposition and the transfer is typically too fast to allow for complete equilibration. Furthermore, post-transfer structural changes cannot be compensated by 2-dimensional mass transport and thus lead always either to hole or island formation in the cases of contraction or expansion. As a result, LB-films often contain a considerable amount of trapped defects (*43*). Since the structure of the molecular crystal is typically the thermodynamically most stable one, there is a high tendency for multilayers to "recrystallize" to the bulk structure. Nevertheless, there exist cases where even non-centrosymmetric multilayers with a thickness in the μ-range can be prepare by the LB-Technique and such films even appear to be stable over a few years.

An interesting example for second harmonic generation is the case of a non-centrosymmetric LB-film of a single amphiphile (2-docosylamino-5-nitropyridine, DCANP) which possesses a non-alternating head-to-head and tail-to-tail structure (Y-type) and a polar axis in the layer plane parallel to the dipping direction (*44,45*). DCANP is also an interesting material, because it allows the detection of monolayer collapse due to its unique spectroscopic properties, even when the pressure/area isotherms indicate that the monolayer seems to be stable at a certain surface pressure (*46*). It is clear that control of monolayer stability is a prerequisite for the fabrication of high quality films.

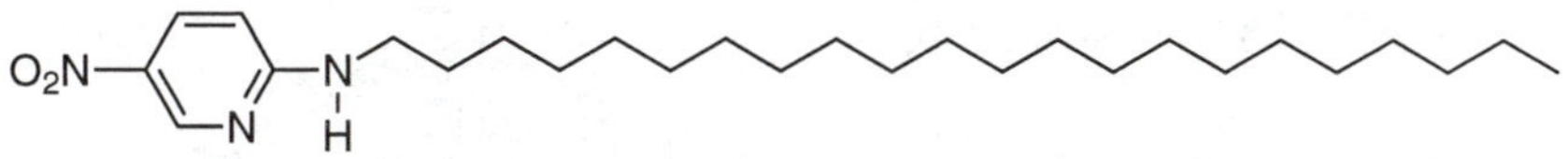

O_2N N N H

2-docosylamino-5-nitropyridine (DCANP)

The in-plane polarity arises from the combination of two effects: a) strongly tilted alkyl chains allow for a chromophore orientation within the layer plain and b) the crystalline domains are aligned preferentially along the dipping direction during transfer (*47*). The nonlinear optical coefficient d_{33} of LB-multilayers of DCANP is 7.8 ± 1.0 pm/V (at 1064 nm). In Cerenkov-type waveguiding setup it is possible to achieve phase matching conditions and values of d_{33} of 27 pm/V were reported for wavelengths of 820 nm (*48,49*). Although it is possible to fabricate transparent films thick enough for optical waveguiding, investigation by atomic force microscopy (AFM) shows that LB-films of DCANP are not molecularly smooth. Instead they exhibit mostly sub-micron size surface defects in the form of holes with a depth of one or several bilayers (*50*), which is not uncommon in LB-films as discussed below. The observation of holes with a depth of more than a bilayer shows that subsequent layers are not able to completely cover holes in the previous layers. On the other hand the average surface roughness of a 15-bilayer film, as estimated by x-ray reflectometry, is only 0.7 ± 0.5 nm (*50*). It should be mentioned that even "high quality" LB-films that can be transferred easily exhibit large numbers of bilayer defects (*51*). Despite these defects, it was demonstrated that LB-films of DCANP are in principle to guide visible light and a loss of about 12 dB/cm was determined at 632.8 nm (*52,53*). Although this seems to be a relatively large number, it should be emphasized that it is very difficult to fabricate LB-film waveguides with a loss of 1-2 dB/cm or less.

It was mentioned above that LB-films are often prone to thermal rearrangements and multilayers of DCANP are a case in which this can be studied in full detail. When LB-films of DCANP are heated up, they rearrange to 2 new phases depending on the age of the film and the annealing conditions. The structural changes are verified independently by X-ray reflectometry, UV/Vis-spectroscopy and optical microscopy. Although the structural rearrangements are drastic, from an original bilayer spacing of 4.42 nm to layer spacings of either 3.80 nm (by annealing at about 60 °C) or 3.09 nm (by annealing at about 75 °C), a large number of well resolved Kiessig fringes are observed in all 3 phases, indicating that the films remain intact and smooth throughout the phase transition. The thermally induced phase transition proceeds so slowly that it can be followed by X-ray with time and a kinetic analysis yields that the nucleation of the 3.80-nm phase is spontaneous and occurs via the edge of a half-sphere. This is in good agreement with the fact that the films rearrange, but are not destroyed. Interestingly the 3.80-nm phase cannot be further converted to the 3.09-nm phase by a second annealing step at 75 °C (*50*).

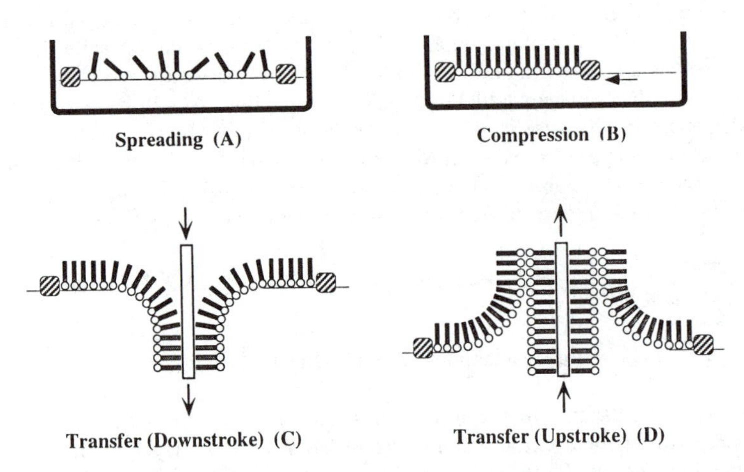

Figure 1. Schematic of the LB-transfer process. (A) After spreading, the monolayer is usually expanded at low surface pressures. (B) Compression with a movable barrier leads typically to a condensed state. (C,D) The condensed monolayer is transferred onto a hydrophobic solid substrate by repeated downstroke/upstroke cycles which result in a head-to-head and tail-to-tail (Y-type) structure.

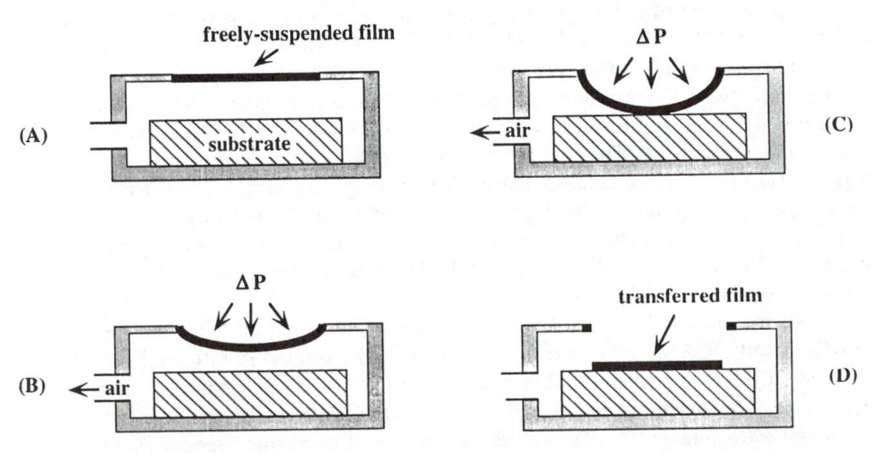

Figure 2. Schematic of the transfer of freely-suspended liquid crystal films onto solid supports. (A) The freely-suspended film spans across an aperture of mm^2 to cm^2 size. (B) Withdrawing air from the inner compartment cause the film to bulge towards the substrate. Note that the drawing is not to scale, the thickness of the film is only a few nm and its actual distance from the substrate is less than a mm. (C) Further reduction of internal pressure brings the film into contact with the substrate. (D) After the transfer is complete the part of the film connecting the aperture and the substrate breaks and transferred FS-films of up to 1-2 cm^2 are obtained.

Molecularly layered systems with built-in control of defects would therefore be highly interesting as an alternative to LB-films and we have developed 2 different approaches that are based on the self-organization of liquid crystals and on layer-by-layer adsorption from solution. Advantages and drawbacks of both techniques are briefly discussed below.

Freely-Suspended and Transferred Freely-Suspended Liquid Crystal Films: Candidates for Optically Homogenous Films

This approach is based on the molecular self-organization of thermotropic liquid crystals in freely-suspended (FS) films, a highly ordered, but very fragile system which was reported for the first time in 1978 (*54,55*). Liquid crystals are interesting materials for low-defect molecular films, because they represent molecular stacks with liquid-like in-plane order in the high temperature smectic A and smectic C phases. In the smectic A phase all molecules are oriented perpendicular to the layer normal (homeotropic alignment) and it is trivial that the absence of in-plane crystallinity automatically excludes the presence of grain boundaries, which are one of the reasons for light scattering. Even more interesting are films of the non-centrosymmetric chiral smectic C^* phase which is polar and ferroelectric. Homogeneously oriented freely-suspended films of this phase are obtained by cooling from the smectic A phase in an electric field.

We have tried to stabilize freely-suspended films in two ways: a) by transfer onto solid substrates and b) by using polymeric liquid crystals.

Transferred Freely-Suspended Films. Transferred freely-suspended (TFS) films are prepared by a simple procedure: At first the freely-suspended film is spread across an aperture on the top of a small box which contains the desired substrate positioned directly below the aperture. The liquid crystalline FS-film is then transferred by lowering the air pressure in the box which causes the film to bulge towards the substrate. Additional reduction of internal pressure brings the FS-film into contact with the substrate after which further transfer occurs by wetting until the film finally ruptures in the vicinity of the edge of the aperture (Figure 2). This way one is able to fabricate supported liquid crystalline films of about 1-2 cm^2 in size which would be large enough for potential optical applications (*14,56*). In the smectic A phase or in the electric field aligned chiral smectic C^* phase, these transferred films represent optical monodomains, which are now stable over long times. When kept at the same temperature, the transferred films can be expected to remain homogenous and essentially free of defects, since they were annealed when freely-suspended and because the transfer does not cause a phase transition or require thermomechanical adaptation.

Freely-suspended Films of Polymeric Liquid Crystals. The stabilization of freely-suspended films by using polymeric liquid crystals is obviously interesting and has been attempted previously. Unfortunately it seems to be extremely difficult to polymerize films of liquid crystalline monomers as these films were reported to always break during polymerization. It seems to be equally difficult to fabricate FS-films of polymeric liquid crystals in their smectic A and smectic C phases, most likely due to their enhanced viscosities. However, if one heats slightly into the isotropic phase it is possible to spread a film across an aperture which thins out to form a truly freely-suspended liquid crystal film after cooling into the smectic phases (*57*). Films of this type are homeotropic in the smectic A phase and show birefringence when cooled to the ferroelectric smectic C^*

phase. The layer structure of these first polymeric films and their phase transition from the smectic A to the smectic C* phase was investigated in some detail by X-ray reflectometry and some differences from the behavior of the bulk was found (*57*). Transmission electron diffraction after transfer onto electron microscope grids shows beautiful diffractograms of a higher ordered phase at lower temperature. Of course it is possible to further increase the stability of polymeric FS-films by transfer onto solid substrates as described above. Here polymeric materials offer a second advantage, namely the possibility to freeze order and orientation of the smectic phases by rapid cooling into the glassy state.

With a different ferroelectric LC-polymer it was even possible to obtain very homogenous FS-films of thicknesses of only 6 to 16 layers as determined by both optical and X-ray reflectometry (*58*). Since these polymeric films contained a small fraction of polymerizable mesogenic side-groups, it was even possible to photocrosslink these systems. It should be mentioned that in this first case the crosslinking caused some changes in film morphology, but the resulting films are very stable and do not even break when small holes are formed by mechanical stress (*58*).

Comparison of Transferred Freely-Suspended Films and Langmuir-Blodgett Films of the Same Liquid Crystalline Amphiphile.

In order to compare structure, defects and stability of both Langmuir-Blodgett and transferred freely-suspended films, we have synthesized an amphiphilic liquid crystal which allows deposition in multilayer films by both techniques. The molecule employed for this purpose was ethyl-4'-n-octyloxybiphenyl-4-carboxylate (**28OBC**) (*40,59*). Its LB-film forming properties are actually quite interesting since LB-transfer initially yields a deposition enforced bilayer phase, which then rearranges, even at room temperature, to a more stable monolayer phase (*40*). This further underlines the fact that LB-films are not necessarily equilibrium structures. X-ray-reflectivity measurements performed on both LB-films and on the transferred FS-films reveal that the layer spacings are the same in both types of films (2.41 ± 0.01 nm and 2.38 ± 0.02 nm respectively) which is very close to the long spacing of the bulk (2.46 ± 0.01 nm), indicating that both films are very close if not identical to the stable bulk structure, given the differences in the experimental set-up for reflectivity and bulk-diffraction experiments.

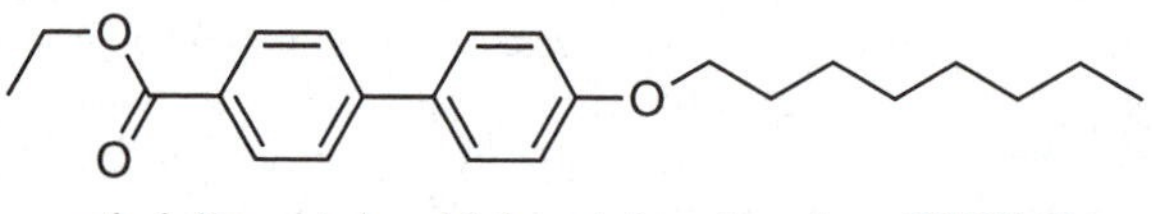

ethyl-4'-n-octyloxybiphenyl-4-carboxylate (**28OBC**)

But what about defects? The TFS-films certainly will possess defects arising from the fact that they were transferred at elevated temperature, but then cooled down to the crystalline bulk phase. Since the thermal expansion coefficients of the substrate and the film should be different, the formation of some defects, even besides the occurrence of grain boundaries due to crystallization, is to be expected. In fact cracks have been observed between very smooth areas from which molecular resolution could be obtained (*51*). However, these films appear perfectly transparent to the naked eye and macroscopic defects (besides dust particles) are not even visible under an optical microscope. LB-films of the same material are also initially transparent but already show some defects visible to the naked eye. After rearrangement to the monolayer phase, the defects become even more pronounced. Nevertheless, comparison of the rearranged LB-films with LB-films of several tens of other materials clearly shows that the LB-films of 28OBC are certainly of low-end quality, but do not really represent a particular poor case.

The structural similarities and expected defect dissimilarities in both kinds of films (TFS- and LB-films), made it interesting to look at both their respective thermal stabilities. When slowly heated the LB-film underwent an irreversible phase transition at 64 °C and started to melt under droplet formation at approximately 80 °C, which is 30 °C below the melting point of the bulk material. At 100 °C the LB-film has almost entirely dewetted the substrate and formed a large number of islands that are clearly visible to the naked eye. As seen for example in the section on LB-films, thermal rearrangements are not uncommon in Langmuir-Blodgett multilayers and melting below the melting point of the bulk is also frequently observed. This further underlines the fact, that ethyl-4'-n-octyloxy-biphenyl-4-carboxylate is not a particularly bad LB-film forming material. In contrast TFS-films of 28OBC can reversibly be heated to 108 °C and cooled back to ambient without noticeable changes in film morphology (*59*). The fact that TFS-films can be heated to 2 degrees below the bulk melting temperature clearly points towards a much lower defect density, even of crystallized transferred freely-suspended films as compared to LB-films of the same material.

Layer-by-Layer Adsorbed Films of Oppositely Charged Synthetic or Biological Polymers or Colloids: Control of Layer Architecture Over Large Surface Areas

Multilayer fabrication by multiple consecutive electrostatically driven adsorption of oppositely charged polyions (*17*) (Figure 3) or colloidal particles (*28*) overcomes two problems: a) the use of charge reduces the steric requirements in comparison to covalent bond formation and b) the use of polyions allows for compensation of defects or small charge densities in underlying layers. The resulting films are, with the exception of very few cases, of high optical quality, but optical loss measurements on films in waveguide configuration have not yet been performed. The reason for the good film quality is probably the fact that polyelectrolyte complexes are generally not crystalline. This means that the films should be rather homogeneous since domain boundaries on the submicron length scale are not present and therefore cannot contribute to light scattering. A consequence of this is, however, that multilayer structural resolution is limited, since it is only defined on the molecular but not on the atomic scale.

The key to a successful multilayer deposition is the surface refunctionalization in each adsorption step (*17*). Each adsorption step is self-regulating and thus leading to a defined thickness of individual layers. In this case the build-up of heterofunctional film architectures in a layer-by-layer fashion is straightforward and easily achieved (Figure 3), since one has complete control of layer sequence (*19,60*). Depending on the application, the multilayers may be fabricated on planar surfaces (e. g. for basic research, optical or sensing applications), on large custom shape surfaces or even on colloidal particles. For strong polyelectrolytes adsorption times may be as fast as 10 seconds, however, such conditions do not always lead to transparent films or to a linear film growth over many layers. For the simplest case of a $(AB)_n$ two-component film in which A is a polyanion and B is a polycation and n is the number of repetitions of this sequence, linear film growth is observed over several tens to hundreds of layers, when adsorption times are in the minute range.

Multilayer Film Structure and Properties. Although the resulting films loose or take up water when heated to different temperatures and re-cooled, they are thermally quite

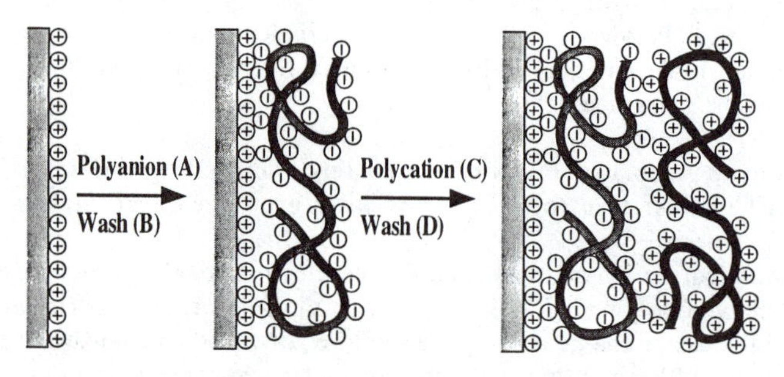

Figure 3. Schematic of the electrostatically driven layer-by-layer adsorption process. It describes the case of the adsorption of a polyanion to a positively charged substrate (A), followed by washing (B), the adsorption of a polycation (C) and another washing step (D). Multilayer films are prepared by repeating steps (A) through (D) in a cyclic fashion. More complicated film architectures are obtained by using additional adsorption/washing steps and applying more than two polyelectrolytes. Note that the drawing is oversimplified with respect to polyion conformation and interpenetration of adjacent layers. Furthermore, any counterions that might be present in the films were omitted for reasons of clarity.

stable. Films were heated to 200 °C in dry air for one week but no significant changes in X-ray reflectivity were observed after re-equilibration at ambient conditions. When heated directly in the X-ray reflectometer the film thickness shrinks, but the surface roughness stays constant.

The conformation of polyelectrolytes in solution depends strongly on the concentration of added electrolyte. At low salt concentrations the polyelectrolytes are rather extended due to the repulsion of charges along the polymer backbone. At high salt concentrations the polyions coil up because the added salt screens this repulsion. Accordingly, layers adsorbed from solutions containing no salt are thin and layers adsorbed from solutions containing large amounts of salt are thick. The control of the average incremental film growth is remarkable. In the case of films composed of poly(styrene sulfonate) (**PSS**) and poly(allylamine) (**PAH**), the thickness of a polyanion/polycation layer pair increases from 1.77 nm to 1.94 nm and to 2.26 nm when the **PSS** is adsorbed from a solution of NaCl with concentrations of 1.0 M to 1.5 M and to 2.0 M respectively and the **PAH** is adsorbed from pure water (*61*). **PSS/PAH** films were also the first on which the spatial confinement of individual polymer layers (= multilayer superlattice architecture) was demonstrated (*62,63*). A film composed of 48 layers of **PSS** and **PAH** in which every sixth layer (= every third layer of **PSS**) was deuterated has a total thickness of 121 nm and a distance of 15.9 nm between the deuterated layers, as determined by neutron reflectivity (*62*). A detailed neutron reflectivity study on the effects of salt and molecular weight on the average layer thicknesses and interfacial roughnesses has been carried out, its results will be published elsewhere (Schmitt, J.; Decher, G.; Bouwman, W.; Kjær, K.; Lösche, M. manuscript in preparation). It is now clear that the overlap of adjacent polyion layers is considerable, but many details such as charge stoichiometry and its relation to the film structure are not yet understood.

Potential Applications. Currently, the technique of consecutive electrostatically driven adsorption of polyanions and polycations has not yet been developed to the point of commercially available devices, but the prospects for future use are quite promising. An advantage over other molecular deposition techniques is, that no dedicated and sensitive equipment (e. g. ultrahigh vacuum apparatus or Langmuir troughs) is needed and that the adsorption is carried out from aqueous solutions, which makes the technique also environmentally attractive and allows incorporation of natural or modified biological materials. This has already prompted considerable research from several groups (e.g. refs. (*20,22,29,30,34-37,64-76*). An overview of the different polyions used in these studies is given in scheme 1.

Certainly one of the most interesting applications lies in the control of architecture (control of hole and electron injecting and active layers) in very thin electroluminescent devices. The existence of a precursor polyelectrolyte **Pre-PPV** (*39,90,91*) of the electroluminescent material poly(p-phenylene-vinylene) (**PPV**) (*92*) makes it possible to fabricate polyion multilayer film architectures, in which the **Pre-PPV** can subsequently be converted to **PPV** by thermal annealing (*19,34-38,87,88,93*). The first electroluminescent devices have been prepared and it was found that devices as thin as 13 nm emit light (*88*). Turn-on voltages of less than 2 V are required to generate light (*35*) and even an influence of the film architecture on the luminescent properties has been observed (*88*). However, the structural details of these films are presently not understood and it is not clear how device properties are influenced by film composition and architecture. We have recently shown that the thermal conversion of **Pre-PPV** to **PPV** can be carried out with preservation of a layered structure. These results were obtained by neutron reflectometry on

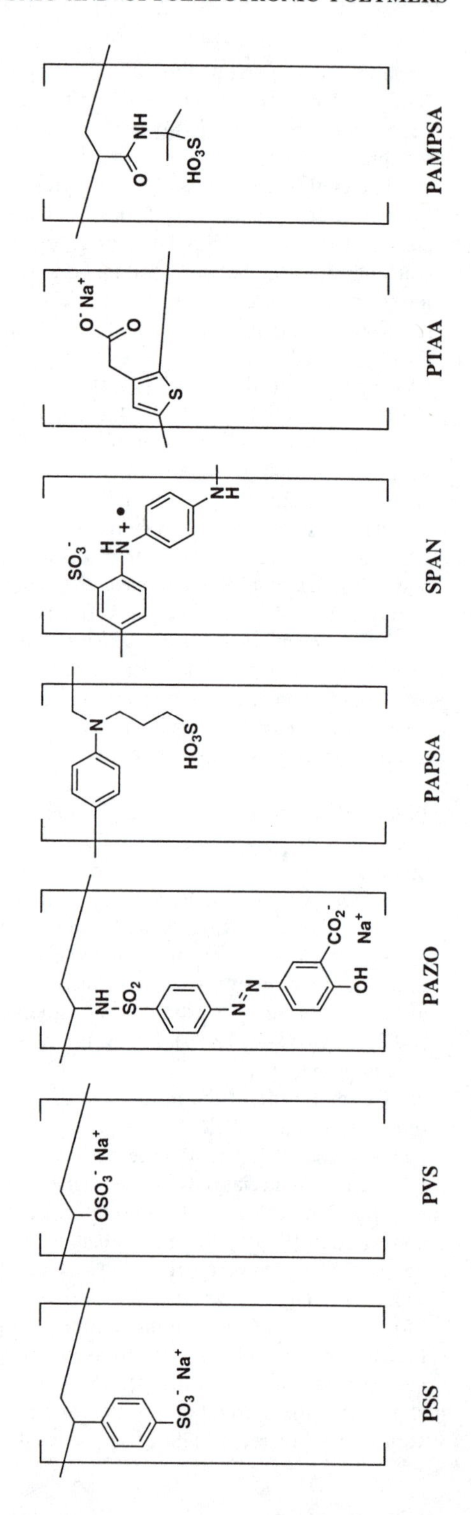
PAMPSA
PTAA
SPAN
PAPSA
PAZO
PVS
PSS

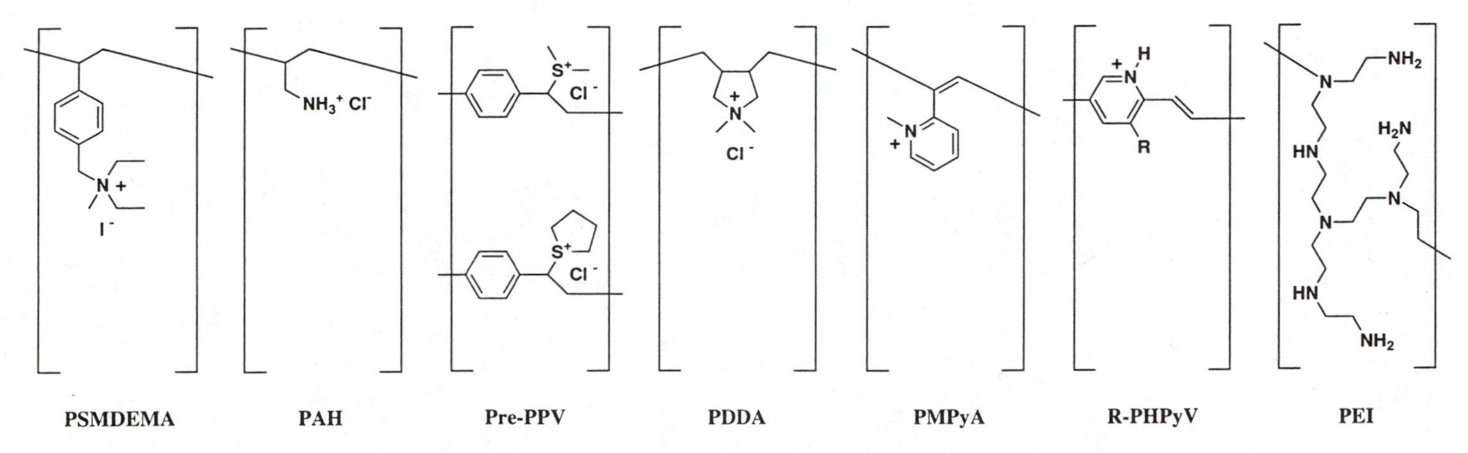

Scheme 1: Chemical structures of synthetic polyelectrolytes used up to now for the fabrication of multilayers. There are considerably more polyelectrolytes whose adsorption onto solid surfaces has been studied previously. Counter-ions are omitted in some cases, most of the abbreviations of the substances follow the original literature. The compounds are referenced as follows: **PSS** (*17,21,24,61-63,71,75,77-79*); **PVS** (*80,81*); **PAZO** (*77*); **PAPSA** (*24*); **SPAN** (*35,65*); **PTAA** (*65*); **PAMPSA** (*82*); **PSMDEMA** (*16,17*); **PAH** (*17,20,24,27,61-63,65,68,71,73-75,79,80,83-86*); **Pre-PPV** (*19,34,35,38,87,88*); **PDDA** (*24,29*); **PMPyA** (*65*); **R-PHpyV** (*36*); **PEI** (*24, 79,81,89*)

multilayer films in which selected layers were deuterated and will be published separately (Lehr, B.; Oeser, R.; Decher, G. manuscript in preparation).

However, the fact that just within 4 years of the first report on polyelectrolyte multilayer films, homogeneous large area light emitting diodes have been described, has produced considerable enthusiasm towards applications (especially in the field of light emitting diodes) of these layered polymeric nanoheterostructures.

Literature Cited.

(1) Swalen, J. D.; Allara, D. L.; Andrade, J. D.; Chandross, E. A.; Garoff, S.; Israelachvili, J.; McCarthy, T. J.; Murray, R.; Pease, R. F.; Rabolt, J. F.; Wynne, K. J.; Yu, H. *Langmuir* **1987**, *3*, 932-950.
(2) Special Issue: Organic Thin Films *Adv. Mater.* **1991**, *3*, 3-180.
(3) Ulman, A. *An Introduction to Ultrathin Organic Films: From Langmuir-Blodgett to Self-Assembly*; Academic Press: Boston, 1991; pp 442.
(4) Ulman, A. *Organic Thin Films and Surfaces: Directions for the Nineties*; Ulman, A., Ed.; Academic Press: San Diego, 1995; Vol. 20, pp 392.
(5) Knoll, W. *Curr. Opinion in Coll. & Interface Sci.* **1996**, *1*, 137-143.
(6) Langmuir, I.; Blodgett, K. B. *Kolloid-Z.* **1935**, *73*, 257-263.
(7) Blodgett, K. B.; Langmuir, I. *Phys. Rev.* **1937**, *51*, 964-982.
(8) Netzer, L.; Iscovici, R.; Sagiv, J. *Thin Solid Films* **1983**, *99*, 235-241.
(9) Maoz, R.; Netzer, L.; Gun, J.; Sagiv, J. *J. de Chim. Phys.* **1988**, *85*, 1059-1065.
(10) Lee, H.; Kepley, L. J.; Hong, H.-G.; Mallouk, T. E. *J. Am. Chem. Soc.* **1987**, *110*, 618-620.
(11) Cao, G.; Hong, H.-G.; Mallouk, T. E. *Acc. Chem. Res.* **1992**, *25*, 420-427.
(12) Kunitake, T.; Shimomura, M.; Kajiyama, T.; Harada, A.; Okuyama, K.; Takayanagi, M. *Thin Solid Films* **1984**, *121*, L89-L91.
(13) Coulon, G.; Russel, T. P.; Deline, V. R.; Green, P. F. *Macromolecules* **1989**, *22*, 2581-2589.
(14) Maclennan, J.; Decher, G.; Sohling, U. *Appl. Phys. Lett.* **1991**, *59*, 917-919.
(15) Decher, G.; Hong, J.-D. *Makromol. Chem., Macromol. Symp.* **1991**, *46*, 321-327.
(16) Decher, G.; Hong, J.-D. *Ber. Bunsenges. Phys. Chem.* **1991**, *95*, 1430-1434.
(17) Decher, G.; Hong, J.-D.; Schmitt, J. *Thin Solid Films* **1992**, *210/211*, 831-835.
(18) Hong, J.-D.; Lowack, K.; Schmitt, J.; Decher, G. *Progr. Colloid Polym. Sci.* **1993**, *93*, 98-102.
(19) Decher, G.; Lehr, B.; Lowack, K.; Lvov, Y.; Schmitt, J. *Biosensors and Bioelectronics* **1994**, *9*, 677-684.
(20) Keller, S. W.; Kim, H.-N.; Mallouk, T. E. *J. Am. Chem. Soc.* **1994**, *116*, 8817-8818.
(21) Lvov, Y.; Ariga, K.; Kunitake, T. *Chemistry Letters* **1994**, 2323-2326.
(22) Kong, W.; Zhang, X.; Gao, M. L.; Zhou, H.; Li, W.; Shen, J. *Makromol. Chem., Rapid Commun.* **1994**, *15*, 405-409.
(23) Kong, W.; Wang, L. P.; Gao, M. L.; Zhou, H.; Zhang, X.; Li, W.; Shen, J. C. *J. Chem. Soc., Chem. Commun.* **1994**, 1297-1298.
(24) Lvov, Y.; Ariga, K.; Kunitake, T. *J. Am. Chem. Soc.* **1995**, *117*, 6117-6123.
(25) Sun, Y.; Zhang, X.; Sun, C.; Wang, B.; Shen, J. *Macromol. Chem. Phys.* **1996**, *197*, 147-153.
(26) Onda, M.; Lvov, Y.; Ariga, K.; Kunitake, T. *Biotechn. and Bioengin.* **1996**, *51*, 163-167.
(27) Lvov, Y.; Decher, G.; Sukhorukov, G. *Macromolecules* **1993**, *26*, 5396-5399.

(28) Iler, R. K. *J. Colloid Interface Sci.* **1966**, *21*, 569-594.
(29) Kleinfeld, E. R.; Ferguson, G. S. *Science* **1994**, *265*, 370-373.
(30) Gao, M.; Gao, M.; Zhang, X.; Yang, Y.; Yang, B.; Shen, J. *J. Chem. Soc., Chem. Commun.* **1994**, 2777-2778.
(31) Lvov, Y.; Ariga, K.; Ichinose, I.; Kunitake, T. *Langmuir* **1996**, *12*, 3038-3044.
(32) Kleinfeld, E. R.; Ferguson, G. S. *Chem. Mater.* **1995**, *7*, 2327-2331.
(33) Sano, M.; Lvov, Y.; Kunitake, T. *Annu. Rev. Mater. Sci.* **1996**, *26*, 153-187.
(34) Ferreira, M.; Rubner, M. F.; Hsieh, B. R. *Mat. Res. Soc. Symp. Proc.* **1994**, *328*, 119-124.
(35) Onoda, M.; Yoshino, K. *Jpn. J. Appl. Phys.* **1995**, *34*, L260-L263.
(36) Tian, J.; Wu, C.-C.; Thompson, M. E.; Sturm, J. C.; Register, R. A.; Marsella, M. J.; Swager, T. M. *Adv. Mater.* **1995**, *7*, 395-398.
(37) Hong, H.; Davidov, D.; Avny, Y.; Chayet, H.; Faraggi, E. Z.; Neumann, R. *Adv. Mater.* **1995**, *7*, 846-849.
(38) Lehr, B.; Seufert, M.; Wenz, G.; Decher, G. *Supramolecular Science* **1996**, *2*, 199-207.
(39) Hörhold, H.-H.; Helbig, M.; Raabe, D.; Opfermann, J.; Scherf, U.; Stockmann, R.; Weiß, D. *Z. Chem.* **1987**, *27*, 126-137.
(40) Decher, G.; Sohling, U. *Ber. Bunsenges. Phys. Chem.* **1991**, *95*, 1538-1542.
(41) Leuthe, A.; Riegler, H. *J. Phys. D: Appl. Phys.* **1992**, *25*, 1786-1797.
(42) Asmussen, A.; Riegler, H. *J. Chem. Phys.* **1996**, *104*, 8151-8158.
(43) Riegler, H.; Spratte, K. In *Organic Thin Films and Surfaces: Directions for the Nineties*; Ulman, A., Ed.; Thin Films; Academic Press: San Diego, 1995, Vol. 20; pp 349-364.
(44) Decher, G.; Tieke, B.; Bosshard, C.; Günter, P. *J. Chem. Soc., Chem. Commun.* **1988**, 933-934.
(45) Decher, G.; Tieke, B.; Bosshard, C.; Günter, P. *Ferroelectrics* **1989**, *91*, 193-207.
(46) Decher, G.; Klinkhammer, F. *Makromol. Chem., Macromol. Symp.* **1991**, *46*, 19-26.
(47) Decher, G.; Klinkhammer, F.; Peterson, I. R.; Steitz, R. *Thin Solid Films* **1989**, *178*, 445-451.
(48) Bosshard, C.; Flörsheimer, M.; Küpfer, M.; Günter, P. *Opt. Commun.* **1991**, *85*, 247-253.
(49) Flörsheimer, M.; Küpfer, M.; Bosshard, C.; Günter, P. In *NONOLINEAR OPTICS: Fundamentals, Materials and Devices*; Miyata, S., Ed.; Elsevier Science Publishers B.V.: 1992, pp 255-269.
(50) Klinkhammer, F. *Herstellung und Untersuchung von dünnen Schichten mit neuartigen Strukturprinzipien für die nichtlineare Optik*; Edition Wissenschaft Reihe Chemie; Tectum Verlag: Marburg, 1995; Vol. 4, (ISBN 3-89608-504-2; printed from Klinkhammer, F. Dissertation, Johannes Gutenberg-Universität Mainz, Germany, 1994).
(51) Overney, R. M.; Meyer, E.; Frommer, J.; Güntherodt, H.-J.; Decher, G.; Reibel, J.; Sohling, U. *Langmuir* **1993**, *9*, 341-346.
(52) Decher, G.; Tieke, B.; Bosshard, C.; Günter, P. *Organische Materialien mit nichtlinearen optischen Eigenschaften*; European Patent No. 89 81 108.4, 1989.
(53) Bosshard, C.; Küpfer, M.; Günter, P.; Pasquier, C.; Zahir, S.; Seifert, M. *Appl. Phys. Lett.* **1990**, *56*, 1204.
(54) Young, C. Y.; Pindak, R.; Clark, N. A.; Meyer, R. B. *Phys. Rev. Lett.* **1978**, *40*, 773-776.
(55) Pindak, R.; Moncton, D. *Physics Today* **1982**, *35*, 57-62.
(56) Decher, G.; Maclennan, J.; Reibel, J.; Sohling, U. *Adv. Mater.* **1991**, *3*, 617-619.

(57) Decher, G.; Honig, M.; Reibel, J.; Voigt-Martin, I.; Dittrich, A.; Poths, H.; Ringsdorf, H.; Zentel, R. *Ber. Bunsenges. Phys. Chem.* **1993**, *97*, 1386-1394.
(58) Reibel, J.; Brehmer, M.; Zentel, R.; Decher, G. *Adv. Mater.* **1995**, *7*, 849-852.
(59) Decher, G.; Maclennan, J.; Sohling, U.; Reibel, J. *Thin Solid Films* **1992**, *210/211*, 504-507.
(60) Decher, G. In *Templating, Self-Assembly and Self-Organization*; Sauvage, J.-P. and Hosseini, M. W., Eds.; Comprehensive Supramolecular Chemistry; Pergamon Press: Oxford, 1996, Vol. 9; pp 507-528.
(61) Decher, G.; Schmitt, J. *Progr. Colloid Polym. Sci.* **1992**, *89*, 160-164.
(62) Schmitt, J.; Grünewald, T.; Kjær, K.; Pershan, P.; Decher, G.; Lösche, M. *Macromolecules* **1993**, *26*, 7058-7063.
(63) Korneev, D.; Lvov, Y.; Decher, G.; Schmitt, J.; Yaradaikin, S. *Physica B* **1995**, *213&214*, 954-956.
(64) Cheung, J. H.; Fou, A. F.; Rubner, M. F. *Thin Solid Films* **1994**, *244*, 985-989.
(65) Ferreira, M.; Cheung, J. H.; Rubner, M. F. *Thin Solid Films* **1994**, *244*, 806-809.
(66) Pommersheim, R.; Schrenzenmeir, J.; Vogt, W. *Macromol. Chem. Phys.* **1994**, *195*, 1557-1567.
(67) Cooper, T. M.; Campbell, A. L.; Crane, R. L. *Langmuir* **1995**, *11*, 2713-2718.
(68) Ferreira, M.; Rubner, M. F. *Macromolecules* **1995**, *28*, 7107-7114.
(69) Keller, S. W.; Johnson, S. A.; Brigham, E. S.; Yonemoto, E. H.; Mallouk, T. E. *J. Am. Chem. Soc.* **1995**, *117*, 12879-12880.
(70) Kotov, N. A.; Dékány, I.; Fendler, J. H. *J. Phys. Chem.* **1995**, *99*, 13065.
(71) Lowack, K.; Helm, C. A. *Macromolecules* **1995**, *28*, 2912-2921.
(72) Mao, G.; Tsao, Y.-H.; Tirrell, M.; Davis, H. T.; Hessel, V.; Ringsdorf, H. *Langmuir* **1995**, *11*, 942-952.
(73) Saremi, F.; Maassen, E.; Tieke, B.; Jordan, G.; Rammensee, W. *Langmuir* **1995**, *11*, 1068-1071.
(74) Hammond, P. T.; Whitesides, G. M. *Macromolecules* **1995**, *28*, 7569-7571.
(75) v. Klitzing, R.; Möhwald, H. *Langmuir* **1995**, *11*, 3554-3559.
(76) Sellergren, B.; Swietlow, A.; Arnebrandt, T.; Unger, K. *Anal. Chem.* **1996**, *68*, 402-407.
(77) Decher, G.; Lvov, Y.; Schmitt, J. *Thin Solid Films* **1994**, *244*, 772-777.
(78) Schlenoff, J. B.; Li, M. *Ber. Bunsenges. Phys. Chem.* **1996**, *100*, 943-947.
(79) Sukhorukov, G. B.; Schmitt, J.; Decher, G. *Ber. Bunsenges. Phys. Chem.* **1996**, *100*, 948-953.
(80) Lvov, Y.; Decher, G.; Möhwald, H. *Langmuir* **1993**, *9*, 481-486.
(81) Laschewsky, A.; Mayer, B.; Wischerhoff, E.; Arys, X.; Jonas, A. *Ber. Bunsenges. Phys. Chem.* **1996**, *100*, 1033-1038.
(82) Mao, G.; Tsao, Y.; Tirrell, M.; Davis, H. T.; Hessel, V.; Ringsdorf, H. *Langmuir* **1993**, *9*, 3461-3470.
(83) Lvov, Y.; Eßler, F.; Decher, G. *J. Phys. Chem.* **1993**, *97*, 13773-13777.
(84) Lvov, Y.; Haas, H.; Decher, G.; Möhwald, H.; Kalachev, A. *J. Phys. Chem.* **1993**, *97*, 12835-12841.
(85) Tronin, A.; Lvov, Y.; Nicolini, C. *Colloid Polym. Sci.* **1994**, *272*, 1317-1321.
(86) Ramsden, J. J.; Lvov, Y. M.; Decher, G. *Thin Solid Films* **1995**, *254*, 246-251 erratum: idem (1995), 261, 343-344.
(87) Lehr, B. *Ultradünne PPV-Filme - Herstellung und Charakterisierung-*; Diploma Thesis; Johannes Gutenberg-Universität: Mainz, Germany, 1993.

(88) Fou, A. C.; Onitsuka, O.; Ferreira, M.; Rubner, M. F.; Hsieh, B. R. *Mat. Res. Soc. Symp. Proc.* **1995**, *369*, 575-580.
(89) Decher, G. *Nachr. Chem. Tech. Lab.* **1993**, *41*, 793-800.
(90) Hörhold, H.-H.; Opfermann, J. *Die Makromolekulare Chemie* **1970**, *131*, 105 - 132.
(91) Wessling, R. A. *J. Polym. Sci.: Polym. Symp.* **1985**, *72*, 55-66.
(92) Burroughes, J. H.; Bradley, D. D. C.; Brown, A. R.; Marks, R. N.; McMackay, K.; Friend, R. H.; Burns, P. L.; Holmes, A. B. *Nature* **1990**, *347*, 539-541.
(93) Tian, J.; Thompson, M. E.; Wu, C.-C.; Sturm, J. C.; Register, R. A.; Marsella, M. J.; Swager, T. M. *Polymer Preprints* **1994**, *35*, 761-762.

Chapter 31

Optical and Photonic Functions of Conjugated Polymer Superlattices and Porphyrin Arrays Connected with Molecular Wires

Takeo Shimidzu[1]

Division of Molecular Engineering, Kyoto University, Kyoto 606–01, Japan

Conjugated polymer superlattices and porphyrin arrays connected with molecular wires are superstructured materials, which exhibit unique optical and photonic functions. The former shows a shift in photoluminescence to higher energy which is interpreted as a quantum size effect. The latter class of materials exhibits photoconductivity by a hole carrier mechanism and photo-information storage by a localized excitation mechanism. The syntheses of these two classes of materials are described.

Both quantum functional materials and molecular devices are considered to be the ultimate functional materials. The former shows a novel property which is specific to the structure and the latter represents the smallest possible functional material. Their properties are closely related to optical and photonic functions. The former shows a quantum size effect, a photoluminescence shift to higher energy which depends on layer thickness, which gives us an idea of a nonlinear optical system. Porphyrin arrays connected with molecular wires show a hole carrier photoconductivity or a photoswitching and a photo-information storage, which suggests an idea of a photoactive neuron model. In this paper, conjugated polymer superlattices and porphyrin arrays connected with molecular wires are described.

Conjugated Polymer Superlattices

Since almost all conjugated polymers are organic semiconductors, structural control, such as compositional control of a copolymer thin film, corresponds

[1]Current address: Kansai Research Institute, Kyoto Research Park, 17 Chudoji-Minami-Machi, Shimogyo-ku, Kyoto 600, Japan

to the manipulation of the electronic band structure of the thin film. Structural control of semiconductor thin films in the 1-10nm scale range can give rise to novel optical and photonic functions which come from carrier confinement due to quantum-size effects. In the past few decades, inorganic semiconductor superlattice and multiple quantum wells have been extremely active research subjects in semiconductor physics and materials science ever since the proposal by Esaki and Tsu in 1970 (*1*). Many new physical phenomena, such as negative differential resistance, have been discovered, and many novel device concepts, such as high electron mobility transistor, have been developed based on these ultra-thin layered structures of semiconductor materials constructed by molecular beam epitaxy and metalorganic chemical vapor deposition. The most important feature, which has produced this great activity, is energy quantization of electronic structure by mesoscopic modulation in materials (*2*). This not only changes the energy band structure but also alters the density of the states and restricts electron motion within the layer planes, leading to a lower dimensional electron system. This feature gives rise to new charge transport, optical, and magnetic properties. Furthermore, because one can control and largely prescribe all these features by adjusting parameters of the heterostructures, periodicities of the layers, and band discontinuities by composition of the constituent layers, new materials with desired physical properties can be essentially designed .

A function which is specific to the overall structure of a heterostrctured material such as superlattice period or a quantum function due to ultrasmall size can be regarded as an ultimate function of the material which can only be creaeted by proper fabrication methods. The design and fabrication of such ultimate function materials has developed into a new field of advanced materials called wave function engineering, quantum technology, etc. Super lattices of inorganic materials are, in general, fabricated by ultrahigh vacuum systems such as molecular beam epitaxy (MBE). However, in order to realize "wave function engineering" in conjugated polymer thin films, and prepare superlattices, a novel method of compositional modulation in conjgated polymer thin films is required. Polymerization followed by the *in-situ* deposition of the resulting polymer onto a substrate makes the fabrication of superlattice structures possible.

Electropolymerization is one of the most interesting methods to control the copolymer composition in molecular or chain sequence. Accordingly, in the case that the electropolymerized material is electroconductive and insoluble, a heterolayered structure and/or a sloped structure with conducting polymers can be constructed on the electrode. The potential-programmed electropolymerization method (PPEP) is utilized for modulating the composition of conducting polymer thin films in the depth direction *(3,4)*. By this method, nanometer scale compositional control of the thin films of conducting polymers are obtained, permitting the fabrication of alternate layered and graded heterostructures. Monomers such as pyrrole, thiophene, and their derivatives, can be electropolymerized, so that the corresponding conducting polymer thin films are obtained on the surface of a working electrode, if they are insoluble. The growth rate of the film thickness is proportional to current *i*. In general, the PPEP method consists of the electropolymerization of a mixture of monomers under potentiostatic control in accordance with an appropriate potential sweep function. The function is programmed in advance from the current fraction curves for each monomer, which leads to a definite copolymer composition and control of layer thickness. The resulting conducting polymer film has a compositionally modulated depth structure corresponding to the applied potential sweep function, for example a layered structure for a rectangular (square wave) function and a graded structure if using a varying function, such as a sawtooth wave.

By applying the PPEP method to the copolymerization of pyrrole and 3-methylthiophene, various kinds of conducting polymer heteromultilayers were fabricated. Figure 1 shows the resulting transmission electron microscopy (TEM) cross section and electron-probe microanalysis (EPMA) line analysis on sulfur reflecting thiophene content. The layer on the electrode side, from which the layer grew, showed a clear and flat interface. The depth profile of the resulting conducting polymer multilayers also was evaluated by SIMS, AES, TEM and EPMA *(5)*. Alternate layered structures were fabricated by a rectangular potential wave and stair-like step sweep function, and triangular sloped structures by a triangular sweep function and triangularly sloped structures resulted from sawtooth potential waves *(4)*.

Figure 2 shows the band structures of several homopolymers and pyrrole-bithiophene copolymers estimated by electrochemical and optical methods as examples. A combination of these homopolymers and/or copolymers implies various kinds of superlattice structures. The electrochemical preparation of both homopolymer multiheterolayers and/or copolymer multiheterolayers results in a superlattices. The electrochemical copolymerization method as used to prepare heterolayers was easier than in the homopolymer heterolayers. The copolymer multiheterolayers are prepared by simply changing the applied electrode potential. On the contrary, the latter needs exchange of the mother solutions. The present electrocopolymerization method which makes compositionally modulated copolymer heterolayers possible is considered to be one of the most fascinating methods to fabricate organic superlattices.

The fabrication of copolymer multiheterolayers was carried out on a rotating highly oriented pyrolytic graphite (HOPG) disk electrode (working electrode; 1000/r.p.m.) which leads to flat and sharp interfaces having a resolution of order 1nm. The electrocopolymerization of a mixture of pyrrole (2.5×10^{-4} M) and bithiopehene (2.5×10^{-2} M) by a rectangular potential sweep having limits of 1.0V and 1.4V, gave superlattice multilayers whose dedoped layers were expected to be a type II superlattice structure whose band structures of well layer and barrier layer are alternative (*1*). The barrier layer was composed of 33% bithiophene and 67% pyrrole while the well layer was composed of 87% bithiophene and 13% pyrrole. In this superlattice, the conduction band edge difference ΔE_C is 0.58V and the

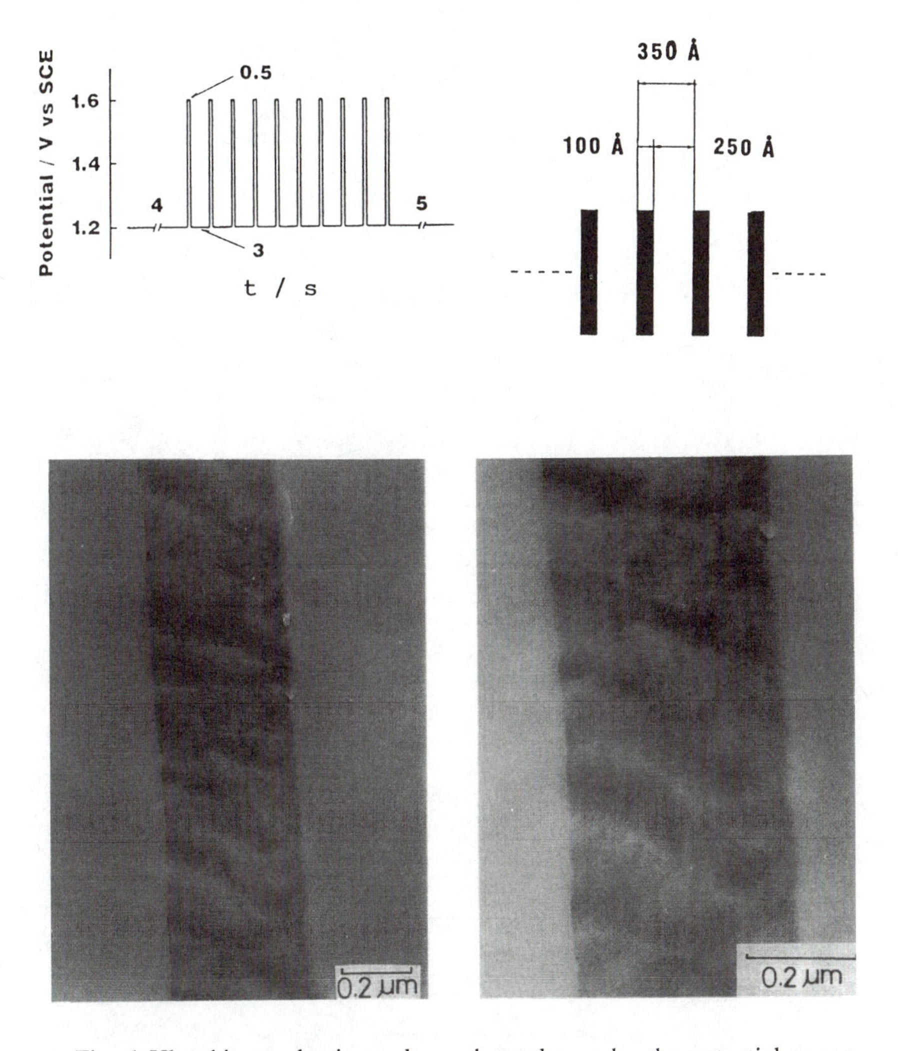

Fig. 1 Ultrathin conducting polymer heterolayers by the potential sweep programmed electropolymerization of pyrrole (25mM) and 3-methylthiophene (50mM) in CH_3CN containing 100mM $LiClO_4$.: Potential sweep programs and TEM pictures of their cross sections.

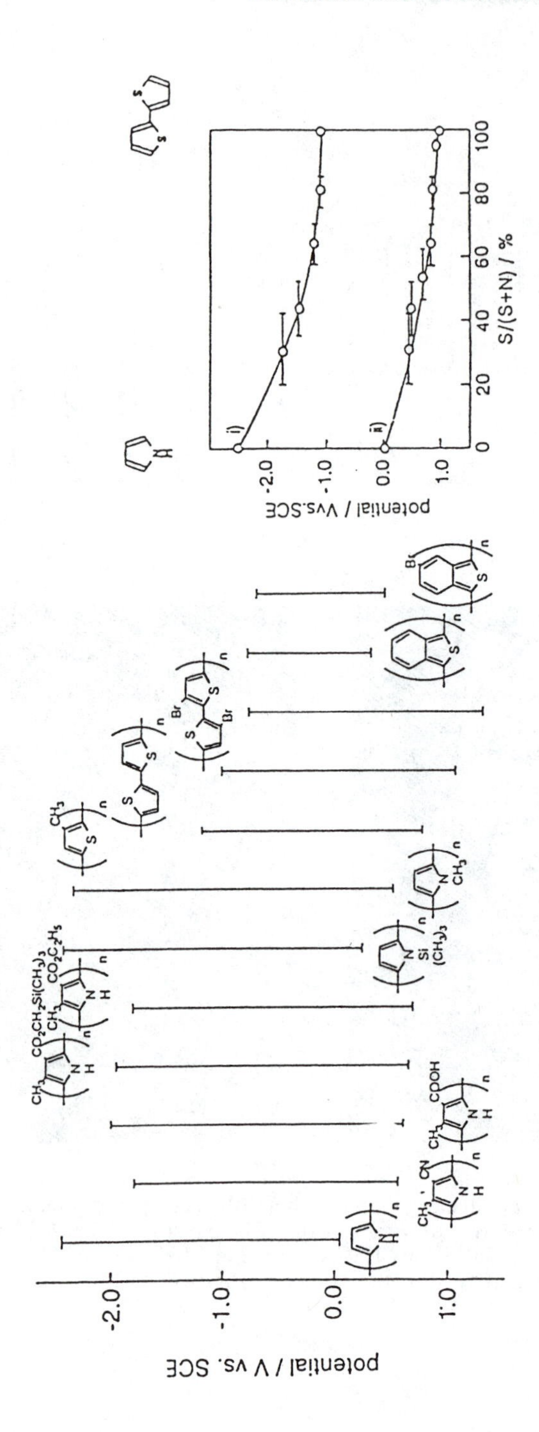

Fig. 2 Band structures, conduction band (E_C) and valence band (E_V) of conjugating homopolymers (left) and copolymer (right).

valence band edge difference ΔE_V is 0.41V, as shown in Figure 3. Both layers can be controlled by copolymer conposition, as shown in Figure 4 (c) in which the barrier layer is composed of 54% bithiophene an 46% pyrrole, giving ΔE_C=0.20eV and ΔEv=0.18eV.

The photoluminescence spectra of the dedoped copolymer (pyrrole/bithiophene) films whose thiophene content was higher than 50%, consisted of three peaks around 2.0, 1.8, and 1.7eV corresponding to phonon side bands at 10K. These peaks correspond to the radiative relaxation of self-trapped excitons. The peak at the highest energy reflects the copolymer's band gap. Actually, the peak positions observed in the spectra of copolymer films shifted to higher energy as thiophene content in the film decreased and peak positions showed good agreement with E_g as estimated in Figure 3. The copolymer containing a thiophene fraction less than 50% did not, however, show photoluminescence. The photoluminescence of the multilayers having 6.0nm of 87% thiophene layer and 10.0nm of 33% thiophene layer (10 layers) $[(87)_{Lw}(33)_{Lb}]_{10}$ (L_b=10.0nm) shifted to higher energy as compared with that of the bulk (87% thiophene content) copolymer film. (Figure 4) (*6*). The photoluminescence of the above-mentioned multilayers shifted to higher energy as the thickness of the well layer (Lw) became smaller than 12.0nm, even when the barrier thickness (Lb) remained constant (10.0nm) and when the ratio, Lw/Lb was constant (0.6). On the other hand, the bulk thin layer did not show a significant energy shift. Such a shift to higher energies is considered to be the result of the confinement of excited electrons in the quantum well layer. We have also found a good fit of experimental results with the Kronig-Penney model, which derives the energy-wave number vector relationship in rectangular type potential profile by assuming that $m^*=0.6\ m_e$ where m_e is the electron mass (depicted as the solid line in Figure 4). The Kronig-Penney model is given as :

$$\cos k\,(Lw + Lb) = \cos \frac{\sqrt{2m^* E}}{h} Lw \cosh \frac{\sqrt{2m^* (Vo - E)}}{h} Lb + \left(\frac{Vo}{2E} - 1\right)\left(\frac{Vo}{E} - 1\right)^{-1/2} \sin \frac{\sqrt{2m^* E}}{h} Lw \sinh \frac{\sqrt{2m^* (Vo - E)}}{h} Lb$$

where Vo and m* are barrier height and effective mass, respectively. It is noteworthy that other multilayers have also shown a similar phenomenon.

These observed photoluminescence spectral properties of conjugated polymer heterolayers fabricated by the present PPEP method appear to be due to quantum size effects. However, additional studies of these materials will be necessary to establish the true origin of their properties. In any case, these results also suggest that many other novel structures of functional materials and devices can be fabricated by this method.

Porphyrin Arrays Connected With Molecular Wires

Nanofabrication of molecular photoelectronic, optical and photonic devices is important *(7-11)*. Incorporation of multiple redox centers into conducting

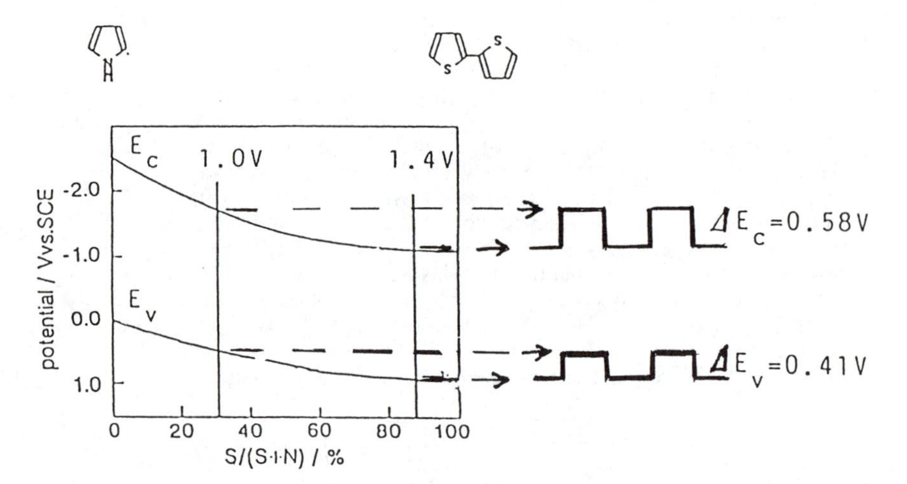

Fig. 3 Fabrication of type II conducting polymer heterolayer superlattice by the electro-copolymerization of pyrrole (2.5×10^{-4}M) and bithiophene (2.5×10^{-2}M) $LiClO_4$ (1.0×10^{-1}M) CH_3CN solution. Sweep potentials, copolymer compositions and ΔE_C, ΔE_V of the resulting heterolayers.

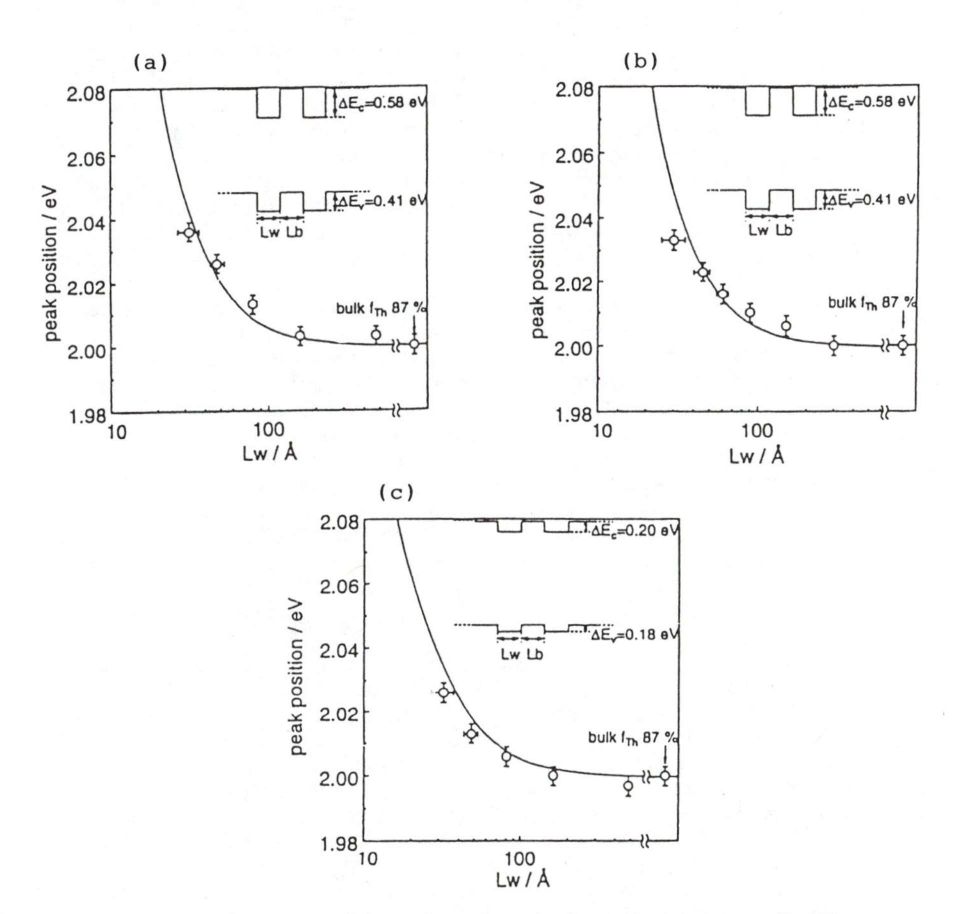

Fig. 4 Structure of type II heterolayer superlattice and emission peak shift as a function of layer thickness. (solid line is estimated from Kronig-Penney model).
(a) $[(87)_{Lw}(33)_{Lb}]_{10}$ Lw/Lb= 0.6
(b) $[(87)_{Lw}(33)_{Lb}]_{10}$ Lb=10.0nm const.
(c) $[(87)_{Lw}(54)_{Lb}]_{10\sim20}$ Lb=10.0nm const.

L_w and L_b denote well layer and barrier layer, and ( ) denotes bithiophene content in pyrrole-bithiophene copolymer.

molecular systems is a useful approach for trial construction of molecular devices. For example, the incorporation of a photosensitizer and a suitable electron donor and/or an acceptor into a polymeric chain has been proposed as a molecular electronic device system based on photoinduced electron transfer *(11)*. However, the production of such polymers containing a number of large aromatic moieties or metal complexes is difficult because of the lack of the solubility and flexibility, which also limit the possibility for controlled fabrication. To overcome these difficulties, electrochemical polymerization is used, since the polymer is deposited directly on the terminal electrode. With this in mind, we have synthesized a series of one- or two-dimensional porphyrin arrays connected with conjugated wires which can be polymerized by normal electrochemical oxidation. On the other hand, porphyrin arrays connected with insulating wires have also been synthesized by esterification.

Construction of intramolecular systems whose photoactive molecules are linked with conducting and insulating molecular wires is an important objective towards the realization of molecular electronic or photonic devices. For such an objective, systematization of donor-photosensitizer-acceptor triad molecules into large molecular systems is one of the feasible approaches to realize a simple model. This is because the incorporation of a photoactive moiety and a suitable electron donor and/or acceptor into a conducting polymeric chain is useful for various molecular electronic systems based on photoinduced electron transfer *(11)*. A symmetrical donor-acceptor-donor triad molecule was polymerized by normal electrochemical oxidation which led to one-dimensional donor-acceptor polymers with porphyrin moieties separated by an ordered oligothienyl molecular wire (1 D porphyrin array) *(12)*.

The oligothiophenes, which are easily polymerized by electrochemical oxidation, were used not only as molecular wires but also as the coupling elements for connecting the phorphyrin moieties. Phosphorus(V)porphyrins (P(V)TPP) which have strong oxidizing powers and are stable to the electrochemical oxidation were used as the photoactive moieties *(13, 14)*. Since P(V)TPP can form two stable axial bonds on the central phosphorus atom, P(V)TPP triads having two oligothiophene moieties in the axial direction can be synthesized easily *(15)*. Three different P(V)TPP derivatives (Figure 5) containing two thienylalkoxy or oligothienylalkoxy groups at the axial positions of the central phosphorus atom were synthesized by the reaction of the dichlorophosphorus-(V)tetraphenylporphyrin and the corresponding thienyl or oligothienyl alcohols (12). The resulting triad molecules give normal P(V)TPP absorption spectra as well as characteristic thienyl or oligothienyl absorption peaks.

All the porphyrin derivatives have similar fluorescence originating from P(V)TPP moiety, but their lifetimes and the relative quantum yields of fluorescence depended on the axial substituents. In particular, the fluorescence was strongly quenched in the latter two cases of molecules (a), (b), and (c) in Figure 5 as compared with diethoxyphosphorus(V)tetraphenylporphyrin which was free of any thienyl

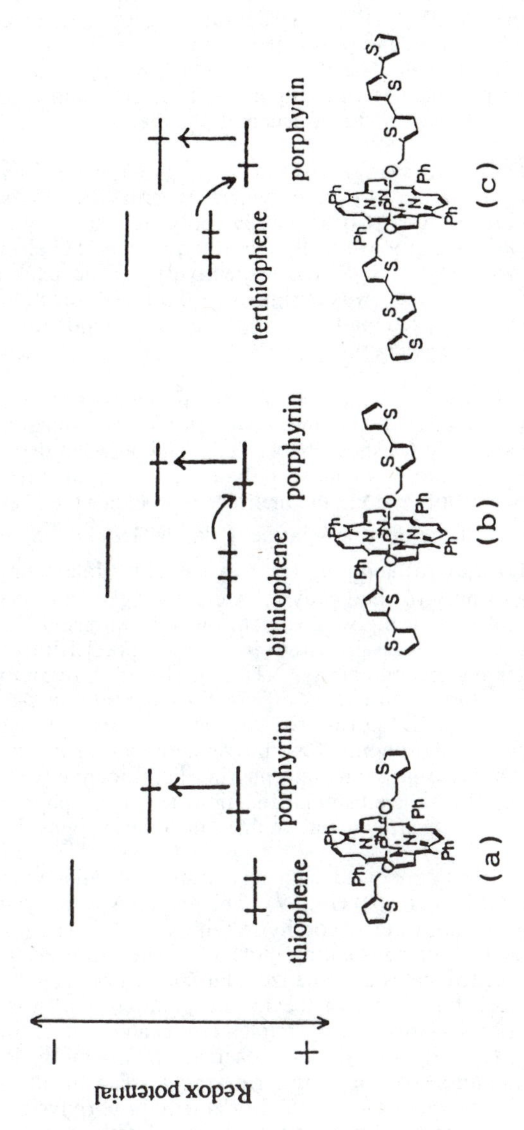

Fig. 5 Schematic representation of relationships between photoinduced electron transfer and corresponding energy levels of phosphorus(V)porphyrin and oligothienyl axial groups.

moieties. Taking into account the energy levels of the P(V)TPP and the oligothienyl moieties, the fluorescence quenching can be attributed to the photoinduced electron transfer from the oligothienyl moieties to the P(V)TPP *(12)*. If fluorescence is quenched, the oxidation potential of the oligothienyl moieties is sufficiently low as compared with the reduction potential of the singlet excited state of the P(V)TPP. These results suggest that the reductive electron transfer occurs in (b) and (c), as depicted in Figure 5. An important point is that the P(V)TPP is able to act as a good photoinduced hole generator in the donor-acceptor molecules with oligothienyl moieties and is therefore expected to play a similar role in donor-acceptor polymers.

Both of the P(V)TPP derivatives (b) and (c) (Figure 5) were polymerized by electrochemical oxidation to give polymers, whereas (a) was scarcely polymerized. (Figure 6 (A)). Consequently, poly-(b) and poly-(c), 1 D porphyrin arrays, were electrochemically deposited on the ITO electrode at potentials >1.2V and 0.9V (vs. SCE), respectively. The peak current observed around -0.4V, which was assigned to the redox reaction of the P(V)TPP moieties, increased and thus signaled the deposition of the porphyrin polymer onto the electrode.

The photoconductivity of the polymers was measured by using a sandwich cell composed of ITO / polymer / Au. Both current-potential (*i*-E) curves of poly-(b) and poly-(c) show that each contact between the polymer and the electrode is ohmic. In these polymers, it is confirmed that a Schottky junction was not formed at the interface with either the ITO or Au. The d.c. conductivity of the polymers in the dark was 1.2×10^{-9} S cm^{-1} and 5.1×10^{-8} S cm^{-1} for poly-(b) and poly-(c), respectively. Interestingly, the conductivity of both poly-(b) and poly-(c) were strongly enhanced upon photoirradiation, and was strongly dependent on light intensity. A little oriented polymer which was prepared in a micropore film, showed significantly greater photoconductivity. This implies that intramolecular photoinduced carrier formation occurs efficiently in these donor-acceptor polymers *(16, 17)*. 2 D porphyrin array was also prepared by electropolymerization of phosphorus(V)porphyrin derivatives containing four oligothienyl groups at meso-position of porphyrin ring. (Figure 6 (C)). These 2 D porphyrin arrays also showed similar functions as 1 D porphyrin arrays. STM image supports the 2D array nature of these materials (Figure 7).

1D porphyrin arrays connected with insulating molecular wires were recently synthesized (Figure 6 (B)) *(18, 19)*. The polymers were synthesized by esterification of dichloro-P(V)) porphyrin with glycols. The porphyrin arrays connected with shorter insulating molecular wires showed that both singlet and triplet excited states are localized, but the arrays connected with longer insulating wires did not show a localized excitation. This was proved by transient absorption spectroscopy by Triplet-Triplet and Singlet-Singlet fs laser pulse excitations. Figure 8 shows a schematic picture of the localized and delocalized excitations of the porphyrin arrays. The results suggest possible photo-information storing capability at certain porphyrin rings of the porphyrin arrays in the porphyrin arrays connected with short insulating

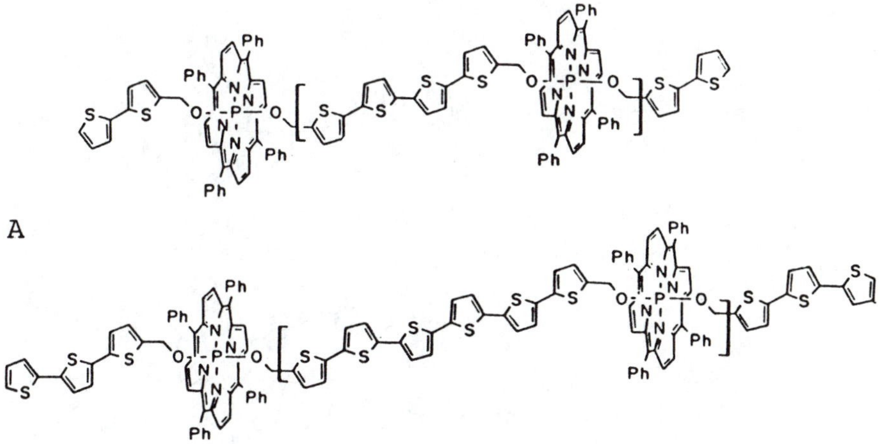

A

B

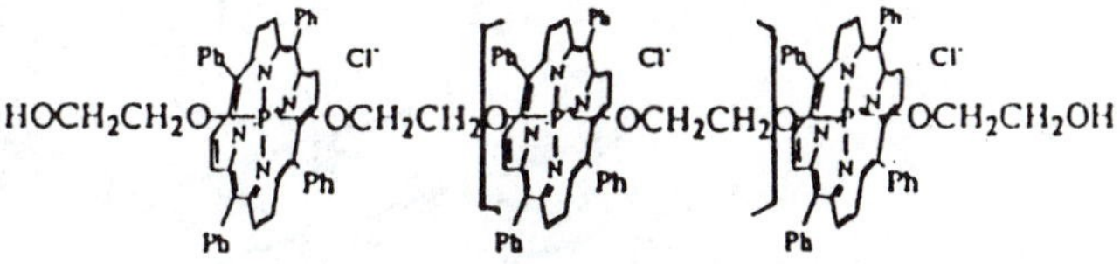

Fig. 6 (A) 1D porphyrin arrays connected with conjugating molecular wire
(B) 1D porphyrin array connected with insulating molecular wire
(C) 2D porphyrin array connected with conjugating molecular wire
(D) A proposed 3D porphyrin array connected with molecular wire

Continued on next page

C

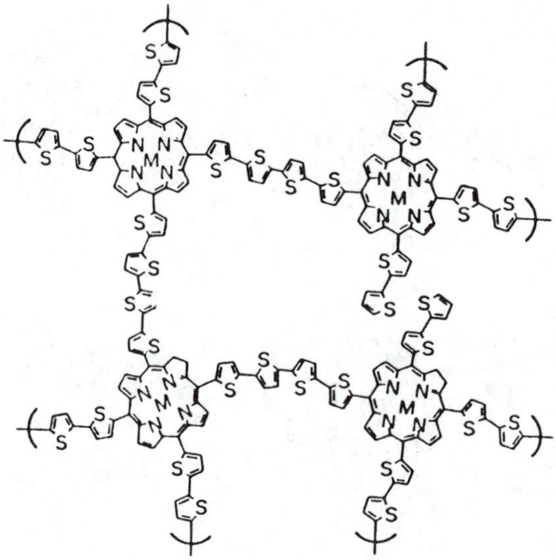

D

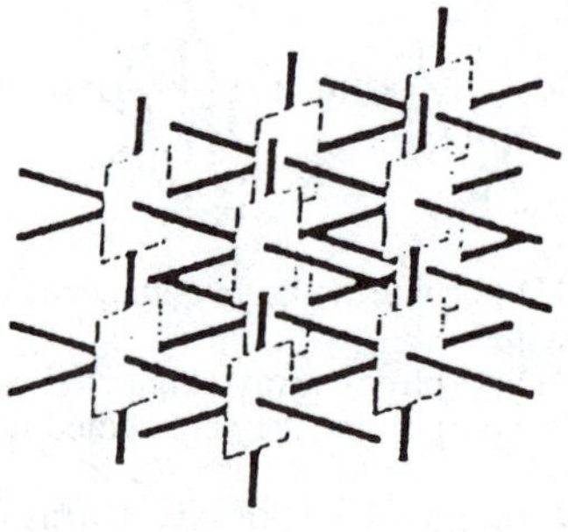

Figure 6. Continued.

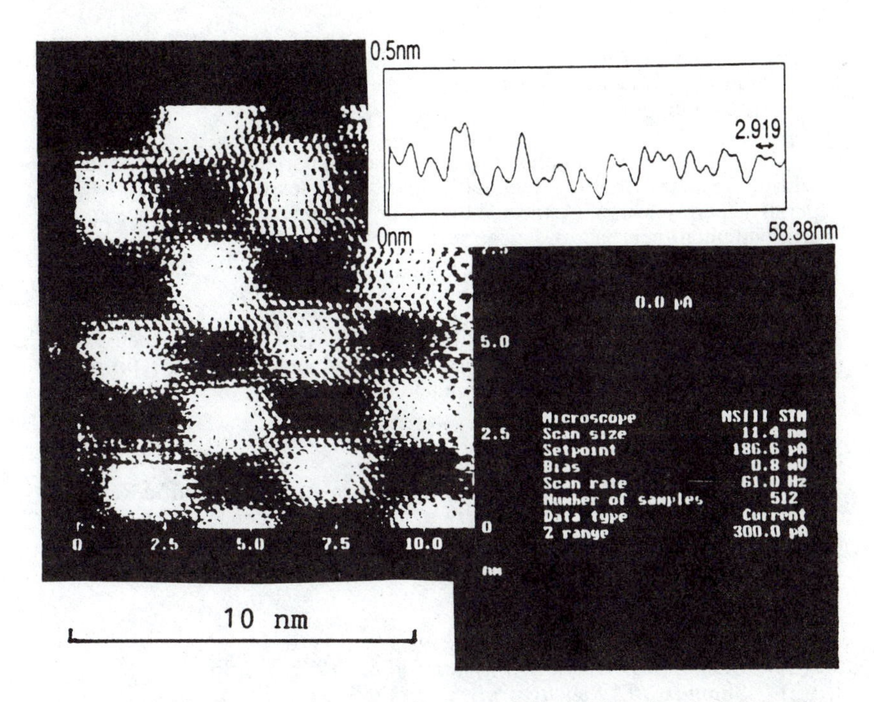

Fig. 7 STM image of 2D porphyrin array (Fig. 6(C)) electrochemically synthesized on the Au(111) substrate. The image was processed with Fourier filter to remove noise. Periodicity of surface level is ca. 2.9nm which is similar to an estimated porphyrin-porphurin distance (ca. 3nm) by CPK model.

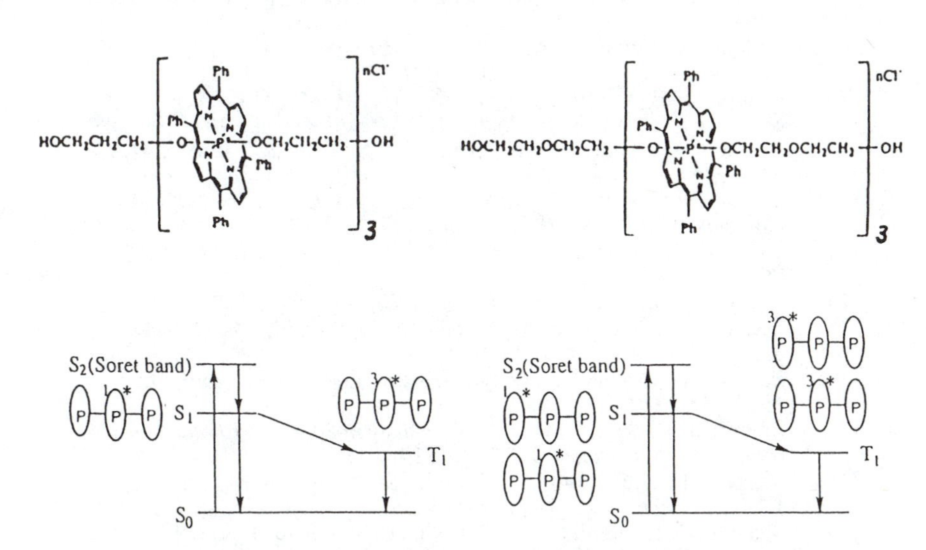

Fig. 8 Singlet and triplet photoexcited states of 1-D porphyrin arrays connected with insulating molecular wires of short and long chains.

wires such porphyrin arrays are thus possible information transduction and storing molecular device systems even if the lifespans of transition status are not sufficiently long.

The present study on photoactive 1D and 2D porphyrin arrays connected with molecular wires together with their syntheses open the way to 3 D porphyrin array which is expected to be a proto-type molecular device and an artificial photo-neuron.

Conclusion

Both conjugated polymer superlattices and porphyrin arrays connected with molecular wires device which are described here represent powerful candidates for optical and photonic materials.

Acknowledgements

This work was partially supported by a Grant-in-Aid for Scientific Research on New Program from the Ministry of Education, Science and Culture of Japan.

Literature Cited

(1) Esaki, L.; Tsu, R. *IBM J. Res. Develop.* **1970**, *14*, 61.
(2) Chang, L.L.; Esaki, L. *Physics Today* **1992**, *45*, 36.
(3) Shimidzu, T. *Reactive Polymers* **1989**, *11*, 177.
(4) Iyoda T.,,Toyoda H., Fujitsuka M., Nakahara R., Tsuchiya H., Honda K.; Shimidzu T., *J. Phys. Chem.,* **1991**, *95*, 5215.
(5) Iyoda, T.; Toyoda, H.; Fujitsuka, M.; Nakahara, R.; Honda, K.; Shimidzu, T.; Tomita, S.; Hatano. Y.; Soeda, F.; Ishitani, A.; Tsuchiya, H *Thin Solid Films*, **1991**, *205*, 258.
(6) Fujitsuka, M.; Nakahara, R.; Iyoda, T.; Shimidzu, T. *Synth. Metals* **1993**, *55-57*, 966.
(7) Aviram, A.; Ratner, M.A. *Chem. Phys. Lett.* **1974**, *29*, 281.
(8) Wrighton, M.S. *Science* **1986**, *231*, 32.
(9) Chidsey, C.E.D.; Murray, R.W. *Science* **1986**, *231*, 25.
(10) Simon. J.; Tournilhac, F.; Andre, J.-J. *New J. Chem.* **1987**, *11*, 383
(11) Hopfield, J.J.; Onuchic, J.N.; Beratan, D.N. *Science* **1988**, *241*, 817
(12) Segawa, H.; Nakayama, N.; Shimidzu, T. *J. Chem. Soc., Chem. Commun.* **1992**; 784.
(13) Sayer, P.; Gouterman, M.; Connell, C.R. *J. Am. Chem. Soc.* **1977**, *99*, 1082.
(14) Carrano, C.J.; Tsutsui, M. *J. Coord. Chem.*, **1977**, *7*, 79.
(15) Segawa, H.; Kunimoto, K.; Nakamoto, A.; Shimidzu T. *J. Chem. Soc. Perkin Trans.* **1992**, *1*, 939.
(16) Segawa, H.; Nakahara, R.; Iyoda, T.; Shimidzu, T. *J. Appl. Phys.* **1993**, *74*, 1283.
(17) Shimidzu, T.; Segawa, H.; Wu, F.; Nakayama, N. *J. Photochem. Photobiol., A. Chem.* **1995**, *92*, 121.
(18) Segawa, H.; Kunimoto, K.; Susumu, K.; Taniguchi, M.; Shimidzu T. J. *Amer. Chem. Soc.* **1994**, *116*, 11193.
(19) Susumu, K.; Kunimoto, K.; Segawa, H.; Shimidzu, T. *J. Phys. Chem.* **1995**, *99*, 29.

Chapter 32

Electroactive and Photoactive Nanostructured Materials from Self-Organizing Rod–Coil Copolymers: Synthesis and Electroluminescent Devices

Richard M. Tarkka[1], X. Linda Chen, and Samson A. Jenekhe[2]

Department of Chemical Engineering and Center for Photoinduced Charge Transfer, University of Rochester, Rochester, NY 14627–0166

Electroluminescent ternary and binary rod-coil copolymer systems have been synthesized, characterized, and used to explore the effects of solid state supramolecular structure and morphology on the mechanism and quantum efficiency of electroluminescence. Enhanced electroluminescence due to efficient electronic energy transfer was observed in ternary rod-coil-rod copolymer systems containing two rodlike segments with different excitation energies, compared to the corresponding binary copolymers. Binary rod-coil copolymers which incorporate, into the rodlike segments, moieties that exhibit excited state intramolecular proton transfer were found to be electroluminescent via a novel mechanism, electrically generated intramolecular proton transfer (EGIPT). The inherent population inversion in the EGIPT process together with the observed stimulated emission imply the feasibility of electrically induced lasing in this class of materials. EGIPT was found to occur in only rod-coil copolymers with low rod-to-coil molar ratios, suggesting the profound effect of supramolecular structure.

The role of non-covalent interaction in dictating overall *supramolecular* structure in biopolymers is well established. Such interactions include, but are not limited to, hydrogen bonding, steric interactions, π-π interactions, dipole-dipole interactions, and van der Waal forces. A well known example is the double helical

[1]Current address: Department of Chemistry, George Washington University, 725 21st Street, NW, Washington, DC 20052

[2]Corresponding author

structure of DNA, which arises from hydrogen bonding. In proteins and other biopolymers, tertiary (supramolecular) structure, arising from non-covalent interaction, dictates function*(1)*. Proteins and related biopolymers can only function properly if folded and packed into the correct geometry. Denaturing of these systems (disrupting the tertiary and secondary structures) disrupts the ability to perform their biological function*(1)*. Nature has built sufficient flexibility into the macromolecular chains that allows folding to the requisite three-dimensional geometry.

Non-covalent interaction has profound effects on the structure and properties of synthetic molecules as well. The tendency of rod-like molecules to align coaxially through self-organization into liquid crystalline phases is just one example(*2*). This phenomenon has both positive and negative consequences. Liquid crystals display systems have successfully exploited the self-organizing features of anisotropic molecules*(2b)*. In contrast, the tendency of conjugated rigid-rod polymers to aggregate in solution and as solids has complicated the investigation and development of these polymers for applications such as molecular composites (*3-6*) and electroluminescent devices *(7-9)*. Because of the strong intermolecular interactions and aggregation of conjugated rigid-rod polymers, solubility in organic solvents is nonexistent, and the glass transition temperatures are extremely high, which makes processing difficult(*4-6*). Perhaps the most dramatic effect of aggregation in these conjugated rigid-rod polymers is the formation of ground state and excited state complexes, such as excimers and exciplexes, which profoundly influences all their photoelectronic properties*(7-10)*. For example, the photoluminescence and electroluminescence spectra of thin films are strongly Stokes shifted compared to solution spectra and the fluorescence quantum yield is significantly reduced in the solid state*(7-9)*.

To increase processibility, control aggregation and supramolecular packing in the solid state, and hence to regulate the electroactive and photoactive properties of conjugated polymer systems, we have adopted a strategy from Nature's example: *self-organizing rod-coil copolymers* that are analogous to proteins in terms of folding(*10*). Folding of the rod-coil copolymer chain and packing of its rigid segments in the solid state governs the photophysical properties*(10)*. Variation of both the rod-to-coil ratio and lengths of the rod-like and coil-like segments, can be used to control the way in which conjugated segments assemble in the solid state. In this way, chromophore aggregation and excimer formation can be controlled and the wavelengths of emission and absorption can also be regulated. Moreover, the rod-coil copolymers have improved solubility in organic solvents relative to the parent conjugated rigid-rod homopolymers.

The efficacy of this "supramolecular engineering" strategy was first demonstrated for heterocyclic rod-coil copolymers, with poly(p-phenylenebenzobisthiazole) (PBZT) and derivatives used as the rod-like segments*(10)*. The wavelength of absorption and emission of rod-coil copolymers containing PBZT moieties could be tuned by the copolymer composition. Photoluminescence (PL) quantum yields could be increased by 600 % over the PBZT homopolymer, and fluorescence lifetimes increased, all of which can be attributed to the regulation of excimer formation through the supramolecular structure. Blending of such rod-coil copolymers with a conjugated homopolymer, into a nanocomposite, was shown to facilitate efficient singlet electronic energy transfer, a phenomenon which further increases the quantum efficiency of emission, and which can be used to regulate the emission wavelength*(11)*.

Our prior studies*(6-11)* have now been extended to include investigation of the effects of supramolecular nanostructure on electroluminescence in rod-coil copolymers. The race to develop commercial light sources from organic polymers faces several challenges, among which are the issues of material stability, processibility, emission efficiency(*14-16*), device durability, and questions of the mechanism of electroluminescence(EL) *(9, 12-15)*. The ease of processing of conjugated heterocyclic polymers, such as PBZT and its derivatives, using the Lewis acid complexation technique*(16)* or solubility in formic acid and related solvents, suggests them as a new class of electroluminescent polymers*(9)*. Herein we report on the effects of supramolecular structure on electroluminescence of several rod-coil copolymers. The goals of this study included improvement of the EL properties of heterocyclic aromatic polymers by incorporating them into self-organizing rod-coil copolymers and exploration of the mechanism of electroluminescence in heterocyclic moieties known to undergo photoinduced intramolecular proton transfer.

Two new electroluminescent rod-coil copolymer systems are reported here. Ternary rod-coil-rod copolymers **1** and **2** (Chart 1), each of which contains two different electroactive and photoactive rigid-rod segments, are used to demonstrate the enhancement of electroluminescence through efficient electronic energy transfer. Comparative studies of binary rod-coil copolymers **3** and **4** (Chart 1), in which there is no energy transfer, confirmed the mechanism of PL and EL enhancement in the ternary copolymers. The second class of electroluminescent rod-coil copolymer system, **5,** is used to explore a novel EL producing chemical reaction, called electrically generated intramolecular proton transfer (EGIPT). The EGIPT process in these binary rod-coil copolymers, incorporating intramolecular hydrogen bonded chromophore **6** which is known to exhibit excited state intramolecular proton transfer, imply the feasibility of electrically induced lasing of such materials.

Experimental Methods

2,5-Diamino-1,4-benzenedithiol (DABDT) (Daychem or TCI) was purified by recrystallization from aqueous HCl as previously reported(*17*). Hydroxyterephthalic acid was prepared by the method reported by Miura *et al.*(*18*). Terephthalic acid (Fluka, > 99 %), 1,10-decanedicarboxylic acid (Aldrich,

99%), 1,4-phenylene diacrylic acid (Aldrich, 97 %), and 2,6-naphthalene dicarboxylic acid (Aldrich, 95 %) were used as received. Poly(phosphoric acid)(PPA) and 85 % phosphoric acid (ACS reagent grade, Aldrich) were used to prepare 77 % PPA, which was used as the polymerization medium. Phosphorus pentoxide (Fluka) was used as received. The copolymers were synthesized by condensation copolymerization in polyphosphoric acid of the tetrafunctional monomer, DABDT, with aromatic and aliphatic diacids, as previously described(*10*). The total molar quantity of the diacid was set to be equal to that of the DABDT, as required for high molecular weight copolymers. In the case of the nonhydroxylated binary(**3**, **4**) and ternary(**1**, **2**) copolymers, the coillike fraction was set at 80 %, with the remaining 20% being composed of equal molar quantities of the remaining rod-like segments. The specific compositions of both **1** and **2** whose results are reported here are: a=0.1, b=0.8, and c=0.1 (see Chart 1). In the case of the intramolecularly hydrogen bonded copolymers **5**, the molar quantity of the rod-like moiety was varied between 1 and 100 %. Polymerization yields were quantitative in all cases.

The names of the rod-coil copolymers (Chart 1) used in this study are: poly(1,4-phenylenebenzobisthiazole-co-decamethylenebenzobisthiazole-co-(1,4-phenylene-divinylene)benzobisthiazole) (PBZT-co-PBTC10-co-PBTPV, **1**); poly(2,6-naphthylenebenzobisthiazole-co-decamethylenebenzobisthiazole-co-(1,4-phenylene-divinylene)benzobisthiazole) (PNBT-co-PBTC10-co-PBTPV, **2**); poly (1,4-phenylenebenzobisthiazole-co-decamethylenebenzobisthiazole) (PBZT-co-PBTC10, **3**); poly((1,4-phenylenedivinylene)benzobisthiazole-co-decamethylenebenzobisthiazole) (PBTPV-co-PBTC10, **4**); and poly (2-hydroxy-1,4-phenylenebenzobisthiazole-co-decamethylene - benzobisthiazole) (HPBT-co-PBTC10, **5**). The molecular structures and compositions of all the rod-coil copolymers **1-5** (Chart 1) were established primarily by ^{1}H NMR and FTIR spectroscopies, thermal analysis (TGA, DSC), intrinsic viscosity and various other spectroscopic measurements, as previously done for related polymers *(10, 11,17)*. These characterizations confirmed the structures and compositions of Chart 1.

Optical absorption spectra of thin films were obtained with a Perkin-Elmer Model Lambda 9 UV/Vis/NIR Spectrophotometer. Steady state photoluminescence studies were performed by using a Spex Fluorolog-2 Spectrofluorometer equipped with Spex DM3000f Spectroscopy computer. The polymer films were positioned such that emission was detected at 22.5° from the incident radiation beam. Thin films of good optical quality were prepared by spin coating of polymer solutions in formic acid onto fused silica, followed by heating under vacuum to remove solvent. Picosecond transient absorption spectroscopy was performed as described elsewhere(*19*). Briefly, the transient absorption system consisted of a Continuum PY61 Series Nd:YAG laser utilizing Kodak QS 5 as the saturable absorber to produce laser light pulses of ~25 ps FWHM. These output pulses were then amplified and the third harmonic generated (355 nm).

Chart 1

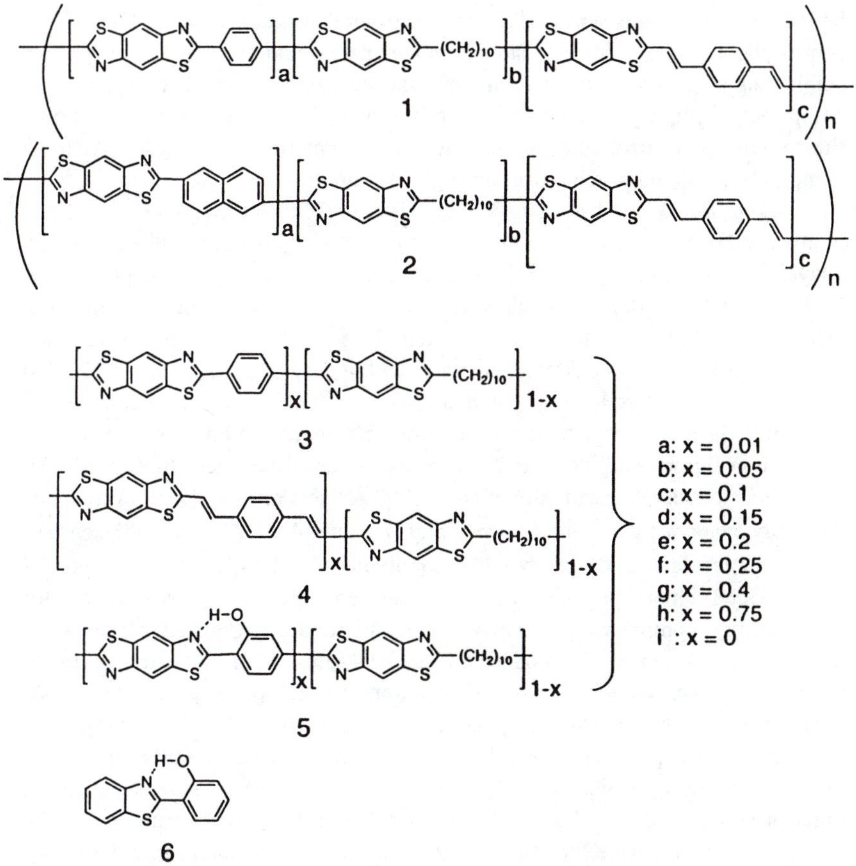

1

2

3

4

5

6

a: x = 0.01
b: x = 0.05
c: x = 0.1
d: x = 0.15
e: x = 0.2
f: x = 0.25
g: x = 0.4
h: x = 0.75
i : x = 0

Dichroic beamsplitters in conjunction with colored glass filters were used to isolate the fundamental (1064 nm) and the harmonic. The fundamental was directed along a variable optical delay and then focused into a 10 cm quartz cell filled with H_2O/D_2O (50:50) to generate a white light continuum probe pulse. The excitation and probe pulses (*ca.* 2 mm diameter) were passed approximately coaxially through the sample. The probe pulse was directed to a Spex 270 M monochromator through a Princeton Instruments fiber optic adapter and dispersed onto a Princeton Instruments dual diode array detector (DPDA 512). This allowed ~350 nm of the visible spectrum to be collected in a single experiment. A ST-121 detector controller/interface was incorporated into a 386/25 MHz PC to control the arrays and for data storage, manipulation and output. Correction for ambient lighting was made by subtracting the spectrum obtained when neither a pump pulse nor a probe pulse passed through the sample, whereas sample fluorescence was corrected for by subtracting the spectrum obtained by passing a pump pulse (with no probe pulse) through the sample.

The EL devices investigated consisted of bilayer polymer thin films sandwiched between two electrodes. To prepare the light-emitting diodes(LEDs), a layer of approximately 500 Å poly(vinyl carbazole) (PVK) was deposited on ITO (indium tin oxide) coated glass substrate (the anode) by spin coating from chloroform solution. The PVK layer functions as the hole transporting and electron blocking layer to confine electrons within the emissive layer. A 500-700 Å emissive layer of rod-coil copolymer was then spin coated from formic acid solution. Since the copolymers were insoluble in chloroform and PVK was insoluble in formic acid, no interfacial mixing occurred between the layers. An aluminum electrode (the cathode) of 500 - 1000 Å thick and *ca.* 7 mm^2 in area was thermally evaporated onto the device at high vacuum (4×10^{-6} torr). In the case of the intramolecularly hydrogen bonded copolymers **5**, EL spectra were recorded with a Spex Fluorolog-2 Spectrofluorometer equipped with Spex DM3000f spectroscopy computer. Current-Voltage and luminance-voltage curves were recorded simultaneously by connecting an HP4155A semiconductor parameter analyzer together with a Grasby S370 optometer equipped with a luminance sensor head. In the case of binary and ternary copolymers **1-4**, EL spectra were obtained by using a calibrated PR-650 colorimeter, whereas brightness was measured by using a Silicon Avalanche Photodiode (APD). The EL quantum efficiencies were calculated using the method proposed by Greenham *et al.*(*20*). No correction was made for losses due to absorption, reflection, or waveguiding effects of the glass. The values given therefore represent lower limits of the efficiency in terms of photons emitted per charge injected. All measurements were performed under ambient conditions.

Results and Discussion

Enhanced Electroluminescence By Energy Transfer. Rod-coil copolymer nanocomposites not only eliminate the formation of excimers, which quench luminescence, but they also allow for efficient singlet electronic energy transfer (EET), a process which leads to increases in the quantum yield of photoluminescence (PL) emission, and can be used in light-harvesting systems(*11*). Such polymer nanocomposites are therefore promising candidates for LED devices since EL and PL emission processes both originate from the same excited states species (*7, 12*). Ternary copolymers consisting of two different conjugated polymer segments and one flexible coil segment, such as **1** and **2** (Chart 1), should form nanostructured polymer assemblies at rod fractions below 0.4-0.5 as illustrated in Figure 1 *(10)*. Efficient exciton transfer, from the rigid-rod component with the higher HOMO-LUMO energy gap, to the component with the lower HOMO-LUMO gap, should occur *via* Förster dipole-dipole interaction(*11, 21*). The component with the higher HOMO-LUMO energy gap thus acts as 'antennae' for harvesting and transferring the electronic energy to the component with the lower HOMO-LUMO gap, from which light is emitted. The singlet excitons involved are initially formed by either photoexcitation or by charge recombination induced by charge injection and transport(*22*).

The optical absorption and emission spectra of thin solid films of binary rod-coil copolymers **3e** and **4e** are shown in Figure 2. The overlap of the emission of **3e** (donor chromophore) with the absorption of **4e** (acceptor chromophore) suggests that a ternary copolymer system incorporating both the PBZT component of **3e** (donor) and the PBTPV component of **4e** (acceptor) will exhibit efficient Förster-type energy transfer*(11)*. Figure 3 shows that this is in fact the case. The PL spectra of **3e** and **4e**, excited at 400 nm, and ternary copolymer **1**, excited at 350, 370, 390, 440 and 460 nm, respectively, are shown. The lineshape and spectral position of the PL emission spectra of **1** are the same, regardless of the excitation wavelength(inset of Figure 3). The emission spectra of **1** resemble that of **4e**, but are vastly different from **3e**. The PL emission of **1** is thus dominated by the PBTPV component. Considering that the extinction coefficient of the PBZT component is higher than that of the PBTPV component between 370 and 390 nm (Figure 2), extensive energy transfer from the PBZT component must occur in this nanocomposite copolymer system. The relative PL quantum efficiency of **1** is 50 % higher than **3e**, suggesting that EET in the ternary copolymer nanocomposite system accounts for the enhanced PL quantum efficiency.

Figure 4 shows the EL spectra of the copolymers **1-3**, which closely resemble the PL spectra shown in Figure 3: green emission occurs from **1** and **2**, whereas blue-green emission occurs from **3e**. (An EL spectrum of **4** could not be obtained due to insufficient light emission). The similarity between the PL and EL spectra for **1** confirms that EET *via* Förster dipole-dipole interaction occurs under conditions leading to EL as well PL.

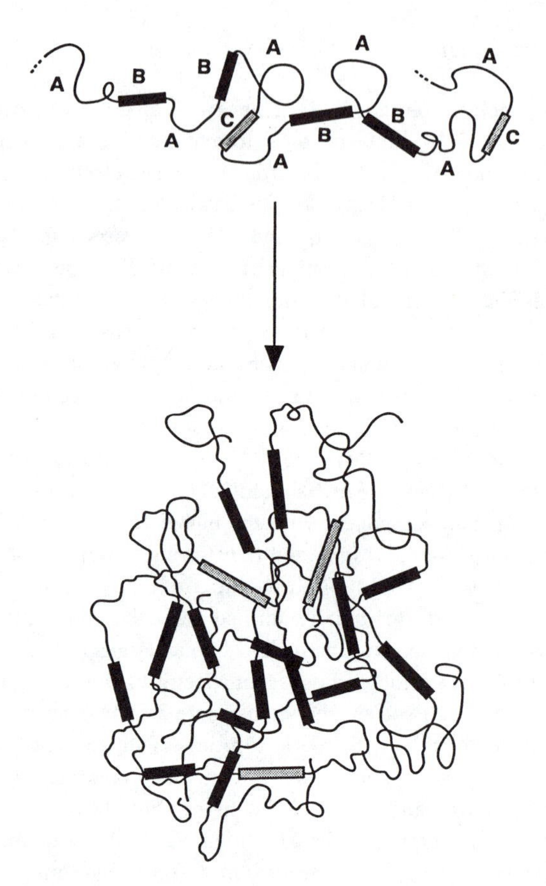

Figure 1. Schematic illustration of the protein-like folding *(1)* of an electroluminescent ternary rod-coil copolymer. The resulting self-organized nanostructure varies with copolymer composition *(10)*.

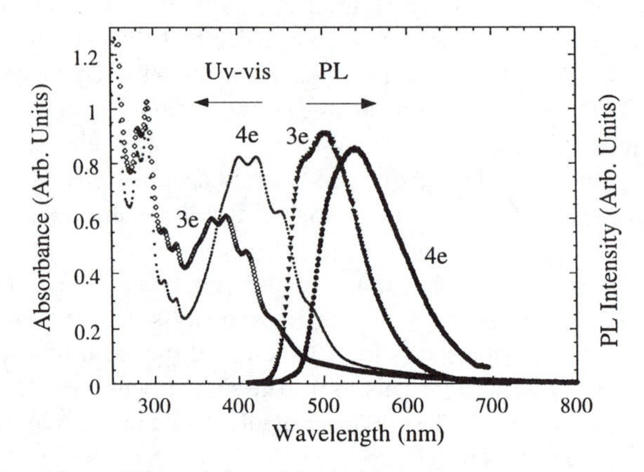

Figure 2. Absorption and fluorescence spectra of thin films of **3e** and **4e**. (λ_{ex} = 400 nm).

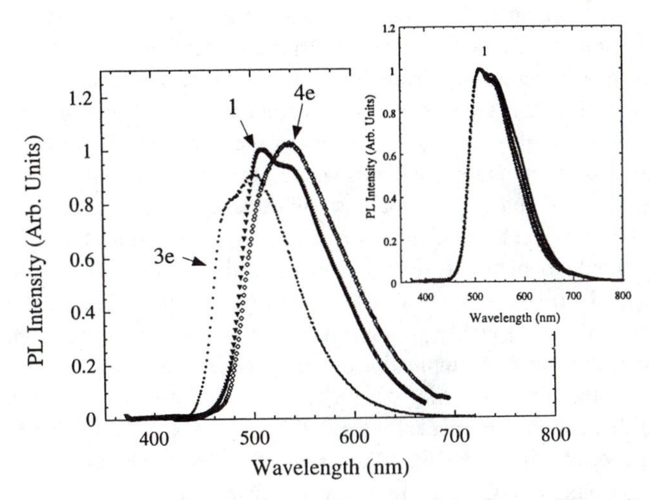

Figure 3. PL spectra of thin films of **1, 3e**, and **4e** (λ_{ex} = 400 nm). Inset: PL spectra of thin films of **1** (λ_{ex} = 350, 370, 390, 440, and 460nm, respectively) are identical at all excitation wavelength.

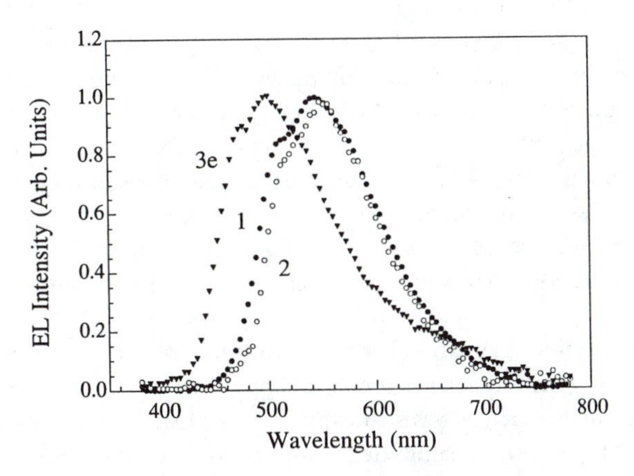

Figure 4. The electroluminescence(EL) spectra of thin films of copolymers **1**, **2**, and **3e.**

Figures 5 and 6 show brightness-voltage and current-voltage (I-V) curves, respectively, for the LED devices whose EL spectra are shown in Figure 4. Under conditions of forward bias, the EL devices have turn-on voltages of 11-13V for the rod-coil copolymers **1-3**. The I-V character of **1** shows two distinct regions as shown in Figure 6. Above the onset of electroluminescence at a field of 1x 10^6 V/cm, the current increases rapidly until saturation is reached at current densities above 170 mA/cm^2. This may arise from a regime of space-charge-limited current due to the finite mobility of charge carriers in the nanocomposites(*23*). This is especially important in nanocomposites where 80 mol% of the solid consists of the non-electroactive and non-photoactive coil-like segments. However, the data strongly show deviation from the tunneling injection mechanism of charge carriers across the Schottky depletion layer at the anode, which was shown to hold for p-type conjugated polymers such as poly (2,5-dialkoxy-phenylenevinylene)(*24*). The brightness of EL is roughly proportional to current, clearly indicating that the EL emission is due to the recombination of charge carriers injected from electrodes into the bulk of the copolymer film. Electroluminescence efficiencies, in terms of photons emitted per charge injected were 0.083% for **1** and 0.04% for **2**. These values are substantially higher than the value of 0.015% obtained for an LED employing the binary copolymer **3e** as the emissive layer. This result suggests that EET in nanostructured rod-coil copolymer polymer systems can significantly enhance EL quantum efficiency. In this regard, it is noteworthy that EL spectra could not be obtained from similarly constructed devices employing the corresponding homopolymers (PBZT, PBTPV, and PNBT) as the emissive layers, due to insufficient emission.

Device architectures other than the bilayer assembly (ITO/PVK/copolymer/Al) were also investigated, but did not give satisfactory results. Single copolymer devices (ITO/Copolymer/Al) had low emission levels, indicating a large imbalance in carrier injection or transport(*25*). When an additional layer of PVK was introduced as a hole transporting and electron blocking layer, device brightness was enhanced by at least one order of magnitude, suggesting that the carrier injection and transport were more balanced, and confirmed that the copolymers were electron-transporting materials.

EGIPT and Proton Transfer Electroluminescence. The second class of electroluminescent self-organizing rod-coil copolymers we have investigated, **5**, incorporates moieties such as **6** which exhibits excited state intramolecular proton transfer (ESIPT)(*26-29*). A simplified representation of the ESIPT process is shown in Scheme 1. Intramolecularly hydrogen bonded molecules such as 2-(2'-hydroxyphenyl) benzothiazole (HBT, **6**), which exist exclusively in the enol form (E) in the ground state, rapidly tautomerization from E* to the keto form (K*) upon photoexcitation to the first singlet excited state. The proton transfer occurs in less than a picosecond(*30*), resulting in a population inversion. The K* form can isomerize to a twisted state, and intersystem crossing to the triplet manifold is

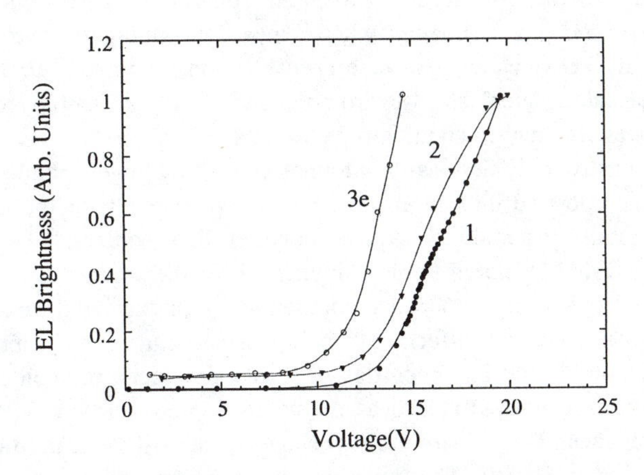

Figure 5. Brightness-Voltage curves for **1**, **2**, and **3e**.

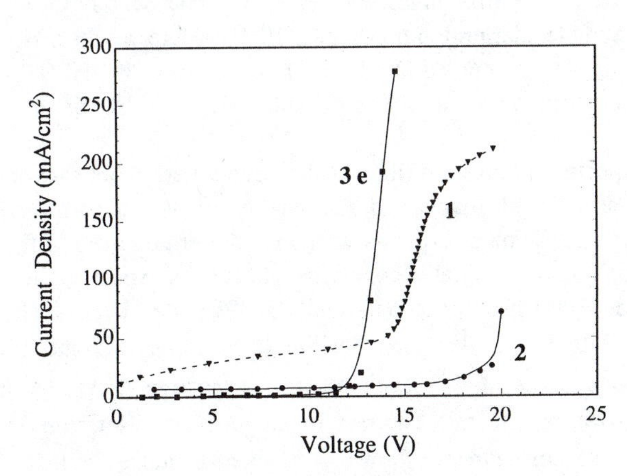

Figure 6. Current-Voltage curves for **1**, **2**, and **3e**.

also possible. Deactivation of the K* states by an emissive pathway occurs with a Stokes shift of approximately 6000 cm^{-1}. Reverse proton transfer in the ground state to regenerate the starting E form completes the cycle. The fact that the process is a closed loop which regenerates the starting material after each cycle means that ESIPT is inherently a photostabilizing process. ESIPT has thus been exploited as a mechanism for protecting polymers from photochemical degradation(*31-34*) (e.g. yellowing when exposed to sunlight). The population inversion of the keto tautomer (Scheme 1) enables some intramolecularly hydrogen bonded molecules, such as 3-hydroxyflavone, to be employed as laser dyes(*35,36*), as was suggested initially by Khan and Kasha(*37*).

Conjugated polymers with intramolecularly hydrogen bonded moieties, such as **6**, incorporated into the main chain were prepared with the objective of developing robust materials for organic polymer light sources. Those studies showed that ESIPT in polymers is complicated by the effects of concentration quenching, extended conjugation, and competition with excimer formation(*38*). In order to investigate the effects of molecular structure and supramolecular nanostructure on PL and EL properties of ESIPT polymers, rod-coil copolymers **5** were synthesized with PBTC10 as the coillike component and HPBT as the rod-like component. Preliminary findings showed that ESIPT occurred in the 5 % HPBT copolymer **5b**(*39*). The emission spectrum was due to the K* form: emission from competing species was not significant. This copolymer(**5b**) was also shown to be electroluminescent via a new chemical reaction, electrically generated intramolecular proton transfer (EGIPT). Here we discuss how changes in the copolymer composition, from 1% to 100 % HPBT, affect the supramolecular structure and the attendant PL and EL of the solid state copolymers.

UV spectra of thin solid films of the copolymers **5** are shown in Figure 7. The variation of band shape and the energy of the absorption band with copolymer composition corresponds well with the effects seen in the case of the corresponding non-hydroxylated copolymers **3**(*10*). As expected, there is a steady increase in the HPBT absorption bands at 375, 395 and 421 nm, for compositions up to 40 %, with λ_{max} at 395 nm. For the 75 % copolymer and 100 % HPBT homopolymer, λ_{max} is at 445 and 450 nm, respectively, with a smaller peak occurring at 479 nm. In stark contrast to the shifting absorption λ_{max} , the PL spectra of these copolymers, shown in Figures 8 and 9, indicate a constant emission maximum. In the case of copolymers with low HPBT content (less than 40 %, Figure 8), the dominant 543-nm emission band, which is significantly Stokes shifted (*ca.* 6000 cm^{-1}) from the absorption, does not shift with composition. This peak clearly arises from emission of the excited keto tautomer (K* in Scheme 1). Evidence for this assignment is as follows. First, the large Stokes shift between the emission band and the absorption band is typical of ESIPT processes leading to the keto state, a process very well known to occur for

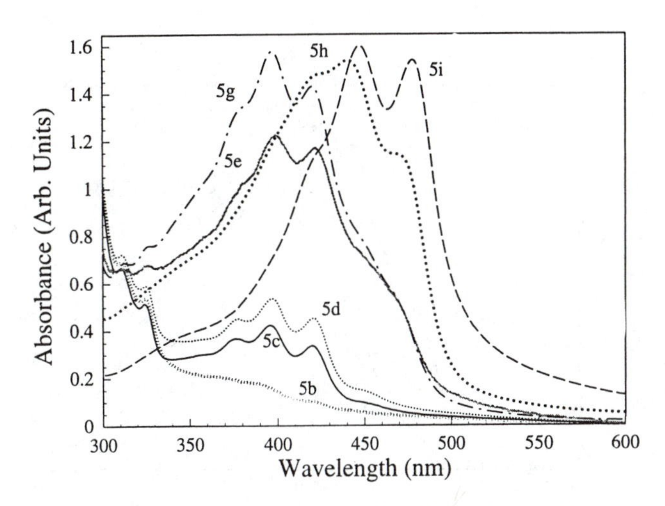

Figure 7. UV spectra of rod-coil copolymers **5**.

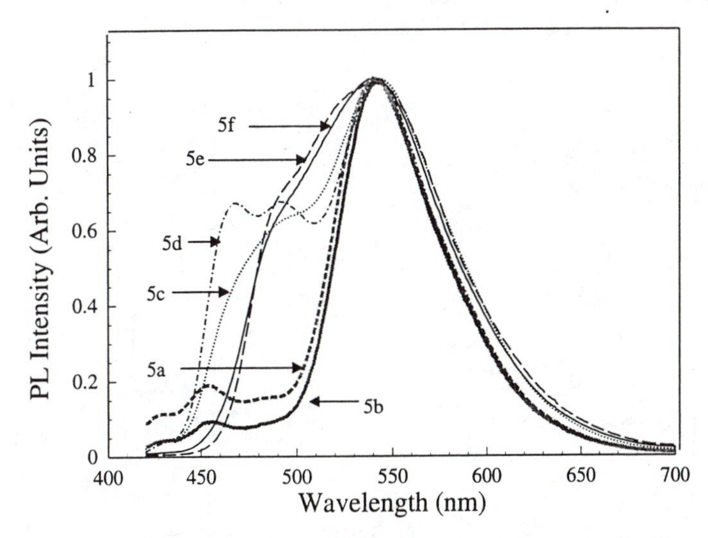

Figure 8. Photoluminescence spectra of copolymers **5a-5f**. All copolymers were excited at 395 nm.

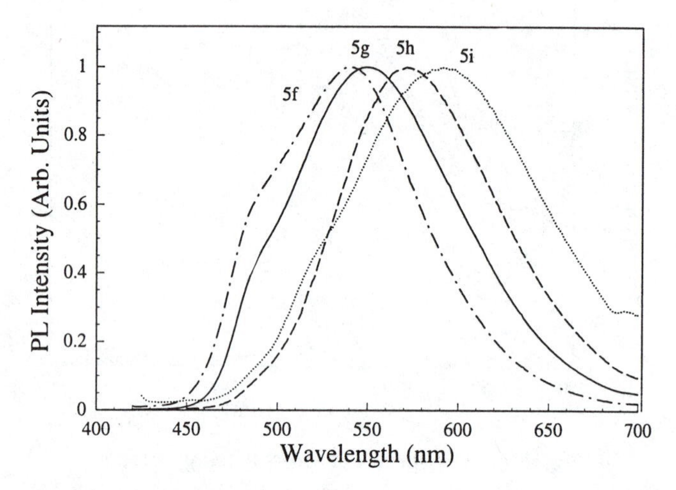

Figure 9. Photoluminescence spectra of copolymers **5g-5i**. All copolymers were excited at 395 nm.

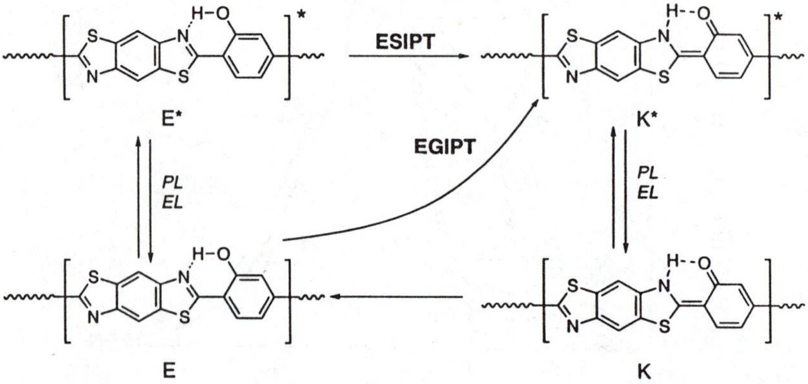

ESIPT

E*

K*

EGIPT

PL
EL

PL
EL

E

K

Scheme 1

intramolecularly hydrogen bonded molecules structurally similar to the HPBT component of **5**(*29*). In contrast, the corresponding copolymers without hydroxy side groups, which are not capable of ESIPT, have a much smaller Stokes shift, with the emission λ_{max} occurring at wavelengths shorter than 500 nm*(10)*. Second, the transient absorption spectrum of **5b** (5% HPBT), shown in Figure 10, has a region of optical gain (stimulated emission) between 540 nm and 640 nm. The keto form K does not initially exist in the ground state, so formation of K* after excitation leads to a population inversion, facilitating observation of stimulated emission. This phenomenon has been well documented in molecules such as **6**(*29*).

Upon increasing the HPBT content of the rod-coil copolymers, a number of trends in the PL properties can be seen. Optical gain, as determined by transient absorption spectroscopy, occurs only in the 1 % (**5a**) and 5 % (**5b**) copolymers. In copolymers with higher HPBT content, competition from transient absorption overwhelms the stimulated emission to the extent that no gain could be seen. Another phenomenon is that the minor bands at the blue end of the emission spectrum increase in magnitude and shift to longer wavelengths, relative to the 543 nm keto emission band (Figure 8). For copolymers with at least 40 % HPBT, the broad emission spectrum appears to be a single broad peak which is red shifted with increasing HPBT content (Figure 9). The origin of these bands is normal emission from E* (Scheme 1) in the case of copolymers with low HPBT content (<40%), which gives way to excimer emission from an (EE)* state in the high HPBT content copolymers. The excitation spectra of the blue end emission bands near 450-470nm match those of the keto emission band, suggesting that both bands arise from a common ground state.

We have previously demonstrated that extended conjugation inhibits ESIPT in polymers, forcing emission to occur from E*, due to an increase in the energy barrier on the excited state potential energy surface*(38)*. It is possible that there is a potential energy barrier to proton transfer on the excited state potential energy surface of copolymers **5**, the magnitude of which increases with increases in conjugation length or HPBT content. This would lead to incomplete proton transfer and emission from both E* and K* (Scheme 1). As the HPBT content in the copolymer increases, the average conjugation length also increases, since the condensation polymerization technique leads to a statistical distribution of molecular weights. This increases the probability that ESIPT will not occur after photoexcitation, due to an unfavorably long conjugation length. The radiative deactivation thus occurs from E*, with a normal Stokes shift. The net result of this effect is an increase in the magnitude, and the shift to lower energies, of the blue end emission peaks. As the average conjugation length increases further, the magnitude of the ESIPT band relative to the E* bands decreases, until the point where ESIPT is completely prohibited. In these copolymers, with high HPBT content, the emission characteristics are similar to those of the corresponding non-hydroxylated copolymers*(10)*. The red shift of the emission band, upon changing from 40 to 70 to 100 % HPBT, is clearly a result of the change of supramolecular

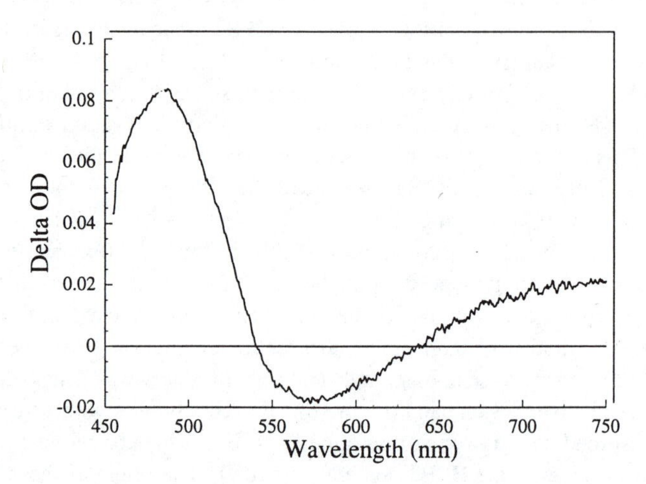

Figure 10. Transient absorption spectrum of **5b** (5 % HPBT). (λ_{ex} = 355 nm; delay = 50 ps).

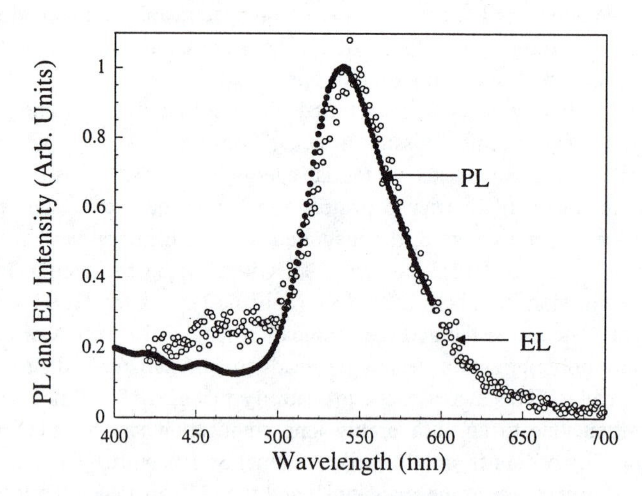

Figure 11. Electroluminescence(EL) and photoluminescence(PL) emission spectra of **5b** (5 % HPBT).

structure and morphology, and hence the increasing contribution of effects of aggregation and excimer emission*(10)*.

The electroluminescent properties of copolymers **5** were also investigated(*39*). Figure 11 shows a representative EL spectrum obtained from a device employing **5b** (5 % HPBT) as the emissive layer. The PL spectrum of the same copolymer is also shown. The PL and EL spectra match very closely, showing that EL emission originates from deactivation of K*. EL spectra for other copolymers **5** are shown in Figure 12. It is seen that the blue end bands, which were found in the PL spectra (Figures 8 and 9) are also found in the case of EL spectra, suggesting that wavelength and peak width of emission in the electroluminescence of the copolymers can be controlled through the composition and supramolecular nanostructure. The origin of the minor bands in the EL spectra cannot be from the PVK layer, which emits at 430 nm(*40*), because the minor bands are also seen in the PL spectra of films taken without a PVK layer. Instead, the origin of the blue-end bands is most likely the same as in the PL spectra. Current-voltage and brightness-voltage curves(not shown) track rather well for the copolymers **5**, indicating that EL emission is due to the recombination of charge carriers injected from electrodes into the copolymer films.

The fact that emission occurs from a K* state, which is generated electrically rather than photochemically, is quite significant: these are the first materials reported to exhibit *electrically generated intramolecular proton transfer*, EGIPT. The mechanism of formation of K* by photoexcitation is well documented to occur by a rapid proton transfer from E*(*29*). However, the mechanism of formation of K* by the new EGIPT reaction is yet to be elucidated. The simplest representation of the overall EGIPT reaction is given in eq. 1.

$$E^{\bullet-} + E^{\bullet+} \rightarrow K^* + E \rightarrow E + E + h\nu \quad (1)$$

Electrical injection of holes and electrons into **5** (at compositions which exhibit photoinduced ESIPT) results in the immediate generation of enol radical anions, $E^{\bullet-}$ and enol radical cations, $E^{\bullet+}$. After this occurs, however, it is not clear whether K* forms simply through tautomerization of E*, or by combination of radical ions of the keto form which may arise from tautomerization of enol radical ions, of which many pathways can be supposed. The detailed mechanism of the observed EGIPT luminescence remains to be established.

Conclusions

The effects of supramolecular structure and morphology on electroluminescence of polymers have been investigated by means of novel ternary and binary rod-coil copolymers. Rod-coil-rod ternary copolymers with two differing rod components were shown to exhibit Förster-type excited state energy transfer, resulting in

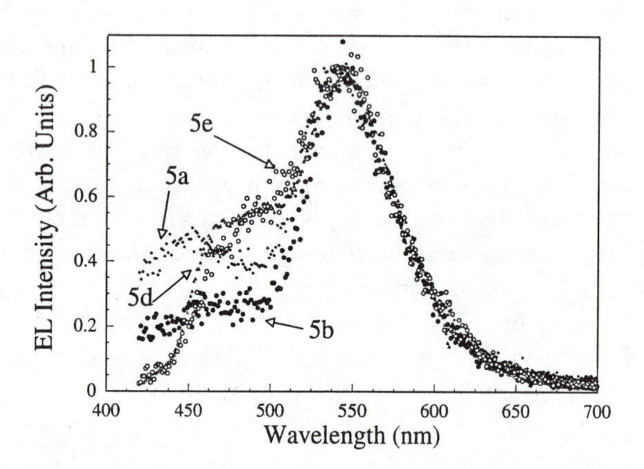

Figure 12. Electroluminescence emission spectra of rod-coil copolymers **5** (1%, 5 %, 10 %, and 15 % HPBT; **5a**, **5b**, **5d**, and **5e**, respectively).

about sixfold enhancement of the quantum efficiency of electroluminescence relative to the corresponding binary rod-coil copolymers. Further improvements of the electroluminescence quantum efficiency can be achieved by selecting the appropriate rigid-rod pairs in ternary copolymers, in which overlap between donor emission and acceptor absorption is maximized. Moreover, the space-charge-limited behavior of the EL devices of these rod-coil copolymers suggests that replacement of the inert nonactive coil by a hole transporting coil segment, the quantum efficiency and the brightness of the LEDs can be further improved. It was also demonstrated that both molecular and supramolecular structures influence the EGIPT reaction and proton transfer electroluminescence of rod-coil copolymers containing intramolecularly hydrogen bonded moieties. Although an electrically pumped diode laser based on EGIPT polymers is yet to constructed, the preliminary EL results and the intrinsic population inversion in such materials suggest that it is a feasible concept.

Acknowledgments

This research was supported by the Office of Naval Research and in part by the National Science Foundation(CHE-9120001, CTS-9311741). We thank Xuejun Zhang for technical assistance with LED fabrication and Elizabeth Gaillard for technical assistance with transient absorption spectroscopy.

Literature Cited

(1) (a) Richards, F. M. *Scientific Am.* **1991**, *January,* 54-63. (b) Kim, P. S.; Baldmin, R. L. *Ann. Rev. Biochem.* **1982**, *51*, 459-489.
(2) (a) Flory, P. J. *Proc. Royal Soc. Lond. A* **1956**, *234,* 60.(b) Chandrasekhar, S. *Liquid Crystals*, 2nd ed; Cambridge University Press: New York, 1992.
(3) Flory, P.J. *Macromolecules* **1978**, *11*, 1138-1141.
(4) Wolfe, J.F. In *Encyclopedia of Polymer Science and Engineering*; Wiley: New York, 1988; Vol. 11, pp 601-635.
(5) Arnold, F.E., Jr.; Arnold, F.E. *Adv. Polym. Sci.* **1994**, *117*, 257-295.
(6) Roberts, M. F.; Jenekhe, S. A. *Chem. Mater.* **1994**, *6*, 135-145.
(7) (a) Jenekhe, S.A.; Osaheni, J.A. *Science* **1994**, *265*, 765-768. (b) Osaheni, J.A.; Jenekhe, S.A. *Macromolecules* **1994**, *27*, 739-741.
(8) (a) Osaheni, J. A.; Jenekhe, S. A. *Macromolecules* **1993,** *26,* 4726-4728. (b) Osaheni, J. A.; Jenekhe, S. A. *Chem. Mater.* **1995**, *7,* 672-682.
(9) (a) Jenekhe, S. A.; Zhang, X.; Chen, X. L.; Choong, V.-E.; Gao, Y.; Hsieh, B. R. *Chem. Mater.* **1997,** *9,* In Press. (b) Zhang, X.; Kale, D. M.; Jenekhe, S. A. *Macromolecules*, Submitted.
(10) (a) Jenekhe, S.A.; Osaheni, J.A. *Chem. Mater.* **1994**, *6,* 1906-1909. (b) Osaheni, J.A.; Jenekhe, S.A. *J. Am. Chem. Soc.* **1995**, *117*, 7389-7398.
(11) Yang, C.J.; Jenekhe, S.A. *Supramolecular Science* **1994**, *1*, 91-101.

(12) (a) Burroghes, J.H.; Bradley, D.D.C.; Brown, A.R.; Marks, R.N.; Mackay, K.; Friend, R.H.; Burn, P.L.; Holmes, A.B. *Nature* **1990**, *347*, 539-541. (b) Burn, P.L.; Holmes, A.B.; Kraft, A.; Bradley, D.D.C.; Brown, A.R.; Friend, R.H.; Gymer, R.W. *Nature* **1992**, *356*, 47-49.

(13) Service, R.F. *Science* **1995**, *269*, 1042.

(14) Zhang, C.; Braun, D.; Heeger, A.J. *J. Appl. Phys.* **1993**, 73, 5177-80.

(15) Zhang, C.; Von Seggern, H.; Pakbaz, K.; Kraabel, B.; Schmidt, H.W.; Heeger, A.J. *Synth. Met.* **1994**, 62, 35-40.

(16) (a) Jenekhe, S. A.; Johnson, P. O.; Agrawal, A. K. *Macromolecules* **1989**, *22*, 3216. (b) Jenekhe, S. A.; Johnson, P. O. *Macromolecules* **1990**, *23*, 4419.

(17) Osaheni, J.A.; Jenekhe, S.A. *Chem. Mater.* **1992**, *4*, 1282-1290.

(18) Miura, Y.; Torres, E.; Panetta, C.A. *J. Org. Chem.* **1988,** *53*, 439-440.

(19) Jenekhe, S.A.; Osaheni, J.A. *J. Phys. Chem.* **1994**, *98,* 12727-12736.

(20) Greenham, N.C.; Friend, R.H.; Bradley, D.D.C. *Adv. Mater.* **1994**, 6, 491-494.

(21) Förster, T. *Ann. Phys.* **1948**, *2*, 55.

(22) Pope, M.; Swenberg, C., *Electronic processes in Organic crystals,* Oxford University Press: New York, 1982.

(23) Gruner, J.; Wittmann, H.F.; Hamer, P.J.; Friend, R.H.; Huber, J,; Scherf, U.; Mullen, K.; Moratti, S.C.; Holmes, A.B. *Synth. Met.* **1994**, *67*, 181-185.

(24) Parker, I.D. *J. Appl. Phys.* **1994**, 75, 1656-1666.

(25) Parker, I.D.; Pei, Q.; Marrocco, M. *Appl. Phys. Lett.* **1994**, *65*, 1272-1274.

(26) Weller, A. *Z. Elektrochem.* **1956**, *60*, 1144-1147.

(27) Barbara, P.F.; Walsh P.K.; Brus, L.E. *J. Phys. Chem.* **1989**, *93*, 29-34.

(28) See the special issues: *Chem. Phys.* **1989**, *136*, 153-360; *J. Phys. Chem.* **1991**, *95*, 10215-10524.

(29) Formosinho, S.J.; Arnault, L.G. *J. Photochem. Photobiol. A: Chem.* **1993**, *75*, 21-48.

(30) Frey, W.; Laermer, F.; Elsaesser, T. *J. Phys. Chem.* **1991**, *95*, 10391-10395.

(31) Heller, H.J.; Blattmann, H.R. *Pure Appl. Chem.* **1973**, *36*, 141-161.

(32) Werner, T. *J. Phys. Chem.* **1979**, *83*, 320-325.

(33) Nir, Z.; Vogl, O. *J. Polym. Sci. Polym. Chem. Ed.* **1982**, *20*, 2735-2754.

(34) O'Connor, D.B.; Scott, G.W.; Coulter, D.R.; Gupta, A.; Webb, S.P.; Yeh, S.W.; Clark, J.H. *Chem. Phys. Lett.* **1985**, *121*, 417-422.

(35) Brucker, G.A.; Swinney, T.C.; Kelley, D.F. *J. Phys. Chem.* **1991**, *95*, 3190- 3195.

(36) Ferrer, M.L.; Acuna, A.U.; Amat-Guerri, F.; Costela, A.; Figuera, J.M.; Florido, F.; Sastre, R. *Appl. Opt.* **1994**, *33*, 2266-2272.

(37) Khan, A.U.; Kasha, M. *Proc. Natl. Acad. Sci. USA.* **1983**, *80*, 1767-1770.

(38) (a) Tarkka, R.M., Jenekhe, S.A. *Mat. Res. Soc. Symp. Proc.* **1996**, *413*, 97-102. (b) Tarkka, R. M.; Jenekhe, S. A. *Chem. Phys. Lett.* **1996**, *260,* 533-538.

(39) Tarkka, R.M.; Zhang, X.; Jenekhe, S.A. *J. Am. Chem. Soc.* **1996**, *118*, 9438-9439.

(40) Hu, B.; Yang, Z., Karasz, F.E. *J. Appl. Phys.* **1994**, *76*, 2419-2422.

Chapter 33

Novel Optically Responsive and Diffracting Materials Derived from Crystalline Colloidal Array Self-Assembly

S. A. Asher, J. Weissman, H. B. Sunkara[1], G. Pan, J. Holtz, Lei Liu, and R. Kesavamoorthy[2]

Department of Chemistry, University of Pittsburgh, Pittsburgh, PA 15260

Novel tunable and switchable diffracting devices useful in the UV, visible and IR region are fabricated through self assembly of monodisperse, highly charged colloidal particles into crystalline colloidal arrays (CCA). The particles form BCC or FCC arrays with spacings that diffract light. We fabricated devices where the particle size or spacing changes with temperature to tune the diffraction. We also fabricated nonlinear devices which switch in ~3 ns. This CCA utilizes dyed particles, which are index matched to the medium. No light is diffracted at low intensities, but at high intensities the particles heat up, their index mismatches from the medium to turn the CCA diffraction on within 5 ns.

Crystalline colloidal arrays (CCA) are mesoscopically periodic fluid materials which efficiently diffract light meeting the Bragg condition (*1-4*). These materials consist of arrays of colloidal particles which self assemble in solution into BCC or FCC crystalline arrays (*1,5*) (Figure 1) with lattice constants in the mesoscale size range (50 to 500 nm). Just as atomic crystals diffract x-rays that meet the Bragg condition, CCA diffract UV, visible, and near IR light (*2-4*); the diffraction phenomena resemble that of opals, which are close-packed arrays of monodisperse silica spheres (*6*).

The CCA, however, can be prepared as macroscopically ordered arrays of non-close-packed spheres. This self assembly is the result of electrostatic repulsions between colloidal particles, each of which has numerous charged surface functional groups. We have concentrated on the development of CCA which diffract light in the visible spectral region, and have generally utilized colloidal particles of ca. 100 nm diameter (*7*). These particles have thousands of charges which result from

[1]Current address: Space Science Laboratory, George C. Marshall Space Flight Center, National Aeronautics and Space Administration, Marshall Space Flight Center, AL 35812

[2]Current address: Materials Science Division, Indira Gandhi Centre for Atomic Research, Kalpakkam, 603102 Tamilnadu, India

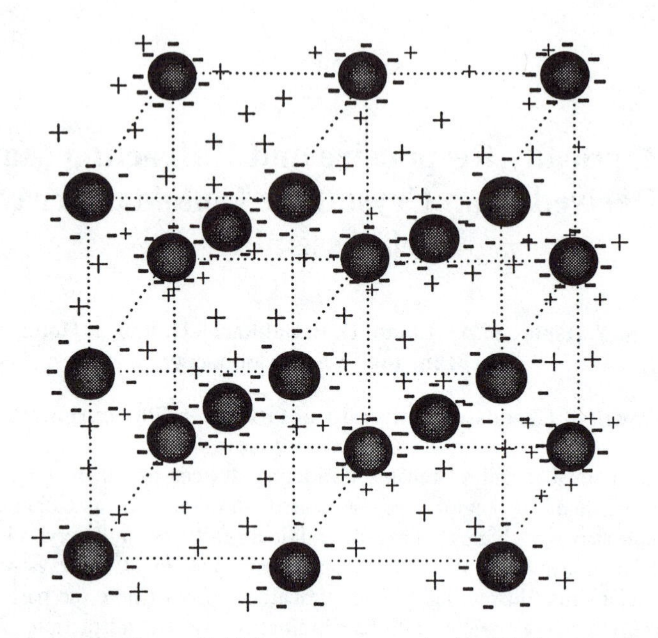

Figure 1. Crystalline colloidal array composed of a BCC array of negatively charged colloidal particles. The colloidal particle charges derive from ionized surface sulfonate groups. The counterions in the surrounding medium maintain the system net neutrality.

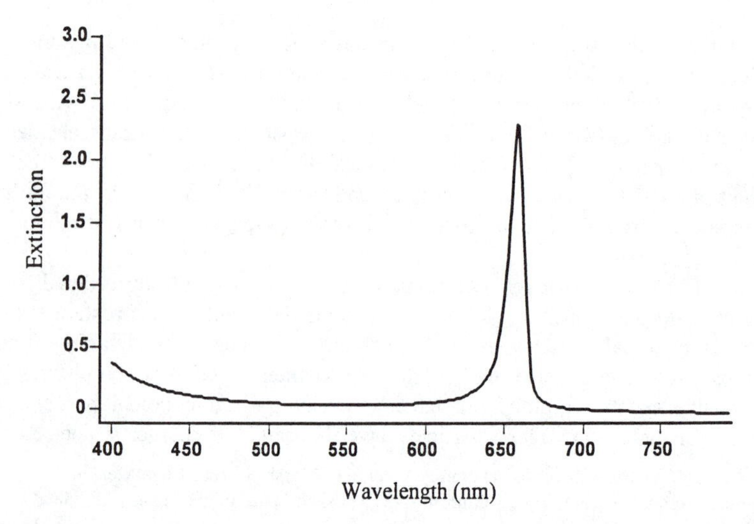

Figure 2. Transmission spectrum of a 1.0 mm thick crystalline colloidal array formed from 131-nm diameter poly(heptafluorobutyl methacrylate) particles. The particle volume fraction was 6.13%.

the ionization of sulfonate groups attached to the particle surfaces. The nearest neighbor distances are often ca. 200 nm.

Optical Diffraction Devices

These BCC or FCC cubic arrays are well ordered and the arrays strongly diffract light in the visible spectral region (Figure 2). All light meeting the Bragg condition is diffracted while adjacent spectral regions freely transmit. We earlier demonstrated the use of these devices as narrow-band optical diffraction filters (*1c,3,8,9*). We more recently developed methods (*10-12*) to solidify and rigidize these arrays by imbedding the CCA cubic lattice in a hydrogel polyacrylamide matrix (Figure 3). This system can be prepared such that the acrylamide and the colloidal particles occupy a small percent of the sample volume, which mostly consists of water. The medium surrounding the spheres can be modified since other solvents can be diffused into the polymerized array to replace the water. Although the hydrogel-linked CCA undergoes swelling and shrinkage as the solvent medium is changed, the array ordering is maintained. Films of this array can be prepared where the (110) plane of the BCC lattice is well oriented and parallel to the surface. The (110) planes of the periodic colloidal structure in the film strongly diffract light meeting the Bragg condition.

Thermally Tunable Diffraction Devices

More recently we utilized the well-known temperature-induced volume phase transition properties of poly(N-isopropylacrylamide) (PNIPAM) (*13-15*) to create novel CCA materials with variable sphere size and variable array periodicity (*16*). In water below ~30°C, PNIPAM is hydrated and swollen, but when heated above its lower critical solution temperature (~32°C) it undergoes a reversible volume phase transition to a collapsed, dehydrated state. The temperature increase causes the polymer to expel water and shrink into a more hydrophobic polymer state.

We developed a synthesis of monodisperse, highly charged colloidal particles of PNIPAM whose diameter depends on temperature (*16*). NIPAM was polymerized with the ionic comonomer 2-acrylamido-2-methyl-1-propane sulfonic acid to increase the colloid surface charge which facilitates CCA self-assembly. Figure 4 shows the temperature dependence of the colloidal sphere diameter, which increases from ~100 nm at 40°C to ~300 nm at 10°C.

These PNIPAM colloids self-assemble in deionized water to form CCA both above and below the polymer phase transition temperature. The ordered array diffracts light almost following Bragg's diffraction law (but not exactly as shown elsewhere (*4*)):

$$m\lambda = 2nd\sin\theta \qquad (1)$$

where m is the order of diffraction, λ is the wavelength of incident light, n is the suspension refractive index, d is the interplanar spacing, and θ is the glancing angle between the incident light and the diffracting crystal planes (*4*), which are oriented parallel to the crystal surface in the CCA we prepare. Figure 5 shows the resulting extinction spectra of a PNIPAM CCA at 10°C and 40°C. At low temperatures, the CCA particles are highly swollen, almost touching, and diffract weakly. Above the

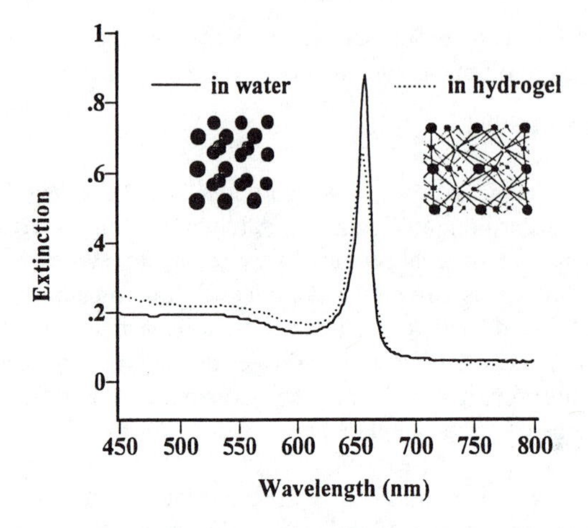

Figure 3. Transmission spectra of 155-nm diameter poly(heptafluorobutyl methacrylate) crystalline colloidal array before and after solidification in a polyacrylamide hydrogel matrix.

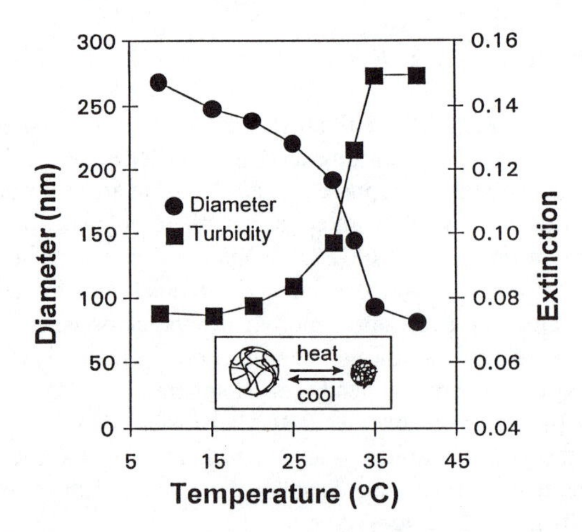

Figure 4. Temperature dependence of the PNIPAM colloid diameter and turbidity. The diameter was determined using a commercial quasielastic light scattering apparatus (Malvern Zetasizer 4). The turbidity was measured for a disordered dilute dispersion of these PNIPAM colloids by measuring light transmission through a 1.0 cm pathlength quartz cell with a UV-visible-near IR spectrophotometer. Solids content of the sample in the turbidity experiment was 0.071%, which corresponds to a particle concentration of 2.49×10^{12} spheres/cc. Also shown is the temperature dependence of the turbidity of this random colloidal dispersion. The light scattering increases as the particle becomes more compact due to its increased refractive index mismatch from the aqueous medium (*16*) (Adapted from ref. 16).

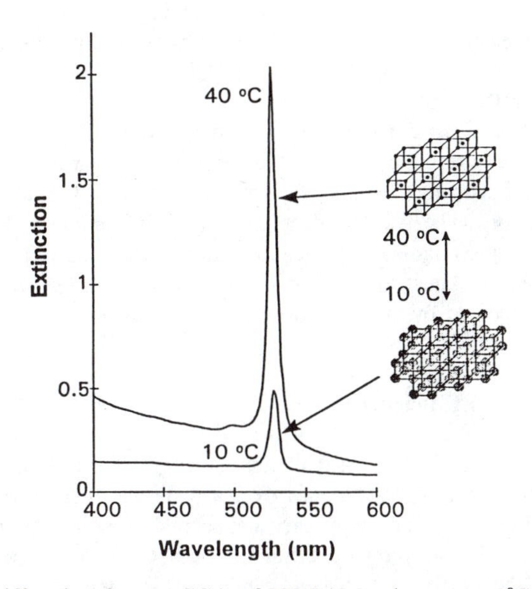

Figure 5. Diffraction from a CCA of PNIPAM spheres at 10°C and at 40°C. The spectra were recorded using a UV-visible-near IR spectrophotometer (Perkin Elmer λ-9). The dispersion was contained in a 1.0 mm quartz cuvette oriented at normal incidence to the incident beam. The observed diffraction switching behavior was reversible; these spectra were recorded after the seventh consecutive heat-cool cycle. Inset: Pictorial representation of the temperature switching between a swollen sphere array below the phase transition temperature and an identical compact sphere array above the transition (Adapted from ref. 16).

phase transition temperature, these particles become compact, increase their refractive index and diffract nearly all incident light at the Bragg wavelength. The temperature change does not affect the lattice spacing. This material acts as a thermally controlled optical switch.

In addition, we used a similar concept (*16*) to fabricate wavelength-tunable diffraction devices using a PNIPAM gel to control the periodicity of the CCA. We polymerized a CCA of polystyrene (PS) spheres in PNIPAM. This polymerized CCA film (PCCA) shrinks and swells continuously and reversibly between 10°C and 35°C; the embedded PS sphere array follows, changing the lattice spacing and thus the diffracted wavelength (Figure 6). The inset to Figure 6 shows the temperature dependence of the diffracted wavelength for this PCCA film, where the incident light is normal to the (110) plane of the lattice.

This PCCA film functions as a tunable optical filter; the diffracted wavelength can be altered by varying either the temperature or the angle of incidence. At a fixed angle to the incident beam this PCCA acts as a tunable wavelength reflector. The width and height of the diffraction peak can be easily controlled by choosing colloidal particles of different sizes and refractive indices or by making different thickness PCCA films (*16*). The tuning range of this device can be widened or narrowed by synthesizing PCCA films with lower or higher cross-linking concentrations.

Improved Diffraction Efficiencies

We also prepared CCA of silica colloidal spheres of ~100-nm diameter and examined their diffraction characteristics over a wide wavelength range. Figure 7 shows the extinction peaks for the primary and secondary diffraction from this CCA for normal incidence. The primary diffraction at 606 nm results from first order diffraction from the (110) BCC planes, while the 315 nm secondary diffraction results from the combination of first order diffraction from numerous higher index planes of the BCC CCA lattice and second order diffraction from the (110) plane. The secondary diffraction efficiency of the BCC CCA is dramatically increased at $\lambda_0/2$ (more than 10-fold) compared to the λ_0 diffraction efficiency (*4*). This anomalous efficiency increase is due to the orientation of numerous higher index planes which fortuitously diffract light at $\lambda_0/2$. This increased diffraction efficiency will increase the response of the nonlinear CCA optical switches described below.

Nonlinear Optical Diffracting Elements

We earlier suggested that we could build nonlinear optical switching materials (*8,17-19*) by creating a CCA of optically nonlinear colloidal particles. One approach would utilize organic polymer spheres containing a dye. We would adjust the composition of the medium such that the real part of the refractive index of the spheres was identical to that of the medium (*18c*). Thus, the array would not diffract light which meets the Bragg condition. At high incident light intensities, however, significant heating would occur within the colloidal particles, the temperature would increase, the refractive index would decrease, and the array

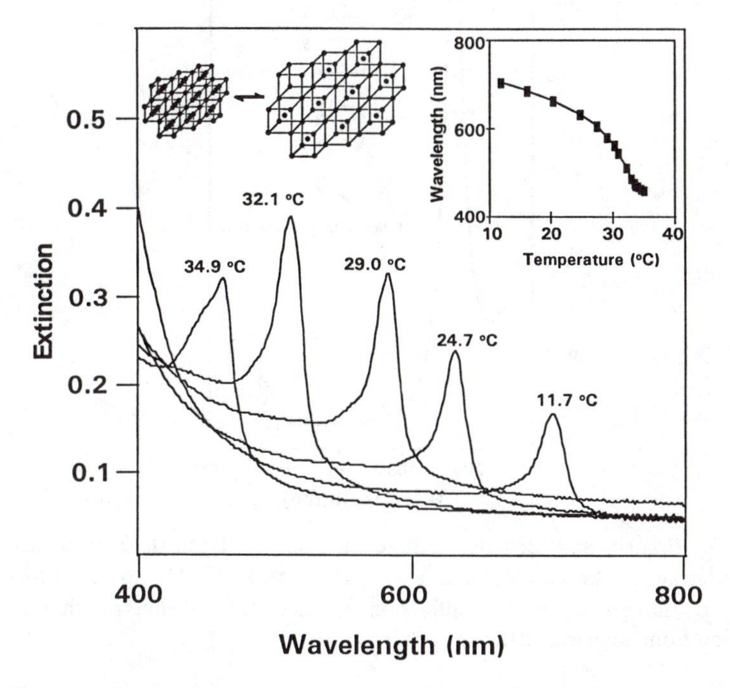

Figure 6. Temperature tuning of Bragg diffraction from a 125-μm-thick PCCA film of 99-nm polystyrene spheres embedded in a PNIPAM gel. The diffraction wavelength shift results from the temperature-induced volume change of the gel, which alters the lattice spacing. Spectra were recorded in a UV-visible-near IR spectrophotometer with the sample placed normal to the incident light beam (Adapted from ref.16).

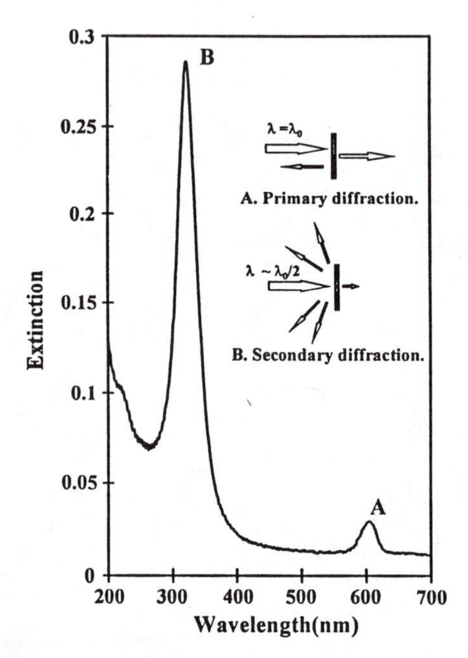

Figure 7. Primary and secondary diffraction from a 6.1-μm thick crystalline colloidal array. A. Primary diffraction at λ_0 from the BCC (110) plane in first order. B. Intense secondary diffraction at $\lambda_0/2$ due to superposition of diffraction from numerous lattice planes.

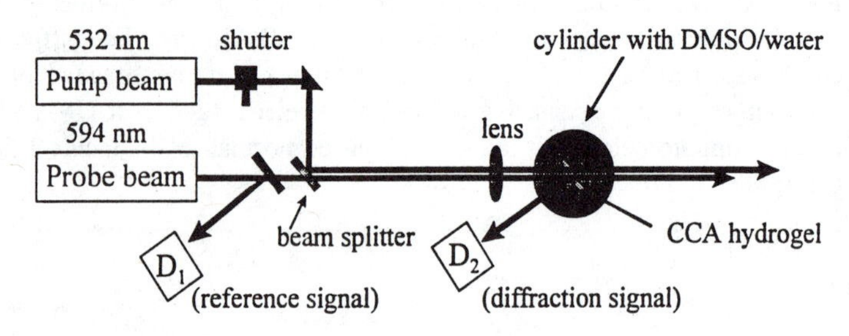

Figure 8. Schematic of the experimental apparatus for optical nonlinear diffraction measurements.

would "pop up" to diffract light. We calculated that the switching time could be within 5 nsec for pulsed laser sources (*18c*).

We synthesized 138-nm diameter monodisperse colloidal particles of heptafluorobutyl methacrylate and covalently attached acylated Oil Blue N dye to these colloidal spheres. These monodisperse highly charged fluorinated colloids have the lowest refractive index (n=1.386) of any monodisperse colloid known. These fluorinated colloids self-assemble into a CCA which was polymerized within an acrylamide hydrogel. Dimethyl-sulfoxide was added to the mainly aqueous medium in order to adjust the medium refractive index to either slightly above or slightly below that of the colloidal particles.

Diffraction by the array was monitored using the experimental apparatus shown in Figure 8. The 532 nm pump beam derived from a Coherent Inc. Infinity frequency doubled YAG laser. The probe beam derived from a dye laser pumped by the 532 nm beam. The pump and the probe beams were ca. 3.5 nsec duration, and were made coincident on the sample, but the probe beam was adjusted temporally such that it was delayed by ca. 2.5 nsec compared to the pump beam. The relative angle of the sample to the probe beam was adjusted such that the Bragg condition was met. If the refractive indices were exactly matched no diffraction should be observed in the absence of the pump beam. If the refractive index of the colloidal particles was adjusted to be slightly greater or less than that of the medium a small fraction of the probe beam will diffract.

Figure 9 shows the extinction spectrum of this dyed crystalline colloidal array hydrogel film measured at normal incidence. The broad 520 nm band results from colloidal particle dye absorption. The 640 nm peak results from the BCC (110) plane Bragg diffraction. The film was oriented such that its normal made an angle of ca. 15° relative to the incident beam, such that light at the dye laser output at 594 nm would be diffracted. Figure 10 shows three measurements where the medium has a refractive index of either 1.3813 or 1.3908, which is either below or above the sphere refractive index, or for a PCCA without dye. Figure 10 plots the relative value of the diffracted light intensity at various pump energy values compared to the value in the absence of the pump laser. As expected, if the colloid refractive index is below that of the medium, incident beam heating further decreases the colloid refractive index below that of the medium and the diffracted intensity increases. In contrast, if the colloid refractive index is above that of the medium, pump beam heating decreases the refractive index towards that of the medium and the diffraction decreases. Without dye no nonlinear response appears. Thus, we have now observed our previously predicted nonlinear optical switching behavior from CCA. We are continuing refinement of the optical materials described here.

Acknowledgments. The authors gratefully acknowledge financial support from the Office of Naval Research through Grant No. N00014-94-1-0592, Air Force Office of Scientific Research through Grant No. F49620-93-1-0008, and from the University of Pittsburgh Materials Research Center through the Air Force Office of Scientific Research Grant No. AFOSR-91-0441.

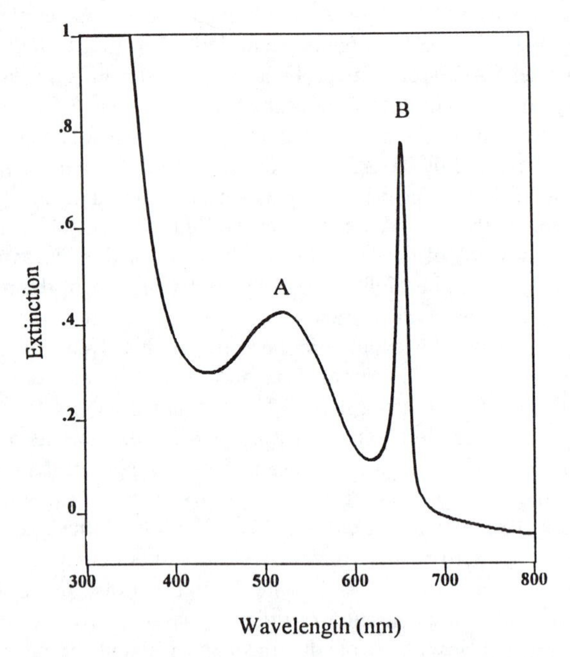

Figure 9. Transmission spectrum of a 138-nm dyed poly(heptafluorobutyl methacrylate) crystalline colloidal array solidified in a polyacrylamide hydrogel matrix. A: Dye absorption peak. B: CCA diffraction peak.

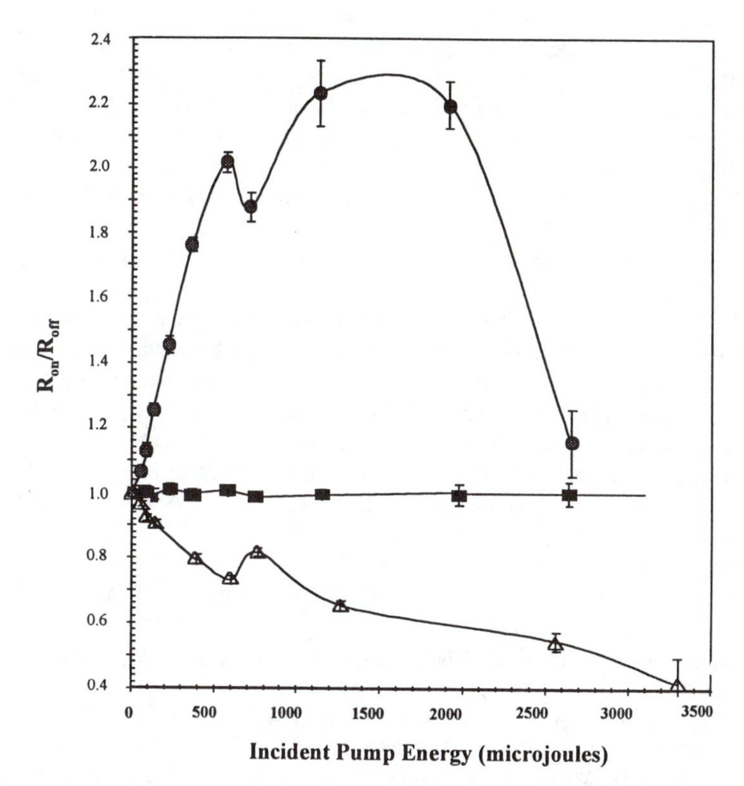

Figure 10. Pump beam energy dependence of Bragg diffraction intensity. The relative value of Bragg diffraction intensities (R_{on}/R_{off}) were monitored at various pump beam energies. • dyed CCA hydrogel in the medium of water and DMSO at n=1.3908; Δ at n=1.3813; ▪ undyed CCA hydrogel in the medium at n=1.3875.

Literature Cited

(1) The physics of crystalline colloidal array ordering and phase transitions is quite extensive with a few hundred references over the last 30 years. The following recent publications are excellent reviews.
(a) Thirumalai, D. *J. Phys. Chem.* **1989,** *93*, 5637. (b) Walsh, A. M.; Coalson, R. D. *J. Chem. Phys.* **1994,** *100*, 1559. (c) Asher, S. A., U.S. Patents **1986,** #4,627,689, 4,632,517 and **1995,** #5,452,123. (d) Hiltner, P.A.; Krieger, I. M. *J. Phys. Chem.* **1969**, *73*, 2386. (e) Clark, N. A.; Hurd, A. J.; Ackerson, B. J. *Nature* **1979**, *281*, 57. (f) Alexander, S.; Chaikin, P. M.; Grant, P.; Morales, G. J.; Pincus, P.; Hone, D. *J. Chem. Phys.* **1984,** *80*, 5776. (g) Monovoukas, Y; and Gast, A. P. *J. Colloid Interface Sci.* **1989,** *128*, 533. (h) Krieger, I. M.; O'Neill, F. M. *J. Am. Chem. Soc.* **1968,** *90*, 3114. (i) Luck, V. W.; Klein, M.; Wesslau, H.; *Ber. Bunsenges, Physik. Chem.* **1963**, *67*, 75. (j) Hone, D.; Alexander, S.; Chaikin, P. M.; Pincus, P. *J. Chem. Phys.* **1983,** *79*, 1474.

(2) Carlson, R. J.; Asher, S. A. *Appl. Spectrosc.* **1984**, *38*, 297.
(3) (a) Flaugh, P. L.; O'Donnell, S. E.; Asher, S. A. *Appl. Spectrosc.* **1984**, *38*, 847. (b) Asher, S. A.; Flaugh, P. L.; Washinger, G. *Spectrosc.* **1986**, *1*, 26.
(4) Rundquist, P. A.; Photinos, P.; Jagannathan, S.; Asher, S. A *J. Chem. Phys.* **1989**, *91*, 4932.
(5) Zahorchak, J. C.; Kesavamoorthy, R.; Coalson, R. D.; Asher, S. A. *J. Chem. Phys.* **1982**, *96*, 6874.
(6) Sanders, J. V. *Nature* **1964**, *204*, 1151.
(7) (a) Rundquist, P. A.; Jagannathan, S.; Kesavamoorthy, R.; Brnardic, C.; Xu, S.; Asher, S. A. *J. Chem. Phys.* **1981**, *94*, 711. (b) Kesavamoorthy, R.; Jagannathan, S.; Rundquist, P. A.; Asher, S. A. *J. Chem. Phys.* **1991**, *94*, 5172. (c) Rundquist, P. A.; Kesavamooorthy, R.; Jagannathan, S.; Asher, S. A. *J. Chem. Phys.* **1991**, *95*, 1249. (d) Rundquist, P. A.; Kesavamoorthy, R.; Jagannathan, S.; Asher, S. A. *J. Chem. Phys.* **1991**, *95*, 8546.
(8) Spry, R. J.; Kosan, D. J. *Appl. Spectrosc.* **1986**, *40*, 782.
(9) (a) Sunkara, H. B.; Jethmalani, J. M.; Ford, W. T. *Chem. Mater.* **1994**, *6*, 362. (b) Sunkara, H. B.; Jethmalani, J. M.; Ford, W. T. *ACS Symp. Ser.* **1995**, *585*, 181.
(10) Asher, S. A.; Holtz, J.; Liu, L.; Wu, Z. *J. Am. Chem. Soc.* **1994**, *116*, 4997.
(11) Asher, S. A.; Jagannathan, S. U. S. Patent, **1994,** #5 281 370.
(12) Haacke, G.; Panzer, H. P.; Magliocco, L. G.; Asher, S. A. U. S. Patent 5 266 238, **1993**.
(13) Hirokawa, Y.; Tanaka, T. *J. Chem. Phys.* **1984**, *81*, 6379.
(14) Schild, H. G. *Prog. Polym. Sci.* **1992**, *17*, 163.
(15) Wu, X. S.; Hoffman, A. S.; Yager, P. *J. Polym Sci. Part A: Polym. Chem.* **1992**, *30*, 2121.
(16) Weissman, J.; Sunkara, H. B.; Tse, A. S.; Asher, S. A. *Science* **1996**, *274*, 959.
(17) (a) Asher, S. A.; Chang, S.-Y.; Tse, A.; Liu, L; Pan, G.; Wu, Z.; Li, P. *Mat. Res. Soc. Symp. Proc.* **1995**, *374*, 305. (b) Asher, S. A.; Kesavamoorthy, R.; Jagannathan, S.; Rundquist, P. *SPIE Vol. 1626 Nonlinear Optics III*, **1992**, 238. (c) Asher, S. A.; Chang, S.-Y.; Jagannathan, S.; Kesavamoorthy, R.; Pan, G., U. S. Patent, **1995,** #5 452 123.
(18) (a) Chang, S.-Y.; Liu, L.; Asher, S. A. *Mat. Res. Soc. Symp. Proc.* **1994**, *346*, 875. (b) Chang, S.-Y.; Liu, L.; Asher, S. A. *J. Am. Chem. Soc.* **1994**, *116*, 6739. (c) Kesavamoorthy, R.; Super, M. S.; Asher, S. A. *J. Appl. Phys.* **1992**, *71*, 1116.
(19) Tse, A. S.; Wu, Z.; Asher, S. A. *Macromolecules* **1995**, *28*, 6533.

Chapter 34

Morphology of Composites of Low-Molar-Mass Liquid Crystals and Polymer Networks

C. V. Rajaram[1], S. D. Hudson[1,3], and L. C. Chien[2]

[1]Department of Macromolecular Science, Case Western Reserve University, Cleveland, OH 44106
[2]Liquid Crystal Institute, Kent State University, Kent, OH 44242

A new class of liquid crystal/polymer network composite with very small amounts of polymer network (3 Wt%) is described. These composites are formed by photopolymerization of the monomers in-situ from a solution of monomer dissolved in low-molar-mass liquid crystals. Several techniques have proven useful to characterize these polymer networks. This review describes polymer network structure and its influence on electro-optic behavior of liquid crystals. Structural formation in these composites begins with the phase separation of polymer micronetworks, which aggregate initially by reaction-limited, and then by diffusion-limited modes. The morphology can be manipulated advantageously by controlling: the crossover condition between such modes, the order of the monomer solution prior to photopolymerization, and the molecular structure of monomers or comonomers.

Liquid Crystalline Polymers are an important class of polymeric materials because they may exhibit optical properties similar to low-molar-mass liquid crystals and high mechanical properties of polymers. These polymers are broadly classified based on their molecular architecture, i.e. attachment of the mesogen to the polymeric backbone, as *main-chain liquid crystal polymers* (*1*) or *side-chain liquid crystal polymers* (*2*). In main-chain liquid crystal polymers, mesogens are incorporated into the backbone. The mesogens may be of different shapes and sizes, and are usually rodlike or disklike. Such polymers have not been used for opto-electronic applications because it is very difficult to reorient these materials by electric field. Instead, these materials find applications that use their exceptional mechanical properties. Even side-chain liquid crystal polymers, whose mesogen is attached to the polymer backbone through a flexible spacer switch too slowly for

[3]Corresponding author

display applications, but are suitable for data storage. Considerable effort has been directed toward understanding the influence of the size and type of flexible spacers on mesomorphic properties (*2*). Another class of materials which are offering promise to be used as optical materials are *side-chain liquid crystal elastomers* (*3*). These are side chain liquid crystal polymers which are lightly crosslinked. These materials can be easily tailored to desired properties based on several modifications concerning molecular architecture, including the amount of crosslinking. These materials may range from translucent to opaque under unperturbed conditions, but can be switched to a transparent state by the application of small amounts of mechanical field.

Another interesting class of materials based on the principle of liquid crystallinity and crosslinking, which is being studied in detail because of its potential use in optical components, is *oriented polymer networks* (*3*). In this class of materials, the monomer, which is usually a diacrylate having a mesogenic moiety and possessing the properties of low-molar-mass liquid crystals, along with a small amount of photoinitiator, is aligned in a desired orientation by either surface alignment or the electric or magnetic field. The entire system is photopolymerized by the application of UV light radiation of suitable intensity and wavelength. The photopolymerization fixes the orientation of the mesogens, resulting in densely crosslinked thermosets with unique anisotropic properties.

A modification of the oriented polymer network systems are *polymer stabilized liquid crystals* (PSLC) (*4*) being studied in detail because of their application in flat-panel displays. In these materials, photopolymerizable diacrylate monomers are usually dissolved at a concentration less than 10% in *non-reactive* low-molar-mass liquid crystal solvents, commonly available, along with a small concentration of photoinitiator. Typically, the addition of small amounts of monomers and photoinitiator reduces the transition temperatures of the pure low-molar-mass liquid crystals slightly, suggesting that the order in the system is not dramatically altered by the addition of monomers or initiator. In application, this solution is aligned in a particular desired state and then photopolymerized. Photopolymerization is preferred to thermal free-radical polymerization, because photopolymerizations are very fast and because the temperature of photopolymerization can be controlled more easily to optimize processing of the display.

These formulations were developed to overcome low viewing angle, a disadvantage of conventional flat-panel displays. The liquid crystal orientation is perturbed by the phase separated polymer network. Since the polymer network is present only in small amounts spanning through the entire cell, the perturbation may be slight in all directions, thereby increasing the viewing angle without sacrificing greatly the contrast of the display. Due to low concentration of the polymer, the electro-optic response of the is dominated by the low-molar-mass liquid crystals in which they are made. These composites combine fast switching and low threshold voltage properties of the conventional low-molar-mass liquid crystals with higher angle of view (haze free) and bistability arising from the polymer network dispersed in the liquid crystal matrix. The polymer networks formed in such systems containing mesogenic monomers have been extensively studied (*5-14*).

To understand this new class of materials requires an interdisciplinary approach: free-radical photopolymerization chemistry, low-molar-mass liquid crystals physics, materials science of thermosets and display technology. This short review will touch on aspects of understanding of the morphology of these polymer networks formed in liquid crystal media.

Materials in Polymer-Stabilized Liquid Crystals

Molecular engineering of the monomers which form the polymer network in these composites is essential for controlling network structure and improving the electro-optic properties of the flat-panel displays. In order to control the polymer networks, diacrylate monomers were designed with varying amounts of flexibility by controlling the spacer between the mesogen and the polymerizable acrylate moiety, by varying the size of the mesogens from biphenyl, triphenyl etc., and by controlling the functionality. Often the building block for the mesogens has been a biphenyl core. A simple reactive monomer is 4,4' bisacryloyloxy 1,1' biphenylene (**BAB**) in which attachment of polymerizable acrylate moieties is on the 4,4' positions of the biphenylic mesogen. The polymer network resulting from these monomers gave displays with low threshold voltage and poor contrast. In order to further increase the alignment of the polymer network and possibly increase the sharpness of the electro-optic curve (transition from transmitting to scattering mode), a hexamethylene spacer was introduced between the biphenyl mesogen and the polymerizable acrylate group of the monomer **BAB**. This monomer, called **BAB6** i.e., 4,4' -bis[6-acryloyloxy)-hexyloxy-] 1,1' biphenylene, forms polymer networks of higher orientation, because the flexible hexamethylene spacer gives more conformational freedom for the monomers during photopolymerization. The displays made from these highly aligned networks (fiber-like morphology) show high contrast ratio and hysteresis which provide gray scale with multiplexing capability (*15*). Further modification of the monomer was done; specifically, the biphenyl core of **BAB6** was extended by another benzylic ester group on both 4,4' positions, resulting in the monomer 4,4'-bis-{4-[6-(acryloyloxy)-hexyloxy]benzoate}-1,1'-biphenylene (**BABB6**). This monomer has a rigid core which is more than twice the size of the mesogen of **BAB**. Unlike **BAB** which melts from crystal to isotropic liquid at 150 ^{0}C, this monomer exhibits several higher order smectic phases (S_{x1} 108.5 ^{0}C; S_{x2} 112.3 ^{0}C; S_c 119.2 ^{0}C and N 124.9 ^{0}C); S_{x1} and S_{x2} are unidentified. **BABB6** materials exhibit high contrast ratio, low threshold voltage and haze-free display devices. Some interesting morphologies of these materials were found to provide a dramatic effect on the display device performance (16). A monomer with higher solubility in common low-molar-mass liquid crystals (such as **5CB** [4'-pentyl 4 biphenyl carbonitrile], **8CB** [4'-octyl 4 biphenyl carbonitrile], or **E48** [a eutectic mixture of several low-molar-mass liquid crystals]) is **C6M** or **RM82** [(1,4-di-(4-(6-acryloyloxyhexyloxy)benzoyloxy)-2-methylbenzene], which was extensively used by Hikmet et al. (*5-14*) in studying these composites. Another monomer used in these studies is **BMB** (*17*) [4,4'-bis(2-methylpropenoyloxy)-biphenyl]. Commonly used photoinitiators are **BME** (benzoin methyl ether) or Irgacure 651 (Ciba Geigy). Photopolymerizations are carried out upon exposure to UV light of wavelength 365 nm. The intensity of the UV lamp is varied depending on the requirements of the experiment. The molecular structures of these materials are shown in Chart I.

Kinetics of Photopolymerization. For each of these monomers, the polymerization rate is relatively low at low monomer concentration (less than 1 Wt%). This is because dilution reduces initiator efficiency and prevents auto-acceleration (Tromsdorff effect), which is typical of bulk photopolymerizations. The polymerization rate increases with increasing monomer concentration (*18*). Similar observation was made by following the double bond conversion of the diacrylate (*18*). The maximum polymerization rate and the conversion at maximum rate increase with increasing monomer concentration, suggesting that, at low monomer concentration, the mobility in the mixtures is high and decrease of rate at later stages results from a depletion of monomers and a decrease in the mobility of the polymer-rich phase as crosslink density increases (*5*). The polymerization rate is also dependent upon the architecture of the monomer. In dilute solution, differences in mobility are less, and factors such as the electronic structure of the monomers is important.

Polymer Network Stabilized Liquid Crystal Displays. The alignment of liquid crystals plays an important role in the operation of a display. The alignment is conventionally induced by the display cell surfaces, but by distributing the surface of a polymer network through out the bulk of the liquid crystal, new properties are possible, and the performance of conventional devices can be improved. This short section will mention some of the conventional liquid crystal display devices modified by these polymer networks.

Cholesteric Texture Displays. Cholesteric liquid crystals have unique optical properties because of their helical structure. When a cholesteric liquid crystal has a planar texture, it exhibits Bragg reflection; a focal-conic texture scatters light; and a homeotropic texture, in which the helical structure is unwound, is transparent. The primary disadvantages in using these materials in flat-panel displays, in spite of their brilliant colors and optical textures, are strong angular dependence of the reflected light intensity, instability of the focal conic texture, and lack of convenient means to switch between focal conic and planar texture. All of these disadvantages are overcome by the inclusion of a polymer network, which functions to perturb the alignment of the planar texture slightly and stabilize the focal conic (*4, 19, 20*). The wide viewing angle in these devices is due to the distribution of the helix axes of the cholesteric liquid crystals around the normal to the surface, resulting in reflected light distributed to a broad range of angles. The reflectivity of the cell is decreased for the above reason, but the contrast of the device remains high because the

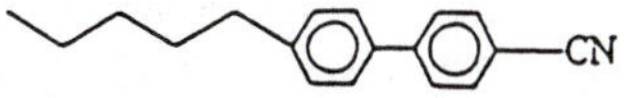

—CN 5CB

—CN 8CB

Eutectic Mixture of Several Low-Molar Mass Liquid Crystals E48

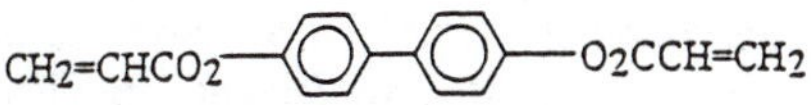

$CH_2{=}CHCO_2$— —$O_2CCH{=}CH_2$ BAB

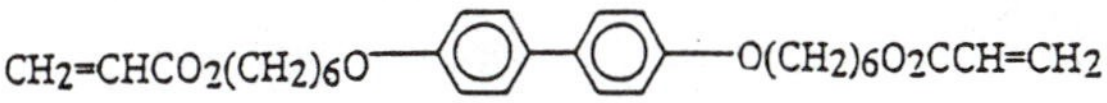

$CH_2{=}CHCO_2(CH_2)_6O$— —$O(CH_2)_6O_2CCH{=}CH_2$ BAB6

BABB6

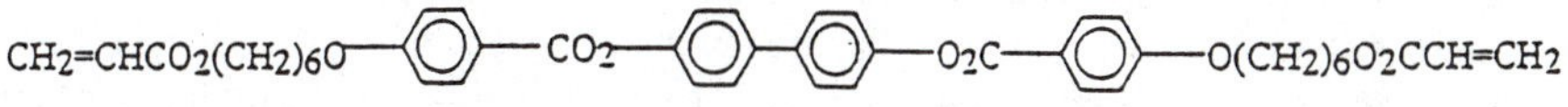

$CH_2{=}CHCO_2(CH_2)_6O$— —CO_2— —O_2C— —$O(CH_2)_6O_2CCH{=}CH_2$

RM82

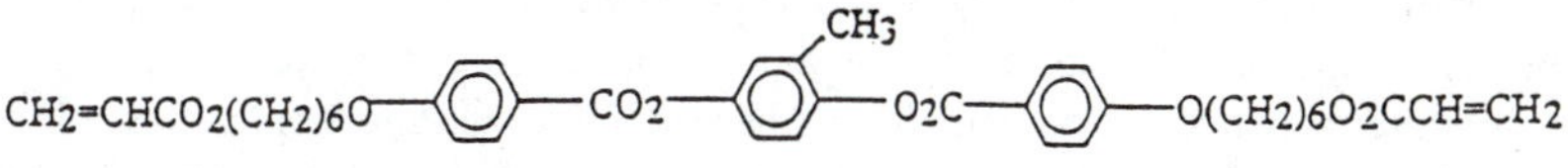

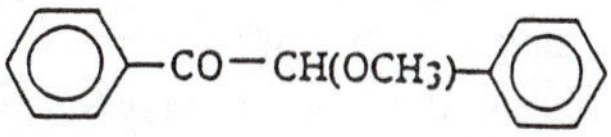

CH_3

$CH_2{=}CHCO_2(CH_2)_6O$— —CO_2— —O_2C— —$O(CH_2)_6O_2CCH{=}CH_2$

—CO—CH(OCH_3)— BME

Chart I: Molecular structure of various solvents and monomers used in the fabrication of polymer-stabilized liquid crystals.

scattering of the focal conic state is high. Efforts are still underway in several groups to improve the response time and driving schemes for these displays (*15*).

These polymer network stabilized cholesteric liquid crystals have also been used successfully as light shutters (*4, 19, 20*). In "normal" mode, the cholesteric mixture containing the diacrylate monomer is polymerized under applied electric field in a cell without surface treatment, producing degenerate planar anchoring (*21*). The network produced by such polymerization is also aligned along the field. When the field is switched off after polymerization, the aligning effect of the polymer network and the surfaces compete, and the highly scattering focal conic texture is stable. On application of electric field, the cell returns to a homeotropic texture, a highly transparent state arising from a monodomain texture of liquid crystals aligned perpendicular to the glass substrate. Since the polymer network is in small concentration, the transparency is not greatly affected, and the contrast of this light shutter valve is high, in excess of 100. To manufacture a "reverse" mode light shutter, polymerization is carried out in the planar state (in a cell rubbed with polyimide), in which the helical axis of the cholesteric is perpendicular to the substrates. The polymer network is formed parallel to the local LC director, and the cell in the "OFF" state is of high transparency (since the pitch of the cholesteric texture is chosen to be significantly greater than the wavelength of visible light). Upon application of electric field, the cell converts to a highly scattering focal conic state, which is stabilized by the polymer network. These cells switch rapidly and have high contrast.

Twisted-Nematic and Super-Twisted-Nematic Displays. In twisted nematic displays, the glass substrates are rubbed with polyimide and assembled with top and bottom glass substrates 90^0 to each other. Strong anchoring of the liquid crystal to the rubbed polyimide layer causes a twist in the nematic liquid crystal mixture filling the cell. The cells are switched from this twisted state to a homeotropic texture on application of electric field. The out-of-plane tilt angle of the liquid crystal molecules at the interface with the rubbed polyimide is an important criteria which determines the operational voltages of these devices. The tilt angle can be varied using special polyimides (yet reproducibility is lacking) or by evaporating a SiO coating obliquely on to the substrates (an expensive process). When polymer networks were formed in these displays during application of an applied field, the operational voltages decreased substantially (*22*). Super-twisted nematic displays are similar to TN displays except their twist angle is much greater, typically between 240^0 and 270^0. A major problem in these displays related to the tilt angle is a striped texture, formed after application of an electric field, in which the helix is parallel, instead of perpendicular, to the substrate. Polymer networks in these displays formed under applied field in homeotropic condition prevent the helix from tipping parallel to the substrate, thereby removing the undesirable stripe texture and at the same time decreasing the operational voltages (*23*).

Ferroelectric Liquid Crystal Displays. Chiral tilted smectic liquid crystals, also known as ferroelectric liquid crystals, exhibit macroscopic dipole density, i.e.

ferroelectric polarization which responds to electric field. Ferroelectric liquid crystal (FLC) displays can be switched from dark to light state between crossed polarizers on reversal of electric field. Addition of monomers into a FLC mixture decreases the net dipole, or ferroelectric polarization, because of increased disorder generated within the mixture. On polymerization of the monomers, the polarization remains at the same level as the monomer dissolved FLC mixture. Increasing the polymer network concentration increases the switching voltage, and switching takes place over a wider voltage range. Also the maximum transmission intensity decreases with increasing network concentration (*24*), since a polymer network creates smaller LC domains with a wide size distribution. Nevertheless, FLC displays modified by a polymer network show enhanced resistance to mechanical shock, because the polymer network fixes the alignment of smectic layers throughout the bulk of the cell, unlike conventional surface stabilized ferroelectric liquid crystal displays with poor shock resistance (*25*).

In all of the devices mentioned above, the polymer network is used to influence the alignment of the liquid crystal, sometimes to stabilize order, other times to introduce disorder. Different types of network structure are desired in each case. Since these polymer networks seem to advantageously alter the electro-optic properties of many types of liquid crystal displays, it is imperative to understand the structure of these polymer networks for modifying, further improving, and even developing entirely new types of liquid crystal displays.

Methods of Analyzing Polymer Network Structure

The polymer network structure can be studied by various means. Optical characterization is particularly versatile, since it can probe the composites directly and test whether, and to what degree, the network is oriented (*15, 27, 30, 31*). Hot-stage cross polarized light microscopy can be used to test the influence of monomer or polymer on LC phase transitions of these composites. Measurement of the birefringence of the bare polymer network, or of the LC composite in the isotropic state, yields information concerning anisotropy of the polymer network and of the type and strength of interaction between the network and LC matrix (*15, 27, 30, 31*).

Spectroscopic methods are also useful. IR and NMR have been used to characterize the polymerization reaction (*32*), as previously mentioned. In addition, dichroic measurements using polarized IR have measured the orientation of polymer networks (*12*). IR and NMR have also measured the order parameter of the LC component, which is effected by polymerization (*12, 32*). NMR and dielectric spectroscopies have been used to determine relaxation times for the polymer and LC solvent (*9, 32*).

Structural characterization of the polymer network itself has been carried out using x-ray and neutron scattering and SEM (*17, 27, 38*). All of the techniques can be applied to LC composites directly, with the exception of SEM, for which the LC matrix must be removed before observation.

Morphology

This section reviews polymer network morphology, its formation processes, and its control for desired electro-optic properties of flat-panel displays.

Phase Separation. All experimental work concerning PSLC's suggests, more or less strongly, that phase separation of a polymer-rich phase occurs upon polymerization. Such polymerization-induced phase separation may be expected, since the total mole fraction of monomer and polymer molecules, and therefore also the entropy of mixing, decreases dramatically upon polymerization. There are changes in the enthalpy of mixing, also as a result of reaction, as the acrylate moiety is converted to its saturated counterpart.

Jakli et al. have observed, on polymerization of 1.5 wt.% **BAB** or **BMB** in **5CB**, paste-like consistency composites (*26*). These composites were white and strongly scattering in the liquid crystal phase and mildly scattering in the isotropic phase suggesting that polymer has phase separated. However, since polymerization was carried out in the liquid crystalline state, it is ambiguous whether the polymer is phase separated or if, through its coupling with the liquid crystal solvent, it induces a disordered liquid crystalline texture that scatters light. Such coupling would persist in the isotropic phase.

To understand phase separation of the polymer network, polymerization was carried out in the isotropic phase (*27*). A droplet of 3 wt.% **BAB** solution in **5CB** was suspended on a wire and heated to 42 ^{0}C, well into the isotropic state. The droplet was clear and remained as such for more than 2 hours. Upon exposure to UV light of intensity 8 mW/cm^2 for 3 minutes, the droplet became turbid, suggesting phase separation. If the polymer remained in solution, the network strands would be isotropic and not scatter light. To demonstrate that there is no favorable coupling between the LC and the network, which would stabilize orientational fluctuations and produce scattering, the nematic-isotropic phase transition temperature of the composite was measured as a function of increasing conversion, both in the liquid crystal (nematic) and isotropic phases. We observed a monotonic decrease in the phase transition temperature, suggesting that the liquid crystal phase is not enhanced by the polymer. Therefore, the turbidity of the droplet on exposure to UV light in the isotropic state is due to phase separation of the polymer from the liquid crystal solvent.

Braun et. al. observed high intensity x-ray scattering at small angles and continuously decreasing intensity with increasing scattering vector (*28*). Such behavior is characteristic of mixtures near a critical point, or of spinodal decomposition. X-ray scattering of the dry networks i.e. after the liquid crystal solvent was removed showed a strong scattering peak corresponding to a periodicity of 10 nm and analysis of the tail of the scattering pattern gave intensity, I, related to the scattering vector q as

$$I \sim q^{-3} \qquad (1)$$

which is characteristic of a two-phase structure, either with sharp boundaries and no fractal type structures, or fractals which are maximally rough. Periodicity of 10 nm is small for a spinodal type decomposition; thus they suggested that phase separation occurs in the gel state.

To test whether the phase separation occurs in the gel state, the experimental geometry of a droplet suspended upon a wire is advantageous, because the shape and volume of the droplet could be monitored during photopolymerization. Volume transitions of gels are well-known (*29*) and would accompany phase separation if it occurred after gelation. However, no change in size or shape of the droplet was observed, and turbidity was uniform throughout the droplet. Therefore, phase separation occurs prior to gelation.

The latter, thus occurs by aggregation of phase-separated polymer-microgel particles. These are apparently sufficiently rigid so as to prevent collapse of the network, and so maintain a uniform turbidity throughout the droplet. Immobility of the polymer-rich phase is reasonable, since it is a crosslinked thermoset. This is confirmed by experiments. For a liquid crystal solvent, there is a characteristic discontinuous change in the relaxation time at T_{ni}. The presence of a polymer network causes the liquid crystal to exhibit an additional distribution of relaxation times, and the magnitude of the change at T_{ni} decreases substantially, indicating the existence of two different populations of liquid crystal. One population is unbound and behaves normally, while another population is bound to the fixed polymer network and its relaxation time spectrum does not change at T_{ni}. Further evidence for this model was provided by Crawford et al. (*30*). They observed droplets of liquid crystal solution containing monomers having a radial director configuration, exhibiting a characteristic cross texture by cross-polarized microscopy which disappears in the isotropic phase. However, on polymerization of the monomers in the nematic state, the optical texture was present even in the isotropic phase, albeit more weakly. Therefore, the polymer network captures the orientation of the liquid crystal phase in which it is polymerized, suggesting that the polymer network is immobile, and the network anisotropy is stable.

Optical anisotropy of the polymer network and the orientational coupling between the polymer network and the liquid crystal was explored by studying the effect of the monomer architecture on the optical properties of these polymer stabilized liquid crystals (*27*). Planar cells were observed after selected-area polymerizations. After polymerization in the nematic state, the cells (both of **BAB** and **BAB6**) were observed in the isotropic state. The cell having the network due to **BAB** (monomer without flexible spacer) appeared faintly birefringent indicating the poor orientation of the polymer network, while the cell with **BAB6** showed much higher birefringence, indicating higher order. In order to understand whether this high birefringence is due to the polymer network or to liquid crystal molecules at the interface of the polymer network, the cells were observed in non-liquid crystalline isotropic solvent (hexane or acetone). Figure 1 shows replacement of the liquid crystals by immersing the cell in acetone. Close examination of the diffusing acetone front in the polymerized region reveals a gradient in the intensity of the birefringence from near the liquid crystalline phase (high intensity) to lower values where the concentration of acetone is greater. Assuming that the network is

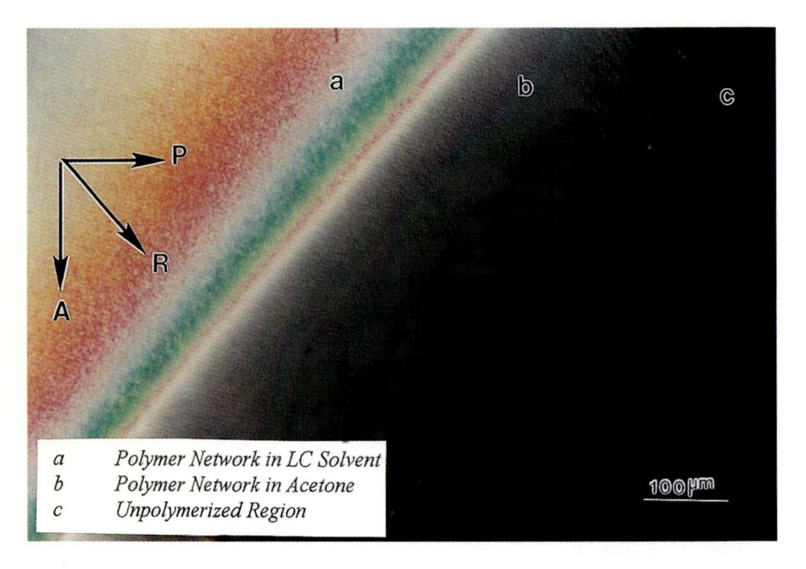

Figure 1: Optical micrograph of a homogenous cell containing p(**BAB6)** during dissolution of the low-molar-mass liquid crystal solvent, **5CB**, by acetone. Crossed polarizers (P and A) are aligned at 45^0 to the rubbing direction (R). Polymerization of the monomer has been carried out under nematic conditions in regions a and b. As acetone diffuses into the cell, the mixture becomes isotropic. Where there is no polymer network, such an isotropic mixture appears black (region c). The polymer network in isotropic liquid is weakly birefringent (region b). (Reproduced, with permission, from ref. 27; copyright 1995 ACS.)

immobile, this gradient indicates the relative contribution of the polymer network and of its coupling with a liquid crystal solvent.

Quantitative measurements of the birefringence and orientational order parameters are consistent with these results. Hikmet et al. (*12*) observed the effect of polymerization on the order parameter of the liquid crystal solution. Before polymerization, the order parameter of the solution falls to zero or vanishes at the clearing point of the solution, similar to the pure liquid crystal solvent. After polymerization of the monomers, however, the order parameter of the composite shows a much different behavior. Most notably, the clearing temperature is reduced. Moreover, the order parameter of the ordered phase is decreased and that of the high temperature, formerly isotropic phase, is finite. Above the clearing temperature, there is a gradual decrease in the order parameter (see Figure 2) (*12*). If the order parameter of the polymer network is fixed, this peculiar behavior can be explained by a two phase model suggesting the presence of two populations of liquid crystal solvents in these composites, as before. The unbound population exhibits a strong decrease in order parameter near the clearing point of the solution, and the (bound) liquid crystals near the polymer network interfaces do not become isotropic even above the clearing point of the solution. The relative fraction of this latter population decreases with increasing temperature, causing the gradual decrease of the order parameter. These results further suggest that the polymer network is immobile and has possibly phase separated from the liquid crystal solvent.

The size scale and the internal structure of the polymer networks were characterized using optical methods, neutron scattering and electron microscopy (*15, 17, 27, 38*). Optical methods involve measuring the birefringence of these composites (formed in a liquid crystal phase) in the isotropic phase of the liquid crystal solvent. A detailed birefringence study on these systems (*15*) assumed a simple model of a cell of thickness 'd' having 'p' fibrils per unit area perpendicular to the networks, schematically represented in Figure 3. This birefringence has contribution both from the polymer network (Δn_p) and the liquid crystal solvent (Δn_{lc}).

$$\Delta n = (\Delta n_p) + (\Delta n_{lc}) \qquad (2)$$

The birefringence of the polymer network was estimated by measuring the birefringence after replacing the liquid crystal solvent with an isotropic solvent such as octane. The birefringence of the liquid crystal composite was modeled as follows. The polymer network was assumed to be a square array of parallel fibrils of radius R and order parameter S_P. The order parameter of the liquid crystal solvent was assumed to be equal to that of the polymer network at the interface, and continuously decreasing away from this interface.

Minimizing the Landau-deGennes free energy equation, and solving for the order parameter S(r), gives an equation, which among other things, is also a function of the fibril radius, R.

$$(\Delta n_{lc}) = K.\ f(R) \qquad (3)$$

A fibril radius of approximately 5 nm provided a reasonable fit to the birefringence data. When a non-uniform spatial distribution of fibrils was considered (*31*), this estimate became even smaller, ~ 2 nm. This analysis ignores interfacial

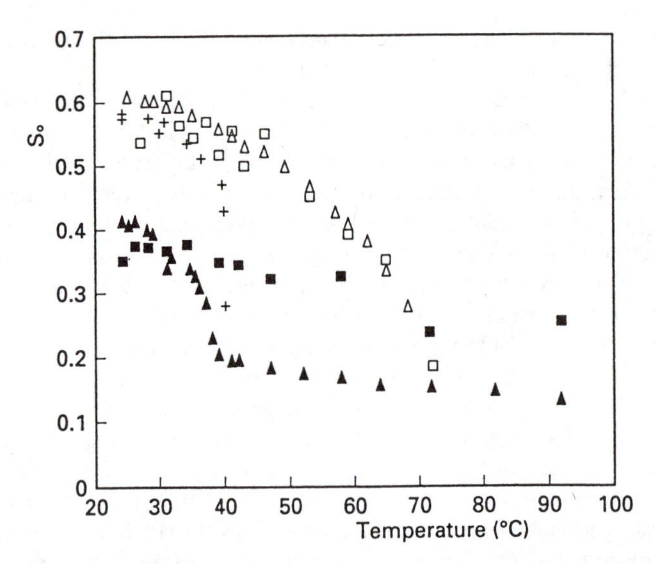

Figure 2: S_o as a function of temperature for a mixture containing 50 wt.% **C6M** in **8CB**. S_o is measured independently for each component: **C6M** (squares) and **8CB** (triangles). Open symbols denote before polymerization; solid symbols denote after polymerization at 71°C; + signs represent pure **8CB** in the bulk. (Reproduced, with permission, from ref. 12; copyright 1993 AIP.)

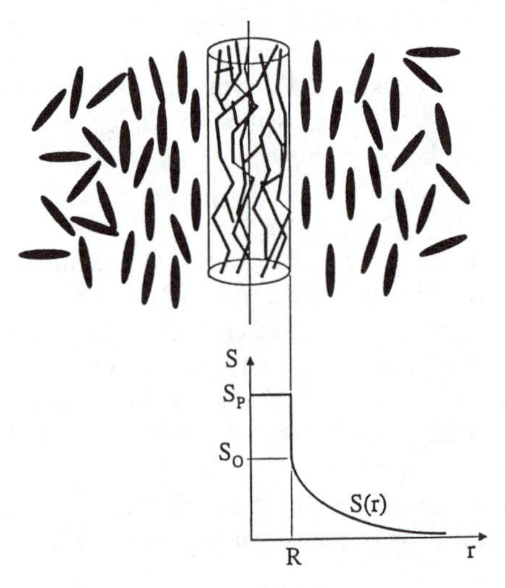

Figure 3: Top: Schematic diagram of a fibril of a polyer network and the nearby nematic liquid crystal molecules in the isotropic state. Bottom: the order paramter of the polymer chains and liquid crystal as a function of position. (Reproduced, with permission, from ref. 15; copyright 1996 Taylor and Francis.)

roughness, and if it is considered, then R actually corresponds to a local interfacial radius of curvature, and such rough polymer fibrils could be much bigger.

Larger sizes have indeed been found using other techniques. Jakli et al. employed neutron scattering experiments to study the shape, size and the surface roughness of the fibrils of the polymer networks in these liquid crystals (*17*). The polymer network fibrils were found to be 30 nm in radius and quite rough, consistent with the preceding argument. The structures measured by electron microscopy range from 100 to 200 nm (*27*). The surfaces of the polymer networks are rough, as observed by neutron scattering, and the roughness could be varied depending on polymerization conditions. It is plausible too, that the structure of the polymer network is microporous. A porous structure within beads or fiber-like morphology is expected by analogy with other polymerization-induced phase separations, in which continuous rejection of the solvent from the polymer-rich phase causes droplets of solvent to nucleate within polymer-rich particles (*33*).

Aggregation process. Since phase separation occurs prior to gelation, gelation occurs by the aggregation of polymer-rich particles. The aggregation process has been studied to understand the evolution of the network morphology in these systems. To determine the structure after phase separation, yet before aggregation, it was necessary to prevent aggregation by using a low concentration of monomer and a short polymerization time. The analysis was made simpler by using a rigid monomer, **BAB**, which forms beads over a wide range of polymerization conditions (*16*). Individual unaggregated smooth beads about 0.1 micron in size were formed by **BAB** at a concentration of 0.28 Wt% exposed to UV light of intensity 8 mW/cm^2 for just six seconds (Figure 4). These are called the primary particles and their size is comparable to the size of nodules, found in nodular polymer beads formed at higher monomer concentrations. Because the nodules are distinct, we conclude that the micronetwork, as it initially precipitates into a primary particle, is sufficiently rigid that it is not deformed during subsequent aggregation. (Similarly, no appreciable influence of specimen preparation procedures is observed). Increasing concentration causes an increase in the bead size, whereas increasing conversion does not. After the nodular beads are formed, subsequent increases in conversion are accomplished mostly within the beads. The bead size also depends significantly on polymerization temperature. When **BAB** is polymerized in **E48** at 28 ^{0}C, nodular beads comprising the network are approximately 0.2 $\pm$ 0.1 μm, while at higher temperature (75 ^{0}C) the morphology is similar, with larger and less nodular beads (~0.4 $\pm$ 0.1 μm in size).

These changes in network morphology with temperature and concentration can be understood based upon a kinetic model, presented below. We hypothesize that the size of nodular beads, and in turn the final morphology, is determined by a balance between the rates of reaction and of bead diffusion. Ideally, the process of morphology evolution in the case of **BAB** monomers may be depicted as in Figure 5, though the processes mentioned might occur at different rates in different regions of the sample, generating fixed fluctuations in the density of the network. Initially, the monomers are uniformly distributed in the liquid crystalline solution. When the UV lamp is turned on, there is initiation of polymerization over the entire sample. The

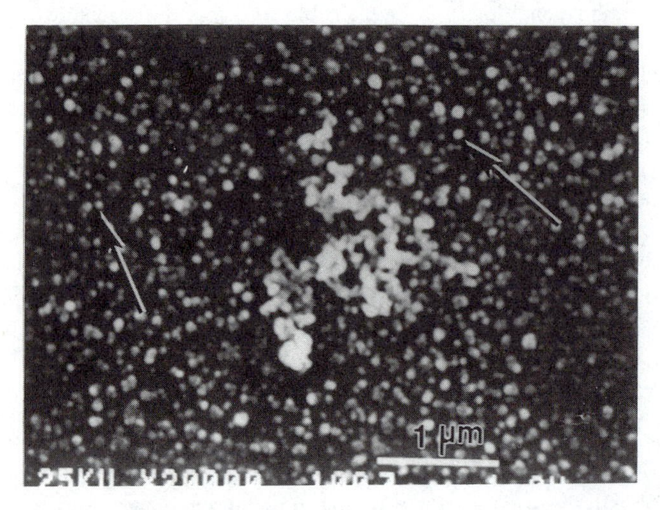

Figure 4: SEM micrograph of the polymer formed by 0.28% **BAB** solution in **5CB** exposed to UV light of intensity about 8 mW/cm^2 for just 6 secs. The brighter object in the center is an aggregate of primary particles; many such primary particles (arrowed) are present in the surrounding field of view. As the exposure time is increased, there are progressively fewer individual beads, and there is an increase in the number of nodular beads and larger network structures comprising nodular beads. (Reproduced, with permission, from ref. 27; copyright 1995 ACS.)

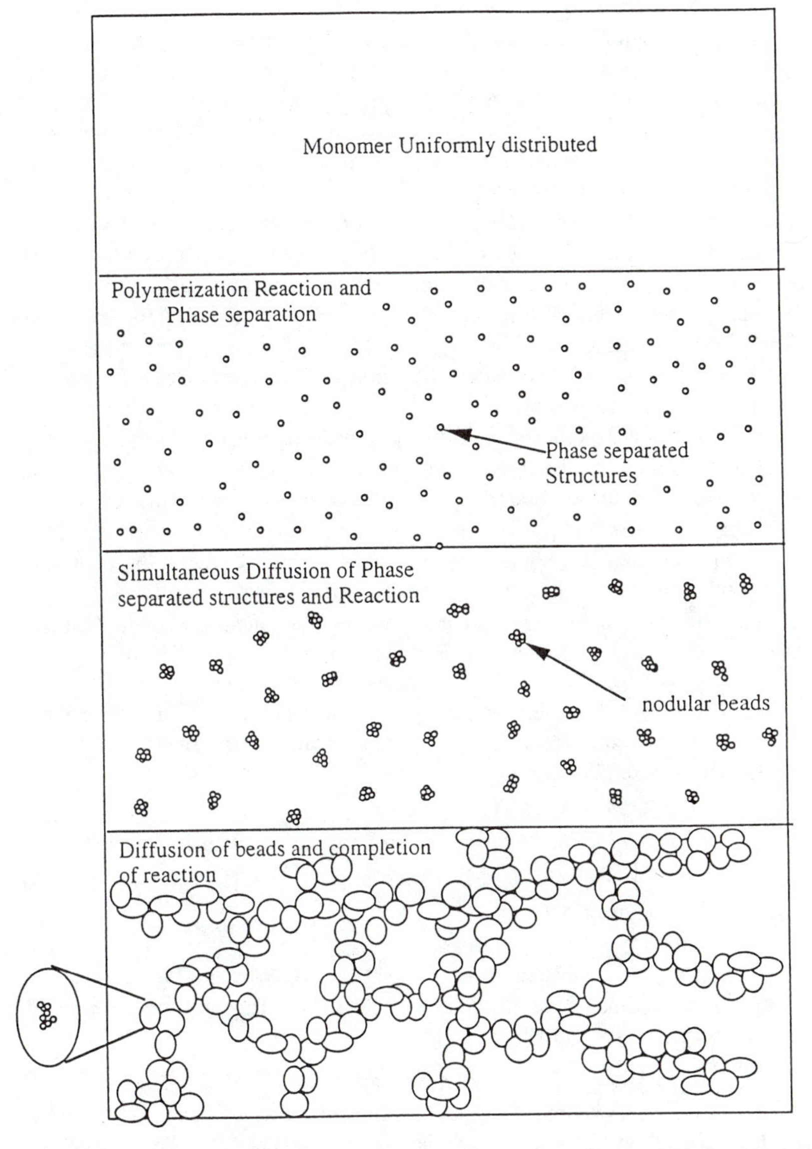

Figure 5: Schematic representation of morphology evolution in polymer stabilized liquid crystals containing p(**BAB**). (Reproduced, with permission, from ref. 16; copyright 1996 ACS.)

monomers migrate from the vicinity and feed the polymerization reaction. As the monomers are being rapidly used up, the polymer molecules eventually reach a sufficiently high molecular weight and/or branch content that they phase separate from the solvent. These primary polymer-rich particles, observed previously after very short polymerization times and low monomer concentration, contain many reactive moieties, a substantial fraction of which are on the surface. These react with other active surface sites on other particles. As the particles grow in size, they become more sluggish, so that eventually the reaction of the surface moieties of different phase separated particles is diffusion- limited. This process leads to final network morphology of the beads.

Based on this model, the characteristic size of the nodular beads depends on the condition of crossover from reaction-limited to diffusion-limited growth. These two types of aggregation processes have been identified since the early 1980's (*34 - 36*). For a given system undergoing aggregation, one of these two processes may occur. These may be distinguished by the compactness of the aggregate, which may be characterized by its fractal dimension. Reaction-limited aggregates are compact, having fractal dimensions in excess of 2. This compares with a fractal dimension of approximately 1.6 to 1.8 for less dense, diffusion-limited aggregates. It has also been pointed out that reaction-limited aggregation processes can, if allowed to continue, become eventually diffusion-limited (*34*). This crossover can occur if there is a fixed amount of aggregating species, which are more likely to react and become more sluggish in their movement as their size increases (*34 - 37*). We extend these ideas to develop an expression for a characteristic dimension of the final network.

At early times, particles are small and close together and the growth is limited by the time for two particles to react with one another, τ_{rxn}. The diffusion time is given by $X^2/D_{particle}$, where X is the average distance between particles and $D_{particle}$ is the diffusion coefficient for the particle or nodular bead of size r. Since the concentration (ϕ) of polymer (monomer units) is constant, the number density, and size, of particles varies to maintain ϕ,

$$\phi = (1/X^3)(4/3\ \pi r^3). \tag{4}$$

Rearranging equation (8) for the average distance between the diffusing particles X is

$$X \sim (\phi)^{-1/3}(4/3\ \pi)^{1/3}\ r. \tag{5}$$

From Einstein's hydrodynamic theory,

$$D_{particle} \sim kT/\eta r, \tag{6}$$

where k is the Boltzmann constant, T is the temperature, η is the viscosity of the medium, and r is the radius of the diffusing particle. The diffusion time can now be expressed using equations 9 and 10.

$$\tau_{Diff} \sim [(\phi)^{-2/3}\ r^3\ \eta]/\ kT. \tag{7}$$

Therefore the diffusion slows markedly as the particle grows in size. The time for two particles to react with one another is inversely related to the rate of the crosslinking reaction:

$$\tau_{rxn} \sim 1/R_X, \tag{8}$$

The crossover from reaction-limited to diffusion-limited aggregation occurs when τ_{Diff} and τ_{rxn} are equivalent. The radius defined by this condition is therefore the size, ξ, of the nodular beads (neglecting internal shrinkage) characteristic of the final network. Replacing ξ for r in equation 7 and combining with equation 8 gives, after rearrangement:

$$\xi \sim (kT/\eta R_X)^{1/3} (\phi)^{2/9} \quad (9)$$

Photocalorimetric measurements demonstrate that the reaction rate does not depend strongly on temperature, so that the strongest temperature dependence is with η. Thus as temperature increases, ξ is predicted to increase, as observed (*16*). The rate of reaction R_x, may be increased by increasing the initiator concentration. Recent experiments further substantiate the model and show that ξ decreases with increasing initiator concentration, holding temperature and monomer concentration fixed.

The above model applies for all crosslinking monomers (with or without flexible spacers) that do not gel before phase separation. These instead separate as microgels and the final macroscopic network is formed only after diffusion and reaction of these particles. The morphology can then be related to monomer concentration, the rate of polymerization, and the solvent viscosity. Moreover, the morphology of anisotropic systems will be influenced by anisotropic diffusion, yet the crossover between reaction-limited and diffusion-limited aggregation applies in general. Polymerization of the monomers in a well ordered smectic A yields polymer networks having an aspect ratio of 1.6, being elongated in the molecular direction, based on small angle neutron scattering patterns (*38*). This value is relatively constant at different scales of scattering, suggesting that the monomers polymerize and aggregate anisotropically.

Controlling the Morphology of the Polymer Stabilized Liquid Crystals. The above examples demonstrate that the size and shape of constituent units of the polymer network are controlled by aggregation kinetics. The morphology of the polymer stabilized liquid crystals can be easily controlled also by the order of the monomer solution prior to photopolymerization (*16,39*), by monomer architecture (*27*) and by copolymerization of different monomers at appropriate ratios (*39*).

Order of monomer solution prior to photopolymerization. The order of the monomer solution prior to photopolymerization can be controlled easily by the temperature of photopolymerization, as increasing temperature typically decreases the order of the system. Cross polarized optical microscopy reveals that a **BAB** network polymerized in the smectic phase produced a much more birefringent PSLC than the network polymerized in the nematic (*40*). Electron microscopy reveals a bead-like morphology in the nematic phase polymerization and a highly ordered fibrous network of rodlike units in smectic phase polymerization suggesting higher orientation of the network when polymerized in smectic phase (*40*). Much more highly aligned networks are produced by more flexible monomers, especially **BAB6** (*27*) and **RM82**.

Monomer size and flexibility. Network anisotropy is much easier to achieve if the monomer is made more flexible (*27*). Therefore **BAB6** was also studied as a function of cure. Figure 6a shows the morphology of **BAB6** exposed to UV light of intensity 1.75 mW/cm^2 for only 30secs. The morphology is a network of beads similar to **BAB**. Figure 6b shows the morphology of **BAB6** exposed to UV light of same intensity but for 3 mins. The network appears much smoother and the beads are less noticeable. The average diameter of the strands have reduced with increase in conversion and also the density of the network increased. Figure 6c shows the morphology of **BAB6** exposed to UV light of same intensity but for a much longer time of 10 mins. The network appears to be made of fine fibers about 0.1 μm in size. The average diameter has decreased and the density of the fibers has increased. Reduction in fiber diameter suggests that the shrinkage of the polymer-rich phase occurs as polymerization continues. Shrinkage, is likely to occur for **BAB** as well, but since **BAB** network is nearly isotropic, the shrinkage is likewise. For **BAB6**, if the network is locally anisotropic, the shrinkage is anisotropic, resulting in a increasingly fibrous morphology with increasing conversion. This fibrous morphology could be changed to bead-like by polymerizing **BAB6** in isotropic phase. A schematic representation of this process is shown in Figure 7. These experiments demonstrate that network anisotropy results from anisotropic shrinkage of polymer fibrils and not just from anisotropic aggregation processes.

Morphology of **BAB6** polymerized in **E48** at 28 ^{0}C is fibrillar in nature with smooth fibers about 0.1 μm in thickness and well aligned along the rubbing direction. Increasing the temperature of photopolymerization to 80 ^{0}C causes the fibers to be rougher, thicker (0.3 $\pm$ 0.1 μm), and less aligned along the rubbing direction while polymerizing at 90 ^{0}C increases this effect further by forming quasi-plate like structures (*16*). When polymerized at still higher temperatures (120 ^{0}C in isotropic state), the morphology is simply an aggregation of small beads (*16*).

The increase in the roughness of fibers with increasing photopolymerization temperature might arise from either decreased order parameter of the solvent or more isotropic diffusion of monomers and phase separated structures. The increase in diameter of the fibers as a function of temperature is consistent with the particle diffusion considerations described for **BAB**. Increasing the length of the mesogenic core of a monomer, increases the stability of smectic phase exhibited by the polymer, and is responsible for some of the unique morphology of polymer networks formed by **BABB6**. The morphology of **BABB6** photopolymerized in **E48** in nematic conditions (28 ^{0}C) is fibrillar with rough, thick fibers about 0.3 $\pm$ 0.1 μm in diameter (*16*). Increasing the temperature of photopolymerization to 75 ^{0}C results in a bead-like network interspersed with plate-like structures, both lacking orientation (*16*). Further increase in temperature of photopolymerization to 140 ^{0}C results in a interconnected plate-like morphology with plates ranging in orientation and size (Figure 8) (*16*). The size of plates typically varies from less than a micron to several μm, and thickness remains approximately 0.2 μm.

It seems that the longer mesogenic monomer, being more likely to form the smectic phase forms roughened and plate-like morphologies more readily. To explain why fibers (presumably nematic like) are formed at the lower temperature, we suggest that insufficient mobility exists within the polymer-rich phase to form the layered structure. Thus the structure characteristic of the initially less concentrated polymer-rich phase is frozen-in. At a much higher temperature of 140 ^{0}C, the polymerization temperature is possibly higher than the glass transition temperature of the polymer microgel, enabling the polymer network to form layered structures leading to a plate-like morphology.

Copolymerization of Monomers forming Different Morphologies. Another method of controlling the morphology of these materials is copolymerization of different monomers resulting in morphologies intermediate to those characteristic of the individual monomers. Copolymerization requires that the reactivity of each of the monomers is similar (*39*). As mentioned in the previous subsection, the fiber-like morphology of **BAB6** and the bead-like morphology of **BAB** are directly related to the presence and absence of flexible spacers between their mesogen and the polymerizable acrylate moiety respectively. Since flexibility can be controlled by copolymerization of these monomers, control of the morphology is also expected.

Indeed, as the relative concentration of **BAB6** in **BAB/BAB6** the copolymer network is increased, the morphology changes from initial beaded network morphology to a fine, smooth fibrillar morphology with rough fibers being produced at intermediate compositions (*39*). By increasing the relative concentration of flexible spacers in the polymer network, there is a corresponding increase in the decoupling of the conformation of mesogen and copolymer backbone. The ability of the copolymer network to be anisotropic and aligned along the nematic director depends on such decoupling. An oriented fibrous morphology is possible for sufficient fraction of flexible units.

BABB6 forms a fibrous morphology in the nematic phase and thus copolymerization with **BAB** in the nematic phase results in a morphology similar to the copolymer network of **BAB** and **BAB6**. At higher temperatures, however **BABB6** forms a plate-like morphology and blends with **BAB** result in some interesting morphologies (*39*). At low relative concentrations (25%) of **BABB6**, there is a single beaded morphology, suggesting copolymerization. At higher concentrations of **BABB6**, however, two morphologies coexist. This result suggests that at least one of the monomer reactivity ratios, under these conditions, is greater than unity. Some of the copolymer product is rich in **BABB6**, and vice versa. The resulting distribution of copolymer composition is sufficiently large, so that the structures that nucleate upon separation are distinct, some forming beads, some forming plates. Although inherent values of monomer reactivity ratios could cause such behavior, the effective values of these ratios also can be influenced strongly by phase separation (*41*). One of the monomers may be more soluble in one or the other of the polymer-rich phases, thus increasing locally the apparent reactivity of that monomer.

The morphology of the blends of **BAB6** and **BABB6**, both having flexible spacers photopolymerized in the nematic state is fibrillar with fibril diameter increasing with increasing **BABB6** content (*39*). Interestingly, in the isotropic state,

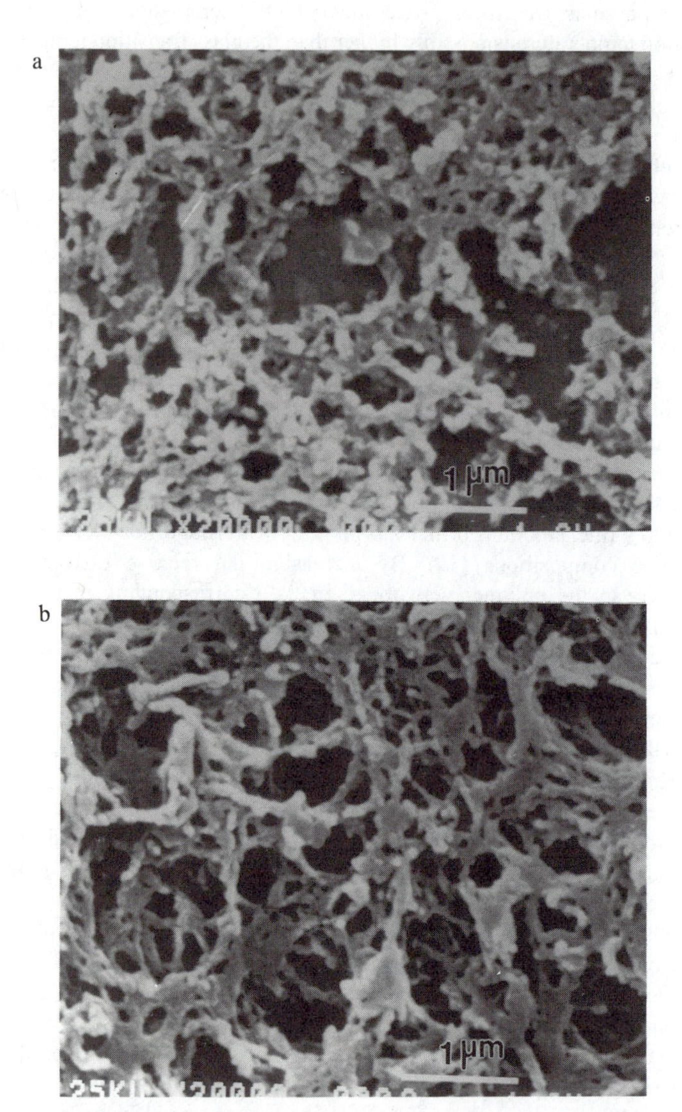

Figure 6: SEM micrograph of p(**BAB6**) polymerized from a solution of 2.8% **BAB6** in **5CB** by exposure to UV light of intensity 1.75 mW/cm^2 for a) 30 secs b) 3 mins c) 10 mins. (Reproduced, with permission, from ref. 27; copyright 1995 ACS.)

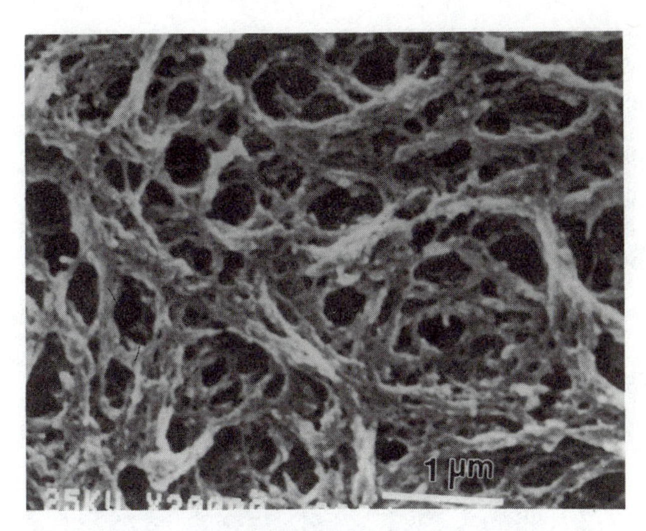

Figure 6. Continued.

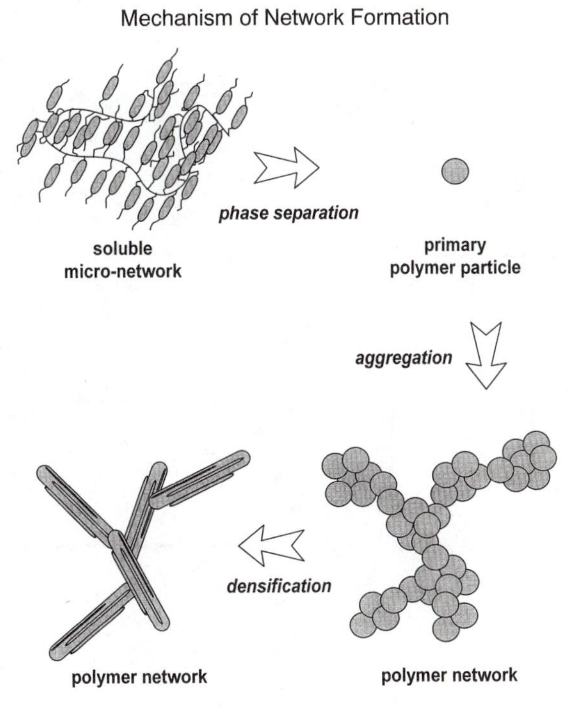

Figure 7: Schematic representation of a proposed model for the evolution of polymer network morphology of p(**BAB6**). (Reproduced, with permission, from ref. 27; copyright 1995 ACS.)

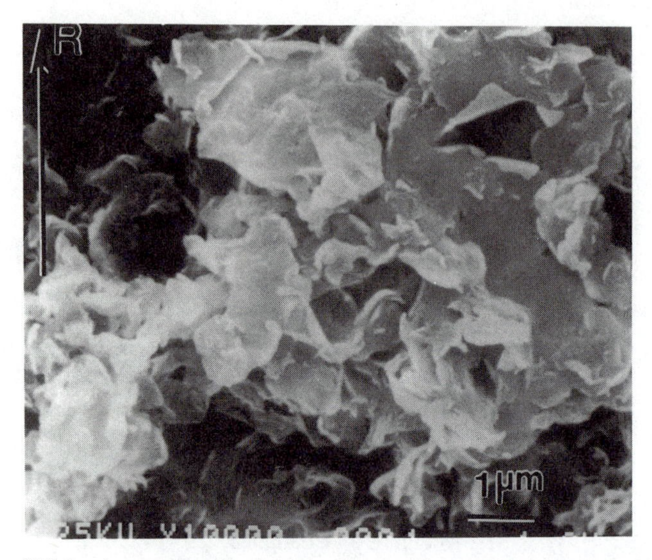

Figure 8: SEM micrograph of a homogenous cell containing 3% BABB6 in E48 polymerized at 140 ^{0}C. (Reproduced, with permission, from ref. 16; copyright 1996 ACS.)

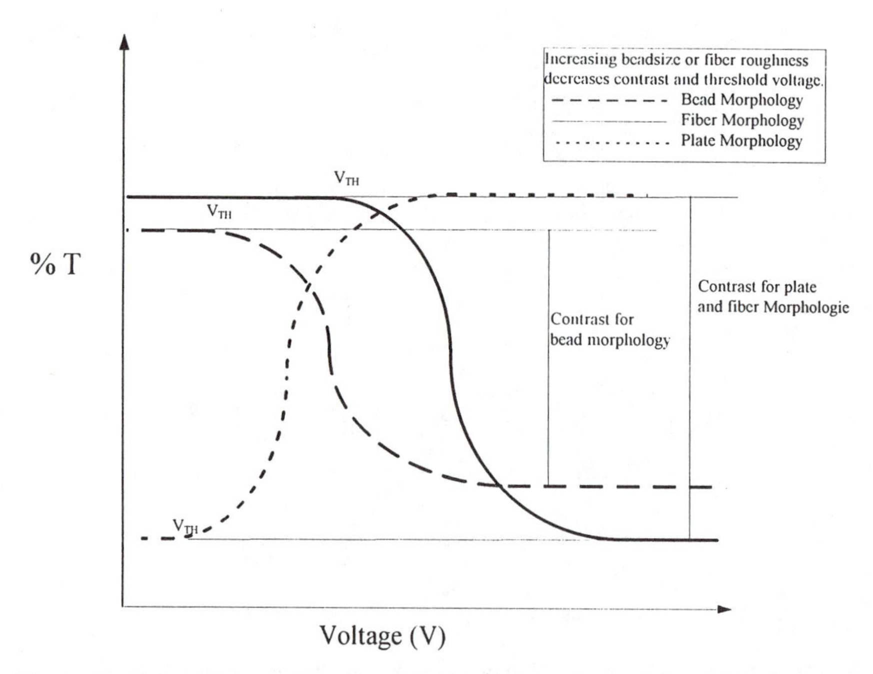

Figure 9: Schematic representation of the electro-optic properties of a "reverse" mode polymer stabilized liquid crystal as a function of the morphology of the polymer network.

at all relative concentrations of **BAB6** and **BABB6**, the plate-like morphology of **BABB6** is formed, suggesting that small amounts of **BABB6** in predominantly **BAB6** polymer network can nucleate formation of layered structures (*39*). This is possible because both **BABB6** and **BAB6** have identical flexible spacers which are attached similarly in the 4,4' positions of the mesogen, making it possible for them to accommodate one another and pack together into layered structures.

Electro-Optic Properties of Polymer Stabilized Liquid Crystals. Polymer networks have been used to stabilize many of the liquid crystal display states in various types of displays quite advantageously. In this section, we present some recent work on correlating the material properties of the liquid crystal/polymer network composite to the electro-optic properties of the flat-panel displays specifically cholesteric texture displays (*15*) and simple nematic birefringent type displays (*16*).

Cholesteric texture displays. In the composite of polymer network and cholesteric liquid crystals, if the pitch of the cholesteric liquid crystal is in the visible or UV region, both the planar and focal-conic textures are stable at zero field with proper surface anchoring condition or dispersed polymer. If the pitch is in the infrared region, the bistability disappears and the planar or focal conic textures can be stabilized at zero-field using polymer networks. Increasing chiral content causes the cell to be unstable at low polymer concentration and stable and scattering at polymer concentration between 1 and 4 wt.% and stable and clear above 4 wt.%. These display cells have wide viewing angle from +55° to -55°. The drive voltage and turn-on time are independent of polymer concentration and the turn-off time increases with increasing polymer concentration in "normal" mode cells from polymer stabilization of the transparent homeotropic 'ON' state of the liquid crystal display (*15*). On the other hand, the drive voltage and turn-on time increases with increasing polymer concentration while the turn-off time decreases with increasing polymer concentration in "reverse" mode cells, since the polymer stabilizes the transparent planar 'OFF' state (*15*).

Simple nematic birefringent displays. In these displays, the monomer solution in a nematic liquid crystal is aligned in a planar orientation and cured without application of electric field. The cell is transparent at zero-field. On application of electric field, the cell becomes scattering as the positive dielectric anisotropic nematic liquid crystal is prevented from aligning completely along the direction of applied field, resulting in a multi-domain scattering state. In this section, we correlate the morphology of the polymer network to the display characteristics in the "reverse" mode cell mainly in terms of its electro-optic curves (Figure 9) (*16*).

Effect of bead-size (*16*). Increased bead size causes greater disruption of the initial LC director and increasing scattering, resulting in lower transparency at zero-field. Secondly, as the bead size increases, the average distance between beads and the corresponding pore size of the polymer network must also increase. Consequently, the network holds the LC orientation more loosely, causing a decrease in the threshold voltage. Moreover, increased disruption of the LC director also reduces the threshold voltage. Increasing the bead-size, however, does not seem to have a significant impact on the contrast or the hysteresis of the cell. Hysteresis suggests the possibility of anchoring transitions at bead surfaces.

Effect of fiber roughness (*16*). Owing to the greater network anisotropy associated with **BAB6**, stronger interaction between the LC solvent and polymer network and corresponding changes in electro-optic properties are expected. The fiber roughness can be varied by varying the order of the liquid crystal solution prior to photopolymerization (by varying the temperature) or by copolymerization with bead forming monomer **BAB**. Increasing the roughness of the fibers causes increasing disruption of the nematic director, in turn causing a decrease in the threshold voltage and contrast of the display cell and an increase in the slope of the transition from transparent to the scattering state. On reversing the voltage it regains transparency with a small hysteresis.

Effect of fiber roughness and the new plat-like morphology (*16*). Increasing the temperature of polymerization of **BABB6** monomers causes a corresponding increase in the amount of plate-like structures along with the decrease in the fiber-like character to beaded structures. This increase in plate-like character is responsible for the unusual electro-optic response of these reverse-mode polymer-stabilized liquid crystal cells, and for decreasing the initial transparency of the cell at zero-field. The reverse-mode cell with plate-like morphology is initially opaque due to the random orientation of the plates. On application of an electric field, the large pore structure permits the liquid crystals to align along the direction of the applied field thereby increasing the transparency of the cell. This effect increases with increasing amounts of plate-like structures. At high concentrations of plates, the electro-optic response of the reverse-mode polymer-stabilized liquid crystal changes into a response which is very similar to that of the normal-mode polymer-stabilized liquid crystal cells which changes from a opaque 'OFF' state due to scattering from the multi-domain nematic state to a clear 'ON' state because of homeotropic alignment of all the liquid crystal molecules along the direction of applied voltage. If the plate-like morphology could be manipulated such that there is an oriented network of plates, it is easily imaginable that the electro-optic properties of the cell would be very different, possibly with very high contrast.

The direct correlation between the morphology of a polymer network and the observed electro-optic properties of these reverse-mode PSLC's is noteworthy. Desirable electro-optic response from these polymer- stabilized liquid crystals can be obtained by manipulating the structure of the polymer network inside the cells.

Conclusions

Polymer stabilized liquid crystals are formed when a small amount of monomer is dissolved in the liquid crystal solvent and photopolymerized in the liquid crystal phase. The resultant polymer network exhibits order, bearing an imprint of the LC template. After photopolymerization, these networks in turn can be used to align the liquid crystals. This aligning effect is a pseudo-bulk effect which is sometimes more effective than conventional surface alignment. Several characterization techniques are employed in studying the morphology of these polymer networks, which depends strongly on the type of the monomer used and the liquid crystal phase in which it is polymerized. In these liquid crystal/polymer composites, formation of the polymer network proceeds through various stages. Initially the monomers crosslink, forming a micronetwork which phase separates after reaching a critical molecular weight and/or branch content. These polymer particles aggregate, initially by a reaction-limited mechanism and finally by a diffusion-limited one. The characteristic size of the polymer particles and network pores can be controlled by the reaction and diffusion kinetics. The internal structure of the phase separated polymer micronetwork determines the final morphology of these systems, and can be controlled by varying the architecture of the monomer. For example, flexible spacers in the monomer permit the polymer to adopt anisotropic configurations which, on aggregation and shrinkage, result in a fibrous morphology. Also, these polymer micronetwork structures could be controlled by copolymerizing different monomers or by controlling the order of the monomer solution prior to photopolymerization. Thus a rich variety of morphologies have been developed, each showing a unique electro-optic response. For example, increasing the bead size of p(**BAB**), or increasing the roughness of the fibers in the fibrous morphology of p(**BAB6**), causes a decrease in the threshold voltage and contrast. While a new interesting plate-like morphology of p(**BABB6**) exhibits dramatically different response, i.e. "normal" mode, instead of the expected "reverse" mode..

Acknowledgments
The authors gratefully acknowledge the financial support of ALCOM grant # DMR89-20147.

Literature Cited

(1) Plate, N. A. *Liquid Crystal Polymers,* Plenum Press, New York, 1995, and references therein.
(2) Finkelmann, H.; Ringsdorf, H.; Wendorff, J. H. *Makromol. Chem.* **1978**, *179*, 273.
(3) Broer, D. J.; Finkelmann, H.; Kondo, K. *Macromol. chem.* **1988**, *189*, 185.
(4) Yang, D. K.;. Doane, J. W. *SID Technical Paper Digest XXIII*, **1992**, 759.
(5) Hikmet, R. A. M. *Liq. Cryst.,* **1991**, *9*, 405.
(6) Hikmet, R. A. M. *Molec. Cryst. Liq. Cryst.*, **1991**, *198*, 357.
(7) Hikmet, R. A. M.; Higgins, J. A. *Liq. Cryst.,* **1992**, *12*, 831.
(8) Hikmet, R. A. M. *J. Appl. Phys.,* **1990**, *68*, 4406.
(9) Hikmet, R. A. M.;. Zwerver, B. H. *Molec. Cryst. Liq. Cryst.,* **1991**, *200*, 197.
(10) Hikmet, R. A. M.;. Zwerver, B. H. *Liq.* Cryst., **1992**, *12*, 319.
(11) Hikmet, R. A. M. Macromolecules, **1992**, *25*, 5759.
(12) Hikmet, R. A. M.; Howard, R. *Phys. Rev.* E **1993**, *48*, 2752.
(13) Hikmet, R. A. M.; Michielsen, M. *Adv. Mater.* **1995**, *7*, 300.
(14) Hikmet, R. A. M. *Adv. Mater.* **1992**, *4*, 679.
(15) Yang, D. K.; Chien, L. C.; Fung, Y. K. In *Liquid Crystals in Complex Geometries;* Crawford, G. P.; Zumer, S., Eds.; Taylor & Francis: Bristol, PA, 1996; Chapter 5.

(16) Rajaram, C. V.; Hudson, S. D.; Chien, L. C. *Chem. Mater.* **1996**, *8*, 2451.
(17) Jakli, A.; Bata, L.; Fodor-Csorba, K.; Rosta, L.; Noirez, L. *Liq. Cryst.* **1994**, *17*, 227.
(18) Guymon, C. A.; Hoggan, E. N.; Walba, D. M.; Clark, N. A.; Bowman, C. N. *Liq. Cryst.* **1995**, *19*, 719.
(19) Yang, D. K.; Chien, L. C.; Doane, J. W. *Proc. Int. Display Research Conf. SID*, **1991**, 49.
(20) Doane, J. W.; Yang, D. K.; Yaniv, Z. *Proc. Int. Display Research Conf., Japan Display*, **1992**, 73.
(21) Jerome, B. *Rep. Prog. Phys.* **1991**, *54*, 391.
(22) Bos, P. J.; Rahman, J.; Doane, J. W. *SID Technical Paper Digest XXIV*, **1993**, 887.
(23) Fredley, D. S.; Quinn, B. M.; Bos, P. J. *Proc. Int. Display Research Conf.*, **1994**, 480.
(24) Hikmet, R. A. M.; Boots, H. M. J.; Michielsen, M. *Liq. Cryst.* **1995**, *19*, 65.
(25) Pirs, J.; Blinc, R.; Martin, B.; Musevic, I.; Pirs, S.; Zumer, S.; Doane, J. W.; *14th International Liquid Crystal Conference*, Pisa, Italy, Poster C-P89.
(26) Jakli, A.; Fodor-Csorba, K.; Vajda, A. In *Liquid Crystals in Complex Geometries;* Crawford, G. P.; Zumer, S., Eds.; Taylor & Francis: Bristol, PA, 1996; Chapter 6.
(27) Rajaram, C. V., Hudson, S. D. and Chien, L. C. *Chem. Mater.* **1995**, *7*, 2300
(28) Braun, D.; Frick, G.; Grell, M.; Klimes, M.; Wendorff, J. H. *Liq. Cryst.* **1992**, *11*, 929
(29) Tanaka, T. *Phys. Rev. Lett.* **1978**, *40*, 820.
(30) Crawford, G. P.; Polak, R. D.; Scharkowski, A.; Chien, L. C.; Zumer, S.; Doane, J. W. *J. Appl. Phys.* **1994**, *75*, 1968.
(31) Crawford, G. P.; Schardowski, A.; Fung, Y. K.; Doane, J. W.; Zumer, S. *Phys. Rev. E.* **1995**, *52*, R1573.
(32) Stannarius, R.; Crawford, G. P.; Chien, L. C.; Doane, J. W. *J. Appl. Phys.* **1991**, *70*, 135.
(33) Kim, Y. J.; Cho, C. H.; Palffy-Muhoray, P.; Mustafa, M.; Kyu, T. *Phys. Rev. Lett.,* **1993**, *71*, 2232.
(34) Kolb, M.; Jullien, R. *J. Physique Lett.* **1984**, *45*, L977.
(35) Family, F.; Meakin, P.; Viesek, T. *J. Chem. Phys.* **1985**, *83*, 4144.
(36) Weitz, D. A.; Huang, J. S.; Lin, M. Y.; Sung, J. *Phys. Rev. Lett.* **1985**, *54*, 1416.
(37) Asnaghi, D.; Carpineti, M.; Giglio, M.; Sozzi, M. *Phys. Rev. A* **1993**, *45*, 1018.
(38) Jakli, A.; Rosta, L.; Noirez, L. *Liq. Cryst.* **1995**, *18*, 601.
(39) Rajaram, C. V., Hudson, S. D. and Chien, L. C. *Polymer*, (Submitted).
(40) Muzic, D. S.; Rajaram, C. V.; Chien, L. C.; Hudson, S. D. *Polym. Adv. Tech.* **1996**, *7*, 737.
(41) Lacik, I., Selb, J. and Candau, F. *Polymer* **1995**, *36*, 3197.

Chapter 35

Novel Multifunctional Polymeric Composites for Photonics

Paras N. Prasad

Photonics Research Laboratory, Department of Chemistry, State University of New York, Buffalo, NY 14260–3000

Multifunctional materials will play an important role in the development of Photonics Technology. This paper describes novel multifunctional polymeric composites for applications in both active and passive photonic components. On the molecular level, we have introduced multifunctionality by design and synthesis of chromophores which by themselves exhibit more than one functionality. At the bulk level, we have introduced the concept of a "multiphasic nanostructured composites" where phase separation is controlled in the nanometer range to produce optically transparent bulk in which each domain produces a specific photonic function. Results are presented from the studies of up-converted two-photon lasing, two-photon confocal microscopy, optical power limiting, photorefractivity and optical channel waveguides to illustrate the application of the multifunctional optical composites.

The development of Photonics or Optical Technology is crucially dependent on the availability of materials which can simultaneously exhibit more than one property. These are multifunctional materials. Multifunctionality is exhibited by a material when it simultaneously performs two or more different functions. Also, a combination of two different functionalities can give rise to a new effect useful for photonics technology. An example is photorefractivity which is produced by the combined action of nonlinear electro-optic effect and photoconductivity. In this paper, the nanostructure control used in our laboratory to introduce multifunctionality will be presented. On the molecular level, we have introduced multifunctionality by design and synthesis of chromophores which by themselves exhibit more than one functionality (*1*). Examples are chromophores which show strong two-photon absorption and strong fluorescence. We have utilized the combined action of these two functions to achieve up-conversion lasing (*1-4*). We have also designed and synthesized segmented co-polymers and side-chain co-polymers consisting of different groups which perform different functions. Thus we have produced multifunctional photorefractive polymers which exhibit both electro-optic effect and hole-conduction (*5*). At the bulk level we have introduced the concept of a "multiphasic nanostructured composite" where phase separation is controlled in the nanometer range. This is accomplished by sol-gel processing (*6*) and by using the cavity of a reverse micelle as a nanoreactor (*7*). The domain size being much smaller than the wavelength of light produces optically transparent bulk. Using this approach we have produced bulk samples consisting of two dyes in two different phases which exhibit lasing from both dyes.

We have designed and processed multiphasic nanostructured composites which exhibit two-photon absorption cross-section more than two orders of magnitude larger than those of commercially available dyes. In these composites efficient two-photon lasing has been achieved at pump energies as low as 1 mJ with 10 nanosecond pulses. Multiphasic composites containing C_{60} together with a two-photon absorbing dye have been prepared and shown to exhibit power limiting behavior derived from both C_{60} and the dye (*6*).

Another area of use of composites is fabrication of Optical Channel waveguides. Using sol-gel processing of inorganic:organic composites we have prepared low loss optical waveguides, the refractive index of which can be selected by adding TiO_2. The thermal characterization of these waveguides has been performed (*7*).

We have prepared multi-component photorefractive polymers for optical data storage applications (*8*). This photorefractivity is a combination of two functionalities: electro-optic effect and photoconductivity. We have achieved holographic diffraction efficiencies up to >30% in our composites.

Materials

Multifunctional Chromophores. Small molecular units can also be designed to be optically multifunctional. In our laboratory we have focused on the designs and synthesis of two types of multifunctional chromophores: (i) efficient up-converters which absorb at longer wavelengths and fluoresce at shorter wavelengths and (ii) electrical-optical bifunctional chromophores which can produce both electro-optic and charge-transporting functions. Up-converted emission can be produced by a combination of multiphoton (such as two-photon absorption) and efficient emission (high quantum yield). The two-photon absorption is a nonlinear optical process which is characterized by the imaginary part of third-order nonlinear polarizability (second hyperpolarizability) of the molecule (*9,10*). These processes have traditionally been assumed to be weak, and, therefore, not of much practical significance. Our recent work based on design of optically nonlinear chromophores have produced structures which exhibit strong two- and even three-photon absorption processes. In addition, they are strongly fluorescent and, therefore, generate strong up-converted emission. Figure 1 shows structure of an example of a chromophore, 4-[N-(2-hydroxyethyl)-N-(methyl)amino phenyl]-4′-(6-hydroxy hexyl sulfonyl stilbene (abbreviated as APSS), which exhibits strong up-converted emission at ~560 nm when pumped with a pulse laser source at 800 nm from a mode-locked Ti-sapphire. This structure contains a donor (dielthylamino) group and an acceptor (sulfonyl) group separated by a conjugated segment and, therefore, also shows second-order nonlinear optical activity such as second-harmonic generation and electro-optic effect when electrically poled in a noncentrosymmetric bulk form (*10*).

Another example of a multifunctional chromophore is given in Figure 2. This chromophore, 4-(N,N′,diphenylamino)-(β-nitrostyrene), developed by Laser Photonics Technology, Inc. contains a charge transporting (triaryl amine) group which can conduct holes (*11*). It also consists of an electron donor group and an electron acceptor group separated by a conjugated segment. Therefore, it exhibits second-order nonlinear optical activity such as electro-optic effect in the electrically poled bulk form. A combination of these two functionalities produce the phenomenon of photorefractivity in which light absorption produces a refractive index change. The mechanism involves a space-charge field generation by the absorption of light which creates charge carriers that are spatially separated by transport of at least one carrier (hole in the present case). This space-charge field produces a refractive index change because of the electro-optic properties of the medium (chromophore in the present case).

Multifunctional Polymers. In addition to the possibility of designing a multifunctional chromophore, multifunctionality can be introduced in a polymeric structure by taking advantage of its tremendous structural flexibility. For example different functionalities can be introduced either by incorporating appropriate functional entities in the side chain or by using segmented units in the form of a co-polymer. The advantage of this approach is that each functionality can be almost independently optimized by a judicious choice of the appropriate structural unit since one structural unit (chromophore) contributes only one major function. The multifunctional polymer approach has the advantage over the guest-host approach (components physically dispersed in a polymer matrix) in that a high level of loading of a desired functional group can be achieved without the risk of immediate or long-term phase separation.

An example of a side-chain multifunctional polymer is a photorefractive polymer shown in Figure 3.

This polymer contains an electro-optic chromophore (labeled as EO), a charge-transporting group (labeled as CT) and an alkyl group to control the Tg of the polymer (labeled as CTG). The relative ratio of these groups can be varied to optimize the photorefractive property (*5*).

Multiphasic Nanostructured Composites Bulks. A novel approach to produce multifunctional polymeric composites is to use a multiphase system with the phase separation at the nanometer scale. Since the domain sizes are much smaller than the wavelength of light, they do not scatter. The result is an optically transparent sample in which each domain may produce a different optical function.

We have developed such a multiphase composite using sol-gel processing which produces large size bulks of excellent optical quality (*6*). In this approach, we prepare highly porous monolith gel (in the present case silica) and thermally process it. The pores in the silica monolith under our processing conditions are in the nanoscale region (~5 nm). This allows various molecules (such as fullerene) to be adsorbed in the pores by diffusion of a solution followed by evaporation of the solvent. Then the pores are filled with a polymerizable liquid such as methyl methacrylate (MMA) which is then

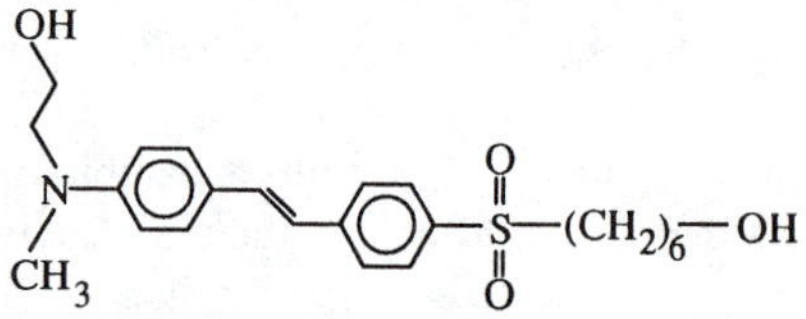

OH

N

CH_3

O

S—$(CH_2)_6$—OH

O

Figure 1. Molecular structure of 4-[N-(2-hydroxyethyl)-N-(methyl)aminophenyl]-4′-(6-hydroxyhexyl sulfonyl) stilbene, from here on abbreviated as APSS.

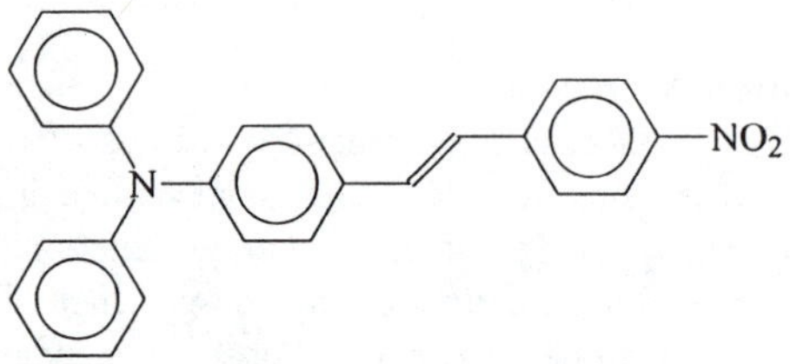

N

—NO_2

Figure 2. Molecular structure of a bifunctional chromophore, 4-(N,N′,diphenylamino)-(β-nitrostyrene).

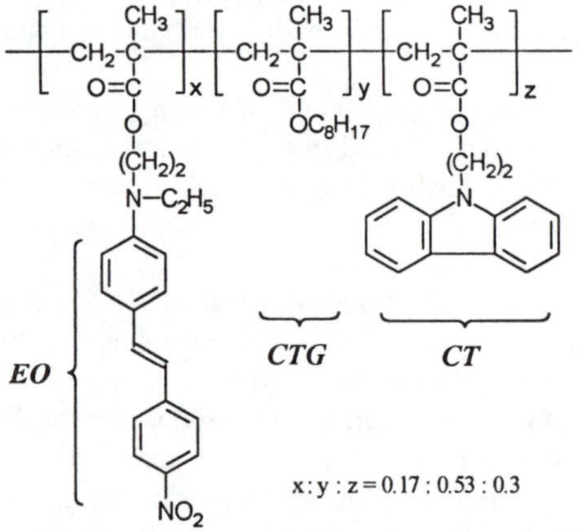

CH_3 CH_3 CH_3

$-CH_2-C-$]x $-CH_2-C-$]y $-CH_2-C-$]z

O=C O=C O=C

O OC_8H_{17} O

$(CH_2)_2$ $(CH_2)_2$

N—C_2H_5 N

NO_2

EO CTG CT

x : y : z = 0.17 : 0.53 : 0.3

Figure 3. The structure of a multifunctional side chain photorefractive polymer.

polymerized in-situ. This MMA procedure first pioneered by Pope, Asami and Mackenzie (*12*) has been used by several groups. We have followed the procedure described by Gvishi et al. (*13*).

The concept of a multiphase nanostructured composite can be used to prepare a wide variety of optical materials. We have been able to dope two (or more) different optically responsive materials, each of which can be in different phases of the matrix (the silica phase, the PMMA phase and the interfacial phase), to make multifunctional bulk materials for photonics. For example, we have doped in addition to the fullerene, which is adsorbed in the interfacial phase, a fluorescent and optically nonlinear chromophore bisbenzothiazole 3,4-didecyloxy thiophene (BBTODT) in the PMMA phase. This nonlinear chromophore was developed by B. Reinhardt and co-workers at the Polymer Branch of U.S. Air Force Wright Laboratory (*14*).

Reverse Micelle Nanoreactors. Reverse micelles are spheroidal aggregates formed by the dispersion of a surfactant[s] in an organic solvent (*15*). They can be formed both in the presence and absence of water. The presence of water is, however, necessary to form large surfactant aggregates. Water can be readily dissolved in the polar core, forming a "water pool". The size of the "water pool" is determined by W_o, the water-surfactant molar ratio ($W_o = [H_2O]/[S]$). The aggregates containing a small amount of water (below $W_o = 15$) are usually called reverse micelles whereas droplets containing a large amount of water molecules (above $W_o = 15$) are called microemulsions. When using iso-octane as a continuous medium, the water pool radius, Rw is found to be linearly dependent on the water content ($Rw(A^0) = 1.5\ W_o$) (*16*) which in turn is related to the size of the particles. As a result, particles of various sizes can be synthesized by varying the size of the water pool which serves as a nanoreactor for carrying out various reactions. Confinement in these small compartments inherently limits particle growth, thereby resulting in controlled size, monodispersed particles.

In the example presented here the reverse micelle was prepared by dispersion of the surfactant, sodium bis(2-ethyl hexyl) sulfosuccinate (AOT) in isooctane. The hydrolysis and sol-gel processing of titanium isopropoxide was carried out in the 5 nm size cavity of the reverse micelle to produce highly uniform TiO_2 nanoparticles. These particles were redispersed in a polymer (polyimide) solution and cast as a film of the polymer composite containing TiO_2 nanoparticles.

Properties

Two-photon Pumped Up-Conversion Lasing. Frequency up-conversion lasing is an important area which is receiving considerable attention. Compared to other coherent frequency up-conversion techniques such as optical parametric mixing or second harmonic generation, the up-conversion lasing provides broad tunability without any phase-matching requirement. Up-conversion lasing has been successfully achieved in inorganic rare-earth ions where sequentially pumped multiphoton transition is used. Direct two-photon pumping to achieve population inversion in organic systems usually requires very high peak power. However, using our multifunctional chromophores, discussed above, we have achieved up-conversion lasing by direct two-photon pumping

(*2-4*). The chromophore APSS was pumped by 5 nanosecond pulses from a dye laser which in turn was pumped by the 532 nm frequency doubled output of a pulse Nd-Yag laser. The lasing wavelength was 560 (*3*). Two-photon pumped lasing has been achieved in the solution phase as well as in a bulk polymer composite phase. Another novel geometry for two-photon pumped lasing is the use of a hollow fiber in which the APSS solution is introduced (*4*). This geometry allows a longer interaction length with beam confinement so that lasing can be produced at a very low pump threshold.

We have demonstrated two-photon pumped up-conversion lasing in another dye, trans-4[p-(N-ethyl-N-(hydroxyethyl) amino)styryl]-N-methylpyridinium tetraphenyl borate (abbreviated as ASPT) (*2*). Its chemical structures is shown in Figure 4. This molecule shows a two-photon pumped lasing at ~620 nm when pumped by the fundamental (1.06 μ) near IR wavelength of a pulse Nd:Yag laser. A lasing efficiency of 3.5% has been achieved. The pump threshold for lasing is as low as 1 μJ of 10 nanosecond pulses. Lasing has been accomplished by using a solution of ASPT or ASPT-doped at the interfacial phase of a multiphasic glass composite discussed above. This result also demonstrates that the multiphasic nanostructured glass composite is of high enough optical quality to be used for intracavity work.

Optical Power Limiting. This is another area of considerable current interest because of its relevance to sensor protection and eye protection from the threat of laser damage (*17,18*). Optical power limiters are devices which limit the optical output to a level below the damage threshold. Power limiting action can be produced by using a variety of physical mechanisms such as nonlinear absorption (for example two-photon absorption), reverse saturable absorption, nonlinear refraction and thermal effects. Each of these mechanisms has its own merits and limitations. Therefore, it may be desirable to use a combination of these mechanisms in order to broaden the scope of optical power limiting. A multiphasic nanostructured composite may be especially suited for this purpose because one can incorporate different power limiting functions in different phase. We have successfully demonstrated this application by using a sol-gel processed glass-polymer composite discussed above in which the fullerene, C_{60}, resides at the walls of the pores and a nonlinear chromophore bisbenzothiazole 3,4-didecyloxythiophene (BBTDOT) is present in the polymer, PMMA, phase (*6*). C_{60} has been shown to exhibit the optical power limiting behavior at 532 nm, mainly by reverse saturable absorption (*17*). BBTDOT is a new nonlinear chromophore synthesized by Bruce Reinhardt et. al at the Polymer Branch of the US Air Force Wright Laboratory (*14*). It exhibits a strong two-photon absorption and, therefore, power limiting behavior at ~602-820 nm (*19*). Our composite which exhibits contributions from both C_{60} and BBTDOT shows optical power limiting at 532 nm and 800 nm.

Two-Photon Confocal Microscopy. Confocal microscopy provides the possibilities of optical sectioning of a sample by the use of a spatial filter to improve axial resolution of the optical microscope (*20*). These optical sections can be used to reconstruct the 3D structure of a polymeric specimen. Confocal microscopy using fluorescence is usually performed by a single-photon excitation (linear absorption) of a fluorophore which may be chemically an integral part of the system (polymer) being investigated or just physically dispersed (solid solution). For the one-photon absorption of these fluorophores emitting in the visible range, one requires shorter (towards UV) wavelength excitation which would have a very short penetration depth in the material, thereby limiting the depth of a material one can probe. Furthermore, although a confocal microscope samples signals from the focal point of the objective, during the image acquisition the entire specimen volume is illuminated, often resulting in reduction of resolution. UV excitation may also lead to chemical damage of the coating or paint.

Multiphoton excitation in confocal microscopy is a nondestructive evaluation technique for polymeric materials and provides a significant advantage in the gain of three-dimensional resolution and image contrast. It also reduces the risk of thermal and photochemical degradation of the material. In multiphoton confocal microscopy, one uses a direct two- or three-photon pumped nonlinear absorption to excite a fluorophore which then emits. In this process, the pump wavelength is longer (towards the IR), while the emission is up-converted to the visible. Since the absorption is highly intensity dependent, strong absorption and subsequent emission occurs very localized at the focal point, therefore, providing one with an opportunity to increase the axial resolution. Since the multiphoton absorption process is much weaker than the linear absorption, and the wavelength is in the near IR, the penetration depth in the sample is very long. Therefore, one can easily study structures with depth profile and investigate the surface, the bulk and any underlying interface of a polymeric material or coating. By virtue of this longer penetration depth, one can use different color fluorophores in different layers of coating to conduct multilayer multiphoton confocal microscopy to selectively probe different layers of a coating. Two-photon microscopy was introduced by Denk et al. in 1990 (*21*), but it has seen limited application. The fluorophores reported in the past did not exhibit strong two-photon induced emission. Therefore, one had to use intense pulse laser sources to study them.

The recent work at the Polymer Branch at US Air Force Wright Laboratory and that at Photonics Research Laboratory at SUNY, Buffalo have produced organic structures that exhibit strong two- and three-photon absorption and exhibit strong up-converted fluorescence. Furthermore, we have shown that with the same pump source for multiphoton excitation, one can excite a number of different fluorophores to generate multicolors to conduct multiphoton multicolor microscopy for the study of microstructures of multilayer coatings, their interfaces and bulk as a function of depth.

We have used a single excitation wavelength of 800 nm from a mode-locked Ti-sapphire laser to pump, by two-photon absorption, a series of new fluorophores and achieve multiple color fluorescence over the entire visible spectrum enabling more than three spectrally separated detection channels. Also, the use of a single excitation wavelength minimizes the axial off-registration due to chromatic aberration. The wavelength of 800 nm can be conveniently reached by a diode laser itself, reducing the cost and size requirement for the pump source for multiphoton confocal microscopy.

We have used a bulk polymer sample to map the microstructure of the surface and the bulk. The sample was a poly-HEMA (poly-hydroxyethyl methacrylate) block containing one of our highly efficient two-photon fluorophores with several fractures infiltrated (*22*). Figure 5 shows a microfracture at the depth of 46 μm from the surface.

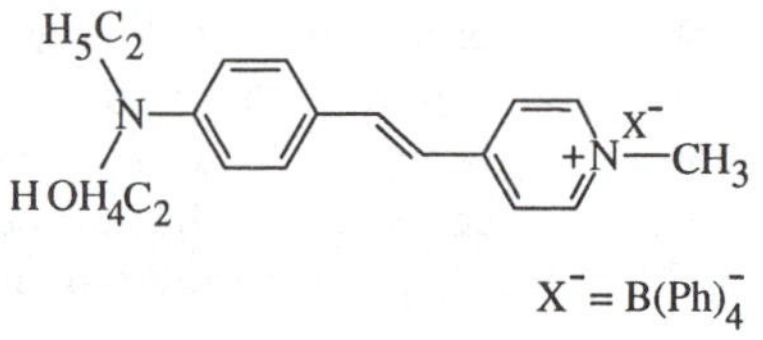

H_5C_2

HOH_4C_2

X^- $-CH_3$

$X^- = B(Ph)_4^-$

Figure 4. Molecular structure of trans-4-[p-(N-ethyl-N-hydroxyethyl-amino)styryl]-N-methylpyridinium tetraphenyl borate, abbreviated as ASPT.

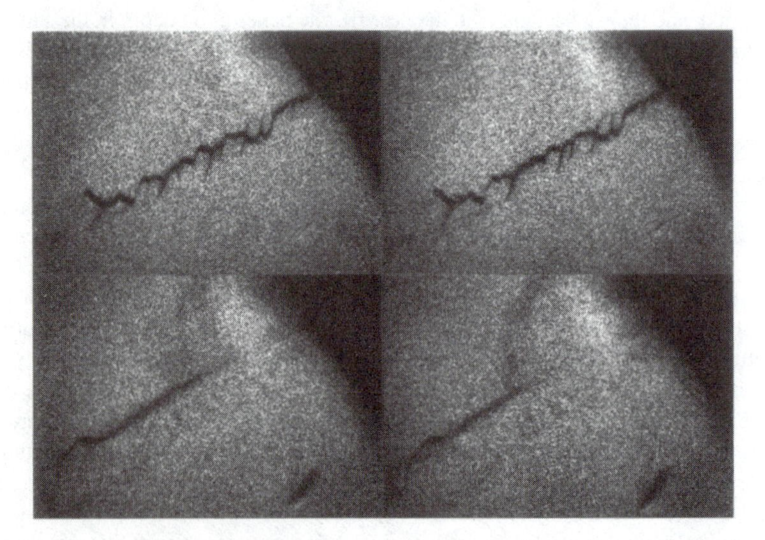

Figure 5. Two-photon confocal image of a doped polymer block at a depth of 46 μm showing a microfracture.

In contrast, a conventional one-photon excited fluorescent confocal microscopy shows no signal beyond about 10 μm in the bulk.

Photorefractive Polymer Composites. Polymeric media have been investigated for both permanent and erasable optical data storage (*23*). For erasable optical data storage, a great deal of interest has been shown in photorefractive materials (*24*). Photorefractivity in inorganic crystals such as $LiNbO_3$, $BaTiO_3$, $KNbO_3$, etc. has been widely investigated over an extensive period (*24*). A more recent entry is by photorefractive polymers which offer advantages of ease of processing and cheaper cost over the traditional inorganic photorefractive materials (*25,26*). Considering the relatively short history of photorefractive polymers, rapid advances have occurred and their photorefractive figure of merit is now comparable and even better in some regard than that of inorganic crystals (*27,28*).

As mentioned above, photorefractivity is a multifunctional property which is produced by the combined action of photoconductivity and nonlinear optical properties of a material (*24*). It is a result of the functions of photogeneration, transport, charge trapping, and the electro-optic effect.

In a holographic experiment, two beams are crossed to produce intensity interference pattern. The action of light produces charge carriers which move to dark regions, get trapped and set up an internal space-charge modulation field. This space charge field then produces a refractive index modulation (refractive index grating) through the electro-optic effect. This is the holographic grating containing the optical information which can be read by the diffraction of a probe beam. The process of photorefractivity in polymeric systems is different from that in inorganic crystals in that the key parameters affecting the photorefractive behavior are electric field dependent for polymeric systems. The role of the applied dc electric field for the polymeric materials is crucial and multifunctional: (i) it enhances the charge-carrier generation; (ii) it stimulates the transport of the carriers via field-dependent mobility; and (iii) it forces a non-centrosymmetric alignment of $\chi^{(2)}$ chromophores. Polymeric materials for the photorefractive effect have to be optimized for charge photo generation, transport, trapping and electro-optic effect. We have found that by using a polymeric composite structure each of these functionalities can be derived from independent components and thus separately optimized (*8,28*) A recent polymeric composite developed in our laboratory which exhibits a very high diffraction efficiency of ~30% at several wavelengths (633 nm, 514.5 nm, 4881 nm) consists of poly[N-vinyl carbazole] as the charge transporting matrix, fullerene C_{60} as photosensitizer, APSS (discussed above) as the electro-optic chromophore and tricresyl phosphate as the plasticizer (28). The plasticizer lowers the Tg of the composite so that it can be electrically poled in situ at room temperature. We have also used a multifunctional polymer containing the charge transporting unit, the electro-optic unit and the Tg lowering unit (its structure given in the earlier section) for photorefractivity. However, the diffraction efficiency is not as high as that of the above composite.

In order to achieve high diffraction efficiency, a very high electric field needs to be applied which is a major hurdle in practical application of these materials. The research effort in our laboratory is now directed towards creating a better understanding of the processes involved in order to design a composite which would work at lower fields. Another objective of our research on photorefractivity is to understand the kinetics of grating formation and decay. The decay process is particularly important for storage of information.

TiO_2 Nanoparticles:Polyimide Composite Waveguides. The TiO_2 nanoparticles can be used to increase the refractive index of a film to make it more suitable as an optical waveguide (*7*). We have used the TiO_2 nanoparticles prepared by the reverse micelle approach described above. The TiO_2 particles prepared with $W_o = 2$ were then extracted from the reverse micellar solution by the addition of a known volume of N-methylpyrolidinone (NMP). Phase separation occurred immediately; the particles were extracted in the NMP phase. To the NMP solution containing TiO_2 particles, a polyimide solution was added. Then a film was cast on a glass slide using this solution. The film was then baked at 200°C for 30 minutes, followed by heat treatment for 30 minutes at 300°C in a nitrogen atmosphere. The concentration of TiO_2 was estimated to be 4% by weight. The optical waveguide loss of this film as a planar waveguide was found to be 1.4 dB/cm at 633 nm. The refractive index of the film was 1.560 compared to 1.550 for a pure polyimide film.

Acknowledgments

This work was supported by the Polymer Branch of the US Air Force Wright Laboratory and the Directorate of Chemistry and Life Sciences of the Air Force Office of Scientific Research. The author acknowledges thanks to Bruce Reinhardt, Professor P. C. Cheng, Dr. J. D. Bhawalkar, Mr. G. S. He, Dr. Deepak Kumar, Mr. G. Ruland, Ms. Manjari Lal and Mr. M. Yoshida for their collaboration.

Literature Cited

(1) Zhao, C. F.; He, G. S.; Bhawalkar, J. D.; Park, C. K.; Prasad, P. N. *Chem. Mat.* **1995**, *7*, 1979.
(2) He, G. S.; Zhao, C. F.; Bhawalkar, J. D.; Prasad, P. N. *Appl. Phys. Lett.* **1995**, *67*, 3703.
(3) Bhawalkar, J. D.; He, G. S.; Park, C. K.; Zhao, C. F.; Ruland, G.; Prasad, P. N. *Opt. Lett.* **1995**, *20*, 2393.
(4) He, G. S.; Bhawalkar, J. D., Zhao, C. F.; Park, C. K.; Prasad, P. N. *Opt. Lett.* **1995**, *20*, 2393.
(5) Zhao, C. F.; Park, C. K.; Prasad, P. N.; Zhang, Y.; Ghosal, S.; Burzynski, R. *Chem. Mat.*, **1995**, *7*, 1237.
(6) Gvishi, R.; Bhawalkar, J. D.; Kumar, N. D.; Ruland, G.; Narang, U.; Prasad, P. N. *Chem. Mat.*, **1995**, *7*, 2199.
(7) Yoshida, M.; Lal, M.; Kumar, N. D.; Prasad, P. N., submitted to Chem. Mat.
(8) Orczyk, M. E.; Swedek, B.; Zieba, J.; Prasad, P. N. *J. Appl. Phys.*, **1994**, *76*, 4995.
(9) Shen, Y. R. *The Principles of Nonlinear Optics*; Wiley: New York, 1984.

(10) Prasad, P. N.; Williams, D. J. *Introduction to Nonlinear Optical Effects in Molecues and Polymers*; Wiley: New York, 1991.
(11) Zhang, Y.; Ghosal, S.; Casstevens, M. K.; Burzynski, R. *Appl. Phys. Lett.*, **1995**, *66*, 256.
(12) Pope, E. J. A.; Asami, M.; Mackenzie, J. D. *J. Mater. Res.*, **1989**, *4*, 1018.
(13) Gvishi, R.; Reisfeld, R.; Biershtein, Z. *J. Sol-Gel Sci. and Tech.*, **1994**, *4*, 49.
(14) Zhao, M.; Samoc, M.; Prasad, P. N.; Reinhardt, B. A.; Unroe, M. R.; Prazak, M.; Evers, R. C.; Kane, J. J.; Jariwalar, C.; Sinsky, M. *Chem. Mat.*, **1990**, 2,670.
(15) Koper, G. J. M.; Sager, W. F. C.; Smelts, J.; Bedeaux, D. *J. Phys. Chem.*, **1995**, *99*, 13291.
(16) Petit, C., Lixon, P.; Pileni, M. P. *J. Phys. Chem.*, **1990**, *94*, 1598.
(17) Tult, L. W.; Boggess, T. F. *Prog. Quant. Electr.*, **1993**, *17*, 299.
(18) Crane, R.; Lewis, K.; Stryland, E. V.,; Khoshnevisan, M. Eds. *Materials for Optical Limiting*, Materials Research Society Symposium Proceedings, Vol. 374, Materials Research Society: Pittsburgh, 1995.
(19) He, G. S.; Gvishi, R.; Prasad, P. N.; Reinhardt, B. A.; Bhatt, J. C.; Dillard, A. G. *Opt. Commun.*, **1995**, *117*, 133.
(20) Wilson, T.; Shepard, C. Theory and Practices of Scanning Optical Microscopy; Academic Press, New York, 1984.
(21) Denk, W.; Stickler, J. H.; Webb, W. W. *Science*, **1990**, *248*, 73.
(22) Bhawalkar, J. D.; Swiatkiewicz, J.; Pan, S. J.; Samarabandu, J. K.; Liou, W. S.; He, G. S.; Berezney, R.; Cheng, P. C.; Prasad, P. N. *J. Scanning Microscopies* (in press).
(23) Mittal, K. L., Ed., *Polymers in Information Storage Technologies*"; Plenum: New York, 1989.
(24) Yen, P. *Introduction to Photorefractive Nonlinear Optics*; Wiley: New York, 1993.
(25) Ducharme, S.; Scott, J. C.; Twieg, R. J.; Moerner, W. E. *Phys. Rev. Lett.*, **1991**, *66*, 1846.
(26) Zhang, Y.; Cui, Y.; Prasad, P. N. *Phys. Rev. B.*, **1992**, *46*, 9900.
(27) Kippelen, B.; Tamura, K.; Peyghambarian, N.; Padias, A. B.; Hall, H. K., Jr. *J. Appl. Phys.*, **1993**, *74*, 3617.
(28) Cui, Y.; Swedek, B.; Kim, K. S.; Prasad, P. N. unpublished results.

INDEXES

Author Index

Affiliation Index

Subject Index